Student Solutions Manual

Essentials of
College Mathematics

Student Solutions Manual

Essentials of College Mathematics

FOR BUSINESS, ECONOMICS, LIFE SCIENCES, AND SOCIAL SCIENCES

THIRD EDITION

RAYMOND A. BARNETT
Merritt College

MICHAEL R. ZIEGLER
Marquette University

Prentice-Hall, Inc., Englewood Cliffs, New Jersey 07632

Production Editor: *Joan Eurell*
Acquisitions Editor: *George Lobell*
Supplement Acquisitions Editor: *Audra Walsh*
Production Coordinator: *Alan Fischer*

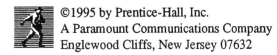

©1995 by Prentice-Hall, Inc.
A Paramount Communications Company
Englewood Cliffs, New Jersey 07632

Printed in the United States of America

10 9 8 7 6 5 4 3 2 1

ISBN 0-02-334387-7

Prentice-Hall International (UK) Limited, *London*
Prentice-Hall of Australia Pty. Limited, *Sydney*
Prentice-Hall Canada Inc., *Toronto*
Prentice-Hall Hispanoamericana, S.A., *Mexico*
Prentice-Hall of India Private Limited, *New Delhi*
Prentice-Hall of Japan, Inc., *Tokyo*
Simon & Schuster Asia Pte. Ltd., *Singapore*
Editora Prentice-Hall do Brasil, Ltda., *Rio de Janeiro*

CONTENTS

PREFACE ix

CHAPTER 1   BASIC ALGEBRAIC OPERATIONS   1

EXERCISE 1-1   Sets   1
EXERCISE 1-2   Algebra and Real Numbers   3
EXERCISE 1-3   Basic Operations on Polynomials   7
EXERCISE 1-4   Factoring Polynomials   10
EXERCISE 1-5   Basic Operations on Rational Expressions   14
EXERCISE 1-6   Integer Exponents and Square Root Radicals   17
EXERCISE 1-7   Rational Exponents and Radicals   21
EXERCISE 1-8   Chapter Review   25

CHAPTER 2   EQUATIONS, GRAPHS, AND FUNCTIONS   31

EXERCISE 2-1   Linear Equations and Inequalities in One Variable   31
EXERCISE 2-2   Quadratic Equations   34
EXERCISE 2-3   Cartesian Coordinate System and Straight Lines   38
EXERCISE 2-4   Functions   45
EXERCISE 2-5   Linear and Quadratic Functions   50
EXERCISE 2-6   Chapter Review   60

CHAPTER 3   EXPONENTIAL AND LOGARITHMIC FUNCTIONS   69

EXERCISE 3-1   Exponential Functions   69
EXERCISE 3-2   The Exponential Function with Base e 74
EXERCISE 3-3   Logarithmic Functions   77
EXERCISE 3-4   Chapter Review   82

CHAPTER 4   MATHEMATICS OF FINANCE   87

EXERCISE 4-1   Simple Interest   87
EXERCISE 4-2   Compound Interest   89
EXERCISE 4-3   Future Value of an Annuity; Sinking Funds   94
EXERCISE 4-4   Present Value of an Annuity; Amortization   98
EXERCISE 4-5   Chapter Review   104

CHAPTER 5	SYSTEMS OF LINEAR EQUATIONS; MATRICES	115
EXERCISE 5-1	Review: Systems of Linear Equations in Two Variables	115
EXERCISE 5-2	Systems of Linear Equations and Augmented Matrices	123
EXERCISE 5-3	Gauss-Jordan Elimination	126
EXERCISE 5-4	Matrices—Addition and Multiplication by a Number	138
EXERCISE 5-5	Matrix Multiplication	142
EXERCISE 5-6	Inverse of a Square Matrix	150
EXERCISE 5-7	Matrix Equations and Systems of Equations	158
EXERCISE 5-8	Chapter Review	167

CHAPTER 6	LINEAR INEQUALITIES AND LINEAR PROGRAMMING	179
EXERCISE 6-1	Systems of Linear Inequalities in Two Variables	179
EXERCISE 6-2	Linear Programming in Two Dimensions—A Geometric Approach	186
EXERCISE 6-3	A Geometric Introduction to the Simplex Method	198
EXERCISE 6-4	The Simplex Method: Maximization with Problem Constraints of the Form ≤	201
EXERCISE 6-5	Chapter Review	221

CHAPTER 7	PROBABILITY	229
EXERCISE 7-1	Basic Counting Principles	229
EXERCISE 7-2	Permutations and Combinations	236
EXERCISE 7-3	Sample Spaces and Events	241
EXERCISE 7-4	Empirical Probability	247
EXERCISE 7-5	Random Variable, Probability Distribution, and Expectation	250
EXERCISE 7-6	Chapter Review	255

CHAPTER 8	ADDITIONAL TOPICS IN PROBABILITY	265
EXERCISE 8-1	Union, Intersection, and Complement of Events; Odds	265
EXERCISE 8-2	Conditional Probability, Intersection, and Independence	271
EXERCISE 8-3	Bayes' Formula	279
EXERCISE 8-4	Chapter Review	286

CHAPTER 9	THE DERIVATIVE	295
EXERCISE 9-1	Limits and Continuity—A Geometric Introduction	295
EXERCISE 9-2	Computation of Limits	301
EXERCISE 9-3	The Derivative	308
EXERCISE 9-4	Derivative of Constants, Power Forms, and Sums	314
EXERCISE 9-5	Derivatives of Products and Quotients	320
EXERCISE 9-6	Chain Rule: Power Form	327
EXERCISE 9-7	Marginal Analysis in Business & Economics	333
EXERCISE 9-8	Chapter Review	338

CHAPTER 10	ADDITIONAL DERIVATIVE TOPICS	349

EXERCISE 10-1	First Derivative and Graphs	349
EXERCISE 10-2	Second Derivative and Graphs	362
EXERCISE 10-3	Curve Sketching Techniques: Unified & Extended	373
EXERCISE 10-4	Optimization; Absolute Maxima and Minima	400
EXERCISE 10-5	The Constant e and Continuous Compound Interest	408
EXERCISE 10-6	Derivatives of Logarithmic and Exponential Functions	411
EXERCISE 10-7	Chain Rule: General Form	422
EXERCISE 10-8	Chapter Review	430

CHAPTER 11	INTEGRATION	449

EXERCISE 11-1	Antiderivative and Indefinite Integrals	449
EXERCISE 11-2	Integration by Substitution	457
EXERCISE 11-3	Definite Integrals	466
EXERCISE 11-4	Area and the Definite Integral	473
EXERCISE 11-5	Definite Integral as a Limit of a Sum	483
EXERCISE 11-6	Consumers' and Producers' Surplus	491
EXERCISE 11-7	Chapter Review	495

APPENDIX A	SPECIAL TOPICS	505

EXERCISE A-1	Sequences, Series, and Summation Notation	505
EXERCISE A-2	Arithmetic Progressions	509
EXERCISE A-3	Geometric Progressions	511
EXERCISE A-4	The Binomial Theorem	513

PREFACE

This supplement accompanies *Essentials of College Mathematics for Business, Economics, Life Sciences, and Social Sciences,* Third Edition, by Raymond A. Barnett and Michael R. Ziegler.

The manual contains the solutions to the odd-numbered problems in each of the exercise sets, and the solutions to all of the problems in the Chapter Reviews. Each of the sections begins with a list of important terms and formulas, given under the heading "Things to Remember." While sufficient details are given for each solution, the first few solutions in each section are more detailed than the remaining ones.

EXERCISE 1-1

Things to remember:

1. $a \in A$ means "a is an element of set A."

2. $a \notin A$ means "a is not an element of set A."

3. $\varnothing$ means "the empty set" or "null set."

4. $S = \{x \mid P(x)\}$ means "S is the set of all x such that $P(x)$ is true."

5. $A \subset B$ means "A is a subset of B."

6. $A = B$ means "A and B have exactly the same elements."

7. $A \not\subset B$ means "A is not a subset of B."

8. $A \neq B$ means "A and B do not have exactly the same elements."

9. $A \cup B = A$ union $B = \{x \mid x \in A \text{ or } x \in B\}$.

10. $A \cap B = A$ intersection $B = \{x \mid x \in A \text{ and } x \in B\}$.

11. $A' = $ complement of $A = \{x \in U \mid x \notin A\}$, where U is a universal set.

1. T   3. T   5. T   7. T   9. $\{1, 3, 5\} \cup \{2, 3, 4\} = \{1, 2, 3, 4, 5\}$

11. $\{1, 3, 4\} \cap \{2, 3, 4\} = \{3, 4\}$ 13. $\{1, 5, 9\} \cap \{3, 4, 6, 8\} = \varnothing$

15. $\{x \mid x - 2 = 0\}$ 17. $x^2 = 49$ is true for $x = 7$ and -7.
    $x - 2 = 0$ is true for $x = 2$.   Hence, $\{x \mid x^2 = 49\} = \{-7, 7\}$.
    Hence, $\{x \mid x - 2 = 0\} = \{2\}$.

19. $\{x \mid x$ is an odd number between 1 and 9 inclusive$\} = \{1, 3, 5, 7, 9\}$.

21. $U = \{1, 2, 3, 4, 5\}$; $A = \{2, 3, 4\}$.   Then $A' = \{1, 5\}$.

23. From the Venn diagram,   25. A' has 60 elements.
    A has 40 elements.

27. $A \cup B$ has 60 elements   29. $A' \cap B$ has 20 elements
    $(35 + 5 + 20)$.   (common elements between A' and B.)

31. $(A \cap B)'$ has 95 elements. 33. $A' \cap B'$ has 40 elements.
    [Note: $A \cap B$ has 5 elements.]

35. (A) $\{x \mid x \in R \text{ or } x \in T\}$.
　　　$= R \cup T$ ("or" translated
　　　　　　as $\cup$, union)
　　　$= \{1, 2, 3, 4\} \cup \{2, 4, 6\}$
　　　$= \{1, 2, 3, 4, 6\}$

(B) $R \cup T = \{1, 2, 3, 4, 6\}$

37. $Q \cap R = \{2, 4, 6\} \cap \{3, 4, 5, 6\}$
　　　$= \{4, 6\}$
$P \cup (Q \cap R) = \{1, 2, 3, 4\} \cup \{4, 6\}$
　　　　　$= \{1, 2, 3, 4, 6\}$

39. Yes. $A \cup B = B$ can be
represented by the Venn diagram

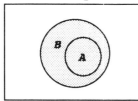

From the diagram, we see that
$A \subset B$. Thus, the given
statement is *true*.

41. Yes. The given statement is always
true. To understand this, see
the following Venn diagram.

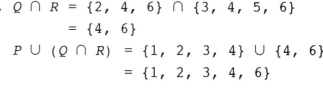

43. Yes. The given statement is
true. To understand this, see
the following Venn diagram.

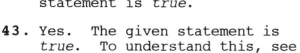

From the diagram, we conclude
that $x \in B$.

45. (A) Set $\{a\}$ has two subsets:
　　$\{a\}$ and $\varnothing$

(B) Set $\{a, b\}$ has four subsets:
　　$\{a, b\}$, $\varnothing$, $\{a\}$, $\{b\}$

(C) Set $\{a, b, c\}$ has eight subsets:
　　$\{a, b, c\}$, $\varnothing$, $\{a\}$, $\{b\}$, $\{c\}$,
　　$\{a, b\}$, $\{a, c\}$, $\{b, c\}$

Parts (A), (B), and (C) suggest
the following formula:
The number of subsets in a set
with n elements $= 2^n$.

47. The Venn diagram that corresponds
to the given information is shown
at the right. We can see that
$N \cup M$ has $300 + 300 + 200 = 800$
students.

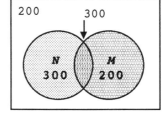

49. $(N \cup M)'$ has 200 students
[because $N \cup M$ has 800
students and $(N \cup M)'$ has $1000 - 800 = 200$].

51. $N' \cap M$ has 200 students.

53. The number of commuters who
listen to either news or music
= number of commuters in the
set $M \cup N$, which is 800.

55. The number of commuters who do
not listen to either news or
music = number of commuters in
set $(N \cup M)'$, which is
　　　$1000 - 800 = 200$.

57. The number of commuters who listen to music but not news = number of commuters in the set $N' \cap M$, which is 200.

59. The six two-person subsets that can be formed from the given set $\{P, V_1, V_2, V_3\}$ are:

$\{P, V_1\}$ $\{P, V_3\}$ $\{V_1, V_3\}$
$\{P, V_2\}$ $\{V_1, V_2\}$ $\{V_2, V_3\}$

61. From the given Venn diagram $A \cap Rh = \{A+, AB+\}$

63. Again, from the given Venn diagram: $A \cup Rh = \{A-, A+, B+, AB-, AB+, O+\}$

65. From the given Venn diagram: $(A \cup B)' = \{O+, O-\}$

67. $A' \cap B = \{B-, B+\}$

69. Statement (2): for every $a, b \in C$, aRb and bRa means that everyone in the clique relates to one another.

EXERCISE 1-2

Things to remember:

1. THE SET OF REAL NUMBERS

SYMBOL	NAME	DESCRIPTION	EXAMPLES
N	Natural numbers	Counting numbers (also called positive integers)	1, 2, 3, ...
Z	Integers	Natural numbers, their negatives, and 0	... -2, -1, 0, 1, 2, ...
Q	Rationals	Any number that can be represented as $\frac{a}{b}$, where a and b are integers and $b \neq 0$. Decimal representations are repeating or terminating.	$-4; 0; 1; 25; \frac{-3}{5}; \frac{2}{3};$ $3.67; -0.333\overline{3}; 5.2727\overline{27}$
I	Irrationals	Any number with a decimal representation that is nonrepeating and non-terminating.	$\sqrt{2}; \pi; \sqrt[3]{7}; 1.414213...;$ $2.718281828...$
R	Reals	Rationals and irrationals	

2. BASIC PROPERTIES OF THE SET OF REAL NUMBERS

Let a, b, and c be arbitrary elements in the set of real numbers R.

ADDITION PROPERTIES

CLOSURE: $a + b$ is a unique element in R

ASSOCIATIVE: $(a + b) + c = a + (b + c)$

COMMUTATIVE: $a + b = b + a$

IDENTITY: 0 is the additive identity; that is, $0 + a = a + 0$ for all a in R, and 0 is the only element in R with this property.

INVERSE: For each a in R, $-a$ is its unique additive inverse; that is, $a + (-a) = (-a) + a = 0$, and $-a$ is the only element in R relative to a with this property.

MULTIPLICATION PROPERTIES

CLOSURE: ab is a unique element in R

ASSOCIATIVE: $(ab)c = a(bc)$

COMMUTATIVE: $ab = ba$

IDENTITY: 1 is the multiplicative identity; that is, $1a = a1 = a$ for all a in R, and 1 is the only element in R with this property.

INVERSE: For each a in R, $a \neq 0$, $\frac{1}{a}$ is its unique multiplicative inverse; that is, $a\left(\frac{1}{a}\right) = \left(\frac{1}{a}\right)a = 1$, and $\frac{1}{a}$ is the only element in R relative to a with this property.

DISTRIBUTIVE PROPERTIES

$$a(b + c) = ab + ac$$
$$(a + b)c = ac + bc$$

3. SUBTRACTION AND DIVISION

For all real numbers a and b.

SUBTRACTION: $a - b = a + (-b)$
$7 - (-5) = 7 + [-(-5)] = 7 + 5 = 12$

DIVISION: $a \div b = a\left(\frac{1}{b}\right)$, $b \neq 0$
$9 \div 4 = 9\left(\frac{1}{4}\right) = \frac{9}{4}$

4. PROPERTIES OF NEGATIVES

For all real numbers a and b.

a. $-(-a) = a$

b. $(-a)b = -(ab) = a(-b) = -ab$

c. $(-a)(-b) = ab$

d. $(-1)a = -a$

e. $\dfrac{-a}{b} = -\dfrac{a}{b} = \dfrac{a}{-b}, \quad b \neq 0$

f. $\dfrac{-a}{-b} = -\dfrac{-a}{b} = -\dfrac{a}{-b} = \dfrac{a}{b}, \quad b \neq 0$

5. ZERO PROPERTIES

For all real numbers a and b.

a. $a \cdot 0 = 0$

b. $ab = 0$ if and only if $a = 0$ or $b = 0$ (or both)

6. FRACTION PROPERTIES

For all real numbers a, b, c, d, and k (division by 0 excluded).

a. $\dfrac{a}{b} = \dfrac{c}{d}$ if and only if $ad = bc$

b. $\dfrac{ka}{kb} = \dfrac{a}{b}$

$\dfrac{7 \cdot 3}{7 \cdot 5} = \dfrac{3}{5}$

c. $\dfrac{a}{b} \cdot \dfrac{c}{d} = \dfrac{ac}{bd}$

$\dfrac{3}{5} \cdot \dfrac{7}{8} = \dfrac{3 \cdot 7}{5 \cdot 8} = \dfrac{21}{40}$

d. $\dfrac{a}{b} \div \dfrac{c}{d} = \dfrac{a}{b} \cdot \dfrac{d}{c}$

$\dfrac{2}{3} \div \dfrac{5}{7} = \dfrac{2}{3} \cdot \dfrac{7}{5}$

e. $\dfrac{a}{b} + \dfrac{c}{b} = \dfrac{a + c}{b}$

$\dfrac{3}{6} + \dfrac{5}{6} = \dfrac{3 + 5}{6} = \dfrac{4}{3}$

f. $\dfrac{a}{b} - \dfrac{c}{b} = \dfrac{a - c}{b}$

$\dfrac{7}{8} - \dfrac{3}{8} = \dfrac{7 - 3}{8} = \dfrac{4}{8} = \dfrac{1}{2}$

g. $\dfrac{a}{b} + \dfrac{c}{d} = \dfrac{ad + bc}{bd}$

$\dfrac{2}{3} + \dfrac{3}{5} = \dfrac{2 \cdot 5 + 3 \cdot 3}{3 \cdot 5} = \dfrac{19}{15}$

h. $\dfrac{a}{b} - \dfrac{c}{d} = \dfrac{ad - bc}{bd}$

$\dfrac{2}{3} - \dfrac{3}{5} = \dfrac{2 \cdot 5 - 3 \cdot 3}{3 \cdot 5} = \dfrac{1}{15}$

---

1. $uv = vu$ 3. $3 + (7 + y) = (3 + 7) + y$ 5. $1(u + v) = u + v$

7. Associative property of multiplication

9. Properties of negatives (use 4e and 4a)

11. Subtraction   13. Division

15. Distributive property   17. Zero property (see 5a)

19. Associative property of multiplication

21. Distributive property   **23.** Property of negatives (see <u>4</u>f)

25. Zero property (see <u>5</u>b)

27. No. For example: $2\left(\dfrac{1}{2}\right) = 1$. In general $a\left(\dfrac{1}{a}\right) = 1$ whenever $a \neq 0$.

29. (A) False. For example, -3 is an integer but not a natural number.

(B) True

(C) True. For example, for any natural number n, $n = \dfrac{n}{1}$.

31. $\sqrt{2}$, $\sqrt{3}$, ...; in general, the square root of any rational number that is not a perfect square; π, e (see Chapter 4).

33. (A) $8 \in N,\ Z,\ Q,\ R$ (B) $\sqrt{2} \in R$

(C) $-1.414 = -\dfrac{1414}{1000} \in Q,\ R$ (D) $\dfrac{-5}{2} \in Q,\ R$

35. (A) True. This is the associative property of addition.

(B) False. For example, $(3 - 7) - 4 = -4 - 4 = -8$
$$\neq 3 - (7 - 4) = 3 - 3 = 0.$$

(C) True. This is the associative property of multiplication.

(D) False. For example, $(12 \div 4) \div 2 = 3 \div 2 = \dfrac{3}{2}$
$$\neq 12 \div (4 \div 2) = 12 \div 2 = 6.$$

37.
$$C = 0.090909...$$
$$100C = 9.090909...$$
$$100C - C = (9.090909...) - (0.090909...)$$
$$99C = 9$$
$$C = \frac{9}{99} = \frac{1}{11}$$

39. 1. Commutative property of addition

2. Associative property of addition

3. Additive inverse

4. Additive identity

41. (A) $\dfrac{13}{6} = 2.1666666...$ (repeating decimal)

(B) $\sqrt{21} \approx 4.5825756...$; $\sqrt{21}$ is an irrational number

(C) $\dfrac{7}{16} = 0.4375$ (terminating decimal)

(D) $\dfrac{29}{111} = 0.261261261...$ (repeating decimal)

Things to remember:

<u>1</u>. NATURAL NUMBER EXPONENT

For n a natural number and b any real number,

$b^n = b \cdot b \cdot \ldots \cdot b$, n factors of b.

For example, $2^3 = 2 \cdot 2 \cdot 2$ $(= 8)$,
$3^5 = 3 \cdot 3 \cdot 3 \cdot 3 \cdot 3$ $(= 243)$.

In the expression b^n, n is called the EXPONENT or POWER, and b is called the BASE.

<u>2</u>. FIRST PROPERTY OF EXPONENTS

For any natural numbers m and n, and any real number b,

$b^m \cdot b^n = b^{m+n}$.

For example, $3^3 \cdot 3^4 = 3^{3+4} = 3^7$.

<u>3</u>. POLYNOMIALS

a. A POLYNOMIAL IN ONE VARIABLE x is constructed by adding or subtracting constants and terms of the form ax^n, where a is a real number, called the COEFFICIENT of the term, and n is a natural number.

b. A POLYNOMIAL IN TWO VARIABLES x AND y is constructed by adding or subtracting constants and terms of the form ax^my^n, bx^k, cy^j, where a, b, and c are real numbers, called COEFFICIENTS and m, n, k, and j are natural numbers.

c. Polynomials in more than two variables are defined similarly.

d. A polynomial with only one term is called a MONOMIAL. A polynomial with two terms is called a BINOMIAL. A polynomial with three terms is called a TRINOMIAL.

<u>4</u>. DEGREE OF A POLYNOMIAL

a. A term of the form ax^n, $a \neq 0$, has degree n. A term of the form ax^my^n, $a \neq 0$, has degree $m + n$. A nonzero constant has degree 0.

b. The DEGREE OF A POLYNOMIAL is the degree of the nonzero term with the highest degree. For example, $3x^4 + \sqrt{2}\,x^3 - 2x + 7$ has degree 4; $2x^3y^2 - 3x^2y + 7x^4 - 5y^3 + 6$ has degree 5; the polynomial 4 has degree 0.

c. The constant 0 is a polynomial but it is not assigned a degree.

5. Two terms in a polynomial are called LIKE TERMS if they have exactly the same variable factors raised to the same powers. For example, in

$$7x^5y^2 - 3x^3y + 2x + 4x^3y - 1,$$

$-3x^3y$ and $4x^3y$ are like terms.

6. To multiply two polynomials, multiply each term of one by each term of the other, and then combine like terms.

7. SPECIAL PRODUCTS

a. $(a - b)(a + b) = a^2 - b^2$

b. $(a + b)^2 = a^2 + 2ab + b^2$

c. $(a - b)^2 = a^2 - 2ab + b^2$

---

1. The term of highest degree in $x^3 + 2x^2 - x + 3$ is x^3 and the degree of this term is 3.

3. $(2x^2 - x + 2) + (x^3 + 2x^2 - x + 3) = x^3 + 2x^2 + 2x^2 - x - x + 2 + 3$
$$= x^3 + 4x^2 - 2x + 5$$

5. $(x^3 + 2x^2 - x + 3) - (2x^2 - x + 2) = x^3 + 2x^2 - x + 3 - 2x^2 + x - 2$
$$= x^3 + 1$$

7. Using a vertical arrangement:

$$
\begin{array}{r}
x^3 + 2x^2 - x + 3 \\
2x^2 - x + 2 \\
\hline
2x^5 + 4x^4 - 2x^3 + 6x^2 \\
- x^4 - 2x^3 + x^2 - 3x \\
2x^3 + 4x^2 - 2x + 6 \\
\hline
2x^5 + 3x^4 - 2x^3 + 11x^2 - 5x + 6
\end{array}
$$

9. $2(u - 1) - (3u + 2) - 2(2u - 3) = 2u - 2 - 3u - 2 - 4u + 6$
$$= -5u + 2$$

11. $4a - 2a[5 - 3(a + 2)] = 4a - 2a[5 - 3a - 6]$
$$= 4a - 2a[-3a - 1]$$
$$= 4a + 6a^2 + 2a$$
$$= 6a^2 + 6a$$

13. $(a + b)(a - b) = a^2 - b^2$ (Special product 7a)

15. Using the FOIL method:
$$(3x - 5)(2x + 1) = 6x^2 + 3x - 10x - 5$$
$$= 6x^2 - 7x - 5$$

17. $(2x - 3y)(x + 2y) = 2x^2 + 4xy - 3xy - 6y^2$
$\qquad\qquad\qquad\qquad\quad = 2x^2 + xy - 6y^2$

19. $(3y + 2)(3y - 2) = (3y)^2 - 2^2 = 9y^2 - 4$ (Special product 7a)

21. $(3m + 7n)(2m - 5n) = 6m^2 - 15mn + 14mn - 35n^2$
$\qquad\qquad\qquad\qquad\qquad = 6m^2 - mn - 35n^2$

23. $(4m + 3n)(4m - 3n) = 16m^2 - 9n^2$

25. $(3u + 4v)^2 = 9u^2 + 24uv + 16v^2$ (Special product 7b)

27. $(a - b)(a^2 + ab + b^2) = a(a^2 + ab + b^2) - b(a^2 + ab + b^2)$
$\qquad\qquad\qquad\qquad\qquad = a^3 + a^2b + ab^2 - a^2b - ab^2 - b^3$
$\qquad\qquad\qquad\qquad\qquad = a^3 - b^3$

29. $(4x + 3y)^2 = 16x^2 + 24xy + 9y^2$

31. $m - \{m - [m - (m - 1)]\} = m - \{m - [m - m + 1]\}$
$\qquad\qquad\qquad\qquad\qquad\quad = m - \{m - 1\}$
$\qquad\qquad\qquad\qquad\qquad\quad = m - m + 1$
$\qquad\qquad\qquad\qquad\qquad\quad = 1$

33. $(x^2 - 2xy + y^2)(x^2 + 2xy + y^2) = (x - y)^2(x + y)^2$
$\qquad\qquad\qquad\qquad\qquad\qquad\qquad = [(x - y)(x + y)]^2$
$\qquad\qquad\qquad\qquad\qquad\qquad\qquad = [x^2 - y^2]^2$
$\qquad\qquad\qquad\qquad\qquad\qquad\qquad = x^4 - 2x^2y^2 + y^4$

35. $(3a - b)(3a + b) - (2a - 3b)^2 = (9a^2 - b^2) - (4a^2 - 12ab + 9b^2)$
$\qquad\qquad\qquad\qquad\qquad\qquad\qquad = 9a^2 - b^2 - 4a^2 + 12ab - 9b^2$
$\qquad\qquad\qquad\qquad\qquad\qquad\qquad = 5a^2 + 12ab - 10b^2$

37. $(m - 2)^2 - (m - 2)(m + 2) = m^2 - 4m + 4 - [m^2 - 4]$
$\qquad\qquad\qquad\qquad\qquad\qquad\quad = m^2 - 4m + 4 - m^2 + 4$
$\qquad\qquad\qquad\qquad\qquad\qquad\quad = -4m + 8$

39. $(x - 2y)(2x + y) - (x + 2y)(2x - y)$
$\qquad\qquad = 2x^2 - 4xy + xy - 2y^2 - [2x^2 + 4xy - xy - 2y^2]$
$\qquad\qquad = 2x^2 - 3xy - 2y^2 - 2x^2 - 3xy + 2y^2$
$\qquad\qquad = -6xy$

41. $(u + v)^3 = (u + v)(u + v)^2 = (u + v)(u^2 + 2uv + v^2)$
$\qquad\qquad\qquad\qquad\qquad\qquad\qquad = u^3 + 3u^2v + 3uv^2 + v^3$

43. $(x - 2y)^3 = (x - 2y)(x - 2y)^2 = (x - 2y)(x^2 - 4xy + 4y^2)$
$\qquad\qquad\qquad\qquad\qquad\qquad\qquad = x(x^2 - 4xy + 4y^2) - 2y(x^2 - 4xy + 4y^2)$
$\qquad\qquad\qquad\qquad\qquad\qquad\qquad = x^3 - 4x^2y + 4xy^2 - 2x^2y + 8xy^2 - 8y^3$
$\qquad\qquad\qquad\qquad\qquad\qquad\qquad = x^3 - 6x^2y + 12xy^2 - 8y^3$

45. $[(2x^2 - 4xy + y^2) + (3xy - y^2)] - [(x^2 - 2xy - y^2) + (-x^2 + 3xy - 2y^2)]$
$= [2x^2 - xy] - [xy - 3y^2] = 2x^2 - 2xy + 3y^2$

47. $(2x - 1)^3 - 2(2x - 1)^2 + 3(2x - 1) + 7$
$= (2x - 1)(2x - 1)^2 - 2[4x^2 - 4x + 1] + 6x - 3 + 7$
$= (2x - 1)(4x^2 - 4x + 1) - 8x^2 + 8x - 2 + 6x + 4$
$= 8x^3 - 12x^2 + 6x - 1 - 8x^2 + 14x + 2$
$= 8x^3 - 20x^2 + 20x + 1$

49. $2\{(x - 3)(x^2 - 2x + 1) - x[3 - x(x - 2)]\}$
$= 2\{x^3 - 5x^2 + 7x - 3 - x[3 - x^2 + 2x]\}$
$= 2\{x^3 - 5x^2 + 7x - 3 + x^3 - 2x^2 - 3x\}$
$= 2\{2x^3 - 7x^2 + 4x - 3\}$
$= 4x^3 - 14x^2 + 8x - 6$

51. $m + n$

53. Let x = amount invested at 9%.
Then $10,000 - x$ = amount invested at 12%.
The total annual income I is:
$$I = 0.09x + 0.12(10,000 - x)$$
$$= 1,200 - 0.03x$$

55. Let x = number of tickets at \$10.
Then $3x$ = number of tickets at \$30 and $4,000 - x - 3x = 4,000 - 4x$
= number of tickets at \$50.
The total receipts R are:
$$R = 10x + 30(3x) + 50(4,000 - 4x)$$
$$= 10x + 90x + 200,000 - 200x = 200,000 - 100x$$

57. Let x = number of kilograms of food A.
Then $10 - x$ = number of kilograms of food B.
The total number of kilograms F of fat in the final food mix is:
$$F = 0.02x + 0.06(10 - x)$$
$$= 0.6 - 0.04x$$

EXERCISE 1-4

Things to remember:

1. FACTORED FORMS

    A polynomial is in FACTORED FORM if it is written as the product of two or more polynomials. A polynomial with integer coefficients is FACTORED COMPLETELY if each factor cannot be expressed as the product of two or more polynomials with integer coefficients, other than itself and one.

2. METHODS
   a. Factor out all factors common to all terms, if they are present.
   b. Try grouping terms.
   c. *ac*-Test for polynomials of the form
$$ax^2 + bx + c \quad \text{or} \quad ax^2 + bxy + cy^2$$
   If the product *ac* has two integer factors *p* and *q* whose sum is the coefficient *b* of the middle term, i.e., if integers *p* and *q* exist so that
$$pq = ac \quad \text{and} \quad p + q = b$$
   then the polynomials have first-degree factors with integer coefficients. If no such integers exist then the polynomials will not have first-degree factors with integer coefficients; the polynomials are *not factorable.*

3. SPECIAL FACTORING FORMULAS
   a. $u^2 + 2uv + v^2 = (u + v)^2$ Perfect square
   b. $u^2 - 2uv + v^2 = (u - v)^2$ Perfect square
   c. $u^2 - v^2 = (u - v)(u + v)$ Difference of squares
   d. $u^3 - v^3 = (u - v)(u^2 + uv + v^2)$ Difference of cubes
   e. $u^3 + v^3 = (u + v)(u^2 - uv + v^2)$ Sum of cubes

---

1. $3m^2$ is a common factor: $6m^4 - 9m^3 - 3m^2 = 3m^2(2m^2 - 3m - 1)$

3. $2uv$ is a common factor: $8u^3v - 6u^2v^2 + 4uv^3 = 2uv(4u^2 - 3uv + 2v^2)$

5. $(2m - 3)$ is a common factor: $7m(2m - 3) + 5(2m - 3) = (7m + 5)(2m - 3)$

7. $(3c + d)$ is a common factor: $a(3c + d) - 4b(3c + d)$
$$= (a - 4b)(3c + d)$$

9. $2x^2 - x + 4x - 2 = (2x^2 - x) + (4x - 2)$
$$= x(2x - 1) + 2(2x - 1)$$
$$= (2x - 1)(x + 2)$$

11. $3y^2 - 3y + 2y - 2 = (3y^2 - 3y) + (2y - 2)$
$$= 3y(y - 1) + 2(y - 1)$$
$$= (y - 1)(3y + 2)$$

13. $2x^2 + 8x - x - 4 = (2x^2 + 8x) - (x + 4)$
$$= 2x(x + 4) - (x + 4)$$
$$= (x + 4)(2x - 1)$$

15. $wy - wz + xy - xz = (wy - wz) + (xy - xz)$
$$= w(y - z) + x(y - z)$$
$$= (y - z)(w + x)$$

or $wy - wz + xy - xz = (wy + xy) - (wz + xz)$
$$= y(w + x) - z(w + x)$$
$$= (w + x)(y - z)$$

17. $am - bn - bm + an = (am - bm) + (an - bn)$
$$= m(a - b) + n(a - b)$$
$$= (a - b)(m + n)$$

or $am - bn - bm + an = (am + an) - (bm + bn)$
$$= a(m + n) - b(m + n)$$
$$= (m + n)(a - b)$$

19. $3y^2 - y - 2$

$a = 3, \ b = -1, \ c = -2$

Step 1. Use the ac-test to test for factorability

$ac = (3)(-2) = -6$

$$\underline{pq}$$
$$(1)(-6)$$
$$(-1)(6)$$
$$\boxed{(2)(-3)}$$
$$(-2)(3)$$

Note that $2 + (-3) = -1 = b$. Thus, $3y^2 - y - 2$ has first-degree factors with integer coefficients.

Step 2. Split the middle term using $b = p + q$ and factor by grouping.

$-1 = -3 + 2$

$3y^2 - y - 2 = 3y^2 - 3y + 2y - 2 = (3y^2 - 3y) + (2y - 2)$
$$= 3y(y - 1) + 2(y - 1)$$
$$= (y - 1)(3y + 2)$$

21. $u^2 - 2uv - 15v^2$

$a = 1, \ b = -2, \ c = -15$

Step 1. Use the ac-test

$ac = 1(-15) = -15$

$$\underline{pq}$$
$$(1)(-15)$$
$$(-1)(15)$$
$$\boxed{(3)(-5)}$$
$$(-3)(5)$$

Note that $3 + (-5) = -2 = b$. Thus $u^2 - 2uv - 15v^2$ has first-degree factors with integer coefficients.

Step 2. Factor by grouping

$-2 = 3 + (-5)$

$u^2 + 3uv - 5uv - 15v^2 = (u^2 + 3uv) - (5uv + 15v^2)$
$$= u(u + 3v) - 5v(u + 3v)$$
$$= (u + 3v)(u - 5v)$$

23. $m^2 - 6m - 3$
$a = 1$, $b = -6$, $c = -3$
Step 1. Use the ac-test

$$ac = (1)(-3) = -3$$

$$\underline{pq}$$
$$(1)(-3)$$
$$(-1)(3)$$

None of the factors add up to $-6 = b$. Thus, this polynomial is *not factorable*.

25. $w^2x^2 - y^2 = (wx - y)(wx + y)$ (difference of squares)

27. $9m^2 - 6mn + n^2 = (3m - n)^2$ (perfect square)

29. $y^2 + 16$
$a = 1$, $b = 0$, $c = 16$
Step 1. Use the ac-test

$$ac = (1)(16)$$

$$\underline{pq}$$
$$(1)(16)$$
$$(-1)(-16)$$
$$(2)(8)$$
$$(-2)(-8)$$
$$(4)(4)$$
$$(-4)(-4)$$

None of the factors add up to $0 = b$. Thus this polynomial is *not factorable*.

31. $4z^2 - 28z + 48 = 4(z^2 - 7z + 12) = 4(z - 3)(z - 4)$

33. $2x^4 - 24x^3 + 40x^2 = 2x^2(x^2 - 12x + 20) = 2x^2(x - 2)(x - 10)$

35. $4xy^2 - 12xy + 9x = x(4y^2 - 12y + 9) = x(2y - 3)^2$

37. $6m^2 - mn - 12n^2 = (2m - 3n)(3m + 4n)$

39. $4u^3v - uv^3 = uv(4u^2 - v^2) = uv[(2u)^2 - v^2] = uv(2u - v)(2u + v)$

41. $2x^3 - 2x^2 + 8x = 2x(x^2 - x + 4)$ [Note: $x^2 - x + 4$ is *not factorable*.]

43. $r^3 - t^3 = (r - t)(r^2 + rt + t^2)$ (difference of cubes)

45. $a^3 + 1 = (a + 1)(a^2 - a + 1)$ (sum of cubes)

47. $(x + 2)^2 - 9y^2 = [(x + 2) - 3y][(x + 2) + 3y]$
$$= (x + 2 - 3y)(x + 2 + 3y)$$

49. $5u^2 + 4uv - 2v^2$ is *not factorable*.

51. $6(x - y)^2 + 23(x - y) - 4 = [6(x - y) - 1][(x - y) + 4]$
$$= (6x - 6y - 1)(x - y + 4)$$

53. $y^4 - 3y^2 - 4 = (y^2)^2 - 3y^2 - 4 = (y^2 - 4)(y^2 + 1)$
$$= (y - 2)(y + 2)(y^2 + 1)$$

55. $27a^2 + a^5b^3 = a^2(27 + a^3b^3) = a^2[3^3 + (ab)^3]$
$$= a^2(3 + ab)(9 - 3ab + a^2b^2)$$

Things to remember:

<u>1</u>. FUNDAMENTAL PROPERTY OF FRACTIONS

If *a*, *b*, and *k* are real numbers with *b*, *k* ≠ 0, then
$$\frac{ka}{kb} = \frac{a}{b}.$$

A fraction is in LOWEST TERMS if the numerator and denominator have no common factors other than 1 or −1.

<u>2</u>. MULTIPLICATION AND DIVISION

For *a*, *b*, *c*, and *d* real numbers:

a. $\frac{a}{b} \cdot \frac{c}{d} = \frac{ac}{bd}$, *b*, *d* ≠ 0

b. $\frac{a}{b} \div \frac{c}{d} = \dfrac{\frac{a}{b}}{\frac{c}{d}} = \frac{a}{b} \cdot \frac{d}{c}$, *b*, *c*, *d* ≠ 0

The same procedures are used to multiply or divide two rational expressions.

<u>3</u>. ADDITION AND SUBTRACTION

For *a*, *b*, and *c* real numbers:

a. $\frac{a}{b} + \frac{c}{b} = \frac{a + c}{b}$, *b* ≠ 0

b. $\frac{a}{b} - \frac{c}{b} = \frac{a - c}{b}$, *b* ≠ 0

The same procedures are used to add or subtract two rational expressions (with the same denominator).

<u>4</u>. THE LEAST COMMON DENOMINATOR (LCD)

The LCD of two or more rational expressions is found as follows:

a. Factor each denominator completely, including integer factors.

b. Identify each different factor from all the denominators.

c. Form a product using each different factor to the highest power that occurs in any one denominator. This product is the LCD.

The least common denominator is used to add or subtract rational expressions having different denominators.

1. $\dfrac{d^5}{3a} \div \left(\dfrac{d^2}{6a^2} \cdot \dfrac{a}{4d^3}\right) = \dfrac{d^5}{3a} \div \left(\dfrac{\cancel{a}d^2}{24\cancel{a^2}d^3}\right) = \dfrac{d^5}{3a} \div \dfrac{1}{24ad} = \dfrac{d^5}{\cancel{3a}} \cdot \dfrac{\overset{8}{\cancel{24}}\cancel{a}d}{1} = 8d^6$

3. $\dfrac{x^2}{12} + \dfrac{x}{18} - \dfrac{1}{30} = \dfrac{15x^2}{180} + \dfrac{10x}{180} - \dfrac{6}{180}$ 　　We find the LCD of 12, 18, 30:

$\qquad\qquad = \dfrac{15x^2 + 10x - 6}{180}$ 　　$12 = 2^2 \cdot 3, \ \ 18 = 2 \cdot 3^2, \ \ 30 = 2 \cdot 3 \cdot 5$

$\qquad\qquad\qquad\qquad\qquad$ 　　Thus, LCD $= 2^2 \cdot 3^2 \cdot 5 = 180$.

5. $\dfrac{4m - 3}{18m^3} + \dfrac{3}{4m} - \dfrac{2m - 1}{6m^2}$ 　　Find the LCD of $18m^3$, $4m$, $6m^2$:

$\qquad\qquad\qquad\qquad\qquad\qquad$ 　　$18m^3 = 2 \cdot 3^2 m^3, \ \ 4m = 2^2 m,$

$= \dfrac{2(4m - 3)}{36m^3} + \dfrac{3(9m^2)}{36m^3} - \dfrac{6m(2m - 1)}{36m^3}$ 　　$6m^2 = 2 \cdot 3m^2$

$\qquad\qquad\qquad\qquad\qquad\qquad$ 　　Thus, LCD $= 36m^3$.

$= \dfrac{8m - 6 + 27m^2 - 6m(2m - 1)}{36m^3}$

$= \dfrac{8m - 6 + 27m^2 - 12m^2 + 6m}{36m^3} = \dfrac{15m^2 + 14m - 6}{36m^3}$

7. $\dfrac{x^2 - 9}{x^2 - 3x} \div (x^2 - x - 12) = \dfrac{\cancel{(x - 3)}\cancel{(x + 3)}}{x\cancel{(x - 3)}} \cdot \dfrac{1}{(x - 4)\cancel{(x + 3)}} = \dfrac{1}{x(x - 4)}$

9. $\dfrac{x^2 - 6x + 9}{x^2 - x - 6} \div \dfrac{x^2 + 2x - 15}{x^2 + 2x} = \dfrac{\cancel{(x - 3)^2}}{\cancel{(x - 3)}\cancel{(x + 2)}} \cdot \dfrac{x\cancel{(x + 2)}}{(x + 5)\cancel{(x - 3)}} = \dfrac{x}{x + 5}$

11. $\dfrac{3}{x^2 - 1} - \dfrac{2}{x^2 - 2x + 1} = \dfrac{3}{(x - 1)(x + 1)} - \dfrac{2}{(x - 1)^2}$ 　　LCD $= (x - 1)^2(x + 1)$

$\qquad\qquad\qquad\qquad\qquad = \dfrac{3(x - 1)}{(x - 1)^2(x + 1)} - \dfrac{2(x + 1)}{(x - 1)^2(x + 1)}$

$\qquad\qquad\qquad\qquad\qquad = \dfrac{3x - 3 - 2(x + 1)}{(x - 1)^2(x + 1)} = \dfrac{3x - 3 - 2x - 2}{(x - 1)^2(x + 1)}$

$\qquad\qquad\qquad\qquad\qquad = \dfrac{x - 5}{(x - 1)^2(x + 1)}$

13. $\dfrac{x + 1}{x - 1} - 1 = \dfrac{x + 1}{x - 1} - \dfrac{x - 1}{x - 1} = \dfrac{x + 1 - (x - 1)}{x - 1} = \dfrac{2}{x - 1}$

15. $\dfrac{3}{a - 1} - \dfrac{2}{1 - a} = \dfrac{3}{a - 1} - \dfrac{-2}{-(1 - a)} = \dfrac{3}{a - 1} + \dfrac{2}{a - 1} = \dfrac{5}{a - 1}$

17. $\dfrac{2x}{x^2 - y^2} + \dfrac{1}{x + y} - \dfrac{1}{x - y}$

$= \dfrac{2x}{(x - y)(x + y)} + \dfrac{x - y}{(x - y)(x + y)} - \dfrac{x + y}{(x - y)(x + y)}$

$= \dfrac{2x + x - y - (x + y)}{(x - y)(x + y)} = \dfrac{2x - 2y}{(x - y)(x + y)} = \dfrac{2\cancel{(x - y)}}{\cancel{(x - y)}(x + y)} = \dfrac{2}{x + y}$

19. $\dfrac{x^2}{x^2 + 2x + 1} + \dfrac{x - 1}{3x + 3} - \dfrac{1}{6} = \dfrac{x^2}{(x + 1)^2} + \dfrac{x - 1}{3(x + 1)} - \dfrac{1}{6}$

$\text{LCD} = 6(x + 1)^2 \qquad\qquad = \dfrac{6x^2}{6(x + 1)^2} + \dfrac{2(x + 1)(x - 1)}{6(x + 1)^2} - \dfrac{(x + 1)^2}{6(x + 1)^2}$

$$= \dfrac{6x^2 + 2(x^2 - 1) - (x^2 + 2x + 1)}{6(x + 1)^2}$$

$$= \dfrac{7x^2 - 2x - 3}{6(x + 1)^2}$$

21. $\dfrac{2 - x}{2x + x^2} \cdot \dfrac{x^2 + 4x + 4}{x^2 - 4} = \dfrac{-\cancel{(x - 2)}}{x\cancel{(x + 2)}} \cdot \dfrac{\cancel{(x + 2)^2}}{\cancel{(x + 2)}\cancel{(x - 2)}} = -\dfrac{1}{x}$

23. $\dfrac{c + 2}{5c - 5} - \dfrac{c - 2}{3c - 3} + \dfrac{c}{1 - c} = \dfrac{c + 2}{5(c - 1)} - \dfrac{c - 2}{3(c - 1)} - \dfrac{c}{c - 1}$

$\text{LCD} = 15(c - 1) \qquad\qquad = \dfrac{3(c + 2)}{15(c - 1)} - \dfrac{5(c - 2)}{15(c - 1)} - \dfrac{15c}{15(c - 1)}$

$$= \dfrac{3c + 6 - 5c + 10 - 15c}{15(c - 1)} = \dfrac{-17c + 16}{15(c - 1)}$$

25. $\left(\dfrac{x^3 - y^3}{y^3} \cdot \dfrac{y}{x - y}\right) \div \dfrac{x^2 + xy + y^2}{y^2} = \dfrac{(x^3 - y^3)y}{y^3(x - y)} \cdot \dfrac{y^2}{x^2 + xy + y^2}$

$$= \dfrac{y^3(x - y)(x^2 + xy + y^2)}{y^3(x - y)(x^2 + xy + y^2)} = 1$$

27. $\left(\dfrac{3}{x - 2} - \dfrac{1}{x + 1}\right) \div \dfrac{x + 4}{x - 2} = \left[\dfrac{3(x + 1) - (x - 2)}{(x - 2)(x + 1)}\right] \cdot \dfrac{x - 2}{x + 4}$

$$= \dfrac{3x + 3 - x + 2}{\cancel{(x - 2)}(x + 1)} \cdot \dfrac{\cancel{x - 2}}{x + 4} = \dfrac{2x + 5}{(x + 1)(x + 4)}$$

29. $\dfrac{1 + \dfrac{3}{x}}{x - \dfrac{9}{x}} = \dfrac{\dfrac{x + 3}{x}}{\dfrac{x^2 - 9}{x}} = \dfrac{x + 3}{x} \cdot \dfrac{x}{x^2 - 9} = \dfrac{\cancel{x + 3}}{\cancel{x}} \cdot \dfrac{\cancel{x}}{\cancel{(x + 3)}(x - 3)} = \dfrac{1}{x - 3}$

31. $\dfrac{\dfrac{1}{m^2} - 1}{\dfrac{1}{m} + 1} = \dfrac{\dfrac{1 - m^2}{m^2}}{\dfrac{1 + m}{m}} = \dfrac{(1 - m)\cancel{(1 + m)}}{\cancel{m^2}} \cdot \dfrac{\cancel{m}}{\cancel{1 + m}} = \dfrac{1 - m}{m}$

33. $\dfrac{c - d}{\dfrac{1}{c} - \dfrac{1}{d}} = \dfrac{c - d}{\dfrac{d - c}{cd}} = (c - d) \cdot \dfrac{cd}{d - c} = -cd$

35. $\dfrac{\dfrac{x}{y} - 2 + \dfrac{y}{x}}{\dfrac{x}{y} - \dfrac{y}{x}} = \dfrac{\dfrac{x^2 - 2xy + y^2}{xy}}{\dfrac{x^2 - y^2}{xy}} = \dfrac{\cancel{(x - y)}^{(x - y)}\cancel{(x - y)^2}}{\cancel{xy}} \cdot \dfrac{\cancel{xy}}{\cancel{(x - y)}(x + y)} = \dfrac{x - y}{x + y}$

37. $\dfrac{\dfrac{s^2}{s-t} - s}{\dfrac{t^2}{s-t} + t} = \dfrac{\dfrac{s^2 - s(s-t)}{s-t}}{\dfrac{t^2 + t(s-t)}{s-t}} = \dfrac{st}{s-t} \cdot \dfrac{s-t}{st} = 1$

39. $1 - \dfrac{1}{1 - \dfrac{1}{1 - \dfrac{1}{x}}} = 1 - \dfrac{1}{1 - \dfrac{1}{\dfrac{x-1}{x}}} = 1 - \dfrac{1}{1 - \dfrac{x}{x-1}}$

$= 1 - \dfrac{1}{\dfrac{(x-1) - x}{x-1}} = 1 - \dfrac{1}{\dfrac{-1}{x-1}} = 1 + x - 1 = x$

EXERCISE 1-6

Things to remember:

<u>1</u>. DEFINITION OF a^n, where n is an integer and a is a real number:

a. For n a positive integer,
   $$a^n = a \cdot a \cdot \ \cdots \ \cdot a, \ n \text{ factors of } a.$$

b. For $n = 0$,
   $$a^0 = 1, \ a \neq 0, \ 0^0 \text{ is not defined.}$$

c. For n a negative integer,
   $$a^n = \dfrac{1}{a^{-n}}, \ a \neq 0.$$

[<u>Note</u>: If n is negative, then $-n$ is positive.]

<u>2</u>. PROPERTIES OF EXPONENTS

GIVEN: n and m are integers and a and b are real numbers.

a. $a^m a^n = a^{m+n}$ $\qquad\qquad\qquad$ $a^8 a^{-3} = a^{8+(-3)} = a^5$

b. $(a^n)^m = a^{mn}$ $\qquad\qquad\qquad$ $(a^{-2})^3 = a^{3(-2)} = a^{-6}$

c. $(ab)^m = a^m b^m$ $\qquad\qquad\qquad$ $(ab)^{-2} = a^{-2}b^{-2}$

d. $\left(\dfrac{a}{b}\right)^m = \dfrac{a^m}{b^m}, \ b \neq 0$ $\qquad\qquad$ $\left(\dfrac{a}{b}\right)^5 = \dfrac{a^5}{b^5}$

e. $\dfrac{a^m}{a^n} = a^{m-n} = \dfrac{1}{a^{n-m}}, \ a \neq 0$ $\qquad$ $\dfrac{a^{-3}}{a^7} = \dfrac{1}{a^{7-(-3)}} = \dfrac{1}{a^{10}}$

3. DEFINITION OF SQUARE ROOT

A number x is a SQUARE ROOT of the number y if $x^2 = y$.

Every positive real number has exactly two square roots, each the negative of the other.

The square root of 0 is 0.

Negative real numbers do not have real number square roots.

4. SQUARE ROOT NOTATION

For a a positive number:

$\sqrt{a}$ is the positive square root of a
$-\sqrt{a}$ is the negative square root of a

[Note: $\sqrt{-a}$ is not a real number.]

5. PROPERTIES OF RADICALS

For a and b nonnegative real numbers:

a. $\sqrt{a^2} = a$ b. $\sqrt{a}\sqrt{b} = \sqrt{ab}$ c. $\dfrac{\sqrt{a}}{\sqrt{b}} = \sqrt{\dfrac{a}{b}}$

6. DEFINITION OF THE SIMPLEST RADICAL FORM

An algebraic expression that contains square root radicals is in SIMPLEST RADICAL FORM if all three of the following conditions are satisfied:

a. No radicand (the expression within the radical sign) when expressed in completely factored form contains a factor raised to a power greater than 1. ($\sqrt{x^3}$ violates this condition.)

b. No radical appears in a denominator. $\left(\dfrac{3}{\sqrt{5}} \text{ violates this condition.}\right)$

c. No fraction appears within a radical. $\left(\sqrt{\dfrac{2}{3}} \text{ violates this condition.}\right)$

---

1. $2x^{-9} = \dfrac{2}{x^9}$

3. $\dfrac{3}{2w^{-7}} = \dfrac{3w^7}{2}$

5. $2x^{-8}x^5 = 2x^{-8+5} = 2x^{-3} = \dfrac{2}{x^3}$

7. $\dfrac{w^{-8}}{w^{-3}} = \dfrac{1}{w^{-3+8}} = \dfrac{1}{w^5}$

9. $5v^8v^{-8} = 5v^{8-8} = 5v^0 = 5\cdot 1 = 5$

11. $(a^{-3})^2 = a^{-6} = \dfrac{1}{a^6}$

13. $(x^6y^{-3})^{-2} = x^{-12}y^6 = \dfrac{y^6}{x^{12}}$ **15.** $\sqrt{x^2} = x$

17. $\sqrt{a^5} = \sqrt{a^4 \cdot a} = \sqrt{a^4}\sqrt{a} = a^2\sqrt{a}$ **19.** $\sqrt{18x^4} = \sqrt{3^2 \cdot 2x^4} = \sqrt{3^2x^4}\sqrt{2} = 3x^2\sqrt{2}$

21. $\dfrac{1}{\sqrt{m}} = \dfrac{1}{\sqrt{m}} \cdot \dfrac{\sqrt{m}}{\sqrt{m}} = \dfrac{\sqrt{m}}{m}$ Rationalizing   **23.** $\sqrt{\dfrac{2}{3}} = \dfrac{\sqrt{2}}{\sqrt{3}} = \dfrac{\sqrt{2}}{\sqrt{3}} \cdot \dfrac{\sqrt{3}}{\sqrt{3}} = \dfrac{\sqrt{6}}{3}$

Rationalizing

25. $\sqrt{\dfrac{2}{x}} = \dfrac{\sqrt{2}}{\sqrt{x}} = \dfrac{\sqrt{2}}{\sqrt{x}} \cdot \dfrac{\sqrt{x}}{\sqrt{x}} = \dfrac{\sqrt{2x}}{x}$ Rationalizing

27. $82,300,000,000 = 8.23 \times 10^{10}$ **29.** $0.783 = 7.83 \times 10^{-1}$

31. $0.000\ 034 = 3.4 \times 10^{-5}$ **33.** $(22 + 31)^0 = (53)^0 = 1$

35. $\dfrac{10^{-3} \times 10^4}{10^{-11} \times 10^{-2}} = \dfrac{10^{-3+4}}{10^{-11-2}} = \dfrac{10^1}{10^{-13}} = 10^{1+13} = 10^{14}$

37. $(5x^2y^{-3})^{-2} = 5^{-2}x^{-4}y^6 = \dfrac{y^6}{5^2x^4} = \dfrac{y^6}{25x^4}$

39. $\dfrac{8 \times 10^{-3}}{2 \times 10^{-5}} = \dfrac{8}{2} \times \dfrac{10^{-3}}{10^{-5}} = 4 \times 10^{-3+5} = 4 \times 10^2$

41. $\dfrac{8x^{-3}y^{-1}}{6x^2y^{-4}} = \dfrac{4y^{-1+4}}{3x^{2+3}} = \dfrac{4y^3}{3x^5}$

43. $\left(\dfrac{6xy^{-2}}{3x^{-1}y^2}\right)^{-3} = \left(\dfrac{2x^{1+1}}{y^{2+2}}\right)^{-3} = \left(\dfrac{2x^2}{y^4}\right)^{-3} = \dfrac{2^{-3}x^{-6}}{y^{-12}} = \dfrac{y^{12}}{2^3x^6}$ or $\dfrac{y^{12}}{8x^6}$

45. $\dfrac{1 - x}{x^{-1} - 1} = \dfrac{1 - x}{\dfrac{1}{x} - 1} = \dfrac{1 - x}{\dfrac{1 - x}{x}} = (1 - x) \cdot \dfrac{x}{1 - x} = x$

47. $\dfrac{u + v}{u^{-1} + v^{-1}} = \dfrac{u + v}{\dfrac{1}{u} + \dfrac{1}{v}} = \dfrac{u + v}{\dfrac{v + u}{uv}} = (u + v) \cdot \dfrac{uv}{v + u} = uv$

49. $\dfrac{7x^5 - x^2}{4x^5} = \dfrac{7x^5}{4x^5} - \dfrac{x^2}{4x^5} = \dfrac{7}{4} - \dfrac{1}{4x^3} = \dfrac{7}{4} - \dfrac{1}{4}x^{-3}$

51. $\dfrac{3x^4 - 4x^2 - 1}{4x^3} = \dfrac{3x^4}{4x^3} - \dfrac{4x^2}{4x^3} - \dfrac{1}{4x^3} = \dfrac{3}{4}x - x^{-1} - \dfrac{1}{4}x^{-3}$

53. $\sqrt{18x^8y^5z^2} = \sqrt{3^2 \cdot 2(x^4)^2 \cdot (y^2)^2 \cdot y \cdot z^2} = 3x^4y^2z\sqrt{2y}$

55. $\dfrac{12}{\sqrt{3x}} = \dfrac{12}{\sqrt{3x}} \cdot \dfrac{\sqrt{3x}}{\sqrt{3x}} = \dfrac{12\sqrt{3x}}{3x} = \dfrac{4\sqrt{3x}}{x}$ Rationalizing

57. $\sqrt{\dfrac{6x}{7y}} = \dfrac{\sqrt{6x}}{\sqrt{7y}} = \dfrac{\sqrt{6x}}{\sqrt{7y}} \cdot \dfrac{\sqrt{7y}}{\sqrt{7y}} = \dfrac{\sqrt{42xy}}{7y}$ Rationalizing

59. $\sqrt{\dfrac{4a^3}{3b}} = \dfrac{2a\sqrt{a}}{\sqrt{3b}} = \dfrac{2a\sqrt{a}}{\sqrt{3b}} \cdot \dfrac{\sqrt{3b}}{\sqrt{3b}} = \dfrac{2a\sqrt{3ab}}{3b}$ Rationalizing

61. $\sqrt{18m^3n^4}\sqrt{2m^3n^2} = \sqrt{36m^6n^6} = \sqrt{6^2(m^3)^2(n^3)^2} = 6m^3n^3$

63. $\dfrac{\sqrt{4a^3}}{\sqrt{3b}} = \dfrac{2a\sqrt{a}}{\sqrt{3b}} = \dfrac{2a\sqrt{a}}{\sqrt{3b}} \cdot \dfrac{\sqrt{3b}}{\sqrt{3b}} = \dfrac{2a\sqrt{3ab}}{3b}$

65. $\dfrac{5\sqrt{x}}{3 - 2\sqrt{x}} = \dfrac{5\sqrt{x}}{3 - 2\sqrt{x}} \cdot \dfrac{3 + 2\sqrt{x}}{3 + 2\sqrt{x}} = \dfrac{15\sqrt{x} + 10x}{9 - 4x}$ Rationalizing

67. $\dfrac{3\sqrt{2} - 2\sqrt{3}}{3\sqrt{3} - 2\sqrt{2}} = \dfrac{3\sqrt{2} - 2\sqrt{3}}{3\sqrt{3} - 2\sqrt{2}} \cdot \dfrac{3\sqrt{3} + 2\sqrt{2}}{3\sqrt{3} + 2\sqrt{2}} = \dfrac{9\sqrt{6} + 6\cdot 2 - 6\cdot 3 - 4\sqrt{6}}{27 - 8} = \dfrac{5\sqrt{6} - 6}{19}$

69. $\dfrac{9,600,000,000}{(1,600,000)(0.000\,000\,25)} = \dfrac{9.6 \times 10^9}{(1.6 \times 10^6)(2.5 \times 10^{-7})} = \dfrac{9.6 \times 10^9}{1.6(2.5) \times 10^{6-7}}$

$\qquad\qquad = \dfrac{9.6 \times 10^9}{4.0 \times 10^{-1}} = 2.4 \times 10^{9+1} = 2.4 \times 10^{10}$

$\qquad\qquad = 24,000,000,000$

71. $\dfrac{(1,250,000)(0.000\,38)}{0.0152} = \dfrac{(1.25 \times 10^6)(3.8 \times 10^{-4})}{1.52 \times 10^{-2}} = \dfrac{1.25(3.8) \times 10^{6-4}}{1.52 \times 10^{-2}}$

$\qquad\qquad = 3.125 \times 10^4 = 31,250$

73. $\left[\left(\dfrac{x^{-2}y^3t}{x^{-3}y^{-2}t^2}\right)^2\right]^{-1} = \left[\left(\dfrac{x^{-2+3}y^{3+2}}{t^{2-1}}\right)^2\right]^{-1} = \left[\left(\dfrac{xy^5}{t}\right)^2\right]^{-1} = \left[\dfrac{x^2y^{10}}{t^2}\right]^{-1} = \dfrac{x^{-2}y^{-10}}{t^{-2}} = \dfrac{t^2}{x^2y^{10}}$

75. $\left(\dfrac{2^2x^2y^0}{8x^{-1}}\right)^{-2}\left(\dfrac{x^{-3}}{x^{-5}}\right)^3 = \left(\dfrac{2^2x^{2+1}}{2^3}\right)^{-2}(x^{-3+5})^3 = \left(\dfrac{x^3}{2^{3-2}}\right)^{-2}(x^2)^3 = \left(\dfrac{x^3}{2}\right)^{-2}x^6$

$\qquad\qquad\qquad = \dfrac{x^{-6}}{2^{-2}} \cdot \dfrac{x^6}{1} = \dfrac{x^{-6+6}}{2^{-2}} = 2^2x^0 = 2^2 = 4$

77. $\dfrac{4(x - 3)^{-4}}{8(x - 3)^{-2}} = \dfrac{4}{8(x - 3)^{-2+4}} = \dfrac{1}{2(x - 3)^2}$

79. $\dfrac{b^{-2} - c^{-2}}{b^{-3} - c^{-3}} = \dfrac{\dfrac{1}{b^2} - \dfrac{1}{c^2}}{\dfrac{1}{b^3} - \dfrac{1}{c^3}} = \dfrac{\dfrac{c^2 - b^2}{b^2 c^2}}{\dfrac{c^3 - b^3}{b^3 c^3}} = \dfrac{\cancel{(c-b)}(c+b)}{\cancel{b^2 c^2}} \cdot \dfrac{\overset{bc}{\cancel{b^3 c^3}}}{\cancel{(c-b)}(c^2 + cb + b^2)}$

$$= \dfrac{bc(c+b)}{c^2 + cb + b^2}$$

81. $\dfrac{\sqrt{2x}\sqrt{5}}{\sqrt{20x}} = \dfrac{\sqrt{2x}\sqrt{5}}{\sqrt{2\cdot 5 \cdot 2x}} = \dfrac{\sqrt{2x}\sqrt{5}}{\sqrt{2}\sqrt{5}\sqrt{2x}} = \dfrac{1}{\sqrt{2}} = \dfrac{1}{\sqrt{2}}\cdot\dfrac{\sqrt{2}}{\sqrt{2}} = \dfrac{\sqrt{2}}{2}$ Rationalizing

83. $\dfrac{2}{\sqrt{x-2}} = \dfrac{2}{\sqrt{x-2}}\cdot\dfrac{\sqrt{x-2}}{\sqrt{x-2}} = \dfrac{2\sqrt{x-2}}{x-2}$ Rationalizing

85. $\dfrac{\sqrt{t}-\sqrt{x}}{t-x} = \dfrac{\sqrt{t}-\sqrt{x}}{t-x}\cdot\dfrac{\sqrt{t}+\sqrt{x}}{\sqrt{t}+\sqrt{x}} = \dfrac{\cancel{t-x}}{\cancel{(t-x)}(\sqrt{t}+\sqrt{x})} = \dfrac{1}{\sqrt{t}+\sqrt{x}}$ Rationalizing

87. $\dfrac{\sqrt{x+h}-\sqrt{x}}{h} = \dfrac{\sqrt{x+h}-\sqrt{x}}{h}\cdot\dfrac{\sqrt{x+h}+\sqrt{x}}{\sqrt{x+h}+\sqrt{x}} = \dfrac{x+h-x}{h(\sqrt{x+h}+\sqrt{x})}$

$$= \dfrac{\cancel{h}}{\cancel{h}(\sqrt{x+h}+\sqrt{x})} = \dfrac{1}{\sqrt{x+h}+\sqrt{x}}$$

EXERCISE 1-7

Things to remember:

<u>1</u>. r is an nth ROOT of b if $r^n = b$.

<u>2</u>. $b^{m/n} = \begin{cases} (b^{1/n})^m = (\sqrt[n]{b})^m \\ (b^m)^{1/n} = (\sqrt[n]{b^m}) \end{cases}$

<u>3</u>. $b^{-m/n} = \dfrac{1}{b^{m/n}},\ b \ne 0$

<u>4</u>. PROPERTIES OF RADICALS

c, n, and m are natural numbers 2 or larger, x and y are positive real numbers.

a. $\sqrt[n]{x^n} = x$ $\qquad\qquad\qquad$ $\sqrt[3]{x^3} = x$

b. $\sqrt[n]{xy} = \sqrt[n]{x}\,\sqrt[n]{y}$ $\qquad\qquad$ $\sqrt[5]{xy} = \sqrt[5]{x}\,\sqrt[5]{y}$

c. $\sqrt[n]{\dfrac{x}{y}} = \dfrac{\sqrt[n]{x}}{\sqrt[n]{y}}$ $\qquad$ $\sqrt[4]{\dfrac{x}{y}} = \dfrac{\sqrt[4]{x}}{\sqrt[4]{y}}$

d. $\sqrt[cn]{x^{cm}} = \sqrt[n]{x^m}$ $\qquad$ $\sqrt[12]{x^8} = \sqrt[4\cdot3]{x^{4\cdot2}} = \sqrt[3]{x^2}$

5. SIMPLEST RADICAL FORM

A radicand (the expression within the radical sign) contains no factor to a power greater than or equal to the index of the radical.

($\sqrt[3]{x^5}$ violates this condition.)

The power of the radicand and the index of the radical have no common factor other than 1.

($\sqrt[6]{x^4}$ violates this condition.)

No radical appears in a denominator.

$\left(\dfrac{y}{\sqrt[3]{x}}$ violates this condition.$\right)$

No fraction appears within a radical.

$\left(\sqrt[4]{\dfrac{3}{5}}$ violates this condition.$\right)$

---

1. $6x^{3/5} = 6\sqrt[5]{x^3}$

3. $(4xy^3)^{2/5} = \sqrt[5]{(4xy^3)^2}$

5. $(x^2 + y^2)^{1/2} = \sqrt{x^2 + y^2}$
[Note: $\sqrt{x^2 + y^2} \neq x + y$.]

7. $5\sqrt[4]{x^3} = 5x^{3/4}$

9. $\sqrt[5]{(2x^2y)^3} = (2x^2y)^{3/5}$

11. $\sqrt[3]{x} + \sqrt[3]{y} = x^{1/3} + y^{1/3}$

13. $25^{1/2} = (5^2)^{1/2} = 5$

15. $16^{3/2} = (4^2)^{3/2} = 4^3 = 64$

17. $-36^{1/2} = -(6^2)^{1/2} = -6$

19. $(-36)^{1/2}$ is not a rational number; -36 does not have a real square root; $(-36)^{1/2}$ is not a real number.

21. $\left(\dfrac{4}{25}\right)^{3/2} = \left(\left(\dfrac{2}{5}\right)^2\right)^{3/2} = \left(\dfrac{2}{5}\right)^3 = \dfrac{2^3}{5^3} = \dfrac{8}{125}$

23. $9^{-3/2} = (3^2)^{-3/2} = 3^{-3} = \dfrac{1}{3^3} = \dfrac{1}{27}$

25. $x^{4/5}x^{-2/5} = x^{4/5-2/5} = x^{2/5}$

27. $\dfrac{m^{2/3}}{m^{-1/3}} = m^{2/3-(-1/3)} = m^1 = m$

29. $(8x^3y^{-6})^{1/3} = (2^3x^3y^{-6})^{1/3} = 2^{3/3}x^{3/3}y^{-6/3} = 2xy^{-2} = \dfrac{2x}{y^2}$

31. $\left(\dfrac{4x^{-2}}{y^4}\right)^{-1/2} = \left(\dfrac{2^2x^{-2}}{y^4}\right)^{-1/2} = \dfrac{2^{2(-1/2)}x^{-2(-1/2)}}{y^{4(-1/2)}} = \dfrac{2^{-1}x^1}{y^{-2}} = \dfrac{xy^2}{2}$

33. $\dfrac{8x^{-1/3}}{12x^{1/4}} = \dfrac{2}{3x^{1/4+1/3}} = \dfrac{2}{3x^{7/12}}$

35. $\left(\dfrac{8x^{-4}y^3}{27x^2y^{-3}}\right)^{1/3} = \left(\dfrac{2^3y^{3+3}}{3^3x^{2+4}}\right)^{1/3} = \left(\dfrac{2^3y^6}{3^3x^6}\right)^{1/3} = \dfrac{2y^2}{3x^2}$

37. $3x^{3/4}(4x^{1/4} - 2x^8) = 12x^{3/4+1/4} - 6x^{3/4+8}$
$\qquad\qquad = 12x - 6x^{3/4+32/4} = 12x - 6x^{35/4}$

39. $(3u^{1/2} - v^{1/2})(u^{1/2} - 4v^{1/2}) = 3u - 12u^{1/2}v^{1/2} - u^{1/2}v^{1/2} + 4v$
$\qquad\qquad\qquad = 3u - 13u^{1/2}v^{1/2} + 4v$

41. $(5m^{1/2} + n^{1/2})(5m^{1/2} - n^{1/2}) = (5m^{1/2})^2 - (n^{1/2})^2 = 25m - n$

43. $(3x^{1/2} - y^{1/2})^2 = (3x^{1/2})^2 - 6x^{1/2}y^{1/2} + (y^{1/2})^2 = 9x - 6x^{1/2}y^{1/2} + y$

45. $\dfrac{x^{2/3} + 2}{2x^{1/3}} = \dfrac{x^{2/3}}{2x^{1/3}} + \dfrac{2}{2x^{1/3}} = \dfrac{1}{2}x^{1/3} + \dfrac{1}{x^{1/3}} = \dfrac{1}{2}x^{1/3} + x^{-1/3}$

47. $\dfrac{2x^{3/4} + 3x^{1/3}}{3x} = \dfrac{2x^{3/4}}{3x} + \dfrac{3x^{1/3}}{3x} = \dfrac{2}{3}x^{3/4-1} + x^{1/3-1} = \dfrac{2}{3}x^{-1/4} + x^{-2/3}$

49. $\dfrac{2x^{1/3} - x^{1/2}}{4x^{1/2}} = \dfrac{2x^{1/3}}{4x^{1/2}} - \dfrac{x^{1/2}}{4x^{1/2}} = \dfrac{1}{2}x^{1/3-1/2} - \dfrac{1}{4} = \dfrac{1}{2}x^{-1/6} - \dfrac{1}{4}$

51. $\sqrt[3]{16m^4n^6} = \sqrt[3]{2\cdot 8\cdot m^3m(n^2)^3} = \sqrt[3]{2\cdot 2^3\cdot m^3m(n^2)^3} = 2mn^2\sqrt[3]{2m}$

53. $\sqrt[4]{32m^9n^7} = \sqrt[4]{2\cdot 2^4(m^2)^4mn^4n^3} = 2m^2n\sqrt[4]{2mn^3}$

55. $\dfrac{x}{\sqrt[3]{x}} = \dfrac{x}{x^{1/3}} = x^{1-1/3} = x^{2/3}$ $\qquad$ or $\qquad$ $\dfrac{x}{\sqrt[3]{x}} = \dfrac{x}{\sqrt[3]{x}}\cdot\dfrac{\sqrt[3]{x^2}}{\sqrt[3]{x^2}} = \dfrac{x\sqrt[3]{x^2}}{x}$

$\qquad\qquad\qquad\qquad = \sqrt[3]{x^2}$ $\qquad\qquad\qquad\qquad\qquad\qquad = \sqrt[3]{x^2}$

57. $\dfrac{4a^3b^2}{\sqrt[3]{2ab^2}} = \dfrac{4a^3b^2}{\sqrt[3]{2ab^2}}\cdot\dfrac{\sqrt[3]{2^2a^2b}}{\sqrt[3]{2^2a^2b}} = \dfrac{4a^3b^2\sqrt[3]{4a^2b}}{2ab} = 2a^2b\sqrt[3]{4a^2b}$

59. $\sqrt[4]{\dfrac{3x^3}{4}} = \sqrt[4]{\dfrac{3x^3}{4}\cdot\dfrac{4}{4}} = \sqrt[4]{\dfrac{12x^3}{16}} = \dfrac{\sqrt[4]{12x^3}}{2}$

61. $\sqrt[12]{(x-3)^9} = (x-3)^{9/12}$

$\qquad\qquad\quad = (x-3)^{3/4}$

$\qquad\qquad\quad = \sqrt[4]{(x-3)^3}$

63. $\sqrt{x}\,\sqrt[3]{x^2} = x^{1/2}x^{2/3}$

$\qquad\qquad\quad = x^{1/2+2/3} = x^{7/6}$

$\qquad\qquad\quad = x \cdot x^{1/6} = x\sqrt[6]{x}$

65. $\dfrac{\sqrt{x}}{\sqrt[3]{x^2}} = \dfrac{x^{1/2}}{x^{2/3}} = \dfrac{1}{x^{2/3-1/2}} = \dfrac{1}{x^{1/6}} = \dfrac{1}{x^{1/6}} \cdot \dfrac{x^{5/6}}{x^{5/6}} = \dfrac{x^{5/6}}{x} = \dfrac{\sqrt[6]{x^5}}{x}$

67. $\dfrac{(x-1)^{1/2} - x\left(\frac{1}{2}\right)(x-1)^{-1/2}}{x-1} = \dfrac{(x-1)^{1/2} - \dfrac{x}{2(x-1)^{1/2}}}{x-1}$

$\qquad\qquad\qquad\qquad\qquad\qquad = \dfrac{\dfrac{2(x-1)^{1/2}(x-1)^{1/2}}{2(x-1)^{1/2}} - \dfrac{x}{2(x-1)^{1/2}}}{x-1}$

$\qquad\qquad\qquad\qquad\qquad\qquad = \dfrac{\dfrac{2(x-1) - x}{2(x-1)^{1/2}}}{x-1} = \dfrac{x-2}{2(x-1)^{3/2}}$

69. $\dfrac{(x+2)^{2/3} - x\left(\frac{2}{3}\right)(x+2)^{-1/3}}{(x+2)^{4/3}} = \dfrac{(x+2)^{2/3} - \dfrac{2x}{3(x+2)^{1/3}}}{(x+2)^{4/3}}$

$\qquad\qquad\qquad\qquad\qquad\qquad = \dfrac{\dfrac{3(x+2)^{1/3}(x+2)^{2/3}}{3(x+2)^{1/3}} - \dfrac{2x}{3(x+2)^{1/3}}}{(x+2)^{4/3}}$

$\qquad\qquad\qquad\qquad\qquad\qquad = \dfrac{\dfrac{3(x+2) - 2x}{3(x+2)^{1/3}}}{(x+2)^{4/3}} = \dfrac{x+6}{3(x+2)^{5/3}}$

71. $22^{3/2} = 22^{1.5} \approx 103.2 \qquad$ or $\qquad 22^{3/2} = \sqrt{(22)^3} = \sqrt{10,648} \approx 103.2$

73. $827^{-3/8} = \dfrac{1}{827^{3/8}} = \dfrac{1}{827^{0.375}} \approx \dfrac{1}{12.42} \approx 0.0805$

75. $37.09^{7/3} \approx 37.09^{2.3333} \approx 4,588$

1. (A) $7 \notin \{4, 6, 8\}$ is true (T)   (B) $\{8\} \subset \{4, 6, 8\}$ is true (T)
   (C) $\varnothing \in \{4, 6, 8\}$ is false (F)   (D) $\varnothing \subset \{4, 6, 8\}$ is true (T)

2. (A) Commutative property ($\cdot$): $x(y + z) = (y + z)x$
   (B) Associative property (+): $2 + (x + y) = (2 + x) + y$
   (C) Distributive property: $(2 + 3)x = 2x + 3x$

3. $(3x - 4) + (x + 2) + (3x^2 + x - 8) + (x^3 + 8)$
   $= 3x - 4 + x + 2 + 3x^2 + x - 8 + x^3 + 8 = x^3 + 3x^2 + 5x - 2$

4. $[(x + 2) + (x^3 + 8)] - [(3x - 4) + (3x^2 + x - 8)]$
   $= x^3 + x + 10 - [3x^2 + 4x - 12]$
   $= x^3 + x + 10 - 3x^2 - 4x + 12 = x^3 - 3x^2 - 3x + 22$

5. $(x^3 + 8)(3x^2 + x - 8) = x^3(3x^2 + x - 8) + 8(3x^2 + x - 8)$
   $= 3x^5 + x^4 - 8x^3 + 24x^2 + 8x - 64$

6. $x^3 + 8$ has degree $\underline{\underline{3}}$

7. The coefficient of the second term in $3x^2 + x - 8$ is $\underline{\underline{1}}$

8. $5x^2 - 3x[4 - 3(x - 2)] = 5x^2 - 3x[4 - 3x + 6]$
   $= 5x^2 - 3x(-3x + 10)$
   $= 5x^2 + 9x^2 - 30x$
   $= 14x^2 - 30x$

9. $(3m - 5n)(3m + 5n) = (3m)^2 - (5n)^2 = 9m^2 - 25n^2$

10. $(2x + y)(3x - 4y) = 6x^2 - 8xy + 3xy - 4y^2$
    $= 6x^2 - 5xy - 4y^2$

11. $(2a - 3b)^2 = (2a)^2 - 2(2a)(3b) + (3b)^2$
    $= 4a^2 - 12ab + 9b^2$

12. $9x^2 - 12x + 4 = (3x - 2)(3x - 2) = (3x - 2)^2$

13. $t^2 - 4t - 6$ This polynomial is prime.

14. $6n^3 - 9n^2 - 15n = 3n(2n^2 - 3n - 5) = 3n(2n - 5)(n + 1)$

15. $\dfrac{2}{5b} - \dfrac{4}{3a^3} - \dfrac{1}{6a^2b^2}$ LCD $= 30a^3b^2$

   $= \dfrac{6a^3b}{6a^3b} \cdot \dfrac{2}{5b} - \dfrac{10b^2}{10b^2} \cdot \dfrac{4}{3a^3} - \dfrac{5a}{5a} \cdot \dfrac{1}{6a^2b^2}$

   $= \dfrac{12a^3b}{30a^3b^2} - \dfrac{40b^2}{30a^3b^2} - \dfrac{5a}{30a^3b^2} = \dfrac{12a^3b - 40b^2 - 5a}{30a^3b^2}$

16. $\dfrac{3x}{3x^2 - 12x} + \dfrac{1}{6x} = \dfrac{\cancel{3x}}{\cancel{3x}(x - 4)} + \dfrac{1}{6x} = \dfrac{1}{x - 4} + \dfrac{1}{6x}$

$\qquad\qquad = \dfrac{6x}{6x(x - 4)} + \dfrac{x - 4}{6x(x - 4)} = \dfrac{6x + x - 4}{6x(x - 4)} = \dfrac{7x - 4}{6x(x - 4)}$

17. $\dfrac{y - 2}{y^2 - 4y + 4} \div \dfrac{y^2 + 2y}{y^2 + 4y + 4} = \dfrac{\cancel{y - 2}}{\underset{y - 2}{\cancel{(y - 2)^2}}} \cdot \dfrac{\overset{y + 2}{\cancel{(y + 2)^2}}}{y\cancel{(y + 2)}} = \dfrac{y + 2}{y(y - 2)}$

18. $\dfrac{u - \dfrac{1}{u}}{1 - \dfrac{1}{u^2}} = \dfrac{\dfrac{u^2 - 1}{u}}{\dfrac{u^2 - 1}{u^2}} = \dfrac{\cancel{u^2 - 1}}{\cancel{u}} \cdot \dfrac{\overset{u}{\cancel{u^2}}}{\cancel{u^2 - 1}} = u$

19. $6(xy^3)^5 = 6x^5y^{15}$

20. $\dfrac{9u^8v^6}{3u^4v^8} = \dfrac{3^2u^8v^6}{3u^4v^8} = \dfrac{3^{2-1}u^{8-4}}{v^{8-6}} = \dfrac{3u^4}{v^2}$

21. $(2 \times 10^5)(3 \times 10^{-3}) = 6 \times 10^{5-3} = 6 \times 10^2 = 600$

22. $(x^{-3}y^2)^{-2} = x^6y^{-4} = \dfrac{x^6}{y^4}$

23. $u^{5/3}u^{2/3} = u^{5/3+2/3} = u^{7/3}$

24. $(9a^4b^{-2})^{1/2} = (3^2a^4b^{-2})^{1/2} = (3^2)^{1/2}(a^4)^{1/2}(b^{-2})^{1/2} = 3a^2b^{-1} = \dfrac{3a^2}{b}$

25. $3x^{2/5} = 3\sqrt[5]{x^2}$

26. $-3\sqrt[3]{(xy)^2} = -3(xy)^{2/3}$

27. $3x\sqrt[3]{x^5y^4} = 3x\sqrt[3]{x^3x^2y^3y} = 3x^2y\sqrt[3]{x^2y}$

28. $\sqrt{2x^2y^5}\sqrt{18x^3y^2} = \sqrt{(2x^2y^5)(18x^3y^2)} = \sqrt{36x^5y^7} = \sqrt{6^2x^4xy^6y} = 6x^2y^3\sqrt{xy}$

29. $\dfrac{6ab}{\sqrt{3a}} = \dfrac{6ab}{\sqrt{3a}}\dfrac{\sqrt{3a}}{\sqrt{3a}} = \dfrac{6ab\sqrt{3a}}{3a} = 2b\sqrt{3a}$

30. $\dfrac{\sqrt{5}}{3 - \sqrt{5}} = \dfrac{\sqrt{5}}{3 - \sqrt{5}} \cdot \dfrac{3 + \sqrt{5}}{3 + \sqrt{5}} = \dfrac{3\sqrt{5} + 5}{9 - 5} = \dfrac{3\sqrt{5} + 5}{4}$

31. $\sqrt[8]{y^6} = y^{6/8} = y^{3/4} = \sqrt[4]{y^3}$

32. (A) $A \cup B = \{1,\ 2,\ 3\} \cup \{2,\ 3,\ 4\} = \{1,\ 2,\ 3,\ 4\}$
(B) $\{x \mid x \in A \text{ and } x \in B\} = A \cap B = \{1,\ 2,\ 3\} \cap \{2,\ 3,\ 4\} = \{2,\ 3\}$

33. $u = \{2,\ 4,\ 5,\ 6,\ 8\}$, $M = \{2,\ 4,\ 5\}$, and $N = \{5,\ 6\}$
(A) $M \cup N = \{2,\ 4,\ 5,\ 6\}$ (B) $M \cap N = \{5\}$
(C) $(M \cup N)' = \{8\}$ (D) $M \cap N' = \{2,\ 4,\ 5\} \cap \{2,\ 4,\ 8\} = \{2,\ 4\}$

34. (A) $N \subset M$ is false (F)   (B) $\varnothing \subset u$ is true (T)
(C) $6 \notin M$ is true (T)   (D) $5 \in N$ is true (T)

35. From the Venn diagram:

(A) $M \cup N$ has $10 + 5 + 13 = 28$ elements

(B) $M \cap N$ has 5 elements

(C) $(M \cup N)'$ has 4 elements

(D) $M \cap N'$ has 10 elements

36. Let u = the set of students in the sample (100),
E = the set of students taking English (70),
M = the set of students taking Math (45), and
$E \cap M$ = the set of students taking English and Math (25).

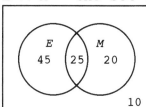

(A) The number of students taking English or Math is the number of elements in $E \cup M$, which is $45 + 25 + 20 = 90$.

(B) The number of students taking English and not Math is the number of elements in $E \cap M'$, which is 45.

37. $-7 - (-5) = -7 + [-(-5)]$: Definition of subtraction

38. $5u + (3v + 2) = (3v + 2) + 5u$: Commutative property (+)

39. $(5m - 2)(2m + 3) = (5m - 2)2m + (5m - 2)3$: Distributive property

40. $9 \cdot (4y) = (9 \cdot 4)y$: Associative property ($\cdot$)

41. $\dfrac{u}{-(v - w)} = -\dfrac{u}{v - w}$: Property of negatives

42. $(x - y) + 0 = (x - y)$: 0 is the additive identity

43. (A) True: $n = \dfrac{n}{1}$ for any natural number n.

(B) False: A repeating decimal represents a rational number.

44. $0, -1, -2, \ldots$ are examples of integers which are not natural numbers.

45. (A) (a) and (d) are second-degree polynomials.   (B) None

46. $(2x - y)(2x + y) - (2x - y)^2 = (2x)^2 - y^2 - (4x^2 - 4xy + y^2)$
$$= 4x^2 - y^2 - 4x^2 + 4xy - y^2$$
$$= 4xy - 2y^2$$

47. $(m^2 + 2mn - n^2)(m^2 - 2mn - n^2)$
$= m^2(m^2 - 2mn - n^2) + 2mn(m^2 - 2mn - n^2) - n^2(m^2 - 2mn - n^2)$
$= m^4 - 2m^3n - m^2n^2 + 2m^3n - 4m^2n^2 - 2mn^3 - m^2n^2 + 2mn^3 + n^4$
$= m^4 - 6m^2n^2 + n^4$

48. $-2x\{(x^2 + 2)(x - 3) - x[x - x(3 - x)]\}$
$= -2x\{x^3 - 3x^2 + 2x - 6 - x[x - 3x + x^2]\}$
$= -2x\{x^3 - 3x^2 + 2x - 6 - x[-2x + x^2]\}$
$= -2x\{x^3 - 3x^2 + 2x - 6 + 2x^2 - x^3\}$
$= -2x\{-x^2 + 2x - 6\} = 2x^3 - 4x^2 + 12x$

49. $(x - 2y)^3 = (x - 2y)(x - 2y)^2$
$$= (x - 2y)(x^2 - 4xy + 4y^2)$$
$$= x(x^2 - 4xy + 4y^2) - 2y(x^2 - 4xy + 4y^2)$$
$$= x^3 - 4x^2y + 4xy^2 - 2x^2y + 8xy^2 - 8y^3$$
$$= x^3 - 6x^2y + 12xy^2 - 8y^3$$

50. $(4x - y)^2 - 9x^2 = (4x - y - 3x)(4x - y + 3x) = (x - y)(7x - y)$

51. $2x^2 + 4xy - 5y^2$ is a prime polynomial relative to the integers.

52. $6x^3y + 12x^2y^2 - 15xy^3 = 3xy(2x^2 + 4xy - 5y^2)$
[Note: $2x^2 + 4xy - 5y^2$ is prime.]

53. $3x^3 + 24y^3 = 3(x^3 + 8y^3) = 3[x^3 + (2y)^3]$
$$= 3(x + 2y)(x^2 - 2xy + 4y^2)$$

54. $\dfrac{m - 1}{m^2 - 4m + 4} + \dfrac{m + 3}{m^2 - 4} + \dfrac{2}{2 - m}$ LCD $= (m + 2)(m - 2)^2$

$$= \frac{(m + 2)(m - 1)}{(m + 2)(m - 2)^2} + \frac{(m + 3)(m - 2)}{(m + 2)(m - 2)^2} - \frac{2(m - 2)(m + 2)}{(m + 2)(m - 2)^2}$$

$$= \frac{m^2 + m - 2 + m^2 + m - 6 - 2(m^2 - 4)}{(m + 2)(m - 2)^2} = \frac{2m}{(m + 2)(m - 2)^2}$$

55. $\dfrac{y}{x^2} \div \left(\dfrac{x^2 + 3x}{2x^2 + 5x - 3} \div \dfrac{x^3y - x^2y}{2x^2 - 3x + 1} \right)$

$$= \frac{y}{x^2} \div \left[\frac{x\cancel{(x + 3)}}{(2x - 1)\cancel{(x + 3)}} \div \frac{x^2y\cancel{(x - 1)}}{(2x - 1)\cancel{(x - 1)}} \right]$$

$$= \frac{y}{x^2} \div \left[\frac{\cancel{x}}{\cancel{2x - 1}} \cdot \frac{\cancel{2x - 1}}{\underset{xy}{\cancel{x^2y}}} \right] = \frac{y}{x^2} \div \frac{1}{xy} = \frac{y}{x^2} \cdot \frac{xy}{1} = \frac{y^2}{x}$$

56. $\dfrac{a^{-1} - b^{-1}}{ab^{-2} - ba^{-2}} = \dfrac{\dfrac{1}{a} - \dfrac{1}{b}}{\dfrac{a}{b^2} - \dfrac{b}{a^2}} = \dfrac{\dfrac{b - a}{ab}}{\dfrac{a^3 - b^3}{a^2b^2}} = \dfrac{b - a}{\cancel{ab}} \cdot \dfrac{\underset{a^2b^2}{\cancel{ab}}^{ab}}{a^3 - b^3}$

$$= (b - a) \cdot \frac{ab}{(a - b)(a^2 + ab + b^2)} = \frac{-ab}{a^2 + ab + b^2}$$

57. $\left(\dfrac{8u^{-1}}{2^2u^2v^0} \right)^{-2} \left(\dfrac{u^{-5}}{u^{-3}} \right)^3 = \left(\dfrac{2^3u^{-1}}{2^2u^2} \right)^{-2} \left(\dfrac{1}{u^{-3+5}} \right)^3 = \left(\dfrac{2}{u^3} \right)^{-2} \left(\dfrac{1}{u^2} \right)^3$

$$= \frac{2^{-2}}{u^{-6}} \cdot \frac{1}{u^6} = \frac{2^{-2}}{u^0} = \frac{1}{2^2} = \frac{1}{4}$$

58. $\dfrac{5^0}{3^2} + \dfrac{3^{-2}}{2^{-2}} = \dfrac{1}{3^2} + \dfrac{\dfrac{1}{3^2}}{\dfrac{1}{2^2}} = \dfrac{1}{9} + \dfrac{1}{9} \cdot \dfrac{4}{1} = \dfrac{5}{9}$

59. $\left(\dfrac{27x^2y^{-3}}{8x^{-4}y^3}\right)^{1/3} = \left(\dfrac{3^3x^{2+4}}{2^3y^{3+3}}\right)^{1/3} = \left(\dfrac{3^3x^6}{2^3y^6}\right)^{1/3} = \dfrac{3x^2}{2y^2}$

60. $(a^{-1/3}b^{1/4})(9a^{1/3}b^{-1/2})^{3/2} = a^{-1/3}b^{1/4}9^{3/2}a^{1/2}b^{-3/4}$

$$= 27a^{-1/3+1/2}b^{1/4-3/4}$$

$$= 27a^{1/6}b^{-1/2} = \dfrac{27a^{1/6}}{b^{1/2}}$$

61. $(x^{1/2} + y^{1/2})^2 = (x^{1/2})^2 + 2x^{1/2}y^{1/2} + (y^{1/2})^2 = x + 2x^{1/2}y^{1/2} + y$

62. $(3x^{1/2} - y^{1/2})(2x^{1/2} + 3y^{1/2}) = 6x + 9x^{1/2}y^{1/2} - 2x^{1/2}y^{1/2} - 3y$

$$= 6x + 7x^{1/2}y^{1/2} - 3y$$

63. $\dfrac{0.000\ 000\ 000\ 52}{(1,300)(0.000\ 002)} = \dfrac{5.2 \times 10^{-10}}{(1.3 \times 10^3)(2 \times 10^{-6})} = \dfrac{5.2 \times 10^{-10}}{2.6 \times 10^{-3}} = 2 \times 10^{-7}$

64. $-2x\sqrt[5]{3^6x^7y^{11}} = -2x\sqrt[5]{3^5 \cdot 3 \cdot x^5 \cdot x^2 \cdot y^{10} \cdot y}$

$$= -2x \cdot 3 \cdot x \cdot y^2 \sqrt[5]{3x^2y}$$

$$= -6x^2y^2\sqrt[5]{3x^2y}$$

65. $\dfrac{2x^2}{\sqrt[3]{4x}} = \dfrac{2x^2}{(2^2x)^{1/3}} = \dfrac{2x^2}{2^{2/3}x^{1/3}} = 2^{1-2/3}x^{2-1/3} = 2^{1/3}x^{5/3}$

$$= \sqrt[3]{2x^5} = \sqrt[3]{2x^3x^2} = x\sqrt[3]{2x^2}$$

66. $\sqrt[5]{\dfrac{3y^2}{8x^2}} = \dfrac{\sqrt[5]{3y^2}}{\sqrt[5]{2^3x^2}}$

$$= \dfrac{3^{1/5}y^{2/5}}{2^{3/5}x^{2/5}} \cdot \dfrac{2^{2/5}x^{3/5}}{2^{2/5}x^{3/5}} = \dfrac{2^{2/5}3^{1/5}x^{3/5}y^{2/5}}{2x} = \dfrac{\sqrt[5]{2^2 3x^3y^2}}{2x} = \dfrac{\sqrt[5]{12x^3y^2}}{2x}$$

67. $\sqrt[9]{8x^6y^{12}} = \sqrt[9]{2^3x^6y^{12}} = (2^3x^6y^{12})^{1/9} = 2^{3/9}x^{6/9}y^{12/9}$

$$= 2^{1/3}x^{2/3}y^{4/3} = 2^{1/3}x^{2/3}y^{1/3}y = y\sqrt[3]{2x^2y}$$

68. $(2\sqrt{x} - 5\sqrt{y})(\sqrt{x} + \sqrt{y}) = 2x + 2\sqrt{x}\sqrt{y} - 5\sqrt{x}\sqrt{y} - 5y = 2x - 3\sqrt{xy} - 5y$

69. $\dfrac{3\sqrt{x}}{2\sqrt{x} - \sqrt{y}} = \dfrac{3\sqrt{x}}{2\sqrt{x} - \sqrt{y}} \cdot \dfrac{2\sqrt{x} + \sqrt{y}}{2\sqrt{x} + \sqrt{y}} = \dfrac{6x + 3\sqrt{xy}}{(2\sqrt{x})^2 - (\sqrt{y})^2} = \dfrac{6x + 3\sqrt{xy}}{4x - y}$

70. $\dfrac{2\sqrt{u} - 3\sqrt{v}}{2\sqrt{u} + 3\sqrt{v}} = \dfrac{2\sqrt{u} - 3\sqrt{v}}{2\sqrt{u} + 3\sqrt{v}} \cdot \dfrac{2\sqrt{u} - 3\sqrt{v}}{2\sqrt{u} - 3\sqrt{v}} = \dfrac{4u - 12\sqrt{uv} + 9v}{4u - 9v}$

71. Yes. If $A \not\subset B$, then there is at least one $x \in A$ such that $x \notin B$. Since $x \notin B$, $x \notin A \cap B$. Thus, we have $x \in A$ and $x \notin A \cap B$, and so $A \neq A \cap B$.

72. $\dfrac{\sqrt{t} - \sqrt{5}}{t - 5} = \dfrac{\sqrt{t} - \sqrt{5}}{t - 5} \cdot \dfrac{\sqrt{t} + \sqrt{5}}{\sqrt{t} + \sqrt{5}} = \dfrac{t - 5}{(t - 5)(\sqrt{t} + \sqrt{5})} = \dfrac{1}{\sqrt{t} + \sqrt{5}}$

73. $\dfrac{4\sqrt{x} - 3}{2\sqrt{x}} = \dfrac{4x^{1/2}}{2x^{1/2}} - \dfrac{3}{2x^{1/2}} = 2 - \dfrac{3}{2}x^{-1/2}$

74. Letting $x = 2 - \sqrt{3}$ in $x^2 - 4x + 1$, we have:
$(2 - \sqrt{3})^2 - 4(2 - \sqrt{3}) + 1 = 4 - 4\sqrt{3} + 3 - 8 + 4\sqrt{3} + 1 = 0$

75. $\begin{aligned} x(2x - 1)(x + 3) - (x - 1)^3 &= x(2x^2 + 5x - 3) - (x - 1)(x - 1)^2 \\ &= 2x^3 + 5x^2 - 3x - (x - 1)(x^2 - 2x + 1) \\ &= 2x^3 + 5x^2 - 3x - (x^3 - 3x^2 + 3x - 1) \\ &= 2x^3 + 5x^2 - 3x - x^3 + 3x^2 - 3x + 1 \\ &= x^3 + 8x^2 - 6x + 1 \end{aligned}$

76. $\begin{aligned} \dfrac{(20{,}410)(0.000\ 003\ 477)}{0.000\ 000\ 022\ 09} &= \dfrac{(2.041 \times 10^4)(3.477 \times 10^{-6})}{2.209 \times 10^{-8}} \\ &\approx 3.213 \times 10^{4+8-6} = 3.213 \times 10^6 \end{aligned}$

77. $(0.1347)^5 \approx 4.434 \times 10^{-5}$

78. $(-60.39)^{-3} = \dfrac{1}{(-60.39)^3} = \dfrac{-1}{(60.39)^3} \approx \dfrac{-1}{220{,}239.44} \approx -4.541 \times 10^{-6}$

79. $(82.45)^{8/3} \approx [(82.45)^{1/3}]^8 \approx [4.3524]^8 \approx 128{,}778 \approx 128{,}800$

80. $\sqrt[5]{0.006\ 604} = (0.006\ 604)^{1/5} = (0.006\ 604)^{0.2} \approx 0.3664$

81. $\sqrt[3]{3 + \sqrt{2}} \approx \sqrt[3]{4.41421} \approx 1.640$

82. Let C = the number of students who smoked (550)
and A = the number of students who drank alcoholic beverages (820).
Then $C \cap A = 470$ students who did both.

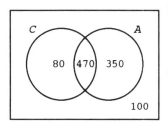

(A) The number of students who smoked or drank is the number of elements in $C \cup A$, which is
80 + 470 + 350 = 900.

(B) The number of students who drank but did not smoke is the number of elements in $A \cap C'$, which is 350.

CHAPTER 2   EQUATIONS, GRAPHS, AND FUNCTIONS

Things to remember:

<u>1</u>. If $a = b$, then   **(a)** $a + c = b + c$

   (b) $a - c = b - c$

   (c) $ac = bc, \quad c \neq 0$

   (d) $\dfrac{a}{c} = \dfrac{b}{c}, \quad c \neq 0$

<u>2</u>. If $a > b$, then   **(a)** $a + c > b + c$

   (b) $a - c > b - c$

   (c) $ac > bc$

   (d) $\dfrac{a}{c} > \dfrac{b}{c}$ $\left.\right\}$ if c is positive

   (e) $ac < bc$

   (f) $\dfrac{a}{c} < \dfrac{b}{c}$ $\left.\right\}$ if c is negative

[<u>Note</u>: Similar properties hold if each inequality is reversed, or if $>$ is replaced by $\geq$ and $<$ is replaced by $\leq$.

<u>3</u>. The double inequality $a \leq x \leq b$ means that $a \leq x$ and $x \leq b$. Other variations, as well as a useful interval notation, are indicated in the following table.

Interval Notation	Inequality Notation	Line Graph
$[a, b]$	$a \leq x \leq b$	
$[a, b)$	$a \leq x < b$	
$(a, b]$	$a < x \leq b$	
(a, b)	$a < x < b$	
$(-\infty, a]$	$x \leq a$	
$(-\infty, a)$	$x < a$	
$[b, \infty)$	$x \geq b$	
(b, ∞)	$x > b$	

[<u>Note</u>: An endpoint on a line graph has a square bracket through it if it is included in the inequality and a parenthesis through it if it is not.]

31

1.
$$2m + 9 = 5m - 6$$
$2m + 9 - 9 = 5m - 6 - 9$ [using $\underline{1}$(b)]
$$2m = 5m - 15$$
$2m - 5m = 5m - 15 - 5m$ [using $\underline{1}$(b)]
$$-3m = -15$$
$$\frac{-3m}{-3} = \frac{-15}{-3}$$ [using $\underline{1}$ (d)]
$$m = 5$$

3.
$$x + 5 < -4$$
$x + 5 - 5 < -4 - 5$ [using $\underline{2}$(b)]
$$x < -9$$

5. $-3x \geq -12$
$$\frac{-3x}{-3} \leq \frac{-12}{-3}$$ [using $\underline{2}$(f)]
$$x \leq 4$$

7. $-4x - 7 > 5$
$$-4x > 5 + 7$$
$$-4x > 12$$
$$x < -3$$
Graph of $x < -3$ is:

9. $2 \leq x + 3 \leq 5$
$$2 - 3 \leq x \leq 5 - 3$$
$$-1 \leq x \leq 2$$
Graph of $-1 \leq x \leq 2$ is:

11. $\frac{y}{7} - 1 = \frac{1}{7}$

Multiply both sides of the equation by 7. We obtain:

$y - 7 = 1$ [using $\underline{1}$(c)]
$$y = 8$$

13. $\frac{x}{3} > -2$

Multiply both sides of the inequality by 3. We obtain:

$x > -6$ [using $\underline{2}$(c)]

15. $\frac{y}{3} = 4 - \frac{y}{6}$

Multiply both sides of the equation by 6. We obtain:

$$2y = 24 - y$$
$$3y = 24$$
$$y = 8$$

17. $10x + 25(x - 3) = 275$
$$10x + 25x - 75 = 275$$
$$35x = 275 + 75$$
$$35x = 350$$
$$x = \frac{350}{35}$$
$$x = 10$$

19. $3 - y \leq 4(y - 3)$
$$3 - y \leq 4y - 12$$
$$-5y \leq -15$$
$$y \geq 3$$

[$\underline{\text{Note}}$: Division by a negative number, -3.]

21. $\frac{x}{5} - \frac{x}{6} = \frac{6}{5}$

Multiply both sides of the equation by 30. We obtain:

$$6x - 5x = 36$$
$$x = 36$$

23. $\frac{m}{5} - 3 < \frac{3}{5} - m$

Multiply both sides of the inequality by 5. We obtain:

$$m - 15 < 3 - 5m$$
$$6m < 18$$
$$m < 3$$

25. $0.1(x - 7) + 0.05x = 0.8$
$$0.1x - 0.7 + 0.05x = 0.8$$
$$0.15x = 1.5$$
$$x = \frac{1.5}{0.15}$$
$$x = 10$$

27. $2 \leq 3x - 7 < 14$

$\quad 7 + 2 \leq 3x < 14 + 7$

$\qquad 9 \leq 3x < 21$

$\qquad 3 \leq x < 7$

Graph of $3 \leq x < 7$ is:

29. $-4 \leq \frac{9}{5}C + 32 \leq 68$

$\quad -36 \leq \frac{9}{5}C \leq 36$

$\quad -36\left(\frac{5}{9}\right) \leq C \leq 36\left(\frac{5}{9}\right)$

$\quad -20 \leq C \leq 20$

Graph of $-20 \leq C \leq 20$ is:

31. $3x - 4y = 12$

$\qquad 3x = 12 + 4y$

$\quad 3x - 12 = 4y$

$\qquad y = \frac{1}{4}(3x - 12)$

$\qquad y = \frac{3}{4}x - 3$

33. $Ax + By = C$

$\qquad By = C - Ax$

$\qquad y = \frac{C}{B} - \frac{Ax}{B}, \; B \neq 0$

$\text{or} \qquad y = -\left(\frac{A}{B}\right)x + \frac{C}{B}$

35. $\qquad F = \frac{9}{5}C + 32$

$\frac{9}{5}C + 32 = F$

$\qquad \frac{9}{5}C = F - 32$

$\qquad C = \frac{5}{9}(F - 32)$

37. $A = Bm - Bn$

$\quad A = B(m - n)$

$\quad B = \dfrac{A}{m - n}$

39. $-3 \leq 4 - 7x < 18$

$\quad -3 - 4 \leq -7x < 18 - 4$

$\qquad -7 \leq -7x < 14.$

Dividing by -7, and recalling $\underline{2}$(f), we have

$1 \geq x > -2 \quad \text{or} \quad -2 < x \leq 1$

The graph is:

41. Let x = number of $6 tickets. Then the number of $10 tickets = $8,000 - x$.

$6x + 10(8,000 - x) = 60,000$

Thus, $6x + 80,000 - 10x = 60,000$

$\qquad\qquad\qquad -4x = 60,000 - 80,000$

$\qquad\qquad\qquad\quad 4x = 20,000$

$\qquad\qquad\qquad\quad\; x = 5,000$

Therefore, $x = 5,000$ $6 tickets and $8,000 - x = 3,000$ $10 tickets were sold.

43. Let x = the amount invested at 10%. Then $12,000 - x$ is the amount invested at 15%.

Required total yield = 12% of $12,000 = 0.12 \cdot 12,000 = \$1,440$. Thus,

$0.10x + 0.15(12,000 - x) = 0.12 \cdot 12,000$

$\quad 10x + 15(12,000 - x) = 12 \cdot 12,000$ (multiply both sides by 100)

$\quad 10x + 180,000 - 15x = 144,000$

$\qquad\qquad\qquad -5x = -36,000$

$\qquad\qquad\qquad\quad x = \$7,200$

Thus, we get $7,200 invested at 10% and $12,000 - 7,200 = \$4,800$ invested at 15%.

45. $\dfrac{\text{Car sold for in 1980}}{\text{Car sold for in 1965}} = \dfrac{\text{Consumer price index in 1980}}{\text{Consumer price index in 1965}}$

$$\frac{x}{3,000} = \frac{247}{95} \quad \text{(refer to Table 2, Example 9)}$$

$$x = \frac{3,000 \times 247}{95}$$

$$= \$7,800$$

47. Let x = the number of rainbow trout in the lake. Then,

$$\frac{x}{200} = \frac{200}{8} \quad \text{(since proportions are the same)}$$

$$x = \frac{200}{8}(200)$$

$$x = 5,000$$

49. $\text{IQ} = \dfrac{\text{Mental age}}{\text{Chronological age}}(100)$

$$\frac{\text{Mental age}}{9}(100) = 140$$

$$\text{Mental age} = \frac{140}{100}(9)$$

$$= 12.6 \text{ years}$$

EXERCISE 2-2

Things to remember:

<u>1.</u> A quadratic equation in one variable is an equation of the form

(A) $ax^2 + bx + c = 0$,

where x is a variable and a, b, and c are constants, $a \neq 0$.

<u>2.</u> Quadratic equations of the form $ax^2 + c = 0$ can be solved by the SQUARE ROOT METHOD. The solutions are:

$$x = \pm\sqrt{\frac{-c}{a}} \quad \text{provided} \quad \frac{-c}{a} \geq 0;$$

otherwise, the equation has no real solutions.

<u>3.</u> If the left side of the quadratic equation (A) can be FACTORED,

$$ax^2 + bx + c = (px + q)(rx + s),$$

then the solutions of (A) are

$$x = \frac{-q}{p} \quad \text{or} \quad x = \frac{-s}{r}.$$

<u>4.</u> The solutions of (A) are given by the QUADRATIC FORMULA:

$$x = \frac{-b \pm \sqrt{b^2 - 4ac}}{2a}$$

The quantity $b^2 - 4ac$ under the radical is called the DISCRIMINANT and:

(i)   (A) has two real solutions if $b^2 - 4ac > 0$;

(ii) (A) has one real solution if $b^2 - 4ac = 0$;

(iii) (A) has no real solution if $b^2 - 4ac < 0$.

<u>5</u>. FACTORABILITY THEOREM

The second-degree polynomial, $ax^2 + bx + c$, with integer coefficients, can be expressed as the product of two first-degree polynomials with integer coefficients if and only if $\sqrt{b^2 - 4ac}$ is an integer.

<u>6</u>. FACTOR THEOREM

If r_1 and r_2 are solutions of $ax^2 + bx + c = 0$, then $ax^2 + bx + c = a(x - r_1)(x - r_2)$.

---

1. $x^2 - 4 = 0$
$$x^2 = 4$$
$$x = \pm\sqrt{4} = \pm 2$$

3. $2x^2 - 22 = 0$
$$x^2 - 11 = 0$$
$$x^2 = 11$$
$$x = \pm\sqrt{11}$$

5. $2u^2 - 8u - 24 = 0$
$$u^2 - 4u - 12 = 0$$
$$(u - 6)(u + 2) = 0$$
$$u - 6 = 0 \text{ or } u + 2 = 0$$
$$u = 6 \text{ or } \quad u = -2$$

7. $x^2 = 2x$
$$x^2 - 2x = 0$$
$$x(x - 2) = 0$$
$$x = 0 \text{ or } x - 2 = 0$$
$$x = 2$$

9. $x^2 - 6x - 3 = 0$
$$x = \frac{-b \pm \sqrt{b^2 - 4ac}}{2a}, \quad a = 1, \ b = -6, \ c = -3$$
$$= \frac{-(-6) \pm \sqrt{(-6)^2 - 4(1)(-3)}}{2(1)}$$
$$= \frac{6 \pm \sqrt{48}}{2} = \frac{6 \pm 4\sqrt{3}}{2} = 3 \pm 2\sqrt{3}$$

11. $3u^2 + 12u + 6 = 0$

Since 3 is a factor of each coefficient, divide both sides by 3.
$$u^2 + 4u + 2 = 0$$
$$u = \frac{-b \pm \sqrt{b^2 - 4ac}}{2a}, \quad a = 1, \ b = 4, \ c = 2$$
$$= \frac{-4 \pm \sqrt{4^2 - 4(1)(2)}}{2(1)} = \frac{-4 \pm \sqrt{8}}{2} = \frac{-4 \pm 2\sqrt{2}}{2} = -2 \pm \sqrt{2}$$

13. $2x^2 = 4x$
$$x^2 = 2x \quad \text{(divide both sides by 2)}$$
$$x^2 - 2x = 0 \quad \text{(solve by factoring)}$$
$$x(x - 2) = 0$$
$$x = 0 \quad \text{or} \quad x - 2 = 0$$
$$x = 2$$

15. $4u^2 - 9 = 0$
$$4u^2 = 9 \quad \text{(solve by square root method)}$$
$$u^2 = \frac{9}{4}$$
$$u = \pm\sqrt{\frac{9}{4}} = \pm\frac{3}{2}$$

17. $8x^2 + 20x = 12$
$$8x^2 + 20x - 12 = 0$$
$$2x^2 + 5x - 3 = 0$$
$$(x + 3)(2x - 1) = 0$$
$$x + 3 = 0 \quad \text{or} \quad 2x - 1 = 0$$
$$x = -3 \text{ or} \qquad 2x = 1$$
$$x = \frac{1}{2}$$

19.
$$x^2 = 1 - x$$
$$x^2 + x - 1 = 0$$
$$x = \frac{-b \pm \sqrt{b^2 - 4ac}}{2a}, \quad a = 1, \; b = 1, \; c = -1$$
$$= \frac{-1 \pm \sqrt{(1)^2 - 4(1)(-1)}}{2(1)} = \frac{-1 \pm \sqrt{5}}{2}$$

21.
$$2x^2 = 6x - 3$$
$$2x^2 - 6x + 3 = 0$$
$$x = \frac{-b \pm \sqrt{b^2 - 4ac}}{2a}, \quad a = 2, \; b = -6, \; c = 3$$
$$= \frac{-(-6) \pm \sqrt{(-6)^2 - 4(2)(3)}}{2(2)} = \frac{6 \pm \sqrt{12}}{4} = \frac{6 \pm 2\sqrt{3}}{4} = \frac{3 \pm \sqrt{3}}{2}$$

23.
$$y^2 - 4y = -8$$
$$y^2 - 4y + 8 = 0$$
$$y = \frac{-b \pm \sqrt{b^2 - 4ac}}{2a}, \quad a, = 1, \; b = -4, \; c = 8$$
$$= \frac{-(-4) \pm \sqrt{(-4)^2 - 4(1)(8)}}{2(1)} = \frac{4 \pm \sqrt{-16}}{2}$$
Since $\sqrt{-16}$ is not a real number, there are no real solutions.

25. $(x + 4)^2 = 11$
$$x + 4 = \pm\sqrt{11}$$
$$x = -4 \pm \sqrt{11}$$

27. $x^2 + 40x - 84$

Step 1. Test for factorability
$$\sqrt{b^2 - 4ac} = \sqrt{(40)^2 - 4(1)(-84)} = \sqrt{1936} = 44$$
Since the result is an integer, the polynomial has first-degree factors with integer coefficients.

Step 2. Use the factor theorem
$$x^2 + 40x - 84 = 0$$
$$x = \frac{-40 \pm 44}{2} = 2, \; -42 \quad \text{(by the quadratic formula)}$$
Thus, $x^2 + 40x - 84 = (x - 2)(x - [-42]) = (x - 2)(x + 42)$

29. $x^2 - 32x + 144$

Step 1. Test for factorability
$$\sqrt{b^2 - 4ac} = \sqrt{(-32)^2 - 4(1)(144)} = \sqrt{448} \approx 21.166$$
Since this is not an integer, the polynomial is not factorable.

31. $2x^2 + 15x - 108$

Step 1. Test for factorability
$$\sqrt{b^2 - 4ac} = \sqrt{(15)^2 - 4(2)(-108)} = \sqrt{1089} = 33$$
Thus, the polynomial has first-degree factors with integer coefficients.

Use the factor theorem

$$2x^2 + 15x - 108$$

$$x = \frac{-15 \pm 33}{4} = \frac{9}{2}, \ -12$$

Thus, $2x^2 + 15x - 108 = 2\left(x - \frac{9}{2}\right)(x - [-12]) = (2x - 9)(x + 12)$

33. $4x^2 + 241x - 434$

Step 1. Test for factorability

$$\sqrt{b^2 - 4ac} = \sqrt{(241)^2 - 4(4)(-434)} = \sqrt{65025} = 255$$

Thus, the polynomial has first-degree factors with integer coefficients.

Step 2. Use the factor theorem

$$4x^2 + 241x - 434$$

$$x = \frac{-241 \pm 255}{8} = \frac{14}{8}, \ -\frac{496}{8} \text{ or } \frac{7}{4}, \ -62$$

Thus, $4x^2 + 241x - 434 = 4\left(x - \frac{7}{4}\right)(x + 62) = (4x - 7)(x + 62)$

35.
$$A = P(1 + r)^2$$

$$(1 + r)^2 = \frac{A}{P}$$

$$1 + r = \sqrt{\frac{A}{P}}$$

$$r = \sqrt{\frac{A}{P}} - 1$$

37. $d = \frac{3,000}{p}$ and $s = 1,000p - 500$

Let $d = s$. Then,

$\frac{3,000}{p} = 1,000p - 500$ or $3,000 = 1,000p^2 - 500p$

and $1,000p^2 - 500p - 3,000 = 0$.
Divide both sides by 500.

$2p^2 - p - 6 = 0$ (solve by factoring)
$(2p + 3)(p - 2) = 0$

$2p + 3 = 0$ or $p - 2 = 0$; thus, $p = -\frac{3}{2}$ or $p = 2$

Since the price, p, must be positive, we have $p = \$2$ as the equilibrium point.

39. $A = P(1 + r)^2 = P(1 + 2r + r^2) = Pr^2 + 2Pr + P$

Let $A = 144$ and $P = 100$. Then,

$100r^2 + 200r + 100 = 144$

$100r^2 + 200r - 44 = 0$

Using the quadratic formula,

$$r = \frac{-200 \pm \sqrt{(200)^2 - 4(100)(-44)}}{200}$$

$$= \frac{-200 \pm 240}{200} = -2.2, \ 0.20$$

Since $r > 0$, we have $r = 0.20$ or 20%.

41. $v^2 = 64h$

For $h = 1$, $v^2 = 64(1) = 64$. Therefore, $v = 8$ ft/sec.
For $h = 0.5$, $v^2 = 64(0.5) = 32$. Therefore, $v = \sqrt{32} = 4\sqrt{2} \approx 5.66$ ft/sec.

EXERCISE 2-3

Things to remember:

<u>1</u>. The graph of any equation of the form $Ax + By = C$ (standard form), where A, B, and C are constants (A and B not both zero), is a straight line. Every straight line in a cartesian coordinate system is the graph of an equation of this type.

<u>2</u>. The slope, m, of a line through the two points (x_1, y_1) and (x_2, y_2) is given by:

$$m = \frac{y_2 - y_1}{x_2 - x_1},$$

$$x_1 \neq x_2.$$

<u>3</u>. In general, the slope of a line may be positive, negative, 0, or not defined. Each of these cases is interpreted geometrically as illustrated in the following table.

Going from left to right:

LINE	SLOPE	EXAMPLE
Rising	Positive	
Falling	Negative	
Horizontal	Zero	
Vertical	Not defined	

<u>4</u>. The equation

$$y = mx + b \qquad\qquad m = \text{slope}$$
$$b = y \text{ intercept}$$

is called the SLOPE-INTERCEPT form of an equation of a line.

<u>5</u>. An equation for the line that passes through (x_1, y_1) with slope m is:

$$y - y_1 = m(x - x_1)$$

This equation is called the POINT-SLOPE FORM.

<u>6</u>. The horizontal line with y intercept c has the equation: $y = c$.

The vertical line with x intercept c has the equation: $x = c$.

<u>7</u>. Let L_1 and L_2 be nonvertical lines with slopes m_1 and m_2, respectively. Then

$L_1 \parallel L_2$ if and only if $m_1 = m_2$;

$L_1 \perp L_2$ if and only if $m_1 \cdot m_2 = -1$ or $m_2 = \dfrac{-1}{m_1}$.

[<u>Note</u>: $\parallel$ means "is parallel to" and $\perp$ means "is perpendicular to."]

---

1. $y = 2x - 3$

x	y
0	-3
1	-1
4	5

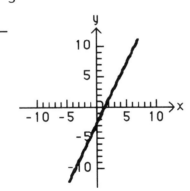

3. $2x + 3y = 12$

x	y
0	4
6	0
9	-2

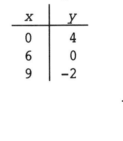

5. Slope $m = 2$

y intercept $b = -3$

7. Slope $m = -\dfrac{2}{3}$

y intercept $b = 2$

9. $m = -2$

$b = 4$

Using <u>4</u>, $y = -2x + 4$.

11. $m = -\dfrac{3}{5}$

$b = 3$

Using <u>4</u>, $y = -\dfrac{3}{5}x + 3$.

13. $y = -\frac{2}{3}x - 2$

$m = -\frac{2}{3}, \; b = -2$

x	y
0	-2
3	-4
-3	0

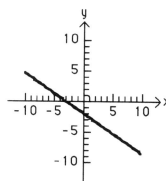

15. $3x - 2y = 10$

x	y
0	-5
10	10
-4	-11

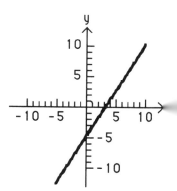

17.

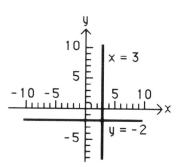

19. $3x + y = 5$
$y = -3x + 5$
$m = -3$ (using $\underline{4}$)

21. $2x + 3y = 12$
$3y = -2x + 12$
Divide both sides by 3:
$y = -\frac{2}{3}x + \frac{12}{3} = -\frac{2}{3}x + 4$
$m = -\frac{2}{3}$ (using $\underline{4}$)

23. $m = -3$
For the point $(4, -1)$, $x_1 = 4$ and $y_1 = -1$. Using $\underline{5}$, we get:
$y - (-1) = -3(x - 4)$
$y + 1 = -3x + 12$
$y = -3x + 11$

25. $m = \frac{2}{3}$
For the point $(-6, -5)$, $x_1 = -6$ and $y_1 = -5$. Using $\underline{5}$, we get:
$y - (-5) = \frac{2}{3}[x - (-6)]$
$y + 5 = \frac{2}{3}(x + 6)$
$y + 5 = \frac{2}{3}x + 4$
$y = \frac{2}{3}x - 1$

27. The points are $(1, 3)$ and $(7, 5)$. Let $x_1 = 1$, $y_1 = 3$, $x_2 = 7$, and $y_2 = 5$. Using $\underline{2}$, we get:
$m = \frac{5 - 3}{7 - 1} = \frac{2}{6} = \frac{1}{3}$

29. Let $x_1 = -5$, $y_1 = -2$, $x_2 = 5$, and $y_2 = -4$. Using $\underline{2}$, we get:
$m = \frac{-4 - (-2)}{5 - (-5)} = \frac{-4 + 2}{5 + 5} = \frac{-2}{10} = -\frac{1}{5}$

31. First, find the slope using $\underline{2}$:
$m = \frac{y_2 - y_1}{x_2 - x_1} = \frac{5 - 3}{7 - 1} = \frac{2}{6} = \frac{1}{3}$

Then, by using $\underline{5}$, $y - y_1 = m(x - x_1)$, where $m = \frac{1}{3}$ and $(x_1, y_1) = (1, 3)$ or $(7, 5)$, we get:
$y - 3 = \frac{1}{3}(x - 1)$ or $y - 5 = \frac{1}{3}(x - 7)$
These two equations are equivalent. After simplifying either one of these, we obtain:
$-x + 3y = 8$ or $x - 3y = -8$

33. First, find the slope using $\underline{2}$:

$m = \dfrac{-4 - (-2)}{5 - (-5)} = \dfrac{-4 + 2}{5 + 5} = \dfrac{-2}{10} = -\dfrac{1}{5}$

By using $\underline{5}$, and either one of these points, we obtain:

$y - (-2) = -\dfrac{1}{5}[x - (-5)]$ [using $(-5, -2)$]

$\quad y + 2 = -\dfrac{1}{5}(x + 5)$

$5(y + 2) = -x - 5$

$\quad 5y + 10 = -x - 5$

$\quad\ \ x + 5y = -15$

35. Using $\underline{6}$ with $c = 3$ for the vertical line and $c = -5$ for the horizontal line, we find that the equation of the vertical line is $x = 3$ and the equation of the horizontal line is $y = -5$.

37. Using $\underline{6}$ with $c = -1$ for the vertical line and $c = -3$ for the horizontal line, we find that the equation of the vertical line is $x = -1$ and the equation of the horizontal line is $y = -3$.

39. $m = -\dfrac{1}{2}$ and $(x_1, y_1) = (-2, 5)$. Using $\underline{5}$:

$y - 5 = -\dfrac{1}{2}[x - (-2)]$

$y - 5 = -\dfrac{1}{2}(x + 2)$

$y - 5 = -\dfrac{1}{2}x - 1$

$\quad\ y = -\dfrac{1}{2}x + 4$

41. The straight line L whose equation is $y = -\dfrac{1}{2}x + 5$ has slope $m = -\dfrac{1}{2}$.

(A) The line through $(-2, 2)$ that is parallel to L has slope $m = -\dfrac{1}{2}$.

Thus, an equation for this line is:

$y - 2 = -\dfrac{1}{2}[x - (-2)]$

$y - 2 = -\dfrac{1}{2}(x + 2)$

$y - 2 = -\dfrac{1}{2}x - 1$

$\quad\ y = -\dfrac{1}{2}x + 1$

(B) The line through $(-2, 2)$ that is perpendicular to L has slope $m = \dfrac{-1}{-\frac{1}{2}} = 2$. Thus, an equation for this line is:

$y - 2 = 2[x - (-2)]$

$y - 2 = 2x + 4$

$\quad\ y = 2x + 6$

43. Writing the equation $x - 2y = 4$ in slope-intercept form, we have $y = \frac{1}{2}x - 2$. Therefore, the slope of this line L is $m = \frac{1}{2}$.

(A) The line through $(-2, -1)$ that is parallel to L has slope $m = \frac{1}{2}$.
Thus, an equation for this line is:

$$y - (-1) = \frac{1}{2}[x - (-2)]$$
$$y + 1 = \frac{1}{2}(x + 2)$$
$$y + 1 = \frac{1}{2}x + 1$$
$$y = \frac{1}{2}x$$

(B) The line through $(-2, -1)$ that is perpendicular to L has slope
$m = \dfrac{-1}{\frac{1}{2}} = -2$. Thus, an equation for this line is:

$$y - (-1) = -2[x - (-2)]$$
$$y + 1 = -2(x + 2)$$
$$y + 1 = -2x - 4$$
$$y = -2x - 5$$

45. $y = mx - 2$.

Using $\underline{3}$ with y intercept = -2 and the given slopes 2, $\frac{1}{2}$, 0, $-\frac{1}{2}$, and -2, we have:

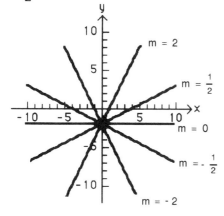

47. $(2, 7)$ and $(2, -3)$
Since each point has the same x-coordinate, the graph of the line formed by these two points will be a *vertical line*. Then, using $\underline{6}$, with $c = 2$, we have $x = 2$ as the equation of the line.

49. $(2, 3)$ and $(-5, 3)$
Since each point has the same y-coordinate, the graph of the line formed by these two points will be a *horizontal line*. Then, using $\underline{6}$, with $c = 3$, we have $y = 3$ as the equation of the line.

51. $y = \frac{1}{2}x - 3$, $y = \frac{1}{2}x$, $y = \frac{1}{2}x + 3$

(A)

(B) Each line has slope $m = \frac{1}{2}$

(C) The lines are parallel since their slopes are equal.

(D) $y = \frac{1}{2}x - 3$; y-intercept $b = -3$

$y = \frac{1}{2}x$; y-intercept $b = 0$

$y = \frac{1}{2}x + 3$; y-intercept $b = 3$

53. (A)

(B) $y = -\frac{1}{2}x + 3$; slope $m = -\frac{1}{2}$

$y = 3$; slope $m = 0$

$y = \frac{1}{2}x + 3$; slope $m = \frac{1}{2}$

(C) The lines are not parallel; the lines do not have the same slopes.

(D) Each line has y-intercept $b = 3$

55. (A)

(B) The lines are perpendicular since their slopes $m_1 = 2$ and

$m_2 = -\frac{1}{2}$ satisfy $m_1 \cdot m_2 = 2\left(-\frac{1}{2}\right) = -1$

57. $A = Prt + P$ (1)

Rate $r = 0.06$

Principal $P = 100$

Substituting in (1), we get:

$A = 6t + 100$ (2)

(A) Let $t = 5$ and $t = 20$ and substitute in (2). We get:

$A = 6(5) + 100 = \$130$

$A = 6(20) + 100 = \$220$

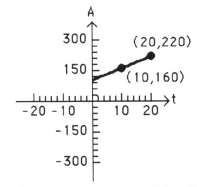

(B)

t	A
0	100
10	160
20	220

(C) Consider two points $(10, 160)$ and $(20, 220)$. Using $\underline{2}$, we have:

$m = \dfrac{220 - 160}{20 - 10} = \dfrac{60}{10} = 6$

59. (A) We find an equation $C = mx + b$ for the line passing through $(0, 200)$ and $(20, 3800)$.

$m = \dfrac{3800 - 200}{20 - 0} = \dfrac{3600}{20} = 180$

Also, since $C = 200$ when $x = 0$, it follows that $b = 200$.

Thus, $C = 180x + 200$.

(B) The total costs at 12 boards per day are:
$$C = 180(12) + 200 = 2,360 \text{ or } \$2,360$$

(C)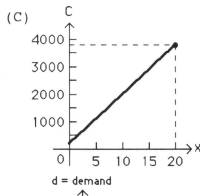

61. (A)

Price	Demand
70	7,800
120	4,800
160	2,400
200	0

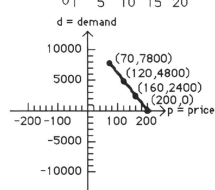

(B) We observe that the points in part (A) lie along a straight line. Consider two points, (120, 4,800) and (160, 2,400). Using 2, we have:
$$m = \frac{2,400 - 4,800}{160 - 120} = -\frac{2,400}{40} = -60$$

The equation of the line which has slope −60 and passes through (120, 4,800) is:
$$(d - 4,800) = -60(p - 120)$$
$$d - 4,800 = -60p + 7,200$$
$$d = -60p + 12,000$$
Slope = −60

The slope of the line indicates that the demand will decrease by 60 power mowers for each dollar increase in price.

63. Mix A contains 20% protein. Mix B contains 10% protein. Let x be the amount of A used, and let y be the amount of B used. Then $0.2x$ is the amount of protein from mix A and $0.1y$ is the amount of protein from mix B. Thus, the linear equation is:
$$0.2x + 0.1y = 20$$

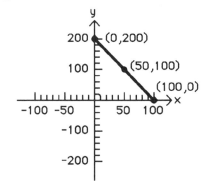

The table shows different combinations of mix A and mix B to provide 20 grams of protein.

[Note: We can get many more combinations. In fact, each point on the graph indicates a combination of mix A and mix B.]

Mix A	Mix B
x	y
100	0
0	200
50	100
10	180

65. $p = -\frac{1}{5}d + 70$, $30 \leq d \leq 175$, where d = distance in centimeters
and p = pull in grams

(A) $d = 30$ $d = 175$

$p = -\frac{1}{5}(30) + 70 = 64$ grams $p = -\frac{1}{5}(175) + 70 = 35$ grams

(B)

d	p
30	64
50	60
175	35

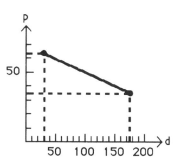

(C) Select two points (30, 64) and
(50, 60) as (x_1, y_1) and (x_2, y_2),
respectively, from part (B).
Using 2:

$$\text{Slope } m = \frac{y_2 - y_1}{x_2 - x_1} = \frac{60 - 64}{50 - 30}$$

$$= -\frac{4}{20} = -\frac{1}{5}$$

EXERCISE 2-4

Things to remember:

1. A FUNCTION is a rule (process or method) that produces a
correspondence between two sets of elements, such that to each
element in the first set there corresponds one and only one
element in the second set. The first set is called the DOMAIN
and the set of all corresponding elements in the second set is
called the RANGE.

2. EQUATIONS AND FUNCTIONS:

Given an equation in two variables. If there corresponds
exactly one value of the dependent variable (output) to each
value of the independent variable (input), then the equation
specifies a function. If there is more than one output for at
least one input, then the equation does not specify a function.

3. VERTICAL LINE TEST FOR A FUNCTION

An equation specifies a function if each vertical line in the
coordinate system passes through at most one point on the graph
of the equation. If any vertical line passes through two or
more points on the graph of an equation, then the equation does
not specify a function.

4. AGREEMENT ON DOMAINS AND RANGES

If a function is specified by an equation and the domain is not
given explicitly, then assume that the domain is the set of all
real number replacements of the independent variable (inputs)
that produce real values for the dependent variable (outputs).
The range is the set of all outputs corresponding to input
values.

5. FUNCTION NOTATION—THE SYMBOL $f(x)$

For any element x in the domain of the function f, the symbol $f(x)$ represents the element in the range of f corresponding to x in the domain of f. If x is an input value, then $f(x)$ is the corresponding output value. If x is an element which is not in the domain of f, then f is NOT DEFINED at x and $f(x)$ DOES NOT EXIST.

---

1. The table specifies a function, since for each domain value there corresponds one and only one range value.

3. The table does not specify a function, since more than one range value corresponds to a given domain value. (Range values 5, 6 correspond to domain value 3; range values 6, 7 correspond to domain value 4.)

5. This is a function.

7. The graph specifies a function; each vertical line in the plane intersects the graph in at most one point.

9. The graph does not specify a function. There are vertical lines which intersect the graph in more than one point. For example, the y-axis intersects the graph in three points.

11. The graph specifies a function.

13. $f(x) = 3x - 2$
$f(2) = 3(2) - 2 = 4$

15. $f(-1) = 3(-1) - 2$
$\quad = -5$

17. $g(x) = x - x^2$
$g(3) = 3 - 3^2 = -6$

19. $f(0) = 3(0) - 2$
$\quad = -2$

21. $g(-3) = -3 - (-3)^2$
$\quad\quad = -12$

23. $f(1) + g(2)$
$\quad = [3(1) - 2] + (2 - 2^2) = -1$

25. $g(2) - f(2)$
$= (2 - 2^2) - [3(2) - 2]$
$= -2 - 4 = -6$

27. $g(3) \cdot f(0) = (3 - 3^2)[3(0) - 2]$
$\quad\quad\quad = (-6)(-2)$
$\quad\quad\quad = 12$

29. $\dfrac{g(-2)}{f(-2)} = \dfrac{-2 - (-2)^2}{3(-2) - 2} = \dfrac{-6}{-8} = \dfrac{3}{4}$

31. The domain of $f(x) = \sqrt{x}$ is all nonnegative real numbers (the square root of a negative number is not a real number).

33. The domain is all real numbers except $x = -3$ and $x = 5$. The function is not defined at $x = -3$ and $x = 5$.

35. Since the square root of a negative number is not real, we must have:
$\quad x + 5 \geq 0$ or $x \geq -5$
Thus, the domain of $f(x) = \sqrt{x + 5}$ is the set of all real numbers such that $x \geq -5$ or $[-5, \infty)$.

37. $f(x) = \dfrac{x^2 + 1}{x^2 - 1} = \dfrac{x^2 + 1}{(x - 1)(x + 1)}$

The domain of f is all real numbers except $x = 1$, $x = -1$.

39. $f(x) = \dfrac{x}{(x^2 + 3x - 4)} = \dfrac{x}{(x + 4)(x - 1)}$

The domain of f is all real numbers except $x = -4$, $x = 1$.

41. $f(x) = \dfrac{x + 4}{x^2 - 4x + 5}$

Since $x^2 - 4x + 5 \neq 0$ for all real numbers x, the domain of f is all real numbers.

43. $f(x) = \dfrac{3}{\sqrt{x + 2}}$

Since the square root of a negative number is not real, and since division by 0 is not defined, we must have $x + 2 > 0$ or $x > -2$. Thus, the domain of f is the set of all real numbers x such that $x > -2$ or $(-2, \infty)$

45. Given $4x - 5y = 20$. Solving for y, we have:

$-5y = -4x + 20$

$y = \dfrac{4}{5}x - 4$

Since each input value x determines a unique output value y, the equation specifies a function. The domain is R, the set of real numbers.

47. Given $x^2 - y = 1$. Solving for y, we have:

$-y = -x^2 + 1 \quad \text{or} \quad y = x^2 - 1$

This equation specifies a function. The domain is R, the set of real numbers.

49. Given $x + y^2 = 10$. Solving for y, we have:

$y^2 = 10 - x$

$y = \pm\sqrt{10 - x}$

This equation does not specify a function since each value of x, $x \leq 10$, determines two values of y. For example, corresponding to $x = 1$, we have $y = 3$ and $y = -3$; corresponding to $x = 6$, we have $y = 2$ and $y = -2$.

51. Given $xy - 4y = 1$. Solving for y, we have:

$(x - 4)y = 1 \quad \text{or} \quad y = \dfrac{1}{x - 4}$

This equation specifies a function. The domain is all real numbers except $x = 4$.

53. Given $x^2 + y^2 = 25$. Solving for y, we have:

$y^2 = 25 - x^2 \quad \text{or} \quad y = \pm\sqrt{25 - x^2}$

Thus, the equation does not specify a function since, for $x = 0$, we have $y = \pm 5$, when $x = 4$, $y = \pm 3$, and so on.

55. Given $F(t) = 4t + 7$. Then:

$$\frac{F(3 + h) - F(3)}{h} = \frac{4(3 + h) + 7 - (4 \cdot 3 + 7)}{h}$$

$$= \frac{12 + 4h + 7 - 19}{h} = \frac{4h}{h} = 4$$

57. Given $g(w) = w^2 - 4$. Then:

$$\frac{g(1 + h) - g(1)}{h} = \frac{(1 + h)^2 - 4 - (1^2 - 4)}{h} = \frac{1 + 2h + h^2 - 4 + 3}{h}$$

$$= \frac{2h + h^2}{h} = \frac{h(2 + h)}{h} = 2 + h$$

59. Given $Q(x) = x^2 - 5x + 1$. Then:

$$\frac{Q(2 + h) - Q(2)}{h} = \frac{(2 + h)^2 - 5(2 + h) + 1 - (2^2 - 5 \cdot 2 + 1)}{h}$$

$$= \frac{4 + 4h + h^2 - 10 - 5h + 1 - (-5)}{h} = \frac{h^2 - h - 5 + 5}{h}$$

$$= \frac{h(h - 1)}{h} = h - 1$$

61. Given $f(x) = 4x - 3$. Then:

$$\frac{f(a + h) - f(a)}{h} = \frac{4(a + h) - 3 - (4a - 3)}{h}$$

$$= \frac{4a + 4h - 3 - 4a + 3}{h} = \frac{4h}{h} = 4$$

63. Given $f(x) = 4x^2 - 7x + 6$. Then:

$$\frac{f(a + h) - f(a)}{h} = \frac{4(a + h)^2 - 7(a + h) + 6 - (4a^2 - 7a - 6)}{h}$$

$$= \frac{4(a^2 + 2ah + h^2) - 7a - 7h + 6 - 4a^2 + 7a - 6}{h}$$

$$= \frac{4a^2 + 8ah + 4h^2 - 7h - 4a^2}{h} = \frac{8ah + 4h^2 - 7h}{h}$$

$$= \frac{h(8a + 4h - 7)}{h} = 8a + 4h - 7$$

65. Given $f(x) = x^3$. Then:

$$\frac{f(a + h) - f(a)}{h} = \frac{(a + h)^3 - a^3}{h} = \frac{a^3 + 3a^2h + 3ah^2 + h^3 - a^3}{h}$$

$$= \frac{h(3a^2 + 3ah + h^2)}{h} = 3a^2 + 3ah + h^2$$

67. Given $f(x) = \sqrt{x}$. Then:

$$\frac{f(a + h) - f(a)}{h} = \frac{\sqrt{a + h} - \sqrt{a}}{h}$$

$$= \frac{\sqrt{a + h} - \sqrt{a}}{h} \cdot \frac{\sqrt{a + h} + \sqrt{a}}{\sqrt{a + h} + \sqrt{a}} \quad \text{(rationalizing the numerator)}$$

$$= \frac{a + h - a}{h(\sqrt{a + h} + \sqrt{a})} = \frac{h}{h(\sqrt{a + h} + \sqrt{a})} = \frac{1}{\sqrt{a + h} + \sqrt{a}}$$

69. Given $A = \ell w = 25$.

Thus, $\ell = \dfrac{25}{w}$. Now $P = 2\ell + 2w$

$$= 2\left(\dfrac{25}{w}\right) + 2w = \dfrac{50}{w} + 2w.$$

The domain is $w > 0$.

71. Given $P = 2\ell w + 2w = 100$ or $\ell + w = 50$ and $w = 50 - \ell$.

Now $A = \ell w = \ell(50 - \ell)$ and $A = 50\ell - \ell^2$.

The domain is $0 \le \ell \le 50$. [<u>Note</u>: $\ell \le 50$ since $\ell > 50$ implies $w < 0$.]

73. Let x = the number of units manufactured. Then:
   $C(x) = 96,000 + 80x$ and
$C(500) = 96,000 + 80(500) = 96,000 + 40,000 = 136,000$
The cost of producing 500 units is \$136,000.

75. The revenue R is given by
   $R = xp$
Solving $x = 8,000 - 40p$ for p, we have:
   $p = \dfrac{8,000 - x}{40}$ or $p = 200 - \dfrac{1}{40}x$
Thus, $R(x) = xp(x)$
$$= x\left(200 - \dfrac{1}{40}x\right)$$
The domain is $0 \le x \le 8,000$. [<u>Note</u>: $x \le 8,000$ since p is nonnegative.]

77. (A) $R(x) = x \cdot p$ where x is the number of units sold and p is the price
   per unit. From the demand equation $x = 48000 - 400p$, we have
   $400p = 48000 - x$
   or $p = 120 - \dfrac{1}{400}x$.

   Thus, $R(x) = x\left(120 - \dfrac{1}{400}x\right) = 120x - \dfrac{1}{400}x^2$

(B) $P(x) = R(x) - C(x)$
$$= 120x - \dfrac{1}{400}x^2 - (100,000 + 20x)$$
$$= 120x - \dfrac{1}{400}x^2 - 100,000 - 20x$$
$$= -\dfrac{1}{400}x^2 + 100x - 100,000$$

(C) $P(10,000) = -\dfrac{1}{400}(10,000)^2 + 100(10,000) - 100,000$
$$= -250,000 + 1,000,000 - 100,000$$
$$= 650,000 \text{ or } \$650,000$$

$P(20,000) = -\dfrac{1}{400}(20,000)^2 + 100(20,000) - 100,000$
$$= -1,000,000 + 2,000,000 - 100,000$$
$$= 900,000 \text{ or } \$900,000$$

$$P(40,000) = -\frac{1}{400}(40,000)^2 + 100(40,000) - 100,000$$
$$= -4,000,000 + 4,000,000 - 100,000$$
$$= -100,000 \text{ or } -\$100,000$$

79. (A)

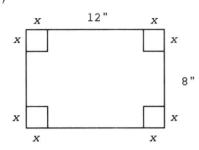

$$V = (\text{length})(\text{width})(\text{height})$$
$$V(x) = (12 - 2x)(8 - 2x)(x)$$
$$\text{or } V(x) = x(8 - 2x)(12 - 2x)$$

(B) Domain $= 0 \le x \le 4$

(C) $V(1) = (12 - 2)(8 - 2)(1)$
$\qquad = (10)(6)(1) = 60$
$\quad V(2) = (12 - 4)(8 - 4)(2)$
$\qquad = (8)(4)(2) = 64$
$\quad V(3) = (12 - 6)(8 - 6)(3)$
$\qquad = (6)(2)(3) = 36$

Thus,

x	$V(x)$
1	60
2	64
3	36

81. Let y = length of each pen. Since the area of each pen is 45 square feet, we have $A = 45 = xy$. Thus, $y = \frac{45}{x}$. There are six lengths y, five widths x, and five widths $x - 3$. Thus, the amount of fencing required is

$$P = 6y + 5x + 5(x - 3) = 6\left(\frac{45}{x}\right) + 5x + 5x - 15$$

and

$$P(x) = \frac{270}{x} + 10x - 15.$$

x	$P(x)$
3	105
4	92.5
5	89
6	90

83. Given $(w + a)(v + b) = c$. Let $a = 15$, $b = 1$, and $c = 90$. Then:
$(w + 15)(v + 1) = 90$
Solving for v, we have
$$v + 1 = \frac{90}{w + 15} \text{ and } v = \frac{90}{w + 15} - 1 = \frac{90 - (w + 15)}{w + 15}, \text{ so that } v = \frac{75 - w}{w + 15}.$$
If $w = 16$, then $v = \frac{75 - 16}{16 + 15} = \frac{59}{31} \approx 1.9032$ cm/sec.

EXERCISE 2-5

Things to remember:

<u>1</u>. GRAPH OF A FUNCTION

The GRAPH of a function f is the graph of the set of ordered pairs (x, y) where x is in the domain of f and $y = f(x)$.

<u>2</u>. The function *f* specified by the equation

$$f(x) = mx + b,$$

where *m* and *b* are constants, is a LINEAR FUNCTION. The graph of $f(x) = mx + b$ is a nonvertical straight line with slope *m* and *y* intercept *b*.

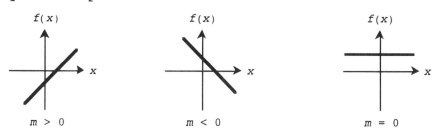

<u>3</u>. The function *f* specified by the equation

$$f(x) = ax^2 + bx + c,$$

where *a*, *b*, and *c* are constants and $a \neq 0$, is called a QUADRATIC FUNCTION. The graph of a quadratic function is a parabola whose AXIS (line of symmetry) is parallel to the vertical axis (or *y* axis). It opens upward if $a > 0$ and downward if $a < 0$. The intersection point of the parabola and its axis is called the VERTEX.

<u>4</u>. PROPERTIES OF $f(x) = ax^2 + bx + c$, $a \neq 0$

a. Axis (of symmetry): $x = -\dfrac{b}{2a}$

b. Vertex: $\left(-\dfrac{b}{2a}, \ f\left(-\dfrac{b}{2a}\right)\right)$

c. Maximum or minimum value of $f(x)$:

$$f\left(-\frac{b}{2a}\right) \quad \begin{cases} \text{Minimum if } a > 0 \\ \text{Maximum if } a < 0 \end{cases}$$

d. Domain: all real numbers
   Range: determine from graph

e. *y* intercept: $f(0) = c$
   x intercepts: solutions of $f(x) = 0$, if any exist

<u>5</u>. ABSOLUTE VALUE

The ABSOLUTE VALUE of a real number *a*, denoted $|a|$, is the (nonnegative) distance on a real number line from *a* to the origin. Equivalently,

$$|a| = \begin{cases} a & \text{if } a \geq 0 \\ -a & \text{if } a < 0 \end{cases}$$

The function $f(x) = |x|$ is called the ABSOLUTE VALUE FUNCTION.

1. $f(x) = 2x - 4$
Slope $m = 2$
y intercept: $b = -4$
x intercept: 2

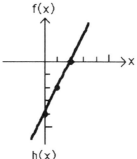

x	$f(x)$
0	-4
1	-2
2	0

3. $h(x) = 4 - 2x$
Slope $m = -2$
y intercept: $b = 4$
x intercept: 2

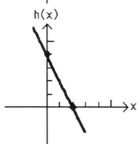

x	$h(x)$
0	4
2	0

5. $g(x) = -\frac{2}{3}x + 4$

Slope $m = -\frac{2}{3}$

y intercept: $b = 4$
x intercept: 6

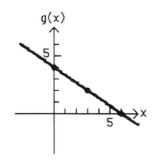

x	$g(x)$
0	4
3	2
6	0

7. Using the slope-intercept form $y = f(x) = mx + b$, we have:
$f(x) = -2x + 6$

9. The slope of the line through $(-1, 5)$ and $(5, 2)$ is:
$m = \dfrac{2 - 5}{5 - (-1)} = -\dfrac{3}{6} = -\dfrac{1}{2}$
Using the point-slope form $y - y_1 = m(x - x_1)$, with $(x_1, y_1) = (5, 2)$,
we have: $y - 2 = -\dfrac{1}{2}(x - 5)$ or $y = -\dfrac{1}{2}x + \dfrac{5}{2} + 2 = -\dfrac{1}{2}x + \dfrac{9}{2}$
Thus, $f(x) = -\dfrac{1}{2}x + \dfrac{9}{2}$.

11. $f(x) = (x - 3)^2 - 1$ Graph:
Axis: $x = 3$
Vertex: $(3, f(3)) = (3, -1)$
Minimum value of f (since $a = 1 > 0$): $f(3) = -1$
y intercept: $f(0) = (-3)^2 - 1 = 9 - 1 = 8$
x intercepts: $f(x) = (x - 3)^2 - 1 \quad = 0$
$\qquad\qquad\qquad\qquad (x - 3)^2 \quad = 1$
$\qquad\qquad\qquad\qquad\quad x - 3 \quad = \pm 1$
$\qquad\qquad\qquad\qquad\qquad x \quad = 2, 4$

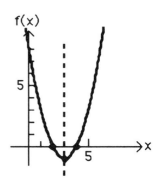

Range: $[f(3), \infty) = [-1, \infty)$

13. $h(x) = -(x + 1)^2 + 9$ Graph:

Axis: $x = -1$

Vertex: $(-1, h(-1)) = (-1, 9)$

Maximum value (since $a = -1 < 0$): $h(-1) = 9$

y intercept: $h(0) = -(1)^2 + 9 = 8$

x intercepts: $h(x) = -(x + 1)^2 + 9 = 0$

$$(x + 1)^2 = 9$$
$$x + 1 = \pm 3$$
$$x = 2, \; -4$$

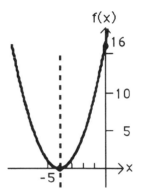

Range: $(-\infty, h(-1)] = (-\infty, 9]$

15. $f(x) = x^2 + 8x + 16 = (x + 4)^2$ Graph:

Axis: $x = -4$

Vertex: $(-4, f(-4)) = (-4, 0)$

Minimum value (since $a = 1 > 0$): $f(-4) = 0$

y intercept: $f(0) = 16$

x intercept: $(x + 4)^2 = 0$
$$x = -4$$

Range: $[f(-4), \infty) = [0, \infty)$

17. $f(u) = u^2 - 2u + 4$; $a = 1$, $b = -2$, $c = 4$.

Axis: $u = -\dfrac{b}{2a} = -\dfrac{(-2)}{2(1)} = 1$ Graph:

Vertex: $\left(-\dfrac{b}{2a}, \; f\left(-\dfrac{b}{2a}\right)\right) = (1, \; f(1)) = (1, \; 3)$

Minimum: $f\left(-\dfrac{b}{2a}\right) = f(1) = 3$

y intercept: $f(0) = 0^2 - 2(0) + 4 = 4$

u intercepts: $u^2 - 2u + 4 = 0$

$$u = \frac{2 \pm \sqrt{4 - 16}}{2} = \frac{2 \pm \sqrt{-12}}{2} \quad \begin{array}{l}\text{(using the quadratic} \\ \text{formula)}\end{array}$$

Thus, there are no u intercepts.

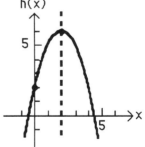

Range: $[f(1), \infty) = [3, \infty)$

19. $h(x) = 2 + 4x - x^2 = -x^2 + 4x + 2$;   Graph:

$a = -1$, $b = 4$, $c = 2$

Axis: $x = -\dfrac{b}{2a} = -\dfrac{(4)}{2(-1)} = 2$

Vertex: $(2, h(2)) = (2, 6)$

Maximum: $h(2) = 6$

y intercept: $h(0) = 2$

x intercepts: $h(x) = -x^2 + 4x + 2 = 0$

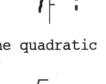

$$x = \frac{-4 \pm \sqrt{16 + 8}}{2(-1)} \quad \begin{array}{l}\text{(using the quadratic} \\ \text{formula)}\end{array}$$
$$= \frac{-4 \pm \sqrt{24}}{-2} = \frac{-4 \pm 2\sqrt{6}}{-2} = 2 \pm \sqrt{6}$$

Range: $(-\infty, 6]$

21. $f(x) = 6x - x^2$; $a = -1$, $b = 6$, $c = 0$.

Axis: $x = -\dfrac{b}{2a} = -\dfrac{6}{2(-1)} = 3$

Vertex: $\left(-\dfrac{b}{2a},\ f\left(-\dfrac{b}{2a}\right)\right) = (3,\ 9)$

Maximum: $f(3) = 9$
y intercept: $f(0) = 0$
x intercepts: $f(x) = 6x - x^2 = 0$
$$x(6 - x) = 0$$
$$x = 0,\ 6$$

Range: $(-\infty,\ 9]$

Graph:

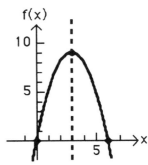

23. $F(s) = s^2 - 4$; $a = 1$, $b = 0$, $c = -4$.

Axis: $s = -\dfrac{b}{2a} = 0$ [the $F(s)$-axis]

Vertex: $(0,\ -4)$
Minimum: $F(0) = -4$
y intercept: $F(0) = -4$
s intercepts: $s^2 - 4 = 0$
$$(s - 2)(s + 2) = 0$$
$$s = 2,\ -2$$
Range: $[-4,\ \infty)$

Graph:

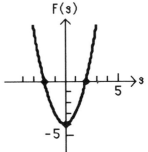

25. $F(x) = 4 - x^2 = -x^2 + 4$; $a = -1$, $b = 0$, $c = 4$.

Axis: $x = -\dfrac{b}{2a} = 0$ [the $F(x)$-axis]

Vertex: $(0,\ 4)$
Maximum: $F(0) = 4$
y intercept: $F(0) = 4$
x intercepts: $4 - x^2 = 0$
$$(2 - x)(2 + x) = 0$$
$$x = 2,\ -2$$
Range: $(-\infty,\ 4]$

Graph:

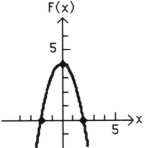

27. $f(x) = \begin{cases} 1 & 0 \le x \le 2 \\ 3 & 2 < x \le 3 \\ 5 & 3 < x \le 5 \end{cases}$

Domain: $[0,\ 5]$
Range: $\{1,\ 3,\ 5\}$
Graph:

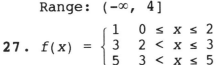

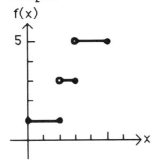

29. $f(x) = \begin{cases} x & -2 \le x < 1 \\ -x + 2 & 1 \le x \le 2 \end{cases}$

Domain: $[-2,\ 2]$
Range: $[-2,\ 1]$
[<u>Note</u>: Use the graph to determine the range.]
Graph:

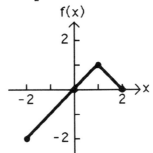

31. $h(x) = \begin{cases} -x^2 - 2 & x < 0 \\ x^2 + 2 & x > 0 \end{cases}$

Domain: All real numbers except 0,
   i.e., $(-\infty, 0) \cup (0, \infty)$

Range: $(-\infty, -2) \cup (2, \infty)$

Graph:

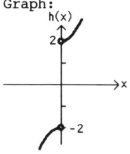

33. $G(x) = \begin{cases} -3 & x < -3 \\ x & -3 \le x \le 3 \\ 3 & x > 3 \end{cases}$

Domain: All real numbers,
   i.e., $(-\infty, \infty)$

Range: $[-3, 3]$

Graph:

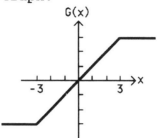

35. $f(x) = x^2 - 7x + 10;\ a = 1,\ b = -7,\ c = 10.$

Axis: $x = -\dfrac{b}{2a} = -\dfrac{(-7)}{2 \cdot 1} = \dfrac{7}{2}$

Vertex: $\left(-\dfrac{b}{2a},\ f\left(-\dfrac{b}{2a}\right)\right) = \left(\dfrac{7}{2},\ f\left(\dfrac{7}{2}\right)\right)$

Now $f\left(\dfrac{7}{2}\right) = \left(\dfrac{7}{2}\right)^2 - 7\left(\dfrac{7}{2}\right) + 10 = \dfrac{49}{4} - \dfrac{49}{2} + 10 = -\dfrac{9}{4}$;

thus, $\left(\dfrac{7}{2},\ -\dfrac{9}{4}\right)$.

Minimum: $f\left(\dfrac{7}{2}\right) = -\dfrac{9}{4}$

y intercept: $f(0) = 10$

x intercepts: $x^2 - 7x + 10 = 0$
$\qquad\qquad (x - 5)(x - 2) = 0$
$\qquad\qquad\qquad\qquad x = 2,\ 5$

Range: $\left[-\dfrac{9}{4},\ \infty\right)$

Graph:

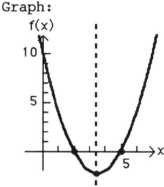

37. $h(x) = 2 - 5x - x^2;\ a = -1,\ b = -5,\ c = 2.$

Axis: $x = -\dfrac{b}{2a} = -\dfrac{(-5)}{2(-1)} = -\dfrac{5}{2}$

Vertex: $\left(-\dfrac{b}{2a},\ h\left(-\dfrac{b}{2a}\right)\right) = \left(-\dfrac{5}{2},\ h\left(-\dfrac{5}{2}\right)\right)$

Now $h\left(-\dfrac{5}{2}\right) = 2 - 5\left(-\dfrac{5}{2}\right) - \left(\dfrac{5}{2}\right)^2 = 2 + \dfrac{25}{2} - \dfrac{25}{4} = \dfrac{33}{4}$;

thus, $\left(-\dfrac{5}{2},\ \dfrac{33}{4}\right)$.

Maximum: $h\left(-\dfrac{5}{2}\right) = \dfrac{33}{4}$

y intercept: $h(0) = 2$

x intercepts: $2 - 5x - x^2 = 0$

$$x = \frac{5 \pm \sqrt{25 + 8}}{2(-1)} = \frac{5 \pm \sqrt{33}}{-2} \approx -5.37,\ 0.372$$

Range: $\left(-\infty,\ \dfrac{33}{4}\right]$

Graph:

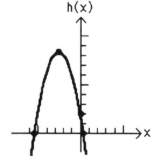

39. Given the linear functions $f(x) = 2 + \frac{1}{2}x$ and $g(x) = 8 - x$.

x	$f(x)$
0	2
2	3

x	$g(x)$
0	8
8	0

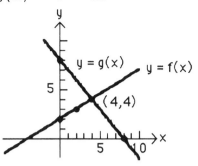

To find the point of intersection, set $f(x) = g(x)$:

$$2 + \frac{1}{2}x = 8 - x$$

$$\frac{3}{2}x = 6$$

$$x = \frac{2}{3}(6) = 4$$

For $x = 4$, $f(4) = g(4) = 4$.

41. Given $f(x) = x^2 - 6x + 8$, $g(x) = x - 2$. The function f is a quadratic function with

Axis: $x = -\dfrac{(-6)}{2(1)} = 3$

Vertex: $(3, f(3)) = (3, -1)$

Minimum: $f(3) = -1$

y intercept: 8

x intercepts: $x^2 - 6x + 8 = 0$

$\qquad\qquad\qquad (x - 4)(x - 2) = 0$

$\qquad\qquad\qquad\qquad\qquad x = 2, 4$

The function g is a linear function with slope 1 and y intercept -2.

To find the points of intersection, set $f(x) = g(x)$:

$$x^2 - 6x + 8 = x - 2$$
$$x^2 - 7x + 10 = 0$$
$$(x - 2)(x - 5) = 0$$
$$x = 2, \ x = 5$$

For $x = 2$, $f(2) = g(2) = 0$; for $x = 5$, $f(5) = g(5) = 3$.

43. Given $f(x) = 11 + 2x - x^2$, $g(x) = x^2 - 1$.

For f:

Axis: $x = -\dfrac{2}{2(-1)} = 1$

Vertex: $(1, f(1)) = (1, 12)$

Maximum: $f(1) = 12$

y intercept: 11

x intercepts: $11 + 2x - x^2 = 0$

$$x = \frac{-2 \pm \sqrt{4 + 44}}{2(-1)}$$

$$= \frac{-2 \pm \sqrt{48}}{-2} = 1 \pm 2\sqrt{3}$$

For g:

Axis: $x = 0$

Vertex: $(0, g(0)) = (0, -1)$

Minimum: $g(0) = -1$

y intercept: -1

x intercepts: $x^2 - 1 = 0$

$$(x - 1)(x + 1) = 0$$

$$x = 1, -1$$

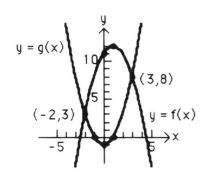

To find the points of intersection, set
$f(x) = g(x)$:
$$11 + 2x - x^2 = x^2 - 1$$
$$-2x^2 + 2x + 12 = 0$$
$$-2(x^2 - x - 6) = 0$$
$$-2(x - 3)(x + 2) = 0$$
$$x = 3, \ -2$$
For $x = -2$, $f(-2) = g(-2) = 3$;
for $x = 3$, $f(3) = g(3) = 8$.

45. $C(x) = \begin{cases} 60 & 0 < x \le 30 \\ 80 & 30 < x \le 45 \\ 100 & 45 < x \le 60 \\ 120 & 60 < x \le 75 \\ 140 & 75 < x \le 90 \end{cases}$

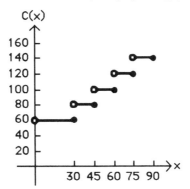

47. Demand equation: $x = 9,000 - 30p$ (1)
Cost equation: $C(x) = 90,000 + 30x$ (2)
Substituting x from (1) into (2), we get:

(A) $C = 90,000 + 30(9,000 - 30p)$
    $= 360,000 - 900p$

(B) Revenue $= xp = (9,000 - 30p)p$
    $R = 9,000p - 30p^2$

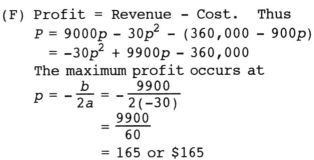

(C) The graphs of the cost function and
the revenue function are shown at
the right; they have been
constructed following the usual graphing technique. Profit and loss
regions are also shown.

(D) At the break-even point, $R = C$:
$$9,000p - 30p^2 = 360,000 - 900p$$
$$30p^2 - 9,900p + 360,000 = 0$$
$$p^2 - 330p + 12,000 = 0$$
$$(p - 42)(p - 288) \approx 0$$
Thus, the break-even points are (approximately) $p = \$42$, $p = \$288$.

(E) The maximum revenue occurs at
$$p = -\frac{b}{2a} = -\frac{9,000}{2(-30)}$$
$$= \frac{-9,000}{-60}$$
$$= 150$$
or $p = \$150$.

(F) Profit = Revenue - Cost. Thus
$$P = 9000p - 30p^2 - (360,000 - 900p)$$
$$= -30p^2 + 9900p - 360,000$$
The maximum profit occurs at
$$p = -\frac{b}{2a} = -\frac{9900}{2(-30)}$$
$$= \frac{9900}{60}$$
$$= 165 \text{ or } \$165$$

49. Demand equation: $x = 8000 - 40p$ (1)   (C)
Cost equation: $C = 100,000 + 20x$ (2)
Substituting x from (1) into (2), we get:

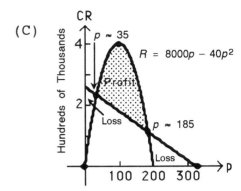

(A) $C = 100,000 + 20(8000 - 40p)$
    $= 260,000 - 800p$

(B) Revenue $R = xp = (8000 - 40p)p$
          $= 8000p - 40p^2$

The graphs of the cost and revenue functions are shown above; they have been constructed using the usual techniques. Profit and loss regions are also shown.

(D) At the break-even points $R = C$:
$$8000p - 40p^2 = 260,000 - 800p$$
$$-40p^2 + 8800p - 260,000 = 0$$
$$p^2 - 220p + 6500 = 0$$

(quadratic formula) $p = \dfrac{220 \pm \sqrt{(-220)^2 - 4(6500)}}{2}$

$$\approx \frac{220 \pm 150}{2} = 185, \ 35$$

Thus, the break-even points are (approximately)
    $p = \$35, \ p = \$185$

(E) $R = -40p^2 + 8000p$. The maximum revenue occurs at
$$p = -\frac{b}{2a} = \frac{-8000}{2(-40)} = \frac{-8000}{-80} = 100$$
or $p = \$100$

(F) Profit = Revenue - Cost. Thus
$$P = -40p^2 + 8000p - (260,000 - 800p)$$
$$= -40p^2 + 8800p - 260,000$$
The maximum profit occurs at
$$p = -\frac{b}{2a} = \frac{-8800}{2(-40)} = \frac{-8800}{-80} = 110$$
or $p = \$110$.

51. From Problem 47(D), Revenue = Cost
$$9000p - 30p^2 = 360,000 - 900p$$
$$-30p^2 + 9900p - 360,000 = 0$$
$$p^2 - 330p + 12,000 = 0$$

Thus, $p = \dfrac{330 \pm \sqrt{(-330)^2 - 4(12,000)}}{2}$ (quadratic formula)

$$= \frac{330 \pm \sqrt{60,900}}{2}$$

$$\approx \frac{330 \pm 246.78}{2} = 288.39, \ 41.61$$

or $p = \$42, \ p = \$288$ to the nearest dollar.

53. From Problem 49(D), Revenue = Cost

$$8000p - 40p^2 = 260{,}000 - 800p$$
$$-40p^2 + 8800p - 260{,}000 = 0$$
$$p^2 - 220p + 6500 = 0$$

$$p = \frac{220 \pm \sqrt{(-220)^2 - 4(6500)}}{2}$$
$$= \frac{220 \pm \sqrt{22{,}400}}{2}$$
$$\approx \frac{220 \pm 149.67}{2} = 184.83,\ 35.17$$

or $p = \$35$, $p = \$185$ to the nearest dollar.

55. Let x = the number of books sold.

(A) $C(x) = 240{,}000 + 20x$

(B) $R(x) = 35x$

(C)

x	$C(x)$
0	240,000
10,000	440,000

x	$R(x)$
0	0
10,000	350,000

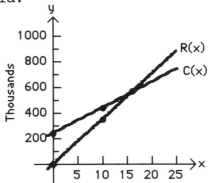

(D) To find the break-even point, set $R(x) = C(x)$: $35x = 240{,}000 + 20x$
$$15x = 240{,}000$$
$$x = 16{,}000 \text{ books}$$

57. Given $v = f(x) = 1{,}000(0.04 - x^2)$, $0 \le x \le 0.2$.

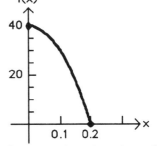

Axis: $x = 0$
Vertex: $(0, 40)$
Maximum: $f(0) = 40$
y intercept: 40
x intercept: $1{,}000(0.04 - x^2) = 0$
$$x^2 = 0.04$$
$$x = 0.2$$

[<u>Note</u>: $x = -0.2$ is not an x intercept since -0.2 is not in the domain of f.]

59. $\dfrac{\Delta s}{s} = k$. For $k = \dfrac{1}{30}$, $\dfrac{\Delta s}{s} = \dfrac{1}{30}$ or $\Delta s = \dfrac{1}{30}s$

(A) When $s = 30$, $\Delta s = \dfrac{1}{30}(30) = 1$ pound.   (B) $\Delta s = \dfrac{1}{30}s$

When $s = 90$, $\Delta s = \dfrac{1}{30}(90) = 3$ pounds.   Slope $m = \dfrac{1}{30}$

y intercept $b = 0$

1. $\frac{u}{5} = \frac{u}{6} + \frac{6}{5}$ Multiply each term by 30: $6u = 5u + 36$
$$6u - 5u = 36$$
$$u = 36$$

2. $2(x + 4) > 5x - 4$
$$2x + 8 > 5x - 4$$
$$2x - 5x > -4 - 8$$
$$-3x > -12 \quad \text{(Divide both sides by -3 and reverse the inequality)}$$
$$x < 4 \quad \text{or } (-\infty, 4)$$

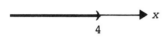

3. $\quad\quad x^2 = 5x$
$$x^2 - 5x = 0 \quad \text{(solve by factoring)}$$
$$x(x - 5) = 0$$
$$x = 0 \text{ or } x - 5 = 0$$
$$x = 5$$

4. $y = \frac{x}{2} - 2 = \frac{1}{2}x - 2$

Slope $m = \frac{1}{2}$

y intercept $b = -2$

x intercept: 4

x	y
0	-2
2	-1
4	0

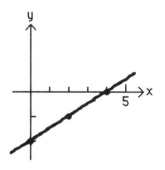

5. Using the point-slope form with $m = \frac{1}{2}$ and $(x_1, y_1) = (4, 3)$, we have:

$y - 3 = \frac{1}{2}(x - 4)$ or $y - 3 = \frac{1}{2}x - 2$ and $y = \frac{1}{2}x + 1$

6. $x - y = 2$. Solving for y, we have:
$-y = -x + 2$ or $y = x - 2$
Slope $m = 1$
y intercept $b = -2$
x intercept: 2

x	y
0	-2
2	0

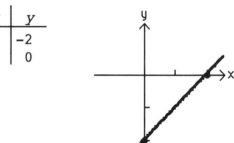

7. $f(x) = 2x - 1$, $g(x) = x^2 - 2x$
$$f(-2) + g(-1) = [2(-2) - 1] + [(-1)^2 - 2(-1)]$$
$$= [-4 - 1] + [1 + 2]$$
$$= -5 + 3 = -2$$

8. $f(x) = \frac{2}{3}x - 1$

Slope $m = \frac{2}{3}$

y intercept $b = -1$

x intercept: $\frac{3}{2}$

x	$f(x)$
0	-1
3	1
$\frac{3}{2}$	0

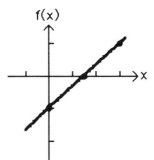

9. Given the quadratic function $f(x) = x^2 - 8x + 7$. We have $a = 1$, $b = -8$, $c = 7$. The x-coordinate of the vertex is:

$x = -\frac{b}{2a} = -\frac{(-8)}{2(1)} = 4$

Thus, vertex: $\left(-\frac{b}{2a},\ f\left(-\frac{b}{2a}\right)\right) = (4,\ f(4)) = (4,\ -9)$.

The minimum value of f is $f(4) = -9$.

10. $\frac{x}{12} - \frac{x-3}{3} = \frac{1}{2}$

Multiply each term by 12: $x - 4(x - 3) = 6$

$$x - 4x + 12 = 6$$
$$-3x = 6 - 12$$
$$-3x = -6$$
$$x = 2$$

11. $1 - \frac{x-3}{3} \leq \frac{1}{2}$

Multiply both sides of the inequality by 6. We do not reverse the direction of the inequality, since $6 > 0$.

$6 - 2(x - 3) \leq 3$
$6 - 2x + 6 \leq 3$
$-2x \leq 3 - 12$
$-2x \leq -9$

Divide both sides by -2 and reverse the direction of the inequality, since $-2 < 0$.

$x \geq \frac{9}{2}$ or $\left[\frac{9}{2},\ \infty\right)$

12. $-2 \leq \frac{x}{2} - 3 < 3$

$-2 + 3 \leq \frac{x}{2} < 3 + 3$

$1 \leq \frac{x}{2} < 6$

$2 \leq x < 12$ or $[2,\ 12)$

13. $2x - 3y = 6$
$-3y = -2x + 6$
$y = \frac{2}{3}x - 2$

14. $xy - y = 3$
$y(x - 1) = 3$
$y = \frac{3}{x - 1}$

15. $3x^2 - 21 = 0$

   $x^2 - 7 = 0$ (solve by the square root method)

   $x^2 = 7$

   $x = \pm\sqrt{7}$

16. $x^2 - x - 20 = 0$ (solve by factoring)

   $(x - 5)(x + 4) = 0$

   $x - 5 = 0$ or $x + 4 = 0$

   $x = 5$ $x = -4$

17. $2x^2 = 3x + 1$

   $2x^2 - 3x - 1 = 0$ (use the quadratic formula with $a = 2$, $b = -3$, $c = -1$)

   $x = \dfrac{-b \pm \sqrt{b^2 - 4ac}}{2a} = \dfrac{-(-3) \pm \sqrt{(-3)^2 - 4(2)(-1)}}{2(2)} = \dfrac{3 \pm \sqrt{9 + 8}}{4} = \dfrac{3 \pm \sqrt{17}}{4}$

18. $4x^2 - 87x - 216$

   <u>Step 1.</u> Test for factorability

   $\sqrt{b^2 - 4ac} = \sqrt{(-87)^2 - 4(4)(-216)} = \sqrt{11025} = 105$

   Since the result is an integer, the polynomial has first-degree factors with integer coefficients.

   <u>Step 2.</u> Use the factor theorem

   $4x^2 - 87x - 216$

   $x = \dfrac{87 \pm 105}{2(4)} = 24, \ -\dfrac{9}{4}$ (by the quadratic formula)

   Thus, $4x^2 - 87x - 216 = 4\left(x - \left[-\dfrac{9}{4}\right]\right)(x - 24)$

   $= (4x + 9)(x - 24)$

19. $2x^2 + 64x - 320$

   <u>Step 1.</u> Test for factorability

   $\sqrt{b^2 - 4ac} = \sqrt{(64)^2 - 4(2)(-320)} = \sqrt{6656} \approx 81.584.$

   Since the result is not an integer, the polynomial does not have first-degree factors with integer coefficients, i.e. the polynomial is *not* factorable.

20. $3x + 6y = 18$

   To find the x intercept, set $y = 0$:

   $3x + 6(0) = 18$

   $x = 6$

   To find the slope and the y intercept, rewrite the equation in the slope-intercept form:

   $6y = -3x + 18$

   $y = -\dfrac{1}{2}x + 3$

x	y
0	3
6	0

   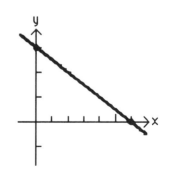

   Thus, the slope $m = -\dfrac{1}{2}$ and the y intercept $b = 3$.

21. Let $(x_1, y_1) = (-2, 3)$ and $(x_2, y_2) = (6, -1)$. The slope of the line is given by:

$$m = \frac{y_2 - y_1}{x_2 - x_1} = \frac{-1 - 3}{6 - (-2)} = -\frac{4}{8} = -\frac{1}{2}$$

Using the point-slope form, we have:

$$y - 3 = -\frac{1}{2}[x - (-2)]$$

$$y - 3 = -\frac{1}{2}(x + 2)$$

$$y - 3 = -\frac{1}{2}x - 1$$

$$y = -\frac{1}{2}x + 2$$

Rewriting this equation in the form $Ax + By = C$, we have:

$2y = -x + 4$ or $x + 2y = 4$

The slope of this line is $m = -\frac{1}{2}$.

22. Given the point P: $(-5, 2)$. The vertical line through P has equation $x = -5$. The horizontal line through P has equation $y = 2$.

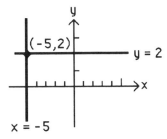

23. Let $(x_1, y_1) = (-2, 5)$ and $(x_2, y_2) = (2, -1)$. The slope of the line is given by:

$$m = \frac{y_2 - y_1}{x_2 - x_1} = \frac{-1 - 5}{2 - (-2)} = -\frac{6}{4} = -\frac{3}{2}$$

Using the point-slope form, we have:

$$y - 5 = -\frac{3}{2}[x - (-2)]$$

$$y - 5 = -\frac{3}{2}(x + 2)$$

$$y - 5 = -\frac{3}{2}x - 3$$

$$y = -\frac{3}{2}x + 2$$

24. $f(x) = 10x - 7$, $g(t) = 6 - 2t$, $F(u) = 3u^2$, and $G(v) = v - v^2$.

(A) $2g(-1) - 3G(-1) = 2[6 - 2(-1)] - 3[-1 - (-1)^2]$
$= 2[6 + 2] - 3[-1 - 1]$
$= 16 + 6$
$= 22$

(B) $4G(-2) - g(-3) = 4[-2 - (-2)^2] - [6 - 2(-3)]$
$= 4[-6] - [12]$
$= -24 - 12$
$= -36$

(C) $\dfrac{f(2) \cdot g(-4)}{G(-1)} = \dfrac{[10(2) - 7] \cdot [6 - 2(-4)]}{-1 - (-1)^2} = \dfrac{(13)(14)}{-2} = -91$

(D) $\dfrac{F(-1)\cdot G(2)}{g(-1)} = \dfrac{[3(-1)^2]\cdot[2 - (2)^2]}{6 - 2(-1)} = \dfrac{(3)(-2)}{8} = \dfrac{-6}{8} = \dfrac{-3}{4}$

25. $f(x) = 2x - x^2$, $g(x) = \dfrac{1}{x - 2}$.

The domain of f is R—the set of real numbers.
The domain of g is all real numbers except $x = 2$; g is not defined at $x = 2$.

26. $f(x) = 2x - 1$

$\dfrac{f(3 + h) - f(3)}{h} = \dfrac{[2(3 + h) - 1] - [2(3) - 1]}{h} = \dfrac{[6 + 2h - 1] - [6 - 1]}{h}$

$= \dfrac{5 + 2h - 5}{h} = \dfrac{2h}{h} = 2$

27. (A) $4x - 3y = 11$
    Solving for y, we have:
    $-3y = -4x + 11$
    $y = \dfrac{4}{3}x - \dfrac{11}{3}$
    This equation *does* specify a function (a linear function). The domain is R, all real numbers.

    (B) $y^2 - 4x = 1$
    Solving for y, we have:
    $y^2 = 4x + 1$
    $y = \pm\sqrt{4x + 1}$
    This equation *does not* specify a function (with independent variable x). For example, if $x = 0$, $y = 1$ or -1; if $x = 2$, $y = 3$ or -3.

    (C) $xy + 3y + 5x = 4$
    Solving for y, we have:
    $(x + 3)y = 4 - 5x$
    $y = \dfrac{4 - 5x}{x + 3}$
    This equation *does* specify a function. The domain is all real numbers except $x = -3$.

28. Given $g(x) = 8x - 2x^2$; $a = -2$, $b = 8$, $c = 0$.

    Axis: $x = -\dfrac{b}{2a} = -\dfrac{8}{2(-2)} = 2$

    Graph:

    Vertex: $\left(-\dfrac{b}{2a},\ g\!\left(-\dfrac{b}{2a}\right)\right) = (2,\ g(2)) = (2,\ 8)$

    Maximum value of g: $g(2) = 8$
    y intercept: $g(0) = 0$
    x intercepts: $8x - 2x^2 = 0$
    $\qquad\qquad\qquad 2x(4 - x) = 0$
    $\qquad\qquad\qquad\qquad\quad x = 0,\ 4$
    Range: $(-\infty,\ 8]$

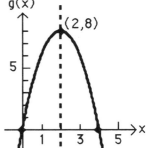

29. $f(x) = \begin{cases} 1 - x^2 & -1 \le x < 0 \\ 1 + x^2 & 0 \le x \le 1 \end{cases}$ Graph:

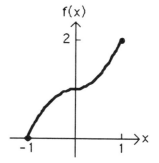

Domain: $[-1, 1]$
Range: $[0, 2]$

30. $x^2 + jx + k = 0$. Using the quadratic formula with $a = 1$, $b = j$, $c = k$, we have:

$$x = \frac{-j \pm \sqrt{j^2 - 4(1)(k)}}{2} = \frac{-j \pm \sqrt{j^2 - 4k}}{2}$$

31. Since both points have the same x-coordinate, the line is vertical. Therefore, the equation is $x = 4$.

32. Given the line L with equation $2x - 4y = 5$. Rewriting this equation in the slope-intercept form, we have:

$-4y = -2x + 5$
$y = \frac{1}{2}x - \frac{5}{4}$

Therefore, the slope of L is $m = \frac{1}{2}$.

(A) Any line parallel to L has slope $m = \frac{1}{2}$. Using the point-slope form, an equation for the line which is parallel to L and passes through $(2, -3)$ is:

$y - (-3) = \frac{1}{2}(x - 2)$
$y + 3 = \frac{1}{2}x - 1$
$y = \frac{1}{2}x - 4$
$2y = x - 8$ or $x - 2y = 8$

(B) Any line perpendicular to L has slope $m = \frac{-1}{\frac{1}{2}} = -2$. Using the point-slope form, an equation for the line which is perpendicular to L and passes through $(2, -3)$ is:

$y + 3 = -2(x - 2)$
$y + 3 = -2x + 4$ or $2x + y = 1$

33. $f(x) = \frac{5}{x - 3}$ and $g(x) = \sqrt{x - 1}$

The domain of f is all real numbers except $x = 3$; f is not defined at $x = 3$.
The domain of g is all real numbers such that $x - 1 \ge 0$, that is, all real numbers x such that $x \ge 1$, or $[1, \infty)$.

34. Given $f(x) = x^2 + 7x - 9$

$$\frac{f(a + h) - f(a)}{h} = \frac{[(a + h)^2 + 7(a + h) - 9] - [a^2 + 7a - 9]}{h}$$

$$= \frac{(a^2 + 2ah + h^2 + 7a + 7h - 9) - a^2 - 7a + 9}{h}$$

$$= \frac{2ah + h^2 + 7h}{h} = \frac{h(2a + h + 7)}{h} = 2a + h + 7$$

35. Given $f(x) = 2x - 7$, $g(x) = x^2 - 6x + 5$.
f is a linear function with slope 2 and
y intercept -7; g is a quadratic function
with vertex

$$\left(-\frac{b}{2a}, \ g\left(-\frac{b}{2a}\right)\right) = (3, -4).$$

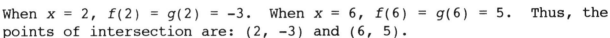

y intercept: 5
x intercepts: $x^2 - 6x + 5 = 0$
$\qquad\qquad\quad (x - 5)(x - 1) = 0$
$\qquad\qquad\qquad\qquad\quad x = 1, \ 5$
Points of intersection: $x^2 - 6x + 5 = 2x - 7$
$\qquad\qquad\qquad\qquad\quad x^2 - 8x + 12 = 0$
$\qquad\qquad\qquad\quad (x - 2)(x - 6) = 0$
$\qquad\qquad\qquad\qquad\qquad\qquad x = 2, \ 6$

When $x = 2$, $f(2) = g(2) = -3$. When $x = 6$, $f(6) = g(6) = 5$. Thus, the
points of intersection are: $(2, -3)$ and $(6, 5)$.

36. Let x = the amount invested at 8%. Then $60,000 - x$ = amount invested at
14%. The interest on $60,000 at 12% for one year is:
$\qquad 0.12(60,000) = 7200$
Thus, we want:
$0.08x + 0.14(60,000 - x) = 7200$
$\qquad 0.08x + 8400 - 0.14x = 7200$
$\qquad\qquad\qquad\qquad -0.06x = -1200$
$\qquad\qquad\qquad\qquad\qquad x = 20,000$
Therefore, $20,000 should be invested at 8% and $40,000 should be
invested at 14%.

37. $\frac{x}{800} = \frac{247}{89}$. Thus, $89x = (247)(800) = 197,600$ and $x = \$2,220.22$.

38. In the formula $A = P(1 + r)^2$, set $A = 1210$ and $P = 1000$. This yields
$1000(1 + r)^2 = 1210$.
Therefore, $(1 + r)^2 = \frac{1210}{1000} = 1.21$. Thus, $1 + r = \pm\sqrt{1.21} = \pm 1.1$
$\qquad\qquad\qquad\qquad\qquad\qquad\qquad$ and $\qquad\qquad r = 0.1$ or -2.1.
Since the interest rate r cannot be negative, $r = 0.1$ or $r = 10\%$.

39. We have $V = 12,000$ when $t = 0$ and $V = 2,000$ when $t = 8$.

(A) We are looking for an equation for the line determined by the two points $(0, 12{,}000)$, $(8, 2{,}000)$. The slope of this line is given by:

$$m = \frac{2{,}000 - 12{,}000}{8 - 0} = \frac{-10{,}000}{8} = -1{,}250$$

Since $V = 12{,}000$ is the V intercept, the linear equation is $V(t) = -1{,}250t + 12{,}000$.

(B) The value of the system after five years is:
$V(5) = -1{,}250(5) + 12{,}000 = -6{,}250 + 12{,}000 = \$5{,}750$

40. (A) We are looking for an equation for the line determined by the two points $(20, 32)$ and $(30, 48)$. The slope of this line is given by:

$$m = \frac{48 - 32}{30 - 20} = \frac{16}{10} = \frac{8}{5}$$

Using the point-slope form with $(C_1, R_1) = (20, 32)$, we have

$$R - 32 = \frac{8}{5}(C - 20) = \frac{8}{5}C - 32.$$

Thus, $R = \frac{8}{5}C$.

(B) For $C = 105$, we have $R = \frac{8}{5}(105) = 8(21) = \$168$.

41. Given $s = 10{,}000p - 25{,}000$ and $d = \frac{90{,}000}{p}$. Set $s = d$ and solve for p.

$10{,}000p - 25{,}000 = \frac{90{,}000}{p}$ (divide both sides of the equation by 5000 and multiply both sides by p)

$$2p^2 - 5p = 18$$
$$2p^2 - 5p - 18 = 0$$
$(2p - 9)(p + 2) = 0$ (factor)
$2p - 9 = 0$ or $p + 2 = 0$
$$p = \frac{9}{2} \qquad\qquad p = -2$$

Since p cannot be negative, we have $p = \frac{9}{2} = 4.5$. Thus, the equilibrium price is \$4.50.

42. Let $x =$ the number of tapes. Then the cost $C(x) = 84{,}000 + 15x$ and the revenue $R(x) = 50x$. To find the break-even point, set $C = R$ and solve for x:

$$84{,}000 + 15x = 50x$$
$$15x - 50x = -84{,}000$$
$$-35x = -84{,}000$$
$$x = 2400$$

Thus, the company must sell 2400 tapes to break even.

43. Given the demand and cost equations $x = 500 - 10p$ and $C = 3{,}000 + 10x$.

(A) Revenue $R = xp = (500 - 10p)p = 500p - 10p^2$.

(B) $C = 3{,}000 + 10x = 3{,}000 + 10(500 - 10p) = 8{,}000 - 100p$.

(C) C is a linear function with y intercept 8,000 and slope -100; R is a quadratic function: the p-coordinate of the vertex is:

$$p = \frac{-500}{-20} = 25,$$

The vertex is $(25, 6,250)$, and the p intercepts are $p = 0$, $p = 50$.

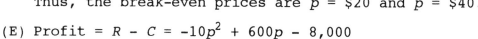

(D) To find the break-even points, set $R = C$. We obtain:

$$500p - 10p^2 = 8,000 - 100p$$
$$-10p^2 + 600p - 8,000 = 0$$
$$p^2 - 60p + 800 = 0$$
$$(p - 40)(p - 20) = 0$$
$$p = 40, \quad p = 20$$

Thus, the break-even prices are $p = \$20$ and $p = \$40$.

(E) Profit $= R - C = -10p^2 + 600p - 8,000$
This is a quadratic function with $a = -10$. The maximum value occurs when

$$p = -\frac{b}{2a} = \frac{-600}{2(-10)} = 30.$$

Thus, the price that produces maximum profit is $p = \$30$.

44. (A) The area of a rectangle is $A = $ (length) $\times$ (width). Thus, $A = 2yx$. Now, the perimeter is $180 = 3x + 4y$. Solving this equation for y, we have:

$$4y = 180 - 3x$$
$$y = 45 - \frac{3}{4}x$$

Therefore, $A = 2\left(45 - \frac{3}{4}x\right)x = 90x - \frac{3}{2}x^2.$

(B) The domain of A is $0 \le x \le 60$ (since $x > 60$ would imply $y < 0$).

(C) We have $A = 90x - \frac{3}{2}x^2$. Since A is a quadratic function with

$a = -\frac{3}{2} < 0$, the maximum value of A occurs when

$$x = -\frac{b}{2a} = -\frac{90}{2\left(-\frac{3}{2}\right)} = 30.$$

Now, when $x = 30$,

$$y = 45 - \frac{3}{4}(30) = 45 - \frac{90}{4} = 22.5.$$

Thus, the dimensions that will maximize the area are $x = 30$, $y = 22.5$.

3 EXPONENTIAL AND LOGARITHMIC FUNCTIONS

EXERCISE 3-1

Things to remember:

1. EXPONENTIAL FUNCTION

   The equation

   $$f(x) = b^x, \; b > 0, \; b \neq 1$$

   defines an EXPONENTIAL FUNCTION for each different constant b, called the BASE. The DOMAIN of f is all real numbers, and the RANGE of f is the set of positive real numbers.

2. BASIC PROPERTIES OF THE GRAPH OF $f(x) = b^x, \; b > 0, \; b \neq 1$

   a. All graphs pass through $(0,1)$; $b^0 = 1$ for any base b.

   b. All graphs are continuous curves; there are no holes or jumps.

   c. The x-axis is a horizontal asymptote.

   d. If $b > 1$, then b^x increases as x increases.

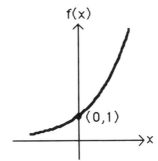

   Graph of $f(x) = b^x, \; b > 1$

   e. If $0 < b < 1$, then b^x decreases as x increases.

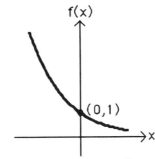

   Graph of $f(x) = b^x, \; 0 < b < 1$

69

3. EXPONENTIAL FUNCTION PROPERTIES

For a, $b > 0$, $a \neq 1$, $b \neq 1$, and x, y real numbers:

a. EXPONENT LAWS

(i) $a^x a^y = a^{x+y}$ (iv) $(ab)^x = a^x b^x$

(ii) $\dfrac{a^x}{a^y} = a^{x-y}$ (v) $\left(\dfrac{a}{b}\right)^x = \dfrac{a^x}{b^x}$

(iii) $(a^x)^y = a^{xy}$

b. $a^x = a^y$ if and only if $x = y$.

c. For $x \neq 0$, $a^x = b^x$ if and only if $a = b$.

4. COMPOUND INTEREST

If a principal P is invested at an annual rate r (expressed as a decimal) compounded m times per year, then the amount A (future value) in the account at the end of t years is given by:

$$A = P\left(1 + \frac{r}{m}\right)^{mt}$$

---

1. $y = 5^x$, $-2 \leq x \leq 2$

x	y
-2	$\frac{1}{25}$
-1	$\frac{1}{5}$
0	1
1	5
2	25

3. $y = \left(\frac{1}{5}\right)^x = 5^{-x}$, $-2 \leq x \leq 2$

x	y
-2	25
-1	5
0	1
1	$\frac{1}{5}$
2	$\frac{1}{25}$

5. $f(x) = -5^x$, $-2 \leq x \leq 2$

x	$f(x)$
-2	$-\frac{1}{25}$
-1	$-\frac{1}{5}$
0	-1
1	-5
2	-25

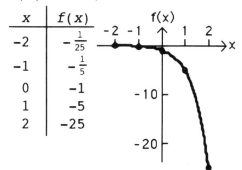

7. $f(x) = 4(5^x)$, $-2 \leq x \leq 2$

x	$f(x)$
-2	$\frac{4}{25}$
-1	$\frac{4}{5}$
0	4
1	20
2	100

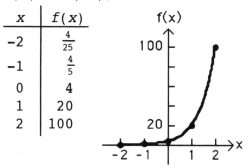

9. $y = 5^{x+2} + 4$, $[-4, 0]$

x	y
-4	$4\frac{1}{25}$
-3	$4\frac{1}{5}$
-2	5
-1	9
0	29

11. $(4^{3x})^{2y} = 4^{6xy}$ [see 3a(iii)]

13. $\dfrac{5^{x-3}}{5^{x-4}} = 5^{(x-3)-(x-4)}$

$\qquad = 5^{x-3-x+4} = 5$ [see 3a(ii)]

15. $(2^x 3^y)^z = 2^{xz} 3^{yz}$

17. $10^{2-3x} = 10^{5x-6}$ implies [see 3b]

$2 - 3x = 5x - 6$

$-8x = -8$

$x = 1$

19. $4^{5x - x^2} = 4^{-6}$ implies

$5x - x^2 = -6$

or $-x^2 + 5x + 6 = 0$

$x^2 - 5x - 6 = 0$

$(x - 6)(x + 1) = 0$

$x = 6, -1$

21. $5^3 = (x + 2)^3$ implies (by property 3c)

$5 = x + 2$

Thus, $x = 3$.

23. $9^{x-1} = 3^x$

$(3^2)^{x-1} = 3^x$ (since $9 = 3^2$)

$3^{2x-2} = 3^x$

Thus, $2x - 2 = x$ and $x = 2$.

25. $f(t) = 2^{t/10}$, $-30 \le t \le 30$

t	$f(t)$
-30	$\frac{1}{8}$
-20	$\frac{1}{4}$
-10	$\frac{1}{2}$
0	1
10	2
20	4
30	8

27. $y = 7(2^{-2x})$, $-2 \le x \le 2$

x	y
-2	112
-1	28
0	7
1	$\frac{7}{4}$
2	$\frac{7}{16}$

29. $f(x) = 2^{|x|}$, $-3 \le x \le 3$

x	$f(x)$
-3	8
-2	4
-1	2
0	1
1	2
2	4
3	8

31. $y = 100(1.03)^x$, $0 \le x \le 20$

x	y
0	100
5	115.9
10	134.4
15	155.8
20	180.6

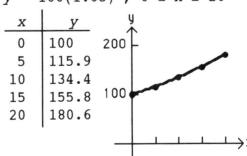

33. $y = 3^{-x^2}$, $-2 \leq x \leq 2$

x	y
-2	$\frac{1}{81}$
-1	$\frac{1}{3}$
0	1
1	$\frac{1}{3}$
2	$\frac{1}{81}$

35. $(3^x - 3^{-x})(3^x + 3^{-x})$
$= (3^x)^2 - (3^{-x})^2 = 3^{2x} - 3^{-2x}$

37. $(3^x - 3^{-x})^2 + (3^x + 3^{-x})^2$
$= (3^x)^2 - 2(3^x)(3^{-x}) + (3^{-x})^2 + (3^x)^2 + 2(3^x)(3^{-x}) + (3^{-x})^2$
$= 3^{2x} - 2 + 3^{-2x} + 3^{2x} + 2 + 3^{-2x}$
$= 2(3^{2x}) + 2(3^{-2x})$

39. $h(x) = x2^x$, $-5 \leq x \leq 0$

x	h(x)
-5	$-\frac{5}{32}$
-4	$-\frac{1}{4}$
-3	$-\frac{3}{8}$
-2	$-\frac{1}{2}$
-1	$-\frac{1}{2}$
0	0

41. $g(x) = \dfrac{3^x + 3^{-x}}{2}$, $-3 \leq x \leq 3$

x	g(x)
-3	≈ 13.5
-2	≈ 4.6
-1	≈ 1.7
0	1
1	≈ 1.7
2	≈ 4.6
3	≈ 13.5

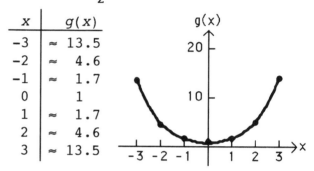

43. $3^{-\sqrt{2}} \approx 0.2115$

45. $\pi^{-\sqrt{3}} \approx 0.1377$

47. $\dfrac{3^\pi - 3^{-\pi}}{2} \approx \dfrac{31.5443 - 0.0317}{2} = 15.7563$

49. Using $\underline{4}$, $A = P\left(1 + \dfrac{r}{m}\right)^{mt}$, we have:

(A) $P = 2,500$, $r = 0.07$, $m = 4$, $t = \dfrac{3}{4}$

$A = 2,500\left(1 + \dfrac{0.07}{4}\right)^{4 \cdot 3/4} = 2,500(1 + 0.0175)^3 = 2,633.56$

Thus, $A = \$2,633.56$.

(B) $A = 2,500\left(1 + \dfrac{0.07}{4}\right)^{4 \cdot 15} = 2,500(1 + 0.0175)^{60} = 7079.54$

Thus, $A = \$7,079.54$.

51. Using $A = P\left(1 + \dfrac{r}{m}\right)^{mt}$, we have:

$A = 15,000$, $r = 0.0975$, $m = 52$, $t = 5$

Thus, $15,000 = P\left(1 + \dfrac{0.0975}{52}\right)^{52 \cdot 5} = P(1 + 0.001875)^{260} \approx P(1.6275)$ and

$P = \dfrac{15,000}{1.6275} \approx 9,217$. Therefore, $P \approx \$9,217$.

53. The doubling time growth model is $P = P_0 2^{t/d}$, where P_0 is the initial population and d is the doubling time. Thus, $P = 23 \cdot 2^{t/19}$ (in millions).

(A) $P = 23 \cdot 2^{10/19} \approx 33.13$
Thus, the population in 10 years will be approximately 33,000,000.

(B) $P = 23 \cdot 2^{30/19} \approx 68.71$
Thus, the population in 30 years will be approximately 69,000,00.

55. Using the half-life decay model $A = A_0 2^{-t/h}$, where A_0 is the initial amount and h is the half-life, we have $A = 12 \cdot 2^{-t/6}$.

(A) $A = 12 \cdot 2^{-3/6} = 12 \cdot 2^{-1/2} \approx 8.49$
Thus, there will be approximately 8.49 mg after 3 hours.

(B) $A = 12 \cdot 2^{-24/6} = 12 \cdot 2^{-4} = \dfrac{12}{16} = 0.75$
Thus, there will be approximately 0.75 mg after 24 hours.

57. From Problem 49, $A = P\left(1 + \dfrac{r}{m}\right)^{mt}$, where $P = 2500$, $r = 0.07$ and $m = 4$. Thus
$A(t) = 2500\left(1 + \dfrac{0.07}{4}\right)^{4t} = 2500(1.0175)^{4t}$ at the end of t years.

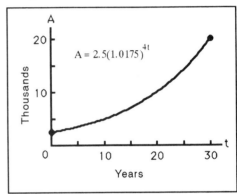

59. From Problem 53, $P = P_0 2^{t/d}$, where $P_0 = 23$ (million) and $d = 19$. Thus $P = 23(2)^{t/19}$ at the end of t years.

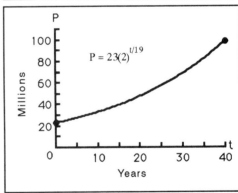

61. From Problem 55, $A = A_0 2^{-t/h}$, where $A_0 = 12$, $h = 6$. Thus $A(t) = 12(2)^{-t/6}$ at the end of t hours.

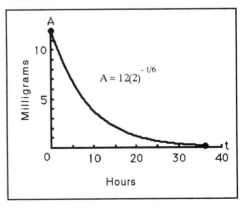

Things to remember:

1. THE IRRATIONAL NUMBER e

   In the expression

   $$\left(1 + \frac{1}{m}\right)^m,$$

   if we let m increase without bound, the value of the expression approaches an irrational number denoted e. To nine decimal places,

   $$e = 2.718\ 281\ 828.$$

2. EXPONENTIAL FUNCTION WITH BASE e

   For x a real number, the equation

   $$f(x) = e^x$$

   defines the EXPONENTIAL FUNCTION WITH BASE e.

3. CONTINUOUS COMPOUND INTEREST FORMULA

   If a principal P is invested at an annual rate r (expressed as a decimal) compounded continuously, then the amount A in the account at the end of t years is given by

   $$A = Pe^{rt}.$$

4. INTEREST FORMULAS

   a. $A = P(1 + rt)$ Simple Interest

   b. $A = P\left(1 + \dfrac{r}{m}\right)^{mt}$ Compound Interest

   c. $A = Pe^{rt}$ Continuous Compound Interest

---

1. $y = -e^{-x}$, $-3 \le x \le 3$

x	y
-3	≈ -20
-2	≈ -7.4
-1	≈ -2.7
0	-1
1	≈ -0.4
2	≈ -0.1
3	≈ -0.05

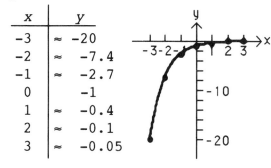

3. $y = 100e^{0.1x}$, $-5 \le x \le 5$

x	y
-5	≈ 60
-3	≈ 74
-1	≈ 90
0	100
1	≈ 111
3	≈ 135
5	≈ 165

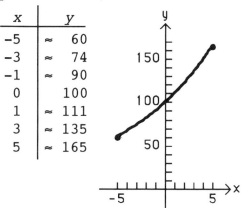

5. $g(t) = 10e^{-0.2t}$, $-5 \le t \le 5$

g	$g(t)$
-5	$\approx$ 27.2
-3	$\approx$ 18.2
-1	$\approx$ 12.2
0	10
1	$\approx$ 8.2
3	$\approx$ 5.5
5	$\approx$ 3.7

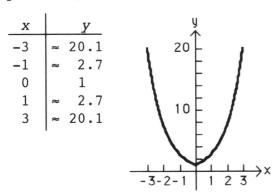

7. $y = -3 + e^{1+x}$, $-4 \le x \le 2$

x	y
-4	$\approx$ -3
-2	$\approx$ -2.6
-1	-2
0	$\approx$ -0.3
1	$\approx$ 4.4
2	$\approx$ 17.1

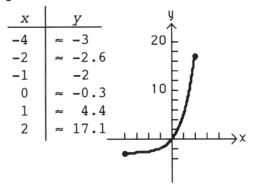

9. $y = e^{|x|}$, $-3 \le x \le 3$

x	y
-3	$\approx$ 20.1
-1	$\approx$ 2.7
0	1
1	$\approx$ 2.7
3	$\approx$ 20.1

11. $C(x) = \dfrac{e^x + e^{-x}}{2}$, $-5 \le x \le 5$

x	$C(x)$
-5	$\approx$ 74
-3	$\approx$ 10
0	1
3	$\approx$ 10
5	$\approx$ 74

13. $e^x(e^{-x} + 1) - e^{-x}(e^x + 1) = e^0 + e^x - e^0 - e^{-x} = e^x - e^{-x}$

15. $\dfrac{e^x(e^x + e^{-x}) - (e^x - e^{-x})e^x}{e^{2x}} = \dfrac{e^{2x} + 1 - (e^{2x} - 1)}{e^{2x}} = \dfrac{2}{e^{2x}}$

17. $(x - 3)e^x = 0$
$x - 3 = 0$ (since $e^x \neq 0$)
$x = 3$

19. $3xe^{-x} + x^2 e^{-x} = 0$
$e^{-x}(3x + x^2) = 0$
$3x + x^2 = 0$ (since $e^{-x} \neq 0$)
$x(3 + x) = 0$
$x = 0, -3$

21. $N = \dfrac{100}{1 + e^{-t}}$, $0 \le t \le 5$

t	N
0	50
1	$\approx$ 73.1
2	$\approx$ 88.1
3	$\approx$ 95.3
5	$\approx$ 99.3

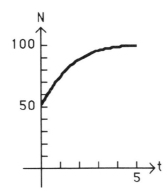

23. Using $\underline{3}$ with $P = 7,500$ and $r = 0.0835$, we have:
$A = 7,500e^{0.0835t}$

(A) $A = 7,500e^{(0.0835)5.5} = 7,500e^{0.45925} \approx 11,871.65$
Thus, there will be $11,871.65 in the account after 5.5 years.

(B) $A = 7,500e^{(0.0835)12} = 7,500e^{1.002} \approx 20,427.93$
Thus, there will be $20,427.93 in the account after 12 years.

25. Alamo Savings:

From Section 3-1, $A = P\left(1 + \dfrac{r}{m}\right)^{mt}$, where P is the principal, r is the annual rate, and m is the number of compounding periods per year. Thus:
$A = 10,000\left(1 + \dfrac{0.0825}{4}\right)^4 = 10,000(1.020625)^4 \approx \$10,850.88$

Lamar Savings:
$A = 10,000e^{0.0805} \approx \$10,838.29$

27. In $A = Pe^{rt}$, we are given $A = 50,000$, $r = 0.1$, and $t = 5.5$. Thus:
$50,000 = Pe^{(0.1)5.5}$ or $P = \dfrac{50,000}{e^{0.55}} \approx 28,847.49$

You should be willing to pay $28,847.49 for the note.

29. Given $N = 2(1 - e^{-0.037t})$, $0 \le t \le 50$

t	N
0	0
10	≈ 0.62
30	≈ 1.34
50	≈ 1.69

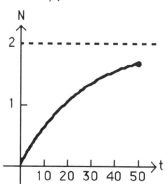

N approaches 2 as t increases without bound.

31. Given $I = I_0 e^{-0.23d}$

(A) $I = I_0 e^{-0.23(10)} = I_0 e^{-2.3} \approx I_0(0.10)$
Thus, about 10% of the surface light will reach a depth of 10 feet.

(B) $I = I_0 e^{-0.23(20)} = I_0 e^{-4.6} \approx I_0(0.010)$
Thus, about 1% of the surface light will reach a depth of 20 feet.

33. Using $A = Pe^{rt}$, and 1989 as the base year, we have:
$A = 24,000e^{0.21t}$

(A) At the end of 1996, $t = 7$ years and
$A = 24,000e^{0.21(7)} = 24,000e^{1.47} \approx 104,382$

(B) At the end of 2000, $t = 11$ years and
$$A = 24,000e^{0.21(11)} = 24,000e^{2.31} \approx 241,786$$

35. Using $A = Pe^{rt}$ with $P = 100$ (million), and $r = 0.023$, we have:
$$A = 100e^{0.023(8)} = 100e^{0.184} \approx 120.20$$
Thus, the population of Mexico will be approximately 120 million.

EXERCISE 3-3

Things to remember:

1. Let $b > 0$, $b \neq 1$. The LOGARITHMIC FUNCTION with base b is defined by
   $$y = \log_b x \text{ if and only if } x = b^y.$$
   In words, the logarithm of a number x to the base b is the exponent to which b must be raised to equal x.

   The domain of a logarithmic function is the set of positive real numbers, and the range is the set of all real numbers.

2. Properties of logarithmic functions: let $b > 0$, $b \neq 1$, $M > 0$, and $N > 0$.
   a. $\log_b b^x = x$
   b. $\log_b (M \cdot N) = \log_b M + \log_b N$
   c. $\log_b \dfrac{M}{N} = \log_b M - \log_b N$
   d. $\log_b M^p = p \log_b M$
   e. $\log_b M = \log_b N$ if and only if $M = N$
   f. $\log_b 1 = 0$

3. $\log_{10} x$ is denoted by $\log x$, and is called the COMMON LOGARITHMIC FUNCTION. $\log x = y$ is equivalent to $x = 10^y$.
   $\log_e x$ is denoted by $\ln x$, and is called the NATURAL LOGARITHMIC FUNCTION. $\ln x = y$ is equivalent to $x = e^y$.

4. CHANGE OF BASE FORMULA
   $$\log_b N = \frac{\log_a N}{\log_a b}$$

---

1. $27 = 3^3$ (using 1)   **3.** $1 = 10^0$ **5.** $8 = 4^{3/2}$

7. $\log_7 49 = 2$ **9.** $\log_4 8 = \dfrac{3}{2}$ **11.** $\log_b A = u$

13. $\log_{10}10^3 = 3$ **15.** $\log_2 2^{-3} = -3$ **17.** $\log_{10}1{,}000 = \log_{10}10^3 =$
(using $\underline{2}$a)

19. $\log_b \dfrac{P}{Q} = \log_b P - \log_b Q$ (using $\underline{2}$c)   **21.** $\log_b L^5 = 5\log_b L$ (using $\underline{2}$d)

23. $\log_b \dfrac{p}{qrs} = \log_b p - \log_b qrs$ (using $\underline{2}$c)
$\phantom{\log_b \dfrac{p}{qrs}} = \log_b p - (\log_b q + \log_b r + \log_b s)$ (using $\underline{2}$b)
$\phantom{\log_b \dfrac{p}{qrs}} = \log_b p - \log_b q - \log_b r - \log_b s$

25. $\log_3 x = 2$ **27.** $\log_7 49 = y$ **29.** $\log_b 10^{-4} = -4$
$x = 3^2$ (using $\underline{1}$) $\log_7 7^2 = y$ $10^{-4} = b^{-4}$
$x = 9$ $2 = y$ This equality implies
                        Thus, $y = 2$. $b = 10$ (since the
                                            exponents are the same).

31. $\log_4 x = \dfrac{1}{2}$ **33.** $\log_{1/3} 9 = y$ **35.** $\log_b 1{,}000 = \dfrac{3}{2}$
$x = 4^{1/2}$ $9 = \left(\dfrac{1}{3}\right)^y$ $\log_b 10^3 = \dfrac{3}{2}$
$x = 2$ $3^2 = (3^{-1})^y$ $3\log_b 10 = \dfrac{3}{2}$
                      $3^2 = 3^{-y}$ $\log_b 10 = \dfrac{1}{2}$
                 This inequality $10 = b^{1/2}$
                 implies that   Square both sides:
                 $2 = -y$ or $y = -2$. $100 = b$, i.e., $b = 100$.

37. $\log_b \dfrac{x^5}{y^3}$ **39.** $\log_b \sqrt[3]{N} = \log_b N^{1/3}$
$= \log_b x^5 - \log_b y^3$ $= \dfrac{1}{3}\log_b N$
$= 5\log_b x - 3\log_b y$

41. $\log_b (x^2 \sqrt[3]{y}) = \log_b x^2 + \log_b y^{1/3} = 2\log_b x + \dfrac{1}{3}\log_b y$

43. $\log_b (50 \cdot 2^{-0.2t}) = \log_b 50 + \log_b 2^{-0.2t} = \log_b 50 - 0.2t\log_b 2$

45. $\log_b P(1 + r)^t = \log_b P + \log_b (1 + r)^t = \log_b P + t\log_b (1 + r)$

47. $\log_e 100e^{-0.01t} = \log_e 100 + \log_e e^{-0.01t}$
$\phantom{\log_e 100e^{-0.01t}} = \log_e 100 - 0.01t\log_e e = \log_e 100 - 0.01t$

49. $\log_b x = \dfrac{2}{3}\log_b 8 + \dfrac{1}{2}\log_b 9 - \log_b 6 = \log_b 8^{2/3} + \log_b 9^{1/2} - \log_b 6$
$ = \log_b 4 + \log_b 3 - \log_b 6 = \log_b \dfrac{4 \cdot 3}{6}$
$\log_b x = \log_b 2$
$x = 2$ (using $\underline{2}$e)

51. $\log_b x = \frac{3}{2}\log_b 4 - \frac{2}{3}\log_b 8 + 2\log_b 2 = \log_b 4^{3/2} - \log_b 8^{2/3} + \log_b 2^2$

$\qquad = \log_b 8 - \log_b 4 + \log_b 4 = \log_b 8$

$\log_b x = \log_b 8$

$\qquad x = 8 \ \text{(using \underline{2}e)}$

53. $\log_b x + \log_b(x - 4) = \log_b 21$

$\qquad\qquad \log_b x(x - 4) = \log_b 21$

Therefore, $x(x - 4) = 21$

$\qquad\quad x^2 - 4x - 21 = 0$

$\qquad (x - 7)(x + 3) = 0$

Thus, $x = 7$.

[Note: $x = -3$ is not a solution since $\log_b(-3)$ is not defined.]

55. $\log_{10}(x - 1) - \log_{10}(x + 1) = 1$

$\qquad\qquad \log_{10}\!\left(\frac{x - 1}{x + 1}\right) = 1$

Therefore, $\frac{x - 1}{x + 1} = 10^1 = 10$

$\qquad\qquad x - 1 = 10(x + 1)$

$\qquad\qquad x - 1 = 10x + 10$

$\qquad\qquad\quad -9x = 11$

$\qquad\qquad\qquad x = -\frac{11}{9}$

There is *no solution*, since

$\log_{10}\!\left(-\frac{11}{9} - 1\right) = \log_{10}\!\left(-\frac{20}{9}\right)$

is not defined. Similarly,

$\log_{10}\!\left(-\frac{11}{9} + 1\right) = \log_{10}\!\left(-\frac{2}{9}\right)$

is not defined.

57. $y = \log_2(x - 2)$

$x - 2 = 2^y$

$x = 2^y + 2$

x	y
$\frac{9}{4}$	-2
$\frac{5}{2}$	-1
3	0
4	1
6	2
18	4

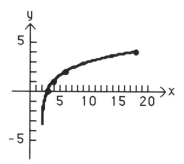

59. (A) 3.54743
(B) −2.16032
(C) 5.62629
(D) −3.19704

61. (A) $\log x = 1.1285$
$\qquad\quad x = 13.4431$
(B) $\log x = -2.0497$
$\qquad\quad x = 0.0089$
(C) $\ln x = 2.7763$
$\qquad\quad x = 16.0595$
(D) $\ln x = -1.8879$
$\qquad\quad x = 0.1514$

63. $n = \dfrac{\log 2}{\log 1.15} \approx \dfrac{0.30103}{0.060698} \approx 4.959$

65. $n = \dfrac{\ln 3}{\ln 1.15} \approx \dfrac{1.09861}{0.13976}$
$\qquad\qquad\qquad \approx 7.861$

67. $x = \dfrac{\ln 0.5}{-0.21} \approx \dfrac{-0.69315}{-0.21}$
$\qquad\qquad\qquad \approx 3.301$

69. $10^x = 12$ (Take common logarithms of both sides)

$\log 10^x = \log 12 \approx 1.0792$

$x \approx 1.0792$ ($\log 10^x = x \log 10 = x$; $\log 10 = 1$)

71. $e^x = 4.304$ (Take natural logarithms of both sides)

$\ln e^x = \ln 4.304 \approx 1.4595$

$x \approx 1.4595$ ($\ln e^x = x \ln e = x$; $\ln e = 1$)

73. $1.03^x = 2.475$ (Take either common or natural logarithms of both sides; we use common logarithms)

$\log(1.03^x) = \log 2.475$

$x \log 1.03 = \log 2.475$

$x = \dfrac{\log 2.475}{\log 1.03} \approx 30.6589$

75. $\log_8 25 = \dfrac{\log 25}{\log 8} \approx 1.548$

or $\log_8 25 = \dfrac{\ln 25}{\ln 8} \approx 1.548$

77. $\log_{32} 4.017 = \dfrac{\log 4.017}{\log 32} \approx 0.401$

79. $\log_{0.5} 0.377 = \dfrac{\ln 0.377}{\ln 0.5} \approx 1.407$

81. $y = \ln x$, $x > 0$

x	y
0.5	≈ -0.69
1	0
2	≈ 0.69
4	≈ 1.39
5	≈ 1.61

83. $y = |\ln x|$, $x > 0$

x	y
0.5	≈ 0.69
1	0
2	≈ 0.69
4	≈ 1.39
5	≈ 1.6

85. $y = 2 \ln(x + 2)$, $x > -2$

x	y
-1.5	≈ -1.39
-1	0
0	≈ 1.39
1	≈ 2.2
5	≈ 3.89
10	≈ 4.61

87. $y = 4 \ln x - 3$, $x > 0$

x	y
0.5	≈ -5.77
1	-3
5	≈ 3.44
10	≈ 6.21

89. Let $\log_b 1 = y$, $b > 0$, $b \neq 1$

Then $b^y = 1$, which implies $y = 0$ (see **2f**). Thus, $\log_b 1 = 0$.

91. $\log_{10} y - \log_{10} c = 0.8x$

$\log_{10} \dfrac{y}{c} = 0.8x$

Therefore, $\dfrac{y}{c} = 10^{0.8x}$ (using **1**)

and $y = c \cdot 10^{0.8x}$.

93. From the compound interest formula $A = P(1 + r)^t$, we have:

$2P = P(1 + .06)^t$ or $(1.06)^t = 2$

Take the natural log of both sides of this equation:

$\ln(1.06)^t = \ln 2$ [Note: The common log could have been used instead of the natural log.]

$t\ln(1.06) = \ln 2$

$$t = \frac{\ln 2}{\ln(1.06)} \approx \frac{.69315}{.05827} = 11.90 \approx 12 \text{ years}$$

95. From the compound interest formula $A = P(1 + r)^t$, we have:

$3P = P(1 + r)^t$ or $(1 + r)^t = 3$

Taking the natural log of both sides of this equation gives:

$\ln(1 + r)^t = \ln 3$ or $t\ln(1 + r) = \ln 3$.

Thus, $t = \dfrac{\ln 3}{\ln(1 + r)}$.

97. (A) $A = P\left(1 + \dfrac{r}{m}\right)^{mt}$, $r = 0.06$, $m = 4$, $P = 1000$, $A = 1800$.

$$1800 = 1000\left(1 + \frac{0.06}{4}\right)^{4t} = 1000(1.015)^{4t}$$

$$(1.015)^{4t} = \frac{1800}{1000} = 1.8$$

$$4t\ln(1.015) = \ln(1.8)$$

$$t = \frac{\ln(1.8)}{4\ln(1.015)} \approx 9.87$$

$1000 at 6% compounded quarterly will grow to $1800 in 9.87 years.

(B) $A = Pe^{rt}$, $r = 0.06$, $P = 1000$, $A = 1800$

$$1000e^{0.06t} = 1800$$

$$e^{0.06t} = 1.8$$

$$0.06t = \ln 1.8$$

$$t = \frac{\ln 1.8}{0.06} \approx 9.80$$

$1000 at 6% compounded continuously will grow to $1800 in 9.80 years.

99. $I = I_0 10^{N/10}$

Take the common log of both sides of this equation. Then:

$\log I = \log(I_0 10^{N/10}) = \log I_0 + \log 10^{N/10}$

$\qquad = \log I_0 + \dfrac{N}{10}\log 10 = \log I_0 + \dfrac{N}{10}$ (since $\log 10 = 1$)

So, $\dfrac{N}{10} = \log I - \log I_0 = \log\left(\dfrac{I}{I_0}\right)$ and $N = 10\log\left(\dfrac{I}{I_0}\right)$.

101. From the compound interest formula $A = P(1 + r)^t$, we have:

$$1.68 \times 10^{14} = 4 \times 10^9(1 + .02)^t \text{ or } (1.02)^t = \frac{1.68 \times 10^{14}}{4 \times 10^9}$$

$$= .42 \times 10^5$$

$$= 42,000$$

Taking the natural log of both sides of this equation gives:

$t \ln(1.02) = \ln(42,000)$ and $t = \dfrac{\ln(42,000)}{\ln(1.02)} \approx \dfrac{10.6454}{.01980} \approx 537.65$

Thus, there will be one square yard of land per person in approximately 538 years.

EXERCISE 3-4 CHAPTER REVIEW

1. $u = e^v$
$v = \ln u$

2. $x = 10^y$
$y = \log x$

3. $\ln M = N$
$M = e^N$

4. $\log u = v$
$u = 10^v$

5. $\dfrac{5^{x+4}}{5^{4-x}} = 5^{x+4-(4-x)} = 5^{2x}$

6. $\left(\dfrac{e^u}{e^{-u}}\right)^u = (e^{u+u})^u = (e^{2u})^u = e^{2u^2}$

7. $\log_3 x = 2$
$x = 3^2$
$x = 9$

8. $\log_x 36 = 2$
$x^2 = 36$
$x = 6$

9. $\log_2 16 = x$
$2^x = 16$
$x = 4$

10. $10^x = 143.7$
$x = \log 143.7$
$x \approx 2.157$

11. $e^x = 503,000$
$x = \ln 503,000 \approx 13.128$

12. $\log x = 3.105$
$x = 10^{3.105} \approx 1273.503$

13. $\ln x = -1.147$
$x = e^{-1.147} \approx 0.318$

14. $\log(x + 5) = \log(2x - 3)$
$x + 5 = 2x - 3$
$-x = -8$
$x = 8$

15. $\begin{aligned} 2\ln(x - 1) &= \ln(x^2 - 5) \\ \ln(x - 1)^2 &= \ln(x^2 - 5) \\ (x - 1)^2 &= x^2 - 5 \\ x^2 - 2x + 1 &= x^2 - 5 \\ -2x &= -6 \\ x &= 3 \end{aligned}$

16. $\begin{aligned} 9^{x-1} &= 3^{1+x} \\ (3^2)^{x-1} &= 3^{1+x} \\ 3^{2x-2} &= 3^{1+x} \\ 2x - 2 &= 1 + x \\ x &= 3 \end{aligned}$

17. $\begin{aligned} e^{2x} &= e^{x^2-3} \\ 2x &= x^2 - 3 \\ x^2 - 2x - 3 &= 0 \\ (x - 3)(x + 1) &= 0 \\ x &= 3, -1 \end{aligned}$

18. $\begin{aligned} 2x^2 e^x &= 3xe^x \\ 2x^2 &= 3x \text{ (divide both sides} \\ 2x^2 - 3x &= 0 \quad \text{by } e^x) \\ x(2x - 3) &= 0 \\ x &= 0, \tfrac{3}{2} \end{aligned}$

19. $\log_{1/3} 9 = x$
$\left(\dfrac{1}{3}\right)^x = 9$
$\dfrac{1}{3^x} = 9$
$3^x = \dfrac{1}{9}$
$x = -2$

20. $\log_x 8 = -3$
$x^{-3} = 8$
$\dfrac{1}{x^3} = 8$
$x^3 = \dfrac{1}{8}$
$x = \dfrac{1}{2}$

21. $\log_9 x = \dfrac{3}{2}$
$9^{3/2} = x$
$x = 27$

22. $x = 3(e^{1.49}) \approx 13.3113$

23. $x = 230(10^{-0.161}) \approx 158.7552$

24. $\log x = -2.0144$
$x \approx 0.0097$

25. $\ln x = 0.3618$
$x \approx 1.4359$

26.
$35 = 7(3^x)$
$3^x = 5$
$\ln 3^x = \ln 5$
$x \ln 3 = \ln 5$
$x = \dfrac{\ln 5}{\ln 3} \approx 1.4650$

27.
$0.01 = e^{-0.05x}$
$\ln(0.01) = \ln(e^{-0.05x}) = -0.05x$
Thus, $x = \dfrac{\ln(0.01)}{-0.05} \approx 92.1034$

28.
$8{,}000 = 4{,}000(1.08)^x$
$(1.08)^x = 2$
$\ln(1.08)^x = \ln 2$
$x \ln 1.08 = \ln 2$
$x = \dfrac{\ln 2}{\ln 1.08} \approx 9.0065$

29.
$5^{2x-3} = 7.08$
$\ln(5^{2x-3}) = \ln 7.08$
$(2x - 3)\ln 5 = \ln 7.08$
$2x \ln 5 - 3 \ln 5 = \ln 7.08$
$x = \dfrac{\ln 7.08 + 3 \ln 5}{2 \ln 5}$
$= \dfrac{\ln 7.08 + \ln 5^3}{2 \ln 5}$
$= \dfrac{\ln[7.08(125)]}{2 \ln 5} \approx 2.1081$

30. $x = \log_2 7 = \dfrac{\log 7}{\log 2} \approx 2.8074$
or $x = \log_2 7 = \dfrac{\ln 7}{\ln 2} \approx 2.8074$

31. $x = \log_{0.2} 5.321 = \dfrac{\log 5.321}{\log 0.2} \approx -1.0387$
or $x = \log_{0.2} 5.321 = \dfrac{\ln 5.321}{\ln 0.2} \approx -1.0387$

32. $e^x(e^{-x} + 1) - (e^x + 1)(e^{-x} - 1) = 1 + e^x - (1 - e^x + e^{-x} - 1)$
$= 1 + e^x + e^x - e^{-x}$
$= 1 + 2e^x - e^{-x}$

33. $(e^x - e^{-x})^2 - (e^x + e^{-x})(e^x - e^{-x})$
$= (e^x)^2 - 2(e^x)(e^{-x}) + (e^{-x})^2 - [(e^x)^2 - (e^{-x})^2]$
$= e^{2x} - 2 + e^{-2x} - [e^{2x} - e^{-2x}]$
$= 2e^{-2x} - 2$

34. $y = 2^{x-1}$, $-2 \le x \le 4$

x	y
-2	$\frac{1}{8}$
-1	$\frac{1}{4}$
0	$\frac{1}{2}$
1	1
2	2
4	8

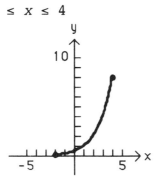

35. $f(t) = 10e^{-0.08t}$, $t \ge 0$

t	$f(t)$
0	10
10	≈ 4.5
20	≈ 2
30	≈ 0.9
40	≈ 0.4

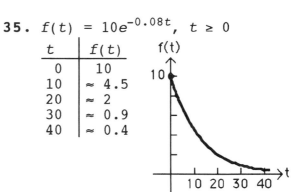

36. $y = \ln(x + 1)$, $-1 < x \leq 10$

x	y
-0.5	$\approx$ -0.7
0	0
4	$\approx$ 1.6
8	$\approx$ 2.2
10	$\approx$ 2.4

37.
$$\log x - \log 3 = \log 4 - \log(x + 4)$$
$$\log \frac{x}{3} = \log \frac{4}{x + 4}$$
$$\frac{x}{3} = \frac{4}{x + 4}$$
$$x(x + 4) = 12$$
$$x^2 + 4x - 12 = 0$$
$$(x + 6)(x - 2) = 0$$
$$x = -6, \ 2$$

Since $\log(-6)$ and $\log(-2)$ are not defined, -6 is not a solution. Therefore, the solution is $x = 2$.

38. $\ln(2x - 2) - \ln(x - 1) = \ln x$
$$\ln\left(\frac{2x - 2}{x - 1}\right) = \ln x$$
$$\ln\left[\frac{2(x - 1)}{x - 1}\right] = \ln x$$
$$\ln 2 = \ln x$$
$$x = 2$$

39. $\ln(x + 3) - \ln x = 2 \ln 2$
$$\ln\left(\frac{x + 3}{x}\right) = \ln(2^2)$$
$$\frac{x + 3}{x} = 4$$
$$x + 3 = 4x$$
$$3x = 3$$
$$x = 1$$

40.
$$\log 3x^2 = 2 + \log 9x$$
$$\log 3x^2 - \log 9x = 2$$
$$\log\left(\frac{3x^2}{9x}\right) = 2$$
$$\log\left(\frac{x}{3}\right) = 2$$
$$\frac{x}{3} = 10^2 = 100$$
$$x = 300$$

41.
$$\ln y = -5t + \ln c$$
$$\ln y - \ln c = -5t$$
$$\ln \frac{y}{c} = -5t$$
$$\frac{y}{c} = e^{-5t}$$
$$y = ce^{-5t}$$

42. Let x be *any* positive real number and suppose $\log_1 x = y$. Then $1^y = x$. But, $1^y = 1$, so $x = 1$, i.e., $x = 1$ for all positive real numbers x. This is clearly impossible.

43. $A = P\left(1 + \frac{r}{m}\right)^{mt}$.

We let $P = 5,000$, $r = 0.12$, $m = 52$, and $t = 6$. Then we have:

$$A = 5,000\left(1 + \frac{0.12}{52}\right)^{52(6)} \approx 5,000(1 + 0.0023)^{312} \approx 10,263.65$$

Thus, there will be \$10,263.65 in the account 6 years from now.

44. $A = Pe^{rt}$. We let $P = 5,000$, $r = 0.12$, and $t = 6$. Then:
$A = 5,000e^{(0.12)6} \approx 10,272.17$
Thus, there will be \$10,272.17 in the account 6 years from now.

45. The compound interest formula for money invested at 15% compounded annually is:
$A = P(1 + 0.15)^t$
To find the tripling time, we set $A = 3P$ and solve for t:
$$3P = P(1.15)^t$$
$$(1.15)^t = 3$$
$$\ln(1.15)^t = \ln 3$$
$$t \ln 1.15 = \ln 3$$
$$t = \frac{\ln 3}{\ln 1.15} \approx 7.86$$
Thus, the tripling time (to the nearest year) is 8 years.

46. The compound interest formula for money invested at 10% compounded continuously is:
$A = Pe^{0.1t}$
To find the doubling time, we set $A = 2P$ and solve for t:
$$2P = Pe^{0.1t}$$
$$e^{0.1t} = 2$$
$$0.1t = \ln 2$$
$$t = \frac{\ln 2}{0.1} \approx 6.93 \text{ years}$$

47. (A) $N(0) = 1$

$N\left(\frac{1}{2}\right) = 2$

$N(1) = 4 = 2^2$

$N\left(\frac{3}{2}\right) = 8 = 2^3$

$N(2) = 16 = 2^4$

$\vdots$

Thus, we conclude that
$N(t) = 2^{2t}$ or $N = 4^t$.

(B) We need to solve:
$$2^{2t} = 10^9$$
$$\log 2^{2t} = \log 10^9 = 9$$
$$2t \log 2 = 9$$
$$t = \frac{9}{2 \log 2} \approx 14.95$$

Thus, the mouse will die in 15 days.

48. Given $I = I_0 e^{-kd}$. When $d = 73.6$, $I = \frac{1}{2}I_0$. Thus, we have:
$$\frac{1}{2}I_0 = I_0 e^{-k(73.6)}$$
$$e^{-k(73.6)} = \frac{1}{2}$$
$$-k(73.6) = \ln\frac{1}{2}$$
$$k = \frac{\ln(0.5)}{-73.6} \approx 0.00942$$
Thus, $k \approx 0.00942$.

To find the depth at which 1% of the surface light remains, we set $I = 0.01I_0$ and solve

$$0.01I_0 = I_0 e^{-0.00942d}$$

for d:

$$0.01 = e^{-0.00942d}$$

$$-0.00942d = \ln 0.01$$

$$d = \frac{\ln 0.01}{-0.00942} \approx 488.87$$

Thus, 1% of the surface light remains at approximately 489 feet.

49. Using the model $P = P_0(1 + r)^t$, we must solve $2P_0 = P_0(1 + 0.03)^t$ for t:

$$2 = (1.03)^t$$

$$\ln(1.03)^t = \ln 2$$

$$t \ln(1.03) = \ln 2$$

$$t = \frac{\ln 2}{\ln 1.03} \approx 23.4$$

Thus, at a 3% growth rate, the population will double in approximately 23.4 years.

50. Using the continuous compounding model, we have:

$$2P_0 = P_0 e^{0.03t}$$

$$2 = e^{0.03t}$$

$$0.03t = \ln 2$$

$$t = \frac{\ln 2}{0.03} \approx 23.1$$

Thus, the model predicts that the population will double in approximately 23.1 years.

4 MATHEMATICS OF FINANCE

Things to remember:

1. SIMPLE INTEREST

$$I = Prt$$

where P = Principal
r = Annual simple interest rate expressed as a decimal
t = Time in years

2. AMOUNT—SIMPLE INTEREST

$$A = P + Prt = P(1 + rt)$$

where P = Principal or *present value*
r = Annual simple interest rate expressed as a decimal
t = Time in years
A = Amount or *future value*

---

1. P = \$500, r = 8% = 0.08, t = 6 months = $\frac{1}{2}$ year

$I = Prt$ (using 1)

$= 500(0.08)\left(\frac{1}{2}\right) = \$20$

3. I = \$80, P = \$500, t = 2 years

$I = Prt$

$r = \dfrac{I}{Pt} = \dfrac{80}{500(2)} = 0.08$ or 8%

5. P = \$100, r = 8% = 0.08, t = 18 months = 1.5 years

$A = P(1 + rt) = 100(1 + 0.08 \cdot 1.5) = \$112$

7. A = \$1000, r = 10% = 0.1, t = 15 months = $\frac{15}{12}$ years

$A = P(1 + rt)$

$P = \dfrac{A}{1 + rt} = \dfrac{1000}{1 + (0.1)\left(\frac{15}{12}\right)} = \$888.89$

9. $I = Prt$

Divide both sides by Pt.

$\dfrac{I}{Pt} = \dfrac{Prt}{Pt}$

$\dfrac{I}{Pt} = r$ or $r = \dfrac{I}{Pt}$

11. $A = P + Prt = P(1 + rt)$

Divide both sides by $(1 + rt)$.

$\dfrac{A}{1 + rt} = \dfrac{P(1 + rt)}{1 + rt}$

$\dfrac{A}{1 + rt} = P$ or $P = \dfrac{A}{1 + rt}$

13. $P = \$3000$, $r = 14\% = 0.14$, $t = 4$ months $= \dfrac{1}{3}$ year

$I = Prt = 3000(0.14)\left(\dfrac{1}{3}\right) = \$140$

15. $P = \$554$, $r = 20\% = 0.2$, $t = 1$ month $= \dfrac{1}{12}$ year

$I = Prt = 554(0.2)\left(\dfrac{1}{12}\right) = \$9.23$

17. $P = \$7250$, $r = 9\% = 0.09$, $t = 8$ months $= \dfrac{2}{3}$ year

$A = 7250\left[1 + 0.09\left(\dfrac{2}{3}\right)\right] = 7250[1.06] = \$7685.00$

19. $P = \$4000$, $A = \$4270$, $t = 8$ months $= \dfrac{2}{3}$ year

The interest on the loan is $I = A - P = \$270$. From Problem 9,
$r = \dfrac{I}{Pt} = \dfrac{270}{4000\left(\dfrac{2}{3}\right)} = 0.10125$. Thus, $r = 10.125\%$.

21. $P = \$1000$, $I = \$30$, $t = 60$ days $= \dfrac{1}{6}$ year

$r = \dfrac{I}{Pt} = \dfrac{30}{1000\left(\dfrac{1}{6}\right)} = 0.18$. Thus, $r = 18\%$.

23. $P = \$1500$. The amount of interest paid is $I = (0.5)(3)(120) = \$180$. Thus, the total amount repaid is $\$1500 + \$180 = \$1680$. To find the annual interest rate, we let $t = 120$ days $= \dfrac{1}{3}$ year. Then

$r = \dfrac{I}{Pt} = \dfrac{180}{1500\left(\dfrac{1}{3}\right)} = 0.36$. Thus, $r = 36\%$.

25. $P = \$9776.94$, $A = \$10,000$, $t = 13$ weeks $= \dfrac{1}{4}$ year

The interest is $I = A - P = \$223.06$.
$r = \dfrac{I}{Pt} = \dfrac{223.06}{9776.94\left(\dfrac{1}{4}\right)} = 0.09126$. Thus, $r = 9.126\%$.

27. $A = \$10,000$, $r = 12.63\% = 0.1263$, $t = 13$ weeks $= \dfrac{1}{4}$ year.

From Problem 11, $P = \dfrac{A}{1 + rt} = \dfrac{10,000}{1 + (0.1263)\dfrac{1}{4}} = \dfrac{10,000}{1.03158} = \$9693.91$.

29. The interest I on a principal $P = \$5500$ at an interest rate $r = 12\% = 0.12$ for $t = 90$ days $= \dfrac{1}{4}$ year is:

$I = Prt = 5500(0.12)\left(\dfrac{1}{4}\right) = \$165$.

The third party will receive $165 in interest on a principal of $5500 for $t = 60$ days in $\frac{1}{6}$ year. Thus,

$$r = \frac{I}{Pt} = \frac{165}{5500\left(\frac{1}{6}\right)} = 0.18 \quad \text{or} \quad r = 18\%.$$

31. The principal P is the cost of the stock plus the broker's commission. The cost of the stock is $500(14.20) = \$7100$ and the commission on this is $62 + (0.003)7100 = \$83.30$. Thus, $P = \$7183.30$. The investor sells the stock for $500(16.84) = \$8420$, and the commission on this amount is $62 + (0.003)8420 = \$87.26$. Thus, the investor has $8420 - 87.26 = \$8332.74$ after selling the stock. We can now conclude that the investor has earned $8332.74 - 7183.30 = \$1149.44$.

Now, $P = \$7183.30$, $I = \$1149.44$, $t = 39$ weeks $= \frac{3}{4}$ year. Therefore,

$$r = \frac{I}{Pt} = \frac{1149.44}{7183.30\left(\frac{3}{4}\right)} = 0.21335 \quad \text{or} \quad r = 21.335\%.$$

33. The principal P is the cost of the stock plus the broker's commission. The cost of the stock is: $2000(23.75) = \$47,500$, and the commission on this is: $84 + 0.002(47,500) = \$179$. Thus $P = \$47,679$. The investor sells this stock for $2000(26.15) = \$52,300$, and the commission on this amount is: $134 + 0.001(52,300) = \$186.30$. Thus, the investor has $52,300 - 186.30 = \$52,113.70$ after selling the stock. We can now conclude that the investor has earned $52,113.70 - 47,679 = \$4,434.70$.

Now, $P = \$47,679$, $I = \$4434.70$, $t = 300$ days $= \frac{300}{360} = \frac{5}{6}$ year.

Therefore, $r = \dfrac{I}{Pt} = \dfrac{4434.70}{(47,679)\left(\frac{5}{6}\right)} \approx 0.11161$ or $r = 11.161\%$

EXERCISE 4-2

Things to remember:

1. AMOUNT—COMPOUND INTEREST

$$A = P(1 + i)^n, \text{ where } i = \frac{r}{m} \text{ and}$$

r = Annual (quoted) rate
m = Number of compounding periods per year
n = Total number of compounding periods
$i = \dfrac{r}{m}$ = Rate per compounding period
P = Principal (present value)
A = Amount (future value) at the end of n periods

2. EFFECTIVE RATE

If principal P is invested at the (nominal) rate r, compounded m times per year, then the effecitve rate, r_e, is given by

$$r_e = \left(1 + \frac{r}{m}\right)^m - 1.$$

---

1. $P = \$100$, $i = 0.01$, $n = 12$
Using 1,
$$\begin{aligned} A = P(1 + i)^n &= 100(1 + 0.01)^{12} \\ &= 100(1.01)^{12} \\ &= \$112.68 \end{aligned}$$

3. $P = \$800$, $i = 0.06$, $n = 25$
Using 1,
$$\begin{aligned} A &= 800(1 + 0.06)^{25} \\ &= 800(1.06)^{25} \\ &= \$3433.50 \end{aligned}$$

5. $A = \$10,000$, $i = 0.03$, $n = 48$
Using 1,
$$A = P(1 + i)^n$$
$$P = \frac{A}{(1 + i)^n} = \frac{10,000}{(1 + 0.03)^{48}}$$
$$= \frac{10,000}{(1.03)^{48}} = \$2419.99$$

7. $A = \$18,000$, $i = 0.01$, $n = 90$
Refer to Problem 5:
$$P = \frac{A}{(1 + i)^n} = \frac{18,000}{(1 + 0.01)^{90}}$$
$$= \frac{18,000}{(1.01)^{90}} = \$7351.04$$

9. $r = 9\%$, $m = 12$. Thus, $i = \dfrac{r}{m} = \dfrac{0.09}{12} = 0.0075$ or 0.75% per month.

11. $r = 7\%$, $m = 4$. Thus, $i = \dfrac{r}{m} = \dfrac{0.07}{4} = 0.0175$ or 1.75% per quarter.

13. $i = 0.8\%$ per month ($m = 12$). Thus, $r = i \cdot m = (0.008)12 = 0.096$ or 9.6% compounded monthly.

15. $i = 4.5\%$ per half year ($m = 2$). Thus, $r = i \cdot m = (0.045)2 = 0.09$ or 9% compounded semiannually.

17. $P = \$100$, $r = 6\% = 0.06$

(A) $m = 1$, $i = 0.06$, $n = 4$
$$\begin{aligned} A &= (1 + i)^n \\ &= 100(1 + 0.06)^4 \\ &= 100(1.06)^4 = \$126.25 \end{aligned}$$
Interest $= 126.25 - 100 = \$26.25$

(B) $m = 4$, $i = \dfrac{0.06}{4} = 0.015$
$$n = 4(4) = 16$$
$$\begin{aligned} A &= 100(1 + 0.015)^{16} \\ &= 100(1.015)^{16} = \$126.90 \end{aligned}$$
Interest $= 126.90 - 100 = \$26.90$

(C) $m = 12$, $i = \dfrac{0.06}{12} = 0.005$, $n = 4(12) = 48$
$$A = 100(1 + 0.005)^{48} = 100(1.005)^{48} = \$127.05$$
Interest $= 127.05 - 100 = \$27.05$

19. $P = \$5000$, $r = 18\%$, $m = 12$

(A) $n = 2(12) = 24$
$$i = \frac{0.18}{12} = 0.015$$
$$\begin{aligned} A &= 5000(1 + 0.015)^{24} \\ &= 5000(1.015)^{24} = \$7147.51 \end{aligned}$$

(B) $n = 4(12) = 48$
$$i = \frac{0.18}{12} = 0.015$$
$$\begin{aligned} A &= 5000(1 + 0.015)^{48} \\ &= 5000(1.015)^{48} = \$10,217.39 \end{aligned}$$

21. $A = \$10,000$, $r = 8\% = 0.08$, $i = \dfrac{0.08}{2} = 0.04$

(A) $n = 2(5) = 10$

$$A = P(1 + i)^n$$
$$10,000 = P(1 + 0.04)^{10}$$
$$= P(1.04)^{10}$$
$$P = \frac{10,000}{(1.04)^{10}} = \$6755.64$$

(B) $n = 2(10) = 20$

$$P = \frac{A}{(1 + i)^n} = \frac{10,000}{(1 + 0.04)^{20}}$$
$$= \frac{10,000}{(1.04)^{20}}$$
$$= \$4563.87$$

23. Use the formula for r_e in 2.

(A) $r = 10\% = 0.1$, $m = 4$

$$r_e = \left(1 + \frac{0.1}{4}\right)^4 - 1 = 0.1038$$
$$\text{or } 10.38\%$$

(B) $r = 12\% = 0.12$, $m = 12$

$$r_e = \left(1 + \frac{0.12}{12}\right)^{12} - 1 = 0.1268$$
$$\text{or } 12.68\%$$

25. We have $P = \$4000$, $A = \$9000$, $r = 15\% = 0.15$, $m = 12$, and $i = \dfrac{0.15}{12} = 0.0125$. Since $A = P(1 + i)^n$, we have:
$$9000 = 4000(1 + 0.0125)^n \quad \text{or} \quad (1.0125)^n = 2.25$$

Method 1: Use Table II. Look down the $(1 + i)^n$ column on the page that has $i = 0.0125$. Find the value of n in this column that is closest to and greater than 2.25. In this case, $n = 66$ months or 5 years and 6 months.

Method 2: Use logarithms and a calculator.
$$\ln(1.0125)^n = \ln 2.25$$
$$n \ln 1.0125 = \ln 2.25$$
$$n = \frac{\ln 2.25}{\ln 1.0125} \approx \frac{0.8109}{0.01242} \approx 65.29$$

Thus, $n = 66$ months or 5 years and 6 months.

27. $A = 2P$, $i = 0.06$

$$A = P(1 + i)^n$$
$$2P = P(1 + 0.06)^n$$
$$(1.06)^n = 2$$
$$\ln(1.06)^n = \ln 2$$
$$n \ln(1.06) = \ln 2$$
$$n = \frac{\ln 2}{\ln 1.06} \approx \frac{0.6931}{0.0583} \approx 11.9 \approx 12$$

29. We have $A = P(1 + i)^n$. To find the doubling time, set $A = 2P$. This yields:
$$2P = P(1 + i)^n \quad \text{or} \quad (1 + i)^n = 2$$
Taking the natural logarithm of both sides, we obtain:
$$\ln(1 + i)^n = \ln 2$$
$$n \ln(1 + i) = \ln 2$$
and
$$n = \frac{\ln 2}{\ln(1 + i)}$$

(A) $r = 10\% = 0.1$, $m = 4$. Thus,

$i = \dfrac{0.1}{4} = 0.025$ and $n = \dfrac{\ln 2}{\ln(1.025)} \approx 28.07$ quarters or $7\frac{1}{4}$ years.

(B) $r = 12\% = 0.12$, $m = 4$. Thus,

$i = \dfrac{0.12}{4} = 0.03$ and $n = \dfrac{\ln 2}{\ln(1.03)} \approx 23.44$ quarters.

That is, 24 quarters or 6 years.

31. $P = \$5000$, $r = 9\% = 0.09$, $m = 4$, $i = \dfrac{0.09}{4} = 0.0225$, $n = 17(4) = 68$

Thus, $A = P(1 + i)^n$

$\qquad = 5000(1 + 0.0225)^{68}$

$\qquad = 5000(1.0225)^{68}$

$\qquad = \$22,702.60$

33. $P = \$110,000$, $r = 6\%$ or 0.06, $m = 1$, $i = 0.06$, $n = 10$

Thus, $A = P(1 + i)^n$

$\qquad = 110,000(1 + 0.06)^{10}$

$\qquad = 110,000(1.06)^{10}$

$\qquad \approx \$196,993.25$

35. $A = \$20$, $r = 7\% = 0.07$, $m = 1$, $i = 0.07$, $n = 5$

$A = P(1 + i)^n$

$P = \dfrac{A}{(1 + i)^n} = \dfrac{20}{(1.07)^5} \approx \$14.26$ per square foot per month

37. From Problem 29, the doubling time is:

$n = \dfrac{\ln 2}{\ln(1 + i)}$

Here $r = i = 0.04$. Thus,

$n = \dfrac{\ln 2}{\ln(1.04)} \approx 17.67$ or 18 years

39. The effective rate, r_e, of $r = 9\% = 0.09$ compounded monthly is:

$r_e = \left(1 + \dfrac{0.09}{12}\right)^{12} - 1 = .0938$ or 9.38%

The effective rate of 9.3% compounded annually is 9.3%. Thus, 9% compounded monthly is better than 9.3% compounded annually.

41. $P = \$7000$, $A = \$9000$, $r = 9\% = 0.09$, $m = 12$, $i = \dfrac{0.09}{12} = 0.0075$

Since $A = P(1 + i)^n$, we have:

$9000 = 7000(1 + 0.0075)^n$ or $(1.0075)^n = \dfrac{9}{7}$

Therefore, $\ln(1.0075)^n = \ln\left(\dfrac{9}{7}\right)$

$\qquad\qquad n \ln(1.0075) = \ln\left(\dfrac{9}{7}\right)$

$\qquad\qquad\qquad n = \dfrac{\ln\left(\dfrac{9}{7}\right)}{\ln(1.0075)} \approx \dfrac{0.2513}{0.0075} \approx 33.6$

Thus, it will take 34 months or 2 years and 10 months.

43. $P = \$20,000$, $r = 8\% = 0.08$, $m = 365$, $i = \dfrac{0.08}{365} \approx 0.0002192$,

$n = (365)35 = 12,775$

Since $A = P(1 + i)^n$, we have:

$$A = 20,000(1.0002192)^{12,775} \approx \$328,791.70$$

45. From Problem 29, the doubling time is:

$$n = \dfrac{\ln 2}{\ln(1 + i)}$$

(A) $r = 14\% = 0.14$, $m = 365$, $i = \dfrac{0.14}{365} \approx 0.0003836$

Thus, $n = \dfrac{\ln 2}{\ln(1.0003836)} \approx 1807.48$ days or 4.952 years.

(B) $r = 15\% = 0.15$, $m = 1$, $i = 0.15$

Thus, $n = \dfrac{\ln 2}{\ln(1.15)} \approx 4.959$ years.

47. The relationship between the effective rate and the annual nominal rate is

$$r_e = \left(1 + \dfrac{r}{m}\right)^m - 1$$

In this case, $r_e = 0.074$ and $m = 365$. Thus, we must solve

$$0.074 = \left(1 + \dfrac{r}{365}\right)^{365} - 1 \text{ for } r$$

$$\left(1 + \dfrac{r}{365}\right)^{365} = 1.074$$

$$1 + \dfrac{r}{365} = (1.074)^{1/365}$$

$$r = 365[(1.074)^{1/365} - 1] \approx 0.0714 \text{ or } r = 7.14\%$$

49. $A = \$30,000$, $r = 10\% = 0.1$, $m = 1$, $i = 0.1$, $n = 17$

From $A = P(1 + i)^n$, we have:

$$P = \dfrac{A}{(1 + i)^n} = \dfrac{30,000}{(1.1)^{17}} \approx \$5935.34$$

51. $A = \$30,000$, $P = \$6844.79$, $r = i$, $n = 17$

Using $A = P(1 + r)^n$, we have:

$$30,000 = 6844.79(1 + r)^{17}$$

$$(1 + r)^{17} = \dfrac{30,000}{6844.79} \approx 4.3829$$

Therefore,

$1 + r \approx (4.3829)^{1/17}$ and $r \approx (4.3829)^{1/17} - 1 \approx 0.0908$ or $r = 9.08\%$

53. From 2, $r_e = \left(1 + \dfrac{r}{m}\right)^m - 1$.

(A) $r = 8.28\% = 0.0828$, $m = 12$

$$r_e = \left(1 + \dfrac{0.0828}{12}\right)^{12} - 1 \approx 0.0860 \quad \text{or} \quad 8.60\%$$

(B) $r = 8.25\% = 0.0825$, $m = 365$

$$r_e = \left(1 + \frac{0.0825}{365}\right)^{365} - 1 \approx 0.0860 \quad \text{or} \quad 8.60\%$$

(C) $r = 8.25\% = 0.0825$, $m = 12$

$$r_e = \left(1 + \frac{0.0825}{12}\right)^{12} - 1 \approx 0.0857 \quad \text{or} \quad 8.57\%$$

55. $A = \$32,456.32$, $P = \$24,766.81$, $m = 1$, $n = 2$

$$A = P(1 + r)^n$$

$$32,456.32 = 24,766.81(1 + r)^2$$

$$(1 + r)^2 = \frac{32,456.32}{24,766.81} = 1.3105$$

Therefore, $1 + r = \sqrt{1.3105} \approx 1.1448$ and $r \approx 0.1448$ or 14.48%

57. The effective rate for 8% compounded quarterly is

$$r_e = \left(1 + \frac{0.08}{4}\right)^4 - 1 = (1.02)^4 - 1 = 0.0824 \text{ or } 8.24\%$$

To find the annual nominal rate compounded monthly which has the effective rate of 8.24%, we solve

$$0.0824 = \left(1 + \frac{r}{12}\right)^{12} - 1 \text{ for } r.$$

$$\left(1 + \frac{r}{12}\right)^{12} = 1.0824$$

$$1 + \frac{r}{12} = (1.0824)^{1/12}$$

$$\frac{r}{12} = (1.0824)^{1/12} - 1$$

$$r = 12[(1.0824)^{1/12} - 1] \approx 0.0795 \text{ or } 7.95\%$$

EXERCISE 4-3

Things to remember:

<u>1</u>. FUTURE VALUE OF AN ORDINARY ANNUITY

$$FV = PMT\,\frac{(1 + i)^n - 1}{i} = PMT\,s_{\overline{n}|i}$$

where PMT = Periodic payment
   i = Rate per period
   n = Number of payments (periods)
   FV = Future value (amount)

(Payments are made at the end of each period.)

2. SINKING FUND PAYMENT

$$PMT = FV \frac{i}{(1 + i)^n - 1} = \frac{FV}{s_{\overline{n}|i}}$$

where PMT = Sinking fund payment
FV = Value of annuity after n payments (future value)
n = Number of payments (periods)
i = Rate per period

(Payments are made at the end of each period.)

---

1. $n = 20$, $i = 0.03$, $PMT = \$500$

$$FV = PMT \frac{(1 + i)^n - 1}{i}$$

$$= PMTs_{\overline{n}|i} \quad \text{(using } \underline{1}\text{)}$$

$$= 500 \frac{(1 + 0.03)^{20} - 1}{0.03} = 500s_{\overline{20}|0.03}$$

$$= 500(26.87037449) = \$13,435.19$$

3. $n = 40$, $i = 0.02$, $PMT = \$1000$

$$FV = 1000 \frac{(1 + 0.02)^{40} - 1}{0.02}$$

$$= 1000s_{\overline{40}|0.02}$$

$$= 1000(60.40198318)$$

$$= \$60,401.98$$

5. $FV = \$3000$, $n = 20$, $i = 0.02$

$$3000 = PMT \frac{(1 + 0.02)^{20} - 1}{0.02}$$

$$= PMTs_{\overline{20}|0.02} \quad \text{(using } \underline{1} \text{ or } \underline{2}\text{)}$$

$$= PMT(24.29736980)$$

$$PMT = \frac{3000}{24.29736980} = \$123.47$$

7. $FV = \$5000$, $n = 15$, $i = 0.01$

$$PMT = FV \frac{i}{(1 + i)^n - 1}$$

$$= \frac{FV}{s_{\overline{n}|i}} \quad \text{(using } \underline{2}\text{)}$$

$$= 5000 \frac{0.01}{(1 + 0.01)^{15} - 1}$$

$$= \frac{5000}{16.09689554} = \$310.62$$

9. $FV = \$4000$, $i = 0.02$, $PMT = 200$, $n = ?$

$$FV = PMT \frac{(1 + i)^n - 1}{i}$$

$$\frac{FVi}{PMT} = (1 + i)^n - 1$$

$$(1 + i)^n = \frac{FVi}{PMT} + 1$$

$$\ln(1 + i)^n = \ln\left[\frac{FVi}{PMT} + 1\right]$$

$$n \ln(1 + i) = \ln\left[\frac{FVi}{PMT} + 1\right]$$

$$n = \frac{\ln\left[\frac{FVi}{PMT} + 1\right]}{\ln(1 + i)} = \frac{\ln\left[\frac{4000(0.02)}{200} + 1\right]}{\ln(1.02)}$$

$$= \frac{\ln(1.4)}{\ln(1.02)} \approx \frac{0.3365}{0.01980} = 16.99 \quad \text{or} \quad 17 \text{ periods}$$

11. $PMT = \$500$, $n = 10(4) = 40$, $i = \dfrac{0.08}{4} = 0.02$

$FV = 500 \, \dfrac{(1 + 0.02)^{40} - 1}{0.02} = 500 s_{\overline{40}|0.02}$

$\qquad = 500(60.40198318) = \$30,200.99$

Total deposits $= 500(40) = \$20,000$.

Interest $= FV - 20,000 = 30,200.99 - 20,000 = \$10,200.99$.

13. $PMT = \$300$, $i = \dfrac{0.06}{12} = 0.005$, $n = 5(12) = 60$

$FV = 300 \, \dfrac{(1 + 0.005)^{60} - 1}{0.005} = 300 s_{\overline{60}|0.005}$ (using $\underline{1}$)

$\qquad = 300(69.77003051) = \$20,931.01$

After five years, $\$20,931.01$ will be in the account.

15. $FV = \$25,000$, $i = \dfrac{0.09}{12} = 0.0075$, $n = 12(5) = 60$

$PMT = \dfrac{FV}{s_{\overline{60}|0.0075}} = \dfrac{25,000}{75.42413693}$ (using the table)

$\qquad = \$331.46$ per month

17. $FV = \$100,000$, $i = \dfrac{0.12}{12} = 0.01$, $n = 8(12) = 96$

$PMT = \dfrac{FV}{s_{\overline{96}|0.01}} = \dfrac{100,000}{159.92729236} = \$625.28$ per month

19. $FV = PMT \, \dfrac{(1 + i)^n - 1}{i} = 100 \, \dfrac{(1 + 0.0075)^{12} - 1}{0.0075}$ (after one year)

$\qquad = 100 \, \dfrac{(1.0075)^{12} - 1}{0.0075}$ $\left[\underline{\text{Note}}: PMT = \$100, \; i = \dfrac{0.09}{12} = 0.0075, \; n = 12\right]$

$\qquad = \underline{\$1250.76}$ $\hfill (1)$

Total deposits in one year $= 12(100) = \$1200$.

Interest earned in first year $= FV - 1200 = 1250.76 - 1200 = \$50.76$.

At the end of the second year:

$FV = 100 \, \dfrac{(1 + 0.0075)^{24} - 1}{0.0075}$ $\quad [\underline{\text{Note}}: n = 24]$

$\qquad = 100 \, \dfrac{(1.0075)^{24} - 1}{0.0075} = \underline{\$2618.85}$ $\hfill (2)$

Total deposits plus interest in the second year $= (2) - (1)$
$\qquad\qquad\qquad\qquad\qquad\qquad\qquad\qquad = 2618.85 - 1250.76$
$\qquad\qquad\qquad\qquad\qquad\qquad\qquad\qquad = \underline{\$1368.09}$ $\hfill (3)$

Interest earned in the second year $= (3) - 1200$
$\qquad\qquad\qquad\qquad\qquad\qquad\qquad = 1368.09 - 1200$
$\qquad\qquad\qquad\qquad\qquad\qquad\qquad = \$168.09$

At the end of the third year,

$FV = 100 \, \dfrac{(1 + 0.0075)^{36} - 1}{0.0075}$ $\quad [\underline{\text{Note}}: n = 36]$

$\qquad = 100 \, \dfrac{(1.0075)^{36} - 1}{0.0075}$

$\qquad = \underline{\$4115.27}$ $\hfill (4)$

Total deposits plus interest in the third year = (4) − (2)
$$= 4115.27 - 2618.85$$
$$= \underline{\$1496.42} \qquad (5)$$

Interest earned in the third year = (5) − 1200
$$= 1496.42 - 1200$$
$$= \$296.42$$

Thus,

Year	Interest earned
1	$ 50.76
2	$168.09
3	$296.42

21. (A) $PMT = \$2000$, $n = 8$, $i = 9\% = 0.09$

$$FV = 2000 \; \frac{(1 + 0.09)^8 - 1}{0.09} = \frac{2000(0.99256)}{0.09} \approx 22{,}056.95$$

Thus, Jane will have $22,056.95 in her account on her 31st birthday. On her 65th birthday, she will have:
$$A = 22{,}056.95(1.09)^{34} \approx \$413{,}092$$

(B) $PMT = \$2000$, $n = 34$, $i = 9\% = 0.09$

$$FV = 2000 \; \frac{(1 + 0.09)^{34} - 1}{0.09} \approx \frac{2000(17.7284)}{0.09} \approx \$393{,}965$$

23. $FV = \$10{,}000$, $n = 48$, $i = \frac{8\%}{12} = \frac{0.08}{12} \approx 0.006667$

From 2, $PMT = \dfrac{10{,}000(0.006667)}{(1 + 0.006667)^{48} - 1} = \dfrac{66.67}{0.3757} \approx \$177.46$

The total of the monthly deposits for 4 years is 48 × 177.46 = 8518.08. Thus, the interest earned is 10,000 − 8518.08 = $1481.92.

25. $PMT = \$150$, $FV = \$7000$, $i = \frac{8.5\%}{12} = \frac{0.085}{12} \approx 0.00708$. From Problem 9:

$$n = \frac{\ln\left[\frac{FVi}{PMT} + 1\right]}{\ln(1 + i)} = \frac{\ln\left[\frac{7000(0.00708)}{150} + 1\right]}{\ln(1.00708)} \approx \frac{0.28548}{0.007055} \approx 40.46$$

Thus, $n = 41$ months or 3 years and 5 months.

27. This problem was done with a graphics calculator.

Start with the equation $\dfrac{(1 + i)^n - 1}{i} - \dfrac{FV}{PMT} = 0$

where $FV = 6300$, $PMT = 1000$, $n = 5$ and $i = \frac{r}{1} = r$

where r is the nominal annual rate with these values, the equation is:

$$\frac{(1 + r)^5 - 1}{r} - \frac{6300}{1000} = 0$$

or $(1 + r)^5 - 1 - 6.3r = 0$

Set $y = (1 + r)^5 - 1 - 6.3r$ and use your calculator to find the zero r of the function y, where $0 < r < 1$. The result is $r = 0.1158$ or 11.58% to two decimal places.

29. Start with the equation $\dfrac{(1 + i)^n - 1}{i} - \dfrac{FV}{PMT} = 0$

where $FV = 620$, $PMT = 50$, $n = 12$ and $i = \dfrac{r}{12}$,

where r is the annual nominal rate. With these values, the equation becomes

$$\frac{(1 + i)^{12} - 1}{i} - \frac{620}{50} = 0$$

or $(1 + i)^{12} - 1 - 12.4i = 0$

Set $y = (1 + i)^{12} - 1 - 12.4i$ and use your calculator to find the zero i of the function y, where $0 < i < 1$. The result is $i = 0.005941$. Thus $r = 12(0.005941) = 0.0713$ or $r = 7.13\%$ to two decimal places.

EXERCISE 4-4

Things to remember:

<u>1</u>. PRESENT VALUE OF AN ORDINARY ANNUITY

$$PV = PMT \frac{1 - (1 + i)^{-n}}{i} = PMTa_{\overline{n}|i}$$

where PMT = Periodic payment
i = Rate per period
n = Number of periods
PV = Present value of all payments

(Payments are made at the end of each period.)

<u>2</u>. AMORTIZATION FORMULA

$$PMT = PV \frac{i}{1 - (1 + i)^{-n}} = PV \frac{1}{a_{\overline{n}|i}}$$

where PV = Amount of loan (present value)
i = Rate per period
n = Number of payments (periods)
PMT = Periodic payment

(Payments are made at the end of each period.)

---

1. $PV = 200 \dfrac{1 - (1 + 0.04)^{-30}}{0.04}$

$= PMTa_{\overline{30}|0.04}$

$= 200(17.29203330)$ (using the table)

$= \$3458.41$

3. $PV = 250 \dfrac{1 - (1 + 0.025)^{-25}}{0.025}$

$= 250a_{\overline{25}|0.025}$

$= 250(18.42437642)$

$= \$4606.09$

5. $PMT = 6000 \dfrac{0.01}{1 - (1 + 0.01)^{-36}}$

$\qquad = \dfrac{PV}{a_{\overline{36}|0.01}}$

$\qquad = \dfrac{6000}{30.10750504} = \$199.29$

7. $PMT = 40{,}000 \dfrac{0.0075}{1 - (1 + 0.0075)^{-96}}$

$\qquad = \dfrac{40{,}000}{a_{\overline{96}|0.0075}}$

$\qquad = \dfrac{40{,}000}{68.25843856} = \$586.01$

9. $PV = \$5000$, $i = 0.01$, $PMT = 200$

We have, $PV = PMT \dfrac{1 - (1 + i)^{-n}}{i}$

$\qquad 5000 = 200 \dfrac{1 - (1 + 0.01)^{-n}}{0.01}$

$\qquad\qquad = 20{,}000[1 - (1.01)^{-n}]$

$\qquad \dfrac{1}{4} = 1 - (1.01)^{-n}$

$\qquad (1.01)^{-n} = \dfrac{3}{4} = 0.75$

$\qquad \ln(1.01)^{-n} = \ln(0.75)$

$\qquad -n \ln(1.01) = \ln(0.75)$

$\qquad\qquad n = \dfrac{-\ln(0.75)}{\ln(1.01)} \approx 29$

11. $PMT = \$4000$, $n = 10(4) = 40$

$\qquad i = \dfrac{0.08}{4} = 0.02$

$\qquad PV = $ Present value

$\qquad\quad = PMT \dfrac{1 - (1 + i)^{-n}}{i}$

$\qquad\quad = PMTa_{\overline{n}|i}$

$\qquad\quad = 4000a_{\overline{40}|0.02}$

$\qquad\quad = 4000(27.35547924)$

$\qquad\quad = \$109{,}421.92$

13. This is a present value problem.

$PMT = \$350$, $n = 4(12) = 48$, $i = \dfrac{0.09}{12} = 0.0075$

Hence, $PV = PMTa_{\overline{n}|i} = 350a_{\overline{48}|0.0075}$

$\qquad\qquad\qquad = 350(40.18478189) = \$14{,}064.67$

They should deposit $14,064.67. The child will receive $350(48) = $16,800.00.

15. (A) $PV = \$600$, $n = 18$, $i = 0.01$

Monthly payment $= PMT = PV \dfrac{i}{1 - (1 + i)^{-n}}$

$\qquad\qquad = \dfrac{PV}{a_{\overline{n}|i}} = \dfrac{600}{a_{\overline{18}|0.01}} = \dfrac{600}{16.39826858}$

$\qquad\qquad = \$36.59$ per month

The amount paid in 18 payments $= 36.59(18) = \$658.62$.
Thus, the interest paid $= 658.62 - 600 = \$58.62$.

(B) $PMT = \dfrac{600}{a_{\overline{18}|0.015}} \quad (i = 0.015)$

$\qquad = \dfrac{600}{15.67256089} = \$38.28$ per month

For 18 payments, the total amount $= 38.28(18) = \$689.04$.
Thus, the interest paid $= 689.04 - 600 = \$89.04$.

17. Amortized amount = $16,000 - (16,000)(0.25) = \$12,000$
Thus, $PV = \$12,000$, $n = 6(12) = 72$, $i = 0.015$

$$PMT = \text{monthly payment} = \frac{PV}{a_{\overline{n}|i}} = \frac{12,000}{a_{\overline{72}|0.015}} = \frac{12,000}{43.84466677} = \$273.69 \text{ per month}$$

The total amount paid in 72 months = $273.69(72) = \$19,705.68$.
Thus, the interest paid = $19,705.68 - 12,000 = \$7705.68$.

19. First, we compute the required quarterly payment for $PV = \$5000$,
$i = 0.045$, and $n = 8$, as follows:

$$PMT = PV \frac{i}{1 - (1 + i)^{-n}} = 5000 \frac{0.045}{1 - (1 + 0.045)^{-8}} = \frac{225}{1 - (1.045)^{-8}}$$
$$= \$758.05 \text{ per quarter}$$

The amortization schedule is as follows:

Payment number	Payment	Interest	Unpaid balance reduction	Unpaid balance
0				$5000.00
1	$758.05	$225.00	$533.05	4466.95
2	758.05	201.01	557.04	3909.91
3	758.05	175.95	582.10	3327.81
4	758.05	149.75	608.30	2719.51
5	758.05	122.38	635.67	2083.84
6	758.05	93.77	664.28	1419.56
7	758.05	63.88	694.17	725.39
8	758.03	32.64	725.39	0.00
Totals	$6064.38	$1064.38	$5000.00	

21. First, we compute the required monthly payment for $PV = \$6000$,
$i = \frac{12}{12(100)} = 0.01$, $n = 3(12) = 36$.

$$PMT = PV \frac{i}{1 - (1 + i)^{-n}} = 6000 \frac{0.01}{1 - (1 + 0.01)^{-36}} = \frac{60}{1 - (1.01)^{-36}}$$
$$= \$199.29$$

Now, compute the unpaid balance after 12 payments by considering 24 unpaid payments: $PMT = \$199.29$, $i = 0.01$, and $n = 24$.

$$PV = PMT \frac{1 - (1 + i)^{-n}}{i} = 199.29 \frac{1 - (1 + 0.01)^{-24}}{0.01}$$
$$= 19,929[1 - (1.01)^{-24}] = \$4233.59$$

Thus, the amount of the loan paid in 12 months is $6000 - 4233.59 = \$1766.41$, and the amount of total payment made during 12 months is $12(199.29) = \$2391.48$. The interest paid during the first 12 months (first year) is:

$2391.48 - 1766.41 = \$625.07$

Similarly, the unpaid balance after two years can be computed by considering 12 unpaid payments: $PMT = \$199.29$, $i = 0.01$, and $n = 12$.

$$PV = 199.29 \frac{1 - (1 + 0.01)^{-12}}{0.01} = 19,929[1 - (1.01)^{-12}] = \$2243.02$$

Thus, the amount of the loan paid during 24 months is $6000 - 2243.02 = \$3756.98$, and the amount of the loan paid during the second year is $3756.98 - 1766.41 = \$1990.57$. The amount of total payment during the second year is $12(199.29) = \$2391.48$. The interest paid during the second year is:

$2391.48 - 1990.57 = \$400.91$

The total amount paid in 36 months is $199.29(36) = \$7174.44$. Thus, the total interest paid is $7174.44 - 6000 = \$1174.44$ and the interest paid during the third year is $1174.44 - (625.07 + 400.91) = 1174.44 - 1025.98 = \$148.46$.

23. PMT = monthly payment = $\$525$, $n = 30(12) = 360$, $i = \dfrac{0.098}{12} \approx 0.0081667$.

Thus, the present value of all payments is:

$$PV = PMT \,\frac{1 - (1 + i)^{-n}}{i} \approx 525 \,\frac{1 - (1 + 0.0081667)^{-360}}{0.0081667} \approx \$60,846.38$$

Hence, selling price = loan + down payment

$$= 60,846.38 + 25,000$$
$$= \$85,846.38$$

The total amount paid in 30 years (360 months) = $525(360) = \$189,000$. The interest paid is: $189,000 - 60,846.38 = \$128,153.62$

25. $P = \$6000$, $n = 2(12) = 24$, $i = \dfrac{0.035}{12} \approx 0.0029167$

The total amount owed at the end of the two years is:

$A = P(1 + i)^n = 6000(1 + 0.0029167)^{24} = 6000(1.0029167)^{24} \approx 6434.39$

Now, the monthly payment is:

$$PMT = PV \,\frac{i}{1 - (1 + i)^{-n}}$$

where $n = 4(12) = 48$, $PV = \$6434.39$, $i = \dfrac{0.035}{12} \approx 0.0029167$. Thus,

$$PMT = 6434.39 \,\frac{0.0029167}{1 - (1 + 0.0029167)^{-48}} = \$143.85 \text{ per month}$$

The total amount paid in 48 payments is $143.85(48) = \$6904.80$. Thus, the interest paid is $6904.80 - 6000 = \$904.80$.

27. First, compute the monthly payment: $PV = \$75,000$, $i = \dfrac{0.132}{12} = 0.011$, $n = 30(12) = 360$.

$$\text{Monthly payment} = PV \,\frac{i}{1 - (1 + i)^{-n}} = 75,000 \,\frac{0.011}{1 - (1 + 0.011)^{-360}}$$

$$= 75,000 \,\frac{0.011}{1 - (1.011)^{-360}} = \$841.39$$

(A) Now, to compute the balance after 10 years (with balance of loan to be paid in 20 years), use $PMT = \$841.39$, $i = 0.011$, and $n = 20(12) = 240$.

$$\text{Balance after 10 years} = PMT \,\frac{1 - (1 + i)^{-n}}{i}$$

$$= 841.39 \,\frac{1 - (1 + 0.011)^{-240}}{0.011}$$

$$= 841.39 \,\frac{1 - (1.011)^{-240}}{0.011} = \$70,952.33$$

(B) Similarly, the balance of the loan after 20 years (with remainder of loan to be paid in 10 years) is:

$$841.39 \frac{1 - (1 + 0.011)^{-120}}{0.011} \quad [\underline{\text{Note}}: n = 12(10) = 120]$$

$$= 841.39 \frac{1 - (1.011)^{-120}}{0.011} = \$55,909.02$$

(C) The balance of the loan after 25 years (with remainder of loan to be paid in 5 years) is:

$$841.39 \frac{1 - (1 + 0.011)^{-60}}{0.011} \quad [\underline{\text{Note}}: n = 12(5) = 60]$$

$$= 841.39 \frac{1 - (1.011)^{-60}}{0.011} = \$36,813.32$$

29. (A) $PV = \$30,000$, $i = \dfrac{0.15}{12} = 0.0125$, $n = 20(12) = 240$.

Monthly payment $PMT = PV \dfrac{i}{1 - (1 + i)^{-n}}$

$$= 30,000 \frac{0.0125}{1 - (1 + 0.0125)^{-240}}$$

$$= 30,000 \frac{0.0125}{1 - (1.0125)^{-240}} = \$395.04$$

The total amount paid in 240 payments is:
395.04(240) = \$94,809.60
Thus, the interest paid is:
\$94,809.60 - \$30,000 = \$64,809.60

(B) New payment = PMT = \$395.04 + \$100.00 = \$495.04. PV = \$30,000, $i = 0.0125$.

$$PMT = PV \frac{i}{1 - (1 + i)^{-n}}$$

$$495.04 = 30,000 \frac{0.0125}{1 - (1 + 0.0125)^{-n}} = \frac{375}{1 - (1.0125)^{-n}}$$

Therefore,

$$1 - (1.0125)^{-n} = \frac{375}{495.04} = 0.7575$$

$$(1.0125)^{-n} = 1 - 0.7575 = 0.2425$$

$$\ln(1.0125)^{-n} = \ln(0.2425)$$

$$-n \ln(1.0125) = \ln(0.2425)$$

$$= \frac{-\ln(0.2425)}{\ln(1.0125)} \approx 114.047 \approx 114 \text{ months or } 9.5 \text{ years}$$

The total amount paid in 114 payments of \$495.04 is:
495.04(114) = \$56,434.56
Thus, the interest paid is:
\$56,434.56 - \$30,000 = \$26,434.56
The savings on interest is:
\$64,809.60 - \$26,434.56 = \$38,375.04

31. $PV = (\$79{,}000)(0.80) = \$63{,}200$, $i = \dfrac{0.12}{12} = 0.01$, $n = 12(30) = 360$.

Monthly payment $PMT = PV \dfrac{i}{1 - (1 + i)^{-n}} = 63{,}200 \dfrac{0.01}{1 - (1 + 0.01)^{-360}}$

$$= \dfrac{632}{1 - (1.01)^{-360}} = \$650.08$$

Next, we find the present value of a \$650.08 per month, 18-year annuity. $PMT = \$650.08$, $i = 0.01$, and $n = 12(18) = 216$.

$$PV = PMT \dfrac{1 - (1 + i)^{-n}}{i} = 650.08 \dfrac{1 - (1.01)^{-216}}{0.01}$$

$$= \dfrac{650.08(0.8834309)}{0.01} = \$57{,}430.08$$

Finally,
Equity = (current market value) − (unpaid loan balance)
$\quad = \$100{,}000 - \$57{,}430.08 = \$42{,}569.92$
The couple can borrow $(\$42{,}569.92)(0.70) = \$29{,}799$.

Problems 33 and 35 start from the equation

$$(*) \quad \dfrac{1 - (1 + i)^{-n}}{i} - \dfrac{PV}{PMT} = 0$$

A graphics calculator was used to solve these problems.

33. $PV = 1000$, $PMT = 90$, $n = 12$, $i = \dfrac{r}{12}$ where r is the annual nominal rate.

With these values, the equation (*) becomes
$$\dfrac{1 - (1 + i)^{-12}}{i} - \dfrac{1000}{90} = 0$$
or $\quad 1 - (1 + i)^{-12} - 11.11i = 0$

Put $y = 1 - (1 + i)^{-12} - 11.11i$ and use your calculator to find the zero i of y, where $0 < i < 1$. The result is $i \approx 0.01204$ and $r = 12(0.01204) = 0.14448$. Thus, $r = 14.45\%$ (two decimal places).

35. $PV = 90{,}000$, $PMT = 1200$, $n = 12(10) = 120$, $i = \dfrac{r}{12}$ where r is the annual nominal rate. With these values, the equation (*) becomes
$$\dfrac{1 - (1 + i)^{-120}}{i} - \dfrac{90{,}000}{1200} = 0$$
or $\quad 1 - (1 + i)^{-120} - 75i = 0$

This equation can be written
$$(1 + i)^{120} - (1 - 75i)^{-1} = 0$$

Put $y = (1 + i)^{120} - (1 - 75i)^{-1}$ and use your calculator to find the zero i of y where $0 < i < 1$. The result is $i \approx 0.00851$ and $r = 12(0.00851) = 0.10212$. Thus $r = 10.21\%$ (two decimal places).

1. $A = 100\left(1 + 0.09 \cdot \dfrac{1}{2}\right)$

$= 100(1.045) = \$104.50$

2. $808 = P\left(1 + 0.12 \cdot \dfrac{1}{12}\right)$

$P = \dfrac{808}{1.01} = \$800$

3. $212 = 200(1 + 0.08 \cdot t)$

$1 + 0.08t = \dfrac{212}{200}$

$0.08t = \dfrac{212}{200} - 1 = \dfrac{12}{200}$

$t = \dfrac{0.06}{0.08} = 0.75$ yr. or 9 mos.

4. $4120 = 4000\left(1 + r \cdot \dfrac{1}{2}\right)$

$1 + \dfrac{r}{2} = \dfrac{4120}{4000}$

$\dfrac{r}{2} = \dfrac{4120}{4000} - 1 = \dfrac{120}{4000} = 0.03$

$r = 0.06$ or   6%

5. $A = 1200(1 + 0.005)^{30}$

$= 1200(1.005)^{30} = \$1393.68$

6. $P = \dfrac{5000}{(1 + 0.0075)^{60}} = \dfrac{5000}{(1.0075)^{60}}$

$= \$3193.50$

7. $FV = 1000 s_{\overline{60}|0.005}$

$= 1000 \cdot 69.77003051$

$= \$69,770.03$

8. $PMT = \dfrac{FV}{s_{\overline{n}|i}} = \dfrac{8000}{s_{\overline{48}|0.015}}$

$= \dfrac{8000}{69.56321929} = \$115.00$

9. $PV = PMT\, a_{\overline{n}|i} = 2500 a_{\overline{16}|0.02}$

$= 2500 \cdot 13.57770931$

$= \$33,944.27$

10. $PMT = \dfrac{PV}{a_{\overline{n}|i}} = \dfrac{8000}{a_{\overline{60}|0.0075}}$

$= \dfrac{8000}{48.17337352} = \$166.07$

11. $2500 = 1000(1.06)^n$

$(1.06)^n = \dfrac{2500}{1000}$

$n = \dfrac{\ln 2.5}{\ln 1.06} \approx 16$

12. $5000 = 100\, \dfrac{(1.01)^n - 1}{0.01}$

$= 10,000[(1.01)^n - 1]$

$(1.01)^n - 1 = \dfrac{5000}{10,000}$

$(1.01)^n = \dfrac{1}{2} + 1$

$n = \dfrac{\ln 1.5}{\ln 1.01} \approx 41$

13. $P = \$3000$, $r = 0.14$, $t = \dfrac{10}{12}$

$A = 3000\left(1 + 0.14 \cdot \dfrac{10}{12}\right)$ [using $A = P(1 + rt)$]

$= \$3350$

Interest $= 3350 - 3000 = \$350$

14. $P = \$635$, $r = 22\% = 0.22$, $t = \dfrac{1}{12}$

$I = Prt = 635(0.22)\dfrac{1}{12} = \$11.64$

15. The interest paid was $2812.50 - $2500 = $312.50. $P = 2500, $t = \frac{10}{12} = \frac{5}{6}$

Solving $I = Prt$ for r, we have:

$r = \dfrac{I}{Pt} = \dfrac{312.50}{2500\left(\frac{5}{6}\right)} = 0.15$ or 15%

16. $P = 1500, $I = 100,
$t = \dfrac{120}{360} = \dfrac{1}{3}$ year
From Problem 15,
$r = \dfrac{I}{Pt} = \dfrac{100}{1500\left(\frac{1}{3}\right)} = 0.20$ or 20%

17. $P = 100, $I = 0.08, $t = \dfrac{1}{360}$
From Problem 15,
$r = \dfrac{I}{Pt} = \dfrac{0.08}{100\left(\frac{1}{360}\right)} = 0.288$ or 28.8%

18. $A = 5000, $P = 4899.08, $t = \dfrac{13}{52} = 0.25$

The interest earned is $I = $5000.00 - $4899.08 = 100.92. Thus:

$r = \dfrac{I}{Pt} = \dfrac{100.92}{(4899.08)(0.25)} \approx 0.0824$ or 8.24%

19. $A = 5000, $r = 10.76\% = 0.1076$, $t = \dfrac{26}{52} = 0.5$

$P = \dfrac{A}{1 + rt} = \dfrac{5000}{1 + (0.1076)(0.5)} = 4744.73

20. $P = 6000, $r = 9\% = 0.09$, $m = 12$, $i = \dfrac{0.09}{12} = 0.0075$, $n = 12(17) = 204$

$A = P(1 + i)^n = 6000(1 + 0.0075)^{204} = 6000(1.0075)^{204} \approx $27{,}551.32$

21. $A = $25{,}000$, $r = 10\% = 0.10$, $m = 2$, $i = \dfrac{0.10}{2} = 0.05$, $n = 2(10) = 20$

$P = \dfrac{A}{(1 + i)^n} = \dfrac{25{,}000}{(1 + 0.05)^{20}} = \dfrac{25{,}000}{(1.05)^{20}} \approx 9422.24

22. $P = 8000, $r = 5\% = 0.05$, $m = 1$, $i = \dfrac{0.05}{1} = 0.05$, $n = 5$

$A = P(1 + i)^n = 8000(1 + 0.05)^5 = 8000(1.05)^5 \approx $10{,}210.25$

23. $A = 8000, $r = 5\% = 0.05$, $m = 1$, $i = \dfrac{0.05}{1} = 0.05$, $n = 5$

$P = \dfrac{A}{(1 + i)^n} = \dfrac{8000}{(1 + 0.05)^5} = \dfrac{8000}{(1.05)^5} \approx 6268.21

24. $P = 2500, $r = 9\% = 0.09$, $m = 4$, $i = \dfrac{0.09}{4} = 0.0225$, $A = 3000

$$A = P(1 + i)^n$$
$$3000 = 2500(1 + 0.0225)^n$$
$$(1.0225)^n = \dfrac{3000}{2500} = 1.2$$
$$\ln(1.0225)^n = \ln 1.2$$
$$n \ln 1.0225 = \ln 1.2$$
$$n = \dfrac{\ln 1.2}{\ln 1.0225} \approx 8.19$$

Thus, it will take 9 quarters, or 2 years and 3 months.

25. (A) $r = 12\% = 0.12$, $m = 12$, $i = \dfrac{0.12}{12} = 0.01$

If we invest P dollars, then we want to know how long it will take to have $2P$ dollars:

$$A = P(1 + i)^n$$
$$2P = P(1 + 0.01)^n$$
$$(1.01)^n = 2$$
$$\ln(1.01)^n = \ln 2$$
$$n \ln 1.01 = \ln 2$$
$$n = \frac{\ln 2}{\ln 1.01} \approx 69.66$$

Thus, it will take 70 months, or 5 years and 10 months, for an investment to double at 12% interest compounded monthly.

(B) $r = 18\% = 0.18$, $m = 12$, $i = \dfrac{0.18}{12} = 0.015$

$$2P = P(1 + 0.015)^n$$
$$(1.015)^n = 2$$
$$\ln(1.015)^n = \ln 2$$
$$n = \frac{\ln 2}{\ln 1.015} \approx 46.56$$

Thus, it will take 47 months, or 3 years and 11 months, for an investment to double at 18% compounded monthly.

26. $r = 9\% = 0.09$, $m = 12$

$$r_e = \left(1 + \frac{r}{m}\right)^m - 1 = \left(1 + \frac{0.09}{12}\right)^{12} - 1$$
$$= (1.0075)^{12} - 1 \approx 0.0938 \quad \text{or} \quad 9.38\%$$

27. The effective rate for 9% compounded quarterly is:

$$r_e = \left(1 + \frac{r}{m}\right)^m - 1, \quad r = 0.09, \quad m = 4$$
$$= \left(1 + \frac{0.09}{4}\right)^4 - 1 = (1.0225)^4 - 1 \approx 0.0931 \quad \text{or} \quad 9.31\%$$

The effective rate for 9.25% compounded annually is 9.25%. Thus, 9% compounded quarterly is the better investment.

28. $PMT = \$200$, $r = 9\% = 0.09$, $m = 12$, $i = \dfrac{0.09}{12} = 0.0075$, $n = 12(8) = 96$

$$FV = PMT \, \frac{(1 + i)^n - 1}{i}$$
$$= 200 \, \frac{(1 + 0.0075)^{96} - 1}{0.0075} = 200 \, \frac{(1.0075)^{96} - 1}{0.0075} \approx \$27,971.23$$

The total amount invested with 96 payments of $200 is:
$96(200) = \$19,200$
Thus, the interest earned with this annuity is:
$I = \$27,971.23 - \$19,200 = \$8771.23$

29. $FV = \$50,000$, $r = 9\% = 0.09$, $m = 12$, $i = \dfrac{0.09}{12} = 0.0075$, $n = 12(6) = 72$

$$PMT = FV \, \frac{i}{(1 + i)^n - 1} = \frac{FV}{s_{\overline{n}|i}} = \frac{50,000}{s_{\overline{72}|0.0075}}$$
$$= \frac{50,000}{95.007028} \quad \text{(from Table II)}$$
$$= \$526.28 \text{ per month}$$

30. Using the sinking fund formula

$$PMT = FV \frac{i}{(1 + i)^n - 1}$$

with $PMT = \$200$, $FV = \$10,000$, and $i = \frac{0.09}{12} = 0.0075$, we have:

$$200 = 10,000 \frac{0.0075}{(1 + 0.0075)^n - 1} = \frac{75}{(1.0075)^n - 1}$$

Therefore,

$$(1.0075)^n - 1 = \frac{75}{200} = 0.375$$
$$(1.0075)^n = 0.375 + 1 = 1.375$$
$$\ln(1.0075)^n = \ln 1.375$$
$$n = \frac{\ln 1.375}{1.0075} \approx 42.62$$

The couple will have to make 43 deposits.

31. $PMT = \$1500$, $r = 8\% = 0.08$, $m = 4$, $i = \frac{0.08}{4} = 0.02$, $n = 2(4) = 8$

We want to find the present value, PV, of this annuity.

$$PV = PMT \frac{1 - (1 + i)^{-n}}{i}$$

$$= 1500 \frac{1 - (1 + 0.02)^{-8}}{0.02} = 1500 \frac{1 - (1.02)^{-8}}{0.02} = \$10,988.22$$

The student will receive $8(\$1500) = \$12,000$.

32. The amount of the loan is $\$3000\left(\frac{2}{3}\right) = \$2000$. The monthly interest rate

is $i = 1.5\% = 0.015$ and $n = 2(12) = 24$.

$$PMT = PV \frac{i}{1 - (1 + i)^{-n}} = 2000 \frac{0.015}{1 - (1 + 0.015)^{-24}} = \frac{30}{1 - (1.015)^{-24}}$$

$$= \$99.85 \text{ per month}$$

The amount paid in 24 payments is
$99.85(24) = \$2396.40$. So, the interest paid is $2396.40 - 2000 = \$396.40$.

33. $PV = \$1000$, $i = 0.025$, $n = 4$

The quarterly payment is:

$$PMT = PV \frac{i}{1 - (1 + i)^{-n}}$$

$$= 1000 \frac{0.025}{1 - (1 + 0.025)^{-4}} = \frac{25}{1 - (1.025)^{-4}} \approx \$265.82$$

Payment number	Payment	Interest	Unpaid balance reduction	Unpaid balance
0				$1000.00
1	$265.82	$25.00	$240.82	759.18
2	265.82	18.98	246.84	512.34
3	265.82	12.81	253.01	259.33
4	265.81	6.48	259.33	0.00
Totals	$1063.27	$63.27	$1000.00	

34. We first compute the monthly payment using $PV = \$10,000$,
$i = \dfrac{0.12}{12} = 0.01$, and $n = 5(12) = 60$.

$$PMT = PV \, \frac{i}{1 - (1 + i)^{-n}}$$

$$= 10,000 \, \frac{0.01}{1 - (1 + 0.01)^{-60}} = \frac{100}{1 - (1.01)^{-60}} = \$222.44 \text{ per month}$$

Now, we calculate the unpaid balance after 24 payments by using
$PMT = \$222.44$, $i = 0.01$, and $n = 60 - 24 = 36$.

$$PV = PMT \, \frac{1 - (1 + i)^{-n}}{i}$$

$$= 222.44 \, \frac{1 - (1 + 0.01)^{-36}}{0.01} = 22,244[1 - (1.01)^{-36}] = \$6697.11$$

Thus, the unpaid balance after 2 years is $6697.11.

35. $PV = \$80,000$, $i = \dfrac{0.15}{12} = 0.0125$, $n = 8(12) = 96$

(A) $PMT = PV \, \dfrac{i}{1 - (1 + i)^{-n}}$

$$= 80,000 \, \frac{0.0125}{1 - (1 + 0.0125)^{-96}} = \frac{1000}{1 - (1.0125)^{-96}}$$

$$= \$1435.63 \text{ monthly payment}$$

(B) Now use $PMT = \$1435.63$, $i = 0.0125$, and $n = 96 - 12 = 84$ to
calculate the unpaid balance.

$$PV = PMT \, \frac{1 - (1 + i)^{-n}}{i}$$

$$= 1435.63 \, \frac{1 - (1 + 0.0125)^{-84}}{0.0125} = 114,850.40[1 - (1.0125)^{-84}]$$

$$= \$74,397.48 \text{ unpaid balance after the first year}$$

(C) Amount of loan paid during the first year:
$\$80,000 - \$74,397.48 = \$5602.52$
Amount of payments during the first year:
$12(\$1435.63) = \$17,227.56$
Thus, the interest paid during the first year is:
$\$17,227.56 - \$5602.52 = \$11,625.04$

36. (A) The present value of an annuity which provides for quarterly
withdrawals of $5000 for 10 years at 12% interest compounded
quarterly is given by:

$$PV = PMT \, \frac{1 - (1 + i)^{-n}}{i} \quad \text{with } PMT = \$5000, \; i = \frac{0.12}{4} = 0.03,$$
$$\text{and } n = 10(4) = 40$$

$$= 5000 \, \frac{1 - (1 + 0.03)^{-40}}{0.03}$$

$$= 166,666.67[1 - (1.03)^{-40}] = \$115,573.86$$

This amount will have to be in the account when he retires.

(B) To determine the quarterly deposit to accumulate the amount in part (A), we use the formula:

$$PMT = FV \frac{i}{(1 + i)^n - 1} \quad \text{where } FV = \$115{,}573.86, \ i = 0.03,$$
$$\text{and } n = 4(20) = 80$$

$$= 115{,}573.86 \frac{0.03}{(1 + 0.03)^{80} - 1}$$

$$= \frac{3467.22}{(1.03)^{80} - 1} = \$359.64 \text{ quarterly payment}$$

(C) The amount collected during the 10-year period is:
($5000)40 = $200,000
The amount deposited during the 20-year period is:
($359.64)80 = $28,771.20
Thus, the interest earned during the 30-year period is:
$200,000 - $28,771.20 = $171,228.80

37. $P = \$10{,}000, \ r = 7\% = 0.07, \ m = 365, \ i = \dfrac{0.07}{365} = 0.0001918,$ and

$n = 40(365) = 14{,}600$

$A = P(1 + i)^n = 10{,}000(1 + 0.0001918)^{14,600}$
$\qquad\qquad\qquad = 10{,}000(1.0001918)^{14,600}$
$\qquad\qquad\qquad = \$164{,}402$

38. To determine how long it will take money to double, we need to solve the equation $2P = P(1 + i)^n$ for n. From this equation, we obtain:

$(1 + i)^n = 2$

$\ln(1 + i)^n = \ln 2$

$n \ln(1 + i) = \ln 2$

$$n = \frac{\ln 2}{\ln(1 + i)}$$

(A) $i = \dfrac{0.10}{365} = 0.000274$

Thus, $n = \dfrac{\ln 2}{\ln(1.000274)} \approx 2530.08$ days or 6.93 years.

(B) $i = 0.10$

Thus, $n = \dfrac{\ln 2}{\ln(1.1)} \approx 7.27$ years.

39. The effective rate for Security S & L is:

$r_e = \left(1 + \dfrac{r}{m}\right)^m - 1$ where $r = 9.38\% = 0.0938$ and $m = 12$

$\quad = \left(1 + \dfrac{0.0938}{12}\right)^{12} - 1 \approx 0.09794$ or 9.794%

The effective rate for West Lake S & L is:

$r_e = \left(1 + \dfrac{r}{m}\right)^m - 1$ where $r = 9.35\% = 0.0935$ and $m = 365$

$\quad = \left(1 + \dfrac{0.0935}{365}\right)^{365} - 1 \approx 0.09799$ or 9.8%

Thus, West Lake S & L is a better investment.

40. $A = \$5000$, $r = i = 9.5\% = 0.095$, $n = 5$

$$P = \frac{A}{(1 + i)^n} = \frac{5000}{(1 + 0.095)^5} = \frac{5000}{(1.095)^5} \approx \$3176.14$$

41. $P = \$4476.20$, $A = \$10,000$, $m = 1$, $r = i$, $n = 10$

$$A = P(1 + i)^n$$
$$10,000 = 4476.20(1 + i)^{10}$$
$$(1 + i)^{10} = \frac{10,000}{4476.20} \approx 2.23404$$
$$10 \ln(1 + i) = \ln(2.23404)$$
$$\ln(1 + i) = \frac{\ln(2.23404)}{10} \approx 0.0803811$$
$$1 + i = e^{0.0803811} \approx 1.0837$$
$$i = 0.0837 \text{ or } 8.37\%$$

42. $A = \$17,388.17$, $P = \$12,903.28$, $m = 1$, $r = i$, $n = 3$

$$A = P(1 + i)^n$$
$$17,388.17 = 12,903.28(1 + i)^3$$
$$(1 + i)^3 = \frac{17,388.17}{12,903.28} \approx 1.3475775$$
$$3 \ln(1 + i) = \ln(1.3475775)$$
$$\ln(1 + i) \approx \frac{0.2983085}{3} \approx 0.0994362$$
$$1 + i = e^{0.0994362} \approx 1.1045$$
$$i = 0.1045 \text{ or } 10.45\%$$

43. (A) $PMT = \$2000$, $m = 1$, $r = i = 7\% = 0.07$, $n = 45$

$$FV = PMT \frac{(1 + i)^n - 1}{i}$$
$$= 2000 \frac{(1 + 0.07)^{45} - 1}{0.07} = 2000 \frac{(1.07)^{45} - 1}{0.07} \approx \$571,499$$

(B) $PMT = \$2000$, $m = 1$, $r = i = 11\% = 0.11$, $n = 45$

$$FV = PMT \frac{(1 + i)^n - 1}{i}$$
$$= 2000 \frac{(1 + 0.11)^{45} - 1}{0.11} = 2000 \frac{(1.11)^{45} - 1}{0.11} \approx \$1,973,277$$

44. $FV = \$850,000$, $r = 8.76\% = 0.0876$, $m = 2$, $i = \frac{0.0876}{2} = 0.0438$,
$n = 2(6) = 12$

$$PMT = FV \frac{i}{(1 + i)^n - 1}$$
$$= 850,000 \frac{0.0438}{(1 + 0.0438)^{12} - 1} = \frac{37,230}{(1.0438)^{12} - 1} \approx \$55,347.48$$

The total amount invested is:
$12(55,347.48) = \$664,169.76$
Thus, the interest earned with this annuity is:
$I = \$850,000 - \$664,169.76 = \$185,830.24$

45. $PMT = \$200$, $FV = \$2500$, $i = \dfrac{0.0798}{12} = 0.00665$

$$FV = PMT\,\frac{(1 + i)^n - 1}{i}$$

$$2500 = 200\,\frac{(1 + 0.00665)^n - 1}{0.00665} = 30{,}075.188[(1.00665)^n - 1]$$

$$(1.00665)^n - 1 = \frac{2500}{30{,}075.188} \approx 0.083125$$

$$(1.0065)^n = 1.083125$$

$$n \ln 1.0065 = \ln 1.083125$$

$$n = \frac{\ln 1.083125}{\ln 1.0065} \approx 12.32 \text{ months}$$

Thus, it will take 13 months, or 1 year and 1 month.

46. The present value, PV, of an annuity of $200 per month for 48 months at 14% interest compounded monthly is given by:

$$PV = PMT\,\frac{1 - (1 + i)^{-n}}{i} \quad \text{where } PMT = \$200, \; i = \frac{0.14}{12} = 0.0116667$$
$$\text{and } n = 48$$

$$= 200\,\frac{1 - (1 + 0.0116667)^{-48}}{0.0116667}$$

$$= 17{,}142.857[1 - (1.0116667)^{-48}] = \$7318.91$$

With the $3000 down payment, the selling price of the car is $10,318.91.
The total amount paid is:
$3000 + 48(\$200) = \$12{,}600$
Thus, the interest paid is:
$I = \$12{,}600 - \$10{,}318.91 = \$2281.09$

47. First, we must calculate the future value of $8000 at 5.5% interest compounded monthly for 2.5 years.

$$A = P(1 + i)^n \quad \text{where } P = \$8000, \; i = \frac{0.055}{12} \text{ and } n = 30$$

$$= 8000\left(1 + \frac{0.055}{12}\right)^{30} = \$9176.33$$

Now, we calculate the monthly payment to amortize this debt at 5.5% interest compounded monthly over 5 years.

$$PMT = PV\,\frac{i}{1 + (1 + i)^{-n}} \quad \text{where } PV = \$9176.33, \; i = \frac{0.055}{12} \approx 0.0045833$$
$$\text{and } n = 12(5) = 60$$

$$= 9176.33\,\frac{0.0045833}{1 - (1 + 0.0045833)^{-60}} = \frac{42.058179}{1 - (1.0045833)^{-60}} \approx \$175.28$$

The total amount paid on the loan is:
$175.28(60) = \$10{,}516.80$
Thus, the interest paid is:
$I = \$10{,}516.80 - \$8000 = \$2516.80$

48. (A) We first calculate the future value of an annuity of $2000 at 8% compounded annually for 9 years.

$$FV = PMT \frac{(1 + i)^n - 1}{i} \text{ where } PMT = \$2000, \ i = 0.08, \text{ and } n = 9$$

$$= 2000 \frac{(1 + 0.08)^9 - 1}{0.08} = 25{,}000[(1.08)^9 - 1] \approx \$24{,}975.12$$

Now, we calculate the future value of this amount at 8% compounded annually for 36 years.

$A = P(1 + i)^n$, where $P = \$24{,}975.12$, $i = 0.08$, and $n = 36$

$= 24{,}975.12(1 + 0.08)^{36} = 24{,}975.12(1.08)^{36} \approx \$398{,}807$

(B) This is the future value of a $2000 annuity at 8% compounded annually for 35 years.

$$FV = PMT \frac{(1 + i)^n - 1}{i} \text{ where } PMT = \$2000, \ i = 0.08, \text{ and } n = 36$$

$$= 2000 \frac{(1 + 0.08)^{36} - 1}{0.08} = 25{,}000[(1.08)^{36} - 1] \approx \$374{,}204$$

49. The amount of the loan is ($100,000)(0.8) = $80,000 and

$$PMT = PV \frac{i}{1 - (1 + i)^{-n}}.$$

(A) First, let $i = \dfrac{0.1075}{12} = 0.0089583$, $n = 12(30) = 360$.   Then,

$$PMT = 80{,}000 \frac{0.0089583}{1 - (1 + 0.0089583)^{-360}} = \frac{716.66667}{0.9596687}$$

$\approx \$746.79$ monthly payment for 30 years.

Next, let $i = \dfrac{0.1075}{12} = 0.0089583$, $n = 12(15) = 180$.   Then

$$PMT = 80{,}000 \frac{0.0089583}{1 - (1 + 0.0089593)^{-180}} = \frac{716.66667}{0.7991735}$$

$\approx \$896.76$ monthly payment for 15 years.

(B) To find the unpaid balance after 10 years, we use

$$PV = PMT \frac{1 - (1 + i)^{-n}}{i}$$

First, for the 30-year mortgage:

$PMT = \$746.79$, $i = \dfrac{0.1075}{12} = 0.0089583$, $n = 12(20) = 240$

$$PV = 746.79 \frac{1 - (1 + 0.0089583)^{-240}}{0.0089583} = 83{,}362.915[1 - (1.0089583)^{-240}]$$

$= \$73{,}558.78$ unpaid balance for the 30-year mortgage

Next, for the 15-year mortgage:

$PMT = \$896.76$, $i = 0.0089583$, $n = 5(12) = 60$

$$PV = 896.76 \frac{1 - (1.0089583)^{-60}}{0.0089583} = 100{,}103.81[1 - (1.0089583)^{-60}]$$

$\approx \$41{,}482.19$ unpaid balance for the 15-year mortgage.

50. The amount of the mortgage is:

($83,000)(0.8) = $66,400

The monthly payment is given by:

$$PMT = PV \frac{i}{1 - (1 + i)^{-n}} \quad \text{where } PV = \$66{,}400, \; i = \frac{0.1125}{12} = 0.009375,$$
$$\text{and } n = 12(30) = 360$$

$$= 66{,}400 \frac{0.009375}{1 - (1 + 0.009375)^{-360}}$$

$$= \frac{622.50}{1 - (1.009375)^{-360}} \approx \$644.92$$

Next, we find the present value of a $644.92 per month, 22-year annuity:

$$PV = PMT \frac{1 - (1 + i)^{-n}}{i} \quad \text{where } PMT = \$644.92, \; i = 0.009375,$$
$$\text{and } n = 12(22) = 264$$

$$= 644.92 \frac{1 - (1 + 0.009375)^{-264}}{0.009375}$$

$$= 68{,}791.467[1 - (1.009375)^{-264}] = \$62{,}934.63$$

Finally,

Equity = (current market value) − (unpaid loan balance)

= $95,000 − $62,934.63 = $32,065.37

The family can borrow up to ($32,065.37)(0.60) = $19,239.

.

EXERCISE 5-1

Things to remember:

1. The system of two linear equations in two variables

    $ax + by = h$
    $cx + dy = k$

    can be solved by:
    (a) graphing;
    (b) substitution;
    (c) elimination by addition.

2. POSSIBLE SOLUTIONS TO A LINEAR SYSTEM

    The linear system

    $ax + by = h$
    $cx + dy = k$

    must have:
    (a) exactly one solution (consistent and independent); or
    (b) no solution (inconsistent); or
    (c) infinitely many solutions (consistent and dependent).

3. Two systems of linear equations are EQUIVALENT if they have exactly the same solution set. A system of linear equations is transformed into an equivalent system if:

    (a) two equations are interchanged;
    (b) an equation is multiplied by a nonzero constant;
    (c) a constant multiple of one equation is added to another equation.

1. $x + y = 5$
   $x - y = 1$
   Point of intersection: (3, 2)
   Solution: $x = 3$; $y = 2$

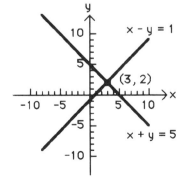

3. $3x - y = 2$
   $x + 2y = 10$
   Point of intersection: (2, 4)
   Solution: $x = 2$; $y = 4$

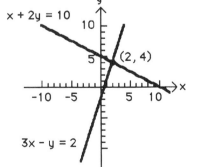

5. $m + 2n = 4$

$2m + 4n = -8$

Since the graphs of the given equations are parallel lines, there is no solution.

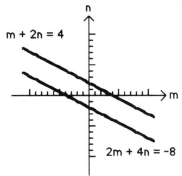

7. $y = 2x - 3$ (1)

$x + 2y = 14$ (2)

By substituting y from (1) into (2), we get:

$x + 2(2x - 3) = 14$

$x + 4x - 6 = 14$

$5x = 20$

$x = 4$

Now, substituting $x = 4$ into (1), we have:

$y = 2(4) - 3$

$y = 5$

Solution: $x = 4$

$y = 5$

9. $2x + y = 6$ (1)

$x - y = -3$ (2)

Solve (2) for y to obtain the system:

$2x + y = 6$ (3)

$y = x + 3$ (4)

Substitute y from (4) into (3):

$2x + x + 3 = 6$

$3x = 3$

$x = 1$

Now, substituting $x = 1$ into (4), we get:

$y = 1 + 3$

$y = 4$

Solution: $x = 1$

$y = 4$

11. $3u - 2v = 12$ (1)

$7u + 2v = 8$ (2)

Add (1) and (2):

$10u = 20$

$u = 2$

Substituting $u = 2$ into (2), we get:

$7(2) + 2v = 8$

$2v = -6$

$v = -3$

Solution: $u = 2$

$v = -3$

13. $2m - n = 10$ (1)

$m - 2n = -4$ (2)

Multiply (1) by -2 and add to (2) to obtain:

$-3m = -24$

$m = 8$

Substituting $m = 8$ into (2), we get:

$8 - 2n = -4$

$-2n = -12$

$n = 6$

Solution: $m = 8$

$n = 6$

15. $9x - 3y = 24$ (1)

$11x + 2y = 1$ (2)

Solve (1) for y to obtain:

$y = 3x - 8$ (3)

and substitute into (2):

$11x + 2(3x - 8) = 1$

$11x + 6x - 16 = 1$

$17x = 17$

$x = 1$

Now, substitute $x = 1$ into (3):

$y = 3(1) - 8$

$y = -5$

Solution: $x = 1$

$y = -5$

17.
$$2x - 3y = -2 \quad (1)$$
$$-4x + 6y = 7 \quad (2)$$
Multiply (1) by 2 and add to (2) to get:
$$0 = 3$$
This implies that the system is inconsistent, and thus there is no solution.

19.
$$3x + 8y = 4 \quad (1)$$
$$15x + 10y = -10 \quad (2)$$
Multiply (1) by -5 and add to (2) to get:
$$-30y = -30$$
$$y = 1$$
Substituting $y = 1$ into (1), we get:
$$3x + 8(1) = 4$$
$$3x = -4$$
$$x = -\frac{4}{3}$$
Solution: $x = -\frac{4}{3}$; $y = 1$

21.
$$-6x + 10y = -30 \quad (1)$$
$$3x - 5y = 15 \quad (2)$$
Multiply (2) by 2 and add to (1). This yields:
$$0 = 0$$
which implies that (1) and (2) are equivalent equations and there are infinitely many solutions. Geometrically, the two lines are coincident. The system is dependent.

23.
$$y = 0.07x \quad (1)$$
$$y = 80 + 0.05x \quad (2)$$
Substitute y from (1) into (2):
$$0.07x = 80 + 0.05x$$
$$0.02x = 80$$
$$x = \frac{80}{0.02}$$
$$x = 4000$$
Next, by substituting $x = 4000$ into (1), we get:
$$y = 0.07(4000) = 280$$
Solution: $x = 4000$; $y = 280$

25.
$$x + y = 1 \quad (1)$$
$$0.3x - 0.4y = 0 \quad (2)$$
Multiply equation (2) by 10 to remove the decimals
$$x + y = 1 \quad (1)$$
$$3x - 4y = 0 \quad (3)$$
Multiply (1) by 4 and add to (2) to get
$$7x = 4$$
$$x = \frac{4}{7}$$
Now substitute $x = \frac{4}{7}$ in (1):
$$\frac{4}{7} + y = 1$$
$$y = 1 - \frac{4}{7} = \frac{3}{7}$$
Solution: $x = \frac{4}{7}$, $y = \frac{3}{7}$

27.
$$0.2x - 0.5y = 0.07 \quad (1)$$
$$0.8x - 0.3y = 0.79 \quad (2)$$
Clear the decimals from (1) and (2) by multiplying each equation by 100.
$$20x - 50y = 7 \quad (3)$$
$$80x - 30y = 79 \quad (4)$$
Multiply (3) by -4 and add to (4) to get:
$$170y = 51$$
$$y = \frac{51}{170}$$
$$y = 0.3$$
Now, substitute $y = 0.3$ into (1):
$$0.2x - 0.5(0.3) = 0.07$$
$$0.2x - 0.15 = 0.07$$
$$0.2x = 0.22$$
$$x = 1.1$$
Solution: $x = 1.1$; $y = 0.3$

29.
$$\frac{2}{5}x + \frac{3}{2}y = 2 \qquad (1)$$

$$\frac{7}{3}x - \frac{5}{4}y = -5 \qquad (2)$$

Multiply (1) by 10 and (2) by 12 to remove the fractions
$$4x + 15y = 20 \qquad (3)$$
$$28x - 15y = -60 \qquad (4)$$

System (3), (4) is equivalent to system (1), (2). Now add equations (3) and (4) to get
$$32x = -40$$
$$x = -\frac{40}{32} = -\frac{5}{4}$$

Now substitute $x = -\frac{5}{4}$ into either (1), (2), (3), or (4) — (3) is probably the easiest

$$4\left(-\frac{5}{4}\right) + 15y = 20$$

$$-5 + 15y = 20$$
$$15y = 25$$
$$y = \frac{25}{15} = \frac{5}{3}$$

Solution: $x = -\frac{5}{4}$, $y = \frac{5}{3}$

31.
$$x - 2y = -6 \qquad (L_1)$$
$$2x + y = 8 \qquad (L_2)$$
$$x + 2y = -2 \qquad (L_3)$$

(A) L_1 and L_2 intersect:
$$x - 2y = -6 \qquad (1)$$
$$2x + y = 8 \qquad (2)$$

Multiply (2) by 2 and add to (1):
$$5x = 10$$
$$x = 2$$

Substitute $x = 2$ in (1) to get
$$2 - 2y = -6$$
$$-2y = -8$$
$$y = 4$$

Solution: $x = 2$, $y = 4$

(B) L_1 and L_3 intersect:
$$x - 2y = -6 \qquad (3)$$
$$x + 2y = -2 \qquad (4)$$

Add (3) and (4):
$$2x = -8$$
$$x = -4$$

Substitute $x = -4$ in (3) to get
$$-4 - 2y = -6$$
$$-2y = -2$$
$$y = 1$$

Solution: $x = -4$, $y = 1$

(C) L_2 and L_3 intersect:
$$2x + y = 8 \qquad (5)$$
$$x + 2y = -2 \qquad (6)$$

Multiply (6) by -2 and add to (5)
$$-3y = 12$$
$$y = -4$$

Substitute $y = -4$ in (5) to get
$$2x - 4 = 8$$
$$2x = 12$$
$$x = 6$$

Solution: $x = 6$, $y = -4$

33. $x + y = 1$ $\qquad$ (L_1)
$\quad\;\;$ $x - 2y = -8$ $\qquad$ (L_2)
$\quad\;\;$ $3x + y = -3$ $\qquad$ (L_3)

$\quad$ (A) L_1 and L_2 intersect
$\qquad$ $x + y = 1$ $\qquad$ (1)
$\qquad$ $x - 2y = -8$ $\qquad$ (2)
$\qquad$ Subtract (2) from (1):
$\qquad$ $3y = 9$
$\qquad$ $\;y = 3$
$\qquad$ Substitute $y = 3$ in (1) to get
$\qquad$ $x + 3 = 1$
$\qquad\;\;$ $x = -2$
$\qquad$ Solution: $x = -2$, $y = 3$

$\quad$ (B) L_1 and L_3 intersect:
$\qquad$ $x + y = 1$ $\qquad$ (3)
$\qquad$ $3x + y = -3$ $\qquad$ (4)
$\qquad$ Subtract (4) from (3):
$\qquad$ $-2x = 4$
$\qquad\;\;$ $x = -2$
$\qquad$ Substitute $x = -2$ in (3) to get
$\qquad$ $-2 + y = 1$
$\qquad\qquad\;$ $y = 3$
$\qquad$ Solution: $x = -2$, $y = 3$

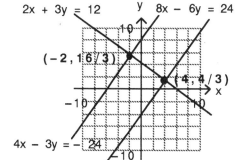

$3x + y = -3$ $\quad$ y
$x - 2y = -8$
$(-2, 3)$
x
$x + y = 1$

$\quad$ (C) It follows from (A) and (B)
$\qquad$ that L_2 and L_3 intersect at
$\qquad$ $x = -2$, $y = 3$.

35. $4x - 3y = -24$ $\qquad$ (L_1)
$\quad\;\;$ $2x + 3y = 12$ $\qquad$ (L_2)
$\quad\;\;$ $8x - 6y = 24$ $\qquad$ (L_3)

$\quad$ (A) L_1 and L_2 intersect:
$\qquad$ $4x - 3y = -24$ $\qquad$ (1)
$\qquad$ $2x + 3y = 12$ $\qquad$ (2)
$\qquad$ Add (1) and (2):
$\qquad$ $6x = -12$
$\qquad\;\;$ $x = -2$
$\qquad$ Substitute $x = -2$ in (2) to get
$\qquad$ $2(-2) + 3y = 12$
$\qquad\qquad\quad\; 3y = 16$
$\qquad\qquad\qquad\; y = \dfrac{16}{3}$

$\qquad$ Solution: $x = -2$, $y = \dfrac{16}{3}$

$2x + 3y = 12$ $\quad$ y $\quad$ $8x - 6y = 24$
$(-2, 16/3)$
$(4, 4/3)$
x
$4x - 3y = -24$

$\quad$ (B) In slope-intercept form, (L_1) and (L_3) have equations

$\qquad$ $y = \dfrac{4}{3}x + 8$ $\qquad$ (L_1)

$\qquad$ $y = \dfrac{4}{3}x - 4$ $\qquad$ (L_2)

$\qquad$ Thus, (L_1) and (L_3) have the same slope and different y-intercepts;
$\qquad$ (L_1) and (L_3) are parallel; they do not intersect.

(C) L_2 and L_3 intersect:

$2x + 3y = 12$ (3)
$8x - 6y = 24$ (4)

Multiply (3) by 2 and add to (4):

$12x = 48$
  $x = 4$

Substitute $x = 4$ in (3) to get

$2(4) + 3y = 12$
        $3y = 4$
          $y = \dfrac{4}{3}$

Solution: $x = 4$, $y = \dfrac{4}{3}$

37. (A) $p = 0.7q + 3$ (1)
      $p = -1.7q + 15$ (2)

Solve the above system for equilibrium price p and the equilibrium quantity q.

     $0.7q + 3 = -1.7q + 15$
$0.7q + 1.7q = 15 - 3$
       $2.4q = 12$
           $q = \dfrac{12}{2.4}$
           $q = 5$ (5 hundreds or 500)

Equilibrium quantity $q = 5$.
Substitute $q = 5$ in (1):

$p = 0.7(5) + 3$
$p = 3.5 + 3$
$p = 6.50$

Equilibrium price $p = \$6.50$

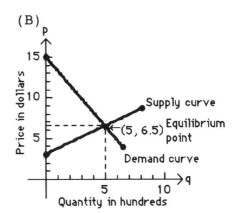

(B)

39. (A) $p = aq + b$, where a and b are to be determined. Now $q = 450$ when $p = 0.6$, and $q = 600$ when $p = 0.75$. This leads to the pair of equations

$450a + b = 0.6$ (1)
$600a + b = 0.75$ (2)

Subtracting (1) from (2), we have

$150a = 0.15$
    $a = 0.001$

Substituting $a = 0.001$ in (2), we get

$600(0.001) + b = 0.75$
            $b = 0.15$

Thus $p = 0.001q + 0.15$ <u>Supply equation</u>

(B) $p = aq + b$; $q = 570$ when $p = 0.6$ and $q = 495$ when $p = 0.75$. This leads to the pair of equations

$570a + b = 0.6$ (3)
$495a + b = 0.75$ (4)

Subtracting (3) from (4), we have
-75a = 0.15
  a = -0.002
Substituting $a = -0.002$ in (3), we get
570(-0.002) + b = 0.6
                  b = 1.74

Thus $p = -0.002q + 1.74$ <u>Demand equation</u>

(C) Equilibrium occurs when supply equals demand. Equating the supply and demand equations yields
0.001q + 0.15 = -0.002q + 1.74
       0.003q = 1.59
            q = 530   <u>Equilibrium quantity</u>

Substituting $q = 530$ into the supply equation (or into the demand equation), we get
  p = 0.001(530) + 0.15
  p = 0.68 or $0.68   <u>Equilibrium price</u>

(D)

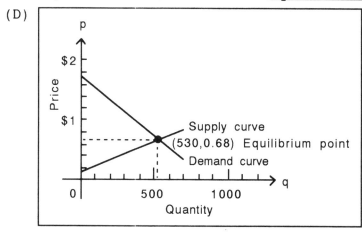

41. (A) The company breaks even when:
             Cost = Revenue
48,000 + 1400x = 1800x
     48,000 = 1800x - 1400x
  or   400x = 48,000
             $x = \dfrac{48,000}{400}$
             x = 120

Thus, 120 units must be manufactured and sold to break even.
Cost = 48,000 + 1400(120)
     = $216,000 = Revenue

(B)

43. Let x = number of tapes marketed per month
(A) Revenue: $R = 19.95x$
Cost: $C = 7.45x + 24,000$
At the break-even point Revenue = Cost, that is,
19.95x = 7.45x + 24,000
12.50x = 24,000
    x = 1920

Thus, 1920 tapes must be sold per month to break even.
Cost = Revenue = $38,304 at the break-even point.

(B)

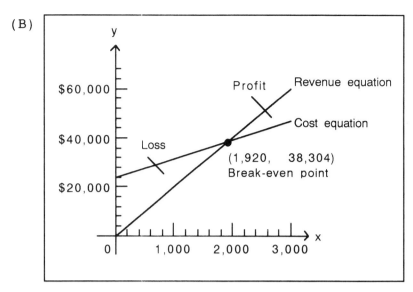

45. Let x = amount of mix A, and
 y = amount of mix B.
We want to solve the following system of equations:
$0.1x + 0.2y = 20$ (1)
$0.06x + 0.02y = 6$ (2)
Clear the decimals from (1) and (2) by multiplying both sides of (1) by 10 and both sides of (2) by 100.
 $x + 2y = 200$ (3)
$6x + 2y = 600$ (4)
Multiply (3) by -1 and add to (4):
$5x = 400$
 $x = 80$
Now substitute $x = 80$ into (3):
$80 + 2y = 200$
     $2y = 120$
      $y = 60$
Solution: x = mix A = 80 grams; y = mix B = 60 grams

47. $p = -\frac{1}{5}d + 70$ [Approach equation]

$p = -\frac{4}{3}d + 230$ [Avoidance equation]

(A) The figure shows the graphs of the two equations.

(B) Setting the two equations equal to each other, we have

$$-\frac{1}{5}d + 70 = -\frac{4}{3}d + 230$$

$$-\frac{1}{5}d + \frac{4}{3}d = 230 - 70$$

$$\frac{17}{15}d = 160$$

$$d = 141 \text{ cm (approx.)}$$

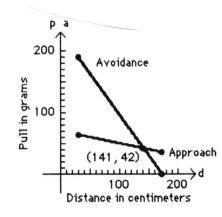

(C) The rat would be very confused (!); it would vacillate.

Things to remember:

1. A system of linear equations is transformed into an equivalent system if:

   (a) two equations are interchanged;
   (b) an equation is multiplied by a nonzero constant;
   (c) a constant multiple of one equation is added to another equation.

2. Associated with the linear system

   $$\begin{aligned} a_1 x_1 + b_1 x_2 &= k_1 \\ a_2 x_1 + b_2 x_2 &= k_2 \end{aligned} \qquad\qquad\qquad (I)$$

   is the AUGMENTED MATRIX of the system

   $$\left[\begin{array}{cc|c} a_1 & b_1 & k_1 \\ a_2 & b_2 & k_2 \end{array}\right]. \qquad\qquad\qquad (II)$$

3. An augmented matrix is transformed into a row-equivalent matrix if:

   (a) two rows are interchanged $(R_i \leftrightarrow R_j)$;
   (b) a row is multiplied by a nonzero constant $(kR_i \rightarrow R_i)$;
   (c) a constant multiple of one row is added to another row $(R_i + kR_j \rightarrow R_i)$.

   (Note: The arrow $\rightarrow$ means "replaces.")

4. Given the system of linear equations (I) and its associated augmented matrix (II). If (II) is row equivalent to a matrix of the form:

   (1) $\left[\begin{array}{cc|c} 1 & 0 & m \\ 0 & 1 & n \end{array}\right]$, then (I) has a unique solution; (consistent and independent);

   (2) $\left[\begin{array}{cc|c} 1 & m & n \\ 0 & 0 & 0 \end{array}\right]$, then (I) has infinitely many solutions (consistent and dependent);

   (3) $\left[\begin{array}{cc|c} 1 & m & n \\ 0 & 0 & p \end{array}\right]$, $p \neq 0$, then (I) has no solution (inconsistent).

1. Interchange row 1 and row 2.

$$\begin{bmatrix} 4 & -6 & | & -8 \\ 1 & -3 & | & 2 \end{bmatrix}$$

3. Multiply row 1 by -4.

$$\begin{bmatrix} -4 & 12 & | & -8 \\ 4 & -6 & | & -8 \end{bmatrix}$$

5. Multiply row 2 by 2.

$$\begin{bmatrix} 1 & -3 & | & 2 \\ 8 & -12 & | & -16 \end{bmatrix}$$

7. Replace row 2 by the sum of row 2 and -4 times row 1.

$$\begin{bmatrix} 1 & -3 & | & 2 \\ 0 & 6 & | & -16 \end{bmatrix}$$

9. Replace row 2 by the sum of row 2 and -2 times row 1.

$$\begin{bmatrix} 1 & -3 & | & 2 \\ 2 & 0 & | & -12 \end{bmatrix}$$

11. Replace row 2 by the sum of row 2 and -1 times row 1

$$\begin{bmatrix} 1 & -3 & | & 2 \\ 3 & -3 & | & -10 \end{bmatrix}$$

13. The corresponding augmented matrix is:

$$\begin{bmatrix} 1 & 1 & | & 5 \\ 1 & -1 & | & 1 \end{bmatrix} \sim \begin{bmatrix} 1 & 1 & | & 5 \\ 0 & -2 & | & -4 \end{bmatrix} \sim \begin{bmatrix} 1 & 1 & | & 5 \\ 0 & 1 & | & 2 \end{bmatrix} \sim \begin{bmatrix} 1 & 0 & | & 3 \\ 0 & 1 & | & 2 \end{bmatrix}$$ Thus, $x_1 = 3$ and $x_2 = 2$.

$$R_2 + (-1)R_1 \rightarrow R_2 \qquad -\frac{1}{2}R_2 \rightarrow R_2 \qquad R_1 + (-1)R_2 \rightarrow R_1$$

15. $$\begin{bmatrix} 1 & -2 & | & 1 \\ 2 & -1 & | & 5 \end{bmatrix} \sim \begin{bmatrix} 1 & -2 & | & 1 \\ 0 & 3 & | & 3 \end{bmatrix} \sim \begin{bmatrix} 1 & -2 & | & 1 \\ 0 & 1 & | & 1 \end{bmatrix} \sim \begin{bmatrix} 1 & 0 & | & 3 \\ 0 & 1 & | & 1 \end{bmatrix}$$ Thus, $x_1 = 3$ and $x_2 = 1$.

$$R_2 + (-2)R_1 \rightarrow R_2 \qquad \frac{1}{3}R_2 \rightarrow R_2 \qquad R_1 + 2R_2 \rightarrow R_1$$

17. $$\begin{bmatrix} 1 & -4 & | & -2 \\ -2 & 1 & | & -3 \end{bmatrix} \sim \begin{bmatrix} 1 & -4 & | & -2 \\ 0 & -7 & | & -7 \end{bmatrix} \sim \begin{bmatrix} 1 & -4 & | & -2 \\ 0 & 1 & | & 1 \end{bmatrix} \sim \begin{bmatrix} 1 & 0 & | & 2 \\ 0 & 1 & | & 1 \end{bmatrix}$$ Thus, $x_1 = 2$ and $x_2 = 1$.

$$R_2 + 2R_1 \rightarrow R_2 \qquad -\frac{1}{7}R_2 \rightarrow R_2 \qquad R_1 + 4R_2 \rightarrow R_1$$

19. $$\begin{bmatrix} 3 & -1 & | & 2 \\ 1 & 2 & | & 10 \end{bmatrix} \sim \begin{bmatrix} 1 & 2 & | & 10 \\ 3 & -1 & | & 2 \end{bmatrix} \sim \begin{bmatrix} 1 & 2 & | & 10 \\ 0 & -7 & | & -28 \end{bmatrix} \sim \begin{bmatrix} 1 & 2 & | & 10 \\ 0 & 1 & | & 4 \end{bmatrix} \sim \begin{bmatrix} 1 & 0 & | & 2 \\ 0 & 1 & | & 4 \end{bmatrix}$$

$$R_1 \leftrightarrow R_2 \qquad R_2 + (-3)R_1 \rightarrow R_2 \qquad -\frac{1}{7}R_2 \rightarrow R_2 \qquad R_1 + (-2)R_2 \rightarrow R_1$$

Thus, $x_1 = 2$ and $x_2 = 4$.

21. $$\begin{bmatrix} 1 & 2 & | & 4 \\ 2 & 4 & | & -8 \end{bmatrix} \sim \begin{bmatrix} 1 & 2 & | & 4 \\ 0 & 0 & | & -16 \end{bmatrix}$$ From **4**, Form (3), the system is inconsistent; there is no solution.

$$R_2 + (-2)R_1 \rightarrow R_2$$

23. $$\begin{bmatrix} 2 & 1 & | & 6 \\ 1 & -1 & | & -3 \end{bmatrix} \sim \begin{bmatrix} 1 & -1 & | & -3 \\ 2 & 1 & | & 6 \end{bmatrix} \sim \begin{bmatrix} 1 & -1 & | & -3 \\ 0 & 3 & | & 12 \end{bmatrix} \sim \begin{bmatrix} 1 & -1 & | & -3 \\ 0 & 1 & | & 4 \end{bmatrix} \sim \begin{bmatrix} 1 & 0 & | & 1 \\ 0 & 1 & | & 4 \end{bmatrix}$$

$$R_1 \leftrightarrow R_2 \qquad R_2 + (-2)R_1 \rightarrow R_2 \qquad \frac{1}{3}R_2 \rightarrow R_2 \qquad R_1 + R_2 \rightarrow R_1$$

Thus, $x_1 = 1$,and $x_2 = 4$.

25. $\begin{bmatrix} 3 & -6 & | & -9 \\ -2 & 4 & | & 6 \end{bmatrix} \sim \begin{bmatrix} 1 & -2 & | & -3 \\ -2 & 4 & | & 6 \end{bmatrix} \sim \begin{bmatrix} 1 & -2 & | & -3 \\ 0 & 0 & | & 0 \end{bmatrix}$

$\quad\quad \frac{1}{3}R_1 \rightarrow R_1 \quad\quad\quad R_2 + 2R_1 \rightarrow R_2$

From $\underline{4}$, Form (2), the system has infinitely many solutions (consistent and dependent). If $x_2 = s$, then $x_1 - 2s = -3$ or $x_1 = 2s - 3$.
Thus, $x_2 = s$, $x_1 = 2s - 3$, for any real number s, are the solutions.

27. $\begin{bmatrix} 4 & -2 & | & 2 \\ -6 & 3 & | & -3 \end{bmatrix} \sim \begin{bmatrix} 1 & -\frac{1}{2} & | & \frac{1}{2} \\ -6 & 3 & | & -3 \end{bmatrix} \sim \begin{bmatrix} 1 & -\frac{1}{2} & | & \frac{1}{2} \\ 0 & 0 & | & 0 \end{bmatrix}$

$\quad\quad \frac{1}{4}R_1 \rightarrow R_1 \quad\quad\quad R_2 + 6R_1 \rightarrow R_2$

Thus, the system has infinitely many solutions (consistent and dependent). Let $x_2 = s$. Then

$$x_1 - \frac{1}{2}s = \frac{1}{2} \quad \text{or} \quad x_1 = \frac{1}{2}s + \frac{1}{2}.$$

The set of solutions is $x_2 = s$, $x_1 = \frac{1}{2}s + \frac{1}{2}$ for any real number s.

29. $\begin{bmatrix} 2 & 1 & | & 1 \\ 4 & -1 & | & -7 \end{bmatrix} \sim \begin{bmatrix} 1 & \frac{1}{2} & | & \frac{1}{2} \\ 4 & -1 & | & -7 \end{bmatrix} \sim \begin{bmatrix} 1 & \frac{1}{2} & | & \frac{1}{2} \\ 0 & -3 & | & -9 \end{bmatrix} \sim \begin{bmatrix} 1 & \frac{1}{2} & | & \frac{1}{2} \\ 0 & 1 & | & 3 \end{bmatrix} \sim \begin{bmatrix} 1 & 0 & | & -1 \\ 0 & 1 & | & 3 \end{bmatrix}$

$\quad\quad \frac{1}{2}R_1 \rightarrow R_1 \quad\quad R_2 + (-4)R_1 \rightarrow R_2 \quad\quad -\frac{1}{3}R_2 \rightarrow R_2 \quad\quad R_1 + \left(-\frac{1}{2}\right)R_2 \rightarrow R_1$

Thus, $x_1 = -1$ and $x_2 = 3$.

31. $\begin{bmatrix} 4 & -6 & | & 8 \\ -6 & 9 & | & -10 \end{bmatrix} \sim \begin{bmatrix} 1 & -\frac{3}{2} & | & 2 \\ -6 & 9 & | & -10 \end{bmatrix} \sim \begin{bmatrix} 1 & -\frac{3}{2} & | & 2 \\ 0 & 0 & | & 2 \end{bmatrix}$

$\quad\quad \frac{1}{4}R_1 \rightarrow R_1 \quad\quad\quad R_2 + 6R_1 \rightarrow R_2$

The second row of the final augmented matrix corresponds to the equation
$0x_1 + 0x_2 = 2$
which has no solution. Thus, the system has no solution; it is inconsistent.

33. $\begin{bmatrix} -4 & 6 & | & -8 \\ 6 & -9 & | & 12 \end{bmatrix} \sim \begin{bmatrix} 1 & -\frac{3}{2} & | & 2 \\ 6 & -9 & | & 12 \end{bmatrix} \sim \begin{bmatrix} 1 & -\frac{3}{2} & | & 2 \\ 0 & 0 & | & 0 \end{bmatrix}$

$\quad\quad -\frac{1}{4}R_1 \rightarrow R_1 \quad\quad\quad R_2 + (-6)R_1 \rightarrow R_2$

The system has infinitely many solutions (consistent and dependent).
If $x_2 = t$, then

$$x_1 - \frac{3}{2}t = 2 \quad \text{or} \quad x_1 = \frac{3}{2}t + 2$$

Thus, the set of solutions is

$$x_2 = t, \quad x_1 = \frac{3}{2}t + 2$$

for any real number t.

35. $\begin{bmatrix} 3 & -1 & | & 7 \\ 2 & 3 & | & 1 \end{bmatrix} \sim \begin{bmatrix} 1 & -\frac{1}{3} & | & \frac{7}{3} \\ 2 & 3 & | & 1 \end{bmatrix} \sim \begin{bmatrix} 1 & -\frac{1}{3} & | & \frac{7}{3} \\ 0 & \frac{11}{3} & | & -\frac{11}{3} \end{bmatrix} \sim \begin{bmatrix} 1 & -\frac{1}{3} & | & \frac{7}{3} \\ 0 & 1 & | & -1 \end{bmatrix} \sim \begin{bmatrix} 1 & 0 & | & 2 \\ 0 & 1 & | & -1 \end{bmatrix}$

$\quad\quad \frac{1}{3}R_1 \to R_1 \quad\quad R_2 + (-2)R_1 \to R_2 \quad\quad \frac{3}{11}R_2 \to R_2 \quad\quad R_1 + \frac{1}{3}R_2 \to R_1 \quad$ Thus, $x_1 = 2$ and $x_2 = -1.$

37. $\begin{bmatrix} 3 & 2 & | & 4 \\ 2 & -1 & | & 5 \end{bmatrix} \sim \begin{bmatrix} 1 & \frac{2}{3} & | & \frac{4}{3} \\ 2 & -1 & | & 5 \end{bmatrix} \sim \begin{bmatrix} 1 & \frac{2}{3} & | & \frac{4}{3} \\ 0 & -\frac{7}{3} & | & \frac{7}{3} \end{bmatrix} \sim \begin{bmatrix} 1 & \frac{2}{3} & | & \frac{4}{3} \\ 0 & 1 & | & -1 \end{bmatrix} \sim \begin{bmatrix} 1 & 0 & | & 2 \\ 0 & 1 & | & -1 \end{bmatrix}$

$\quad\quad \frac{1}{3}R_1 \to R_1 \quad\quad R_2 + (-2)R_1 \to R_2 \quad\quad -\frac{3}{7}R_2 \to R_2 \quad\quad R_1 + \left(-\frac{2}{3}\right)R_2 \to R_1 \quad$ Thus, $x_1 = 2$ and $x_2 = -1.$

39. $\begin{bmatrix} 0.2 & -0.5 & | & 0.07 \\ 0.8 & -0.3 & | & 0.79 \end{bmatrix} \sim \begin{bmatrix} 1 & -2.5 & | & 0.35 \\ 0.8 & -0.3 & | & 0.79 \end{bmatrix} \sim \begin{bmatrix} 1 & -2.5 & | & 0.35 \\ 0 & 1.7 & | & 0.51 \end{bmatrix}$

$\quad\quad\quad \frac{1}{0.2}R_1 \to R_1 \quad\quad\quad R_2 + (-0.8)R_1 \to R_2 \quad\quad\quad \frac{1}{1.7}R_2 \to R_2$

$\sim \begin{bmatrix} 1 & -2.5 & | & 0.35 \\ 0 & 1 & | & 0.3 \end{bmatrix} \sim \begin{bmatrix} 1 & 0 & | & 1.1 \\ 0 & 1 & | & 0.3 \end{bmatrix}$ Thus, $x_1 = 1.1$ and $x_2 = 0.3.$

$\quad\quad R_1 + 2.5R_2 \to R_1$

EXERCISE 5-3

Things to remember:

1. A matrix is in REDUCED FORM if

   (a) each row consisting entirely of zeros is below any row having at least one nonzero element;
   (b) the left-most nonzero element in each row is 1;
   (c) all other elements in the column containing the left-most 1 of a given row are zeros;
   (d) the left-most 1 in any row is to the right of the left-most 1 in any row above.

2. GAUSS-JORDAN ELIMINATION

   (a) Choose the left-most nonzero column and use appropriate row operations to get a 1 at the top.
   (b) Use multiples of the first row to get zeros in all places below the 1 obtained in part (a).
   (c) Delete (mentally) the top row and first nonzero column of the matrix. Repeat parts (a)—(c) with the submatrix (the matrix remaining after deleting the top row and first nonzero column). Continue the process given above until it is not possible to go further.

(d) Now consider the whole matrix. Begin with the bottom nonzero row and use appropriate multiples of it to get zeros above the left-most 1. Continue this process, moving up row by row until the matrix is finally in reduced form.

Note: If at any point in this process we obtain a row with all zeros to the left of the vertical line and a nonzero number n to the right, then we can stop, since we will have a contradiction: $0 = n$, $n \neq 0$. We can conclude that the system has no solution.

1. $\left[\begin{array}{cc|c} 1 & 0 & 2 \\ 0 & 1 & -1 \end{array}\right]$
Is in reduced form. Use 1.

3. $\left[\begin{array}{ccc|c} 1 & 0 & 2 & 3 \\ 0 & 0 & 0 & 0 \\ 0 & 1 & -1 & 4 \end{array}\right]$
Is not in reduced form. Condition (a) has been violated. The second row should be at the bottom.

5. $\left[\begin{array}{ccc|c} 0 & 1 & 0 & 2 \\ 0 & 0 & 3 & -1 \\ 0 & 0 & 0 & 0 \end{array}\right]$
Is not in reduced form. Condition (b) has been violated. The left-most nonzero element in the second row should be 1, not 3.

7. $\left[\begin{array}{cccc|c} 1 & 2 & 0 & 3 & 2 \\ 0 & 0 & 1 & -1 & 0 \end{array}\right]$
Is in reduced form.

9. $x_1 \qquad = -2$
$\qquad x_2 \quad = 3$
$\qquad\qquad x_3 = 0$

11. $x_1 \qquad - 2x_3 = 3 \quad (1)$
$\qquad x_2 + x_3 = -5 \quad (2)$
Let $x_3 = t$. From (2), $x_2 = -5 - t$. From (1), $x_1 = 3 + 2t$. Thus, the solution is
$x_1 = 2t + 3$
$x_2 = -t - 5$
$x_3 = t$
t any real number.

13. $x_1 \quad = 0$
$\quad x_2 = 0$
$\quad 0 = 1$

Inconsistent; no solution.

15. $x_1 - 2x_2 \qquad - 3x_4 = -5$
$\qquad\qquad x_3 + 3x_4 = 2$
Let $x_2 = s$ and $x_4 = t$. Then
$x_1 = 2s + 3t - 5$
$x_2 = s$
$x_3 = -3t + 2$
$x_4 = t$
s and t any real numbers.

17. $\begin{bmatrix} 1 & 2 & | & -1 \\ 0 & 1 & | & 3 \end{bmatrix} \sim \begin{bmatrix} 1 & 0 & | & -7 \\ 0 & 1 & | & 3 \end{bmatrix}$

$R_1 + (-2)R_2 \rightarrow R_1$

19. $\begin{bmatrix} 1 & 0 & -3 & | & 1 \\ 0 & 1 & 2 & | & 0 \\ 0 & 0 & 3 & | & -6 \end{bmatrix} \sim \begin{bmatrix} 1 & 0 & -3 & | & 1 \\ 0 & 1 & 2 & | & 0 \\ 0 & 0 & 1 & | & -2 \end{bmatrix} \sim \begin{bmatrix} 1 & 0 & 0 & | & -5 \\ 0 & 1 & 0 & | & 4 \\ 0 & 0 & 1 & | & -2 \end{bmatrix}$

$\frac{1}{3}R_3 \rightarrow R_3$ $\qquad\qquad R_1 + 3R_3 \rightarrow R_1$
$\qquad\qquad\qquad\qquad R_2 + (-2)R_3 \rightarrow R_2$

21. $\begin{bmatrix} 1 & 2 & -2 & | & -1 \\ 0 & 3 & -6 & | & 1 \\ 0 & -1 & 2 & | & -\frac{1}{3} \end{bmatrix} \sim \begin{bmatrix} 1 & 2 & -2 & | & -1 \\ 0 & 1 & -2 & | & \frac{1}{3} \\ 0 & -1 & 2 & | & -\frac{1}{3} \end{bmatrix} \sim \begin{bmatrix} 1 & 2 & -2 & | & -1 \\ 0 & 1 & -2 & | & \frac{1}{3} \\ 0 & 0 & 0 & | & 0 \end{bmatrix} \sim \begin{bmatrix} 1 & 0 & 2 & | & -\frac{5}{3} \\ 0 & 1 & -2 & | & \frac{1}{3} \\ 0 & 0 & 0 & | & 0 \end{bmatrix}$

$\frac{1}{3}R_2 \rightarrow R_2$ $\qquad\qquad R_3 + R_2 \rightarrow R_3$ $\qquad\qquad R_1 + (-2)R_2 \rightarrow R_1$

23. The corresponding augmented matrix is:

$\begin{bmatrix} 2 & 4 & -10 & | & -2 \\ 3 & 9 & -21 & | & 0 \\ 1 & 5 & -12 & | & 1 \end{bmatrix} \sim \begin{bmatrix} 1 & 2 & -5 & | & -1 \\ 3 & 9 & -21 & | & 0 \\ 1 & 5 & -12 & | & 1 \end{bmatrix} \sim \begin{bmatrix} 1 & 2 & -5 & | & -1 \\ 0 & 3 & -6 & | & 3 \\ 0 & 3 & -7 & | & 2 \end{bmatrix} \sim \begin{bmatrix} 1 & 2 & -5 & | & -1 \\ 0 & 1 & -2 & | & 1 \\ 0 & 3 & -7 & | & 2 \end{bmatrix}$

$\frac{1}{2}R_1 \rightarrow R_1$ $\qquad\qquad R_2 + (-3)R_1 \rightarrow R_2$ $\qquad\qquad \frac{1}{3}R_2 \rightarrow R_2$ $\qquad\qquad R_3 + (-3)R_2 \rightarrow R_3$
$\qquad\qquad\qquad\qquad R_3 + (-1)R_1 \rightarrow R_3$

$\sim \begin{bmatrix} 1 & 2 & -5 & | & -1 \\ 0 & 1 & -2 & | & 1 \\ 0 & 0 & -1 & | & -1 \end{bmatrix} \sim \begin{bmatrix} 1 & 2 & -5 & | & -1 \\ 0 & 1 & -2 & | & 1 \\ 0 & 0 & 1 & | & 1 \end{bmatrix} \sim \begin{bmatrix} 1 & 2 & 0 & | & 4 \\ 0 & 1 & 0 & | & 3 \\ 0 & 0 & 1 & | & 1 \end{bmatrix} \sim \begin{bmatrix} 1 & 0 & 0 & | & -2 \\ 0 & 1 & 0 & | & 3 \\ 0 & 0 & 1 & | & 1 \end{bmatrix}$

$(-1)R_3 \rightarrow R_3$ $\qquad\qquad R_2 + 2R_3 \rightarrow R_2$ $\qquad\qquad R_1 + (-2)R_2 \rightarrow R_1$
$\qquad\qquad\qquad\qquad R_1 + 5R_3 \rightarrow R_1$

Thus, $x_1 = -2$; $x_2 = 3$; $x_3 = 1$.

25. The corresponding augmented matrix is:

$\begin{bmatrix} 3 & 8 & -1 & | & -18 \\ 2 & 1 & 5 & | & 8 \\ 2 & 4 & 2 & | & -4 \end{bmatrix} \sim \begin{bmatrix} 2 & 4 & 2 & | & -4 \\ 2 & 1 & 5 & | & 8 \\ 3 & 8 & -1 & | & -18 \end{bmatrix} \sim \begin{bmatrix} 1 & 2 & 1 & | & -2 \\ 2 & 1 & 5 & | & 8 \\ 3 & 8 & -1 & | & -18 \end{bmatrix}$

$R_1 \leftrightarrow R_3$ $\qquad\qquad \frac{1}{2}R_1 \rightarrow R_1$ $\qquad\qquad R_2 + (-2)R_1 \rightarrow R_2$
$\qquad\qquad\qquad\qquad R_3 + (-3)R_1 \rightarrow R_3$

$\sim \begin{bmatrix} 1 & 2 & 1 & | & -2 \\ 0 & -3 & 3 & | & 12 \\ 0 & 2 & -4 & | & -12 \end{bmatrix} \sim \begin{bmatrix} 1 & 2 & 1 & | & -2 \\ 0 & 1 & -1 & | & -4 \\ 0 & 2 & -4 & | & -12 \end{bmatrix} \sim \begin{bmatrix} 1 & 2 & 1 & | & -2 \\ 0 & 1 & -1 & | & -4 \\ 0 & 0 & -2 & | & -4 \end{bmatrix}$

$-\frac{1}{3}R_2 \rightarrow R_2$ $\qquad\qquad R_3 + (-2)R_2 \rightarrow R_3$ $\qquad\qquad -\frac{1}{2}R_3 \rightarrow R_3$

$$\sim \begin{bmatrix} 1 & 2 & 1 & | & -2 \\ 0 & 1 & -1 & | & -4 \\ 0 & 0 & 1 & | & 2 \end{bmatrix} \sim \begin{bmatrix} 1 & 2 & 0 & | & -4 \\ 0 & 1 & 0 & | & -2 \\ 0 & 0 & 1 & | & 2 \end{bmatrix} \sim \begin{bmatrix} 1 & 0 & 0 & | & 0 \\ 0 & 1 & 0 & | & -2 \\ 0 & 0 & 1 & | & 2 \end{bmatrix}$$

Thus, $x_1 = 0$
$x_2 = -2$
$x_3 = 2.$

$R_2 + R_3 \rightarrow R_2$ $\qquad R_1 + (-2)R_2 \rightarrow R_1$

$R_1 + (-1)R_3 \rightarrow R_1$

27. $\begin{bmatrix} 2 & -1 & -3 & | & 8 \\ 1 & -2 & 0 & | & 7 \end{bmatrix} \sim \begin{bmatrix} 1 & -2 & 0 & | & 7 \\ 2 & -1 & -3 & | & 8 \end{bmatrix} \sim \begin{bmatrix} 1 & -2 & 0 & | & 7 \\ 0 & 3 & -3 & | & -6 \end{bmatrix} \sim \begin{bmatrix} 1 & -2 & 0 & | & 7 \\ 0 & 1 & -1 & | & -2 \end{bmatrix}$

$\qquad R_1 \leftrightarrow R_2 \qquad R_2 + (-2)R_1 \rightarrow R_2 \qquad \frac{1}{3}R_2 \rightarrow R_2 \qquad R_1 + 2R_2 \rightarrow R_1$

$\sim \begin{bmatrix} 1 & 0 & -2 & | & 3 \\ 0 & 1 & -1 & | & -2 \end{bmatrix}$

Thus, $x_1 \quad - 2x_3 = 3 \quad (1)$
$x_2 - x_3 = -2 \quad (2)$

Let $x_3 = t$, where t is any real number. Then:
$x_1 = 2t + 3$
$x_2 = t - 2$
$x_3 = t$

29. $\begin{bmatrix} 2 & 3 & -1 & | & 1 \\ 1 & -2 & 2 & | & -2 \end{bmatrix} \sim \begin{bmatrix} 1 & -2 & 2 & | & -2 \\ 2 & 3 & -1 & | & 1 \end{bmatrix} \sim \begin{bmatrix} 1 & -2 & 2 & | & -2 \\ 0 & 7 & -5 & | & 5 \end{bmatrix} \sim \begin{bmatrix} 1 & -2 & 2 & | & -2 \\ 0 & 1 & -\frac{5}{7} & | & \frac{5}{7} \end{bmatrix}$

$\qquad R_1 \leftrightarrow R_2 \qquad R_2 + (-2)R_1 \rightarrow R_2 \qquad \frac{1}{7}R_2 \rightarrow R_2 \qquad R_1 + 2R_2 \rightarrow R_1$

$\sim \begin{bmatrix} 1 & 0 & \frac{4}{7} & | & -\frac{4}{7} \\ 0 & 1 & -\frac{5}{7} & | & \frac{5}{7} \end{bmatrix}$

Thus, $x_1 \quad + \frac{4}{7}x_3 = -\frac{4}{7} \quad (1)$
$x_2 - \frac{5}{7}x_3 = \frac{5}{7} \quad (2)$

Let $x_3 = t$, where t is any real number. Then:
$x_1 = \dfrac{(-4t - 4)}{7}$
$x_2 = \dfrac{(5t + 5)}{7}$
$x_3 = t$

31. $\begin{bmatrix} 2 & 2 & | & 2 \\ 1 & 2 & | & 3 \\ 0 & -3 & | & -6 \end{bmatrix} \sim \begin{bmatrix} 1 & 2 & | & 3 \\ 2 & 2 & | & 2 \\ 0 & -3 & | & -6 \end{bmatrix} \sim \begin{bmatrix} 1 & 2 & | & 3 \\ 0 & -2 & | & -4 \\ 0 & -3 & | & -6 \end{bmatrix} \sim \begin{bmatrix} 1 & 2 & | & 3 \\ 0 & 1 & | & 2 \\ 0 & -3 & | & -6 \end{bmatrix}$

$\qquad R_1 \leftrightarrow R_2 \qquad R_2 + (-2)R_1 \rightarrow R_2 \qquad -\frac{1}{2}R_2 \rightarrow R_2 \qquad R_3 + 3R_2 \rightarrow R_3$

$\sim \begin{bmatrix} 1 & 2 & | & 3 \\ 0 & 1 & | & 2 \\ 0 & 0 & | & 0 \end{bmatrix} \sim \begin{bmatrix} 1 & 0 & | & -1 \\ 0 & 1 & | & 2 \\ 0 & 0 & | & 0 \end{bmatrix}$

Thus, $x_1 = -1$
$x_2 = 2.$

$R_1 + (-2)R_2 \rightarrow R_1$

33. $\begin{bmatrix} 2 & -1 & | & 0 \\ 3 & 2 & | & 7 \\ 1 & -1 & | & -2 \end{bmatrix} \sim \begin{bmatrix} 1 & -1 & | & -2 \\ 3 & 2 & | & 7 \\ 2 & -1 & | & 0 \end{bmatrix} \sim \begin{bmatrix} 1 & -1 & | & -2 \\ 0 & 5 & | & 13 \\ 0 & 1 & | & 4 \end{bmatrix} \sim \begin{bmatrix} 1 & -1 & | & -2 \\ 0 & 1 & | & 4 \\ 0 & 5 & | & 13 \end{bmatrix} \sim \begin{bmatrix} 1 & -1 & | & -2 \\ 0 & 1 & | & 4 \\ 0 & 0 & | & -7 \end{bmatrix}$

$\quad R_1 \leftrightarrow R_3 \qquad R_2 + (-3)R_1 \to R_2 \qquad R_2 \leftrightarrow R_3 \qquad R_3 + (-5)R_2 \to R_3$

$\qquad\qquad\qquad\qquad R_3 + (-2)R_1 \to R_3$

From the last row, we conclude that there is no solution; the system is inconsistent.

35. $\begin{bmatrix} 3 & -4 & -1 & | & 1 \\ 2 & -3 & 1 & | & 1 \\ 1 & -2 & 3 & | & 2 \end{bmatrix} \sim \begin{bmatrix} 1 & -2 & 3 & | & 2 \\ 2 & -3 & 1 & | & 1 \\ 3 & -4 & -1 & | & 1 \end{bmatrix} \sim \begin{bmatrix} 1 & -2 & 3 & | & 2 \\ 0 & 1 & -5 & | & -3 \\ 0 & 2 & -10 & | & -5 \end{bmatrix} \sim \begin{bmatrix} 1 & -2 & 3 & | & 2 \\ 0 & 1 & -5 & | & -3 \\ 0 & 0 & 0 & | & 1 \end{bmatrix}$

$\quad R_1 \leftrightarrow R_3 \qquad R_2 + (-2)R_1 \to R_2 \qquad R_3 + (-2)R_2 \to R_3$

$\qquad\qquad\qquad\qquad R_3 + (-3)R_1 \to R_3$

From the last row, we conclude that there is no solution; the system is inconsistent.

37. $\begin{bmatrix} 3 & -2 & 1 & | & -7 \\ 2 & 1 & -4 & | & 0 \\ 1 & 1 & -3 & | & 1 \end{bmatrix} \sim \begin{bmatrix} 1 & 1 & -3 & | & 1 \\ 2 & 1 & -4 & | & 0 \\ 3 & -2 & 1 & | & -7 \end{bmatrix} \sim \begin{bmatrix} 1 & 1 & -3 & | & 1 \\ 0 & -1 & 2 & | & -2 \\ 0 & -5 & 10 & | & -10 \end{bmatrix}$

$\quad R_1 \leftrightarrow R_3 \qquad\qquad R_2 + (-2)R_1 \to R_2 \qquad\qquad (-1)R_2 \to R_2$

$\qquad\qquad\qquad\qquad R_3 + (-3)R_1 \to R_3$

$\begin{bmatrix} 1 & 1 & -3 & | & 1 \\ 0 & 1 & -2 & | & 2 \\ 0 & -5 & 10 & | & -10 \end{bmatrix} \sim \begin{bmatrix} 1 & 1 & -3 & | & 1 \\ 0 & 1 & -2 & | & 2 \\ 0 & 0 & 0 & | & 0 \end{bmatrix} \sim \begin{bmatrix} 1 & 0 & -1 & | & -1 \\ 0 & 1 & -2 & | & 2 \\ 0 & 0 & 0 & | & 0 \end{bmatrix}$

$\quad R_3 + 5R_2 \to R_3 \qquad R_1 + (-1)R_2 \to R_1$

From this matrix, $x_1 - x_3 = -1$ and $x_2 - 2x_3 = 2$. Let $x_3 = t$ be any real number, then $x_1 = t - 1$, $x_2 = 2t + 2$, and $x_3 = t$.

39. $\begin{bmatrix} 2 & 4 & -2 & | & 2 \\ -3 & -6 & 3 & | & -3 \end{bmatrix} \sim \begin{bmatrix} 1 & 2 & -1 & | & 1 \\ -3 & -6 & 3 & | & -3 \end{bmatrix} \sim \begin{bmatrix} 1 & 2 & -1 & | & 1 \\ 0 & 0 & 0 & | & 0 \end{bmatrix}$

$\quad \frac{1}{2}R_1 \to R_1 \qquad\qquad R_2 + 3R_1 \to R_2$

From this matrix, $x_1 + 2x_2 - x_3 = 1$. Let $x_2 = s$ and $x_3 = t$. Then $x_1 = -2s + t + 1$, $x_2 = s$, and $x_3 = t$, s and t any real numbers.

41. $\begin{bmatrix} 2 & -3 & 3 & | & -15 \\ 3 & 2 & -5 & | & 19 \\ 5 & -4 & -2 & | & -2 \end{bmatrix} \sim \begin{bmatrix} 1 & -\frac{3}{2} & \frac{3}{2} & | & -\frac{15}{2} \\ 3 & 2 & -5 & | & 19 \\ 5 & -4 & -2 & | & -2 \end{bmatrix} \sim \begin{bmatrix} 1 & -\frac{3}{2} & \frac{3}{2} & | & -\frac{15}{2} \\ 0 & \frac{13}{2} & -\frac{19}{2} & | & \frac{83}{2} \\ 0 & \frac{7}{2} & -\frac{19}{2} & | & \frac{71}{2} \end{bmatrix} \sim \begin{bmatrix} 1 & -\frac{3}{2} & \frac{3}{2} & | & -\frac{15}{2} \\ 0 & 1 & -\frac{19}{13} & | & \frac{83}{13} \\ 0 & \frac{7}{2} & -\frac{19}{2} & | & \frac{71}{2} \end{bmatrix}$

$\quad \frac{1}{2}R_1 \to R_1 \qquad\qquad R_2 + (-3)R_1 \to R_2 \qquad\qquad \frac{2}{13}R_2 \to R_2 \qquad R_3 + \left(-\frac{7}{2}\right)R_2 \to R_3$

$\qquad\qquad\qquad\qquad R_3 + (-5)R_1 \to R_3$

$$\begin{bmatrix} 1 & -\frac{3}{2} & \frac{3}{2} & | & -\frac{15}{2} \\ 0 & 1 & -\frac{19}{13} & | & \frac{83}{13} \\ 0 & 0 & -\frac{114}{26} & | & \frac{342}{26} \end{bmatrix} \sim \begin{bmatrix} 1 & -\frac{3}{2} & \frac{3}{2} & | & -\frac{15}{2} \\ 0 & 1 & -\frac{19}{13} & | & \frac{83}{13} \\ 0 & 0 & 1 & | & -3 \end{bmatrix} \sim \begin{bmatrix} 1 & -\frac{3}{2} & 0 & | & -3 \\ 0 & 1 & 0 & | & 2 \\ 0 & 0 & 1 & | & -3 \end{bmatrix} \sim \begin{bmatrix} 1 & 0 & 0 & | & 0 \\ 0 & 1 & 0 & | & 2 \\ 0 & 0 & 1 & | & -3 \end{bmatrix}$$

$\dfrac{-26}{114} R_3 \to R_3$ $\qquad$ $R_2 + \dfrac{19}{13} R_3 \to R_2$ $\qquad$ $R_1 + \dfrac{3}{2} R_2 \to R_1$ $\qquad$ Thus, $x_1 = 0,$

$\qquad\qquad\qquad\qquad\qquad\qquad\qquad\qquad\qquad\qquad\qquad\qquad\qquad\qquad$ $x_2 = 2,$

$\qquad\qquad\qquad\qquad\qquad R_1 + \left(-\dfrac{3}{2}\right) R_3 \to R_1$ $\qquad\qquad\qquad\qquad$ and $x_3 = -3.$

43. $\begin{bmatrix} 5 & -3 & 2 & | & 13 \\ 2 & 4 & -3 & | & -9 \\ 4 & -2 & 5 & | & 13 \end{bmatrix} \sim \begin{bmatrix} 2 & 4 & -3 & | & -9 \\ 5 & -3 & 2 & | & 13 \\ 4 & -2 & 5 & | & 13 \end{bmatrix} \sim \begin{bmatrix} 1 & 2 & -\frac{3}{2} & | & -\frac{9}{2} \\ 5 & -3 & 2 & | & 13 \\ 4 & -2 & 5 & | & 13 \end{bmatrix} \sim \begin{bmatrix} 1 & 2 & -\frac{3}{2} & | & -\frac{9}{2} \\ 0 & -13 & \frac{19}{2} & | & \frac{71}{2} \\ 0 & -10 & 11 & | & 31 \end{bmatrix}$

$\qquad R_2 \leftrightarrow R_1$ $\qquad\qquad\qquad$ $\dfrac{1}{2} R_1 \to R_1$ $\qquad\qquad$ $R_2 + (-5) R_1 \to R_2$ $\qquad$ $-\dfrac{1}{13} R_2 \to R_2$

$\qquad\qquad\qquad\qquad\qquad\qquad\qquad\qquad\qquad$ $R_3 + (-4) R_1 \to R_3$

$\sim \begin{bmatrix} 1 & 2 & -\frac{3}{2} & | & -\frac{9}{2} \\ 0 & 1 & -\frac{19}{26} & | & -\frac{71}{26} \\ 0 & -10 & 11 & | & 31 \end{bmatrix} \sim \begin{bmatrix} 1 & 2 & -\frac{3}{2} & | & -\frac{9}{2} \\ 0 & 1 & -\frac{19}{26} & | & -\frac{71}{26} \\ 0 & 0 & \frac{48}{13} & | & \frac{48}{13} \end{bmatrix} \sim \begin{bmatrix} 1 & 2 & -\frac{3}{2} & | & -\frac{9}{2} \\ 0 & 1 & -\frac{19}{26} & | & -\frac{71}{26} \\ 0 & 0 & 1 & | & 1 \end{bmatrix}$

$\qquad R_3 + 10 R_2 \to R_3$ $\qquad\qquad$ $\dfrac{13}{48} R_3 \to R_3$ $\qquad\qquad$ $R_2 + \dfrac{19}{26} R_3 \to R_2$

$\qquad\qquad\qquad\qquad\qquad\qquad\qquad\qquad\qquad\qquad\qquad\qquad\qquad$ $R_1 + \dfrac{3}{2} R_3 \to R_1$

$\qquad\qquad\qquad\qquad\qquad\qquad\qquad\qquad\qquad\qquad\qquad$ Thus, $x_1 = 1,$

$\sim \begin{bmatrix} 1 & 2 & 0 & | & -3 \\ 0 & 1 & 0 & | & -2 \\ 0 & 0 & 1 & | & 1 \end{bmatrix} \sim \begin{bmatrix} 1 & 0 & 0 & | & 1 \\ 0 & 1 & 0 & | & -2 \\ 0 & 0 & 1 & | & 1 \end{bmatrix}$ $\qquad$ $x_2 = -2,$

$\qquad\qquad\qquad\qquad\qquad\qquad\qquad\qquad\qquad\qquad$ and $x_3 = 1.$

$\qquad R_1 + (-2) R_2 \to R_1$

45. $\begin{bmatrix} 1 & 2 & -4 & -1 & | & 7 \\ 2 & 5 & -9 & -4 & | & 16 \\ 1 & 5 & -7 & -7 & | & 13 \end{bmatrix} \sim \begin{bmatrix} 1 & 2 & -4 & -1 & | & 7 \\ 0 & 1 & -1 & -2 & | & 2 \\ 0 & 3 & -3 & -6 & | & 6 \end{bmatrix} \sim \begin{bmatrix} 1 & 2 & -4 & -1 & | & 7 \\ 0 & 1 & -1 & -2 & | & 2 \\ 0 & 0 & 0 & 0 & | & 0 \end{bmatrix}$

$\qquad R_2 + (-2) R_1 \to R_2$ $\qquad\qquad\qquad$ $R_3 + (-3) R_2 \to R_3$ $\qquad\qquad\qquad$ $R_1 + (-2) R_2 \to R_1$

$\qquad R_3 + (-1) R_1 \to R_3$

$\sim \begin{bmatrix} 1 & 0 & -2 & 3 & | & 3 \\ 0 & 1 & -1 & -2 & | & 2 \\ 0 & 0 & 0 & 0 & | & 0 \end{bmatrix}$

Thus, $x_1 - 2x_3 + 3x_4 = 3$ and $x_2 - x_3 - 2x_4 = 2.$ Let $x_3 = s$ and $x_4 = t.$
Then $x_1 = 2s - 3t + 3,$ $x_2 = s + 2t + 2,$ $x_3 = s,$ $x_4 = t,$ where $s,$ t are
any real numbers.

47. Let x_1 = number of one-person boats,
  x_2 = number of two-person boats,
and x_3 = number of four-person boats.

We have the following system of linear equations:
$$0.5x_1 + x_2 + 1.5x_3 = 380$$
$$0.6x_1 + 0.9x_2 + 1.2x_3 = 330$$
$$0.2x_1 + 0.3x_2 + 0.5x_3 = 120$$

$$\begin{bmatrix} 0.5 & 1 & 1.5 & | & 380 \\ 0.6 & 0.9 & 1.2 & | & 330 \\ 0.2 & 0.3 & 0.5 & | & 120 \end{bmatrix} \sim \begin{bmatrix} 1 & 2 & 3 & | & 760 \\ 0.6 & 0.9 & 1.2 & | & 330 \\ 0.2 & 0.3 & 0.5 & | & 120 \end{bmatrix} \sim \begin{bmatrix} 1 & 2 & 3 & | & 760 \\ 0 & -0.3 & -0.6 & | & -126 \\ 0 & -0.1 & -0.1 & | & -32 \end{bmatrix}$$

$\quad 2R_1 \rightarrow R_1 \qquad\qquad\qquad R_2 + (-0.6)R_1 \rightarrow R_2 \qquad\qquad -\dfrac{1}{0.3}R_2 \rightarrow R_2$

$\qquad\qquad\qquad\qquad\qquad\qquad R_3 + (-0.2)R_1 \rightarrow R_3$

$$\sim \begin{bmatrix} 1 & 2 & 3 & | & 760 \\ 0 & 1 & 2 & | & 420 \\ 0 & -0.1 & -0.1 & | & -32 \end{bmatrix} \sim \begin{bmatrix} 1 & 2 & 3 & | & 760 \\ 0 & 1 & 2 & | & 420 \\ 0 & 0 & 0.1 & | & 10 \end{bmatrix} \sim \begin{bmatrix} 1 & 2 & 3 & | & 760 \\ 0 & 1 & 2 & | & 420 \\ 0 & 0 & 1 & | & 100 \end{bmatrix}$$

$\quad R_3 + (0.1)R_2 \rightarrow R_3 \qquad\qquad 10R_3 \rightarrow R_3 \qquad\qquad R_1 + (-3)R_3 \rightarrow R_1$

$\qquad\qquad\qquad\qquad\qquad\qquad\qquad\qquad\qquad\qquad R_2 + (-2)R_3 \rightarrow R_2$

$$\sim \begin{bmatrix} 1 & 2 & 0 & | & 460 \\ 0 & 1 & 0 & | & 220 \\ 0 & 0 & 1 & | & 100 \end{bmatrix} \sim \begin{bmatrix} 1 & 0 & 0 & | & 20 \\ 0 & 1 & 0 & | & 220 \\ 0 & 0 & 1 & | & 100 \end{bmatrix}$$

$\quad R_1 + (-2)R_2 \rightarrow R_1$

Thus, $x_1 = 20$, $x_2 = 220$, and $x_3 = 100$, or 20 one-person boats, 220 two-person boats, and 100 four-person boats.

49. Referring to Problem 47, we now have the following system of equations to solve:
$$0.5x_1 + x_2 + 1.5x_3 = 380$$
$$0.6x_1 + 0.9x_2 + 1.2x_3 = 330$$

$$\begin{bmatrix} 0.5 & 1 & 1.5 & | & 380 \\ 0.6 & 0.9 & 1.2 & | & 330 \end{bmatrix} \sim \begin{bmatrix} 1 & 2 & 3 & | & 760 \\ 0.6 & 0.9 & 1.2 & | & 330 \end{bmatrix} \sim \begin{bmatrix} 1 & 2 & 3 & | & 760 \\ 0 & -0.3 & -0.6 & | & -126 \end{bmatrix}$$

$\quad\quad 2R_1 \rightarrow R_1 \qquad\qquad R_2 + (-0.6)R_1 \rightarrow R_2 \qquad\qquad -\dfrac{1}{0.3}R_2 \rightarrow R_2$

$$\sim \begin{bmatrix} 1 & 2 & 3 & | & 760 \\ 0 & 1 & 2 & | & 420 \end{bmatrix} \sim \begin{bmatrix} 1 & 0 & -1 & | & -80 \\ 0 & 1 & 2 & | & 420 \end{bmatrix}$$

$\quad R_1 + (-2)R_2 \rightarrow R_1$

Thus, $x_1 - x_3 = -80 \qquad$ (1)
$\qquad\qquad x_2 + 2x_3 = 420 \qquad$ (2)
Let $x_3 = t \qquad$ (t any real number)

Then, $x_2 = 420 - 2t$ [from (2)]
$x_1 = t - 80$ [from (1)]

In order to keep x_1 and x_2 positive, $t \leq 210$ and $t \geq 80$.

Thus, $x_1 = t - 80$ (one-person boats)
$x_2 = 420 - 2t$ (two-person boats)
$x_3 = t$ (four-person boats)

where $80 \leq t \leq 210$ and t is an integer.

51. Again referring to Problem 47, we have have the following system:

$0.5x_1 + x_2 = 380$
$0.6x_1 + 0.9x_2 = 330$
$0.2x_1 + 0.3x_2 = 120$

$$\begin{bmatrix} 0.5 & 1 & | & 380 \\ 0.6 & 0.9 & | & 330 \\ 0.2 & 0.3 & | & 120 \end{bmatrix} \sim \begin{bmatrix} 1 & 2 & | & 760 \\ 0.6 & 0.9 & | & 330 \\ 0.2 & 0.3 & | & 120 \end{bmatrix} \sim \begin{bmatrix} 1 & 2 & | & 760 \\ 0 & -0.3 & | & -126 \\ 0 & -0.1 & | & -32 \end{bmatrix} \sim \begin{bmatrix} 1 & 2 & | & 760 \\ 0 & 1 & | & 420 \\ 0 & -0.1 & | & -32 \end{bmatrix}$$

$2R_1 \to R_1 R_2 + (-0.6R_1) \to R_2 -\dfrac{1}{0.3}R_2 \to R_2 R_3 + 0.1R_2 \to R_3$

$R_3 + (-0.2R_1) \to R_3$

$$\sim \begin{bmatrix} 1 & 2 & | & 760 \\ 0 & 1 & | & 420 \\ 0 & 0 & | & 10 \end{bmatrix}$$ From this matrix, we conclude that there is no solution; there is no production schedule that will use all the labor-hours in all departments.

53. Let x_1 = number of 6000 gallon tank cars
x_2 = number of 8000 gallon tank cars
x_3 = number of 18,000 gallon tank cars

Then
$x_1 + x_2 + x_3 = 24$
and $6000x_1 + 8000x_2 + 18,000x_3 = 250,000$

Dividing the second equation by 2000, we get the system
$x_1 + x_2 + x_3 = 24$
$3x_1 + 4x_2 + 9x_3 = 125$

The augmented matrix corresponding to this system is:

$$\begin{pmatrix} 1 & 1 & 1 & | & 24 \\ 3 & 4 & 9 & | & 125 \end{pmatrix} \sim \begin{pmatrix} 1 & 1 & 1 & | & 24 \\ 0 & 1 & 6 & | & 53 \end{pmatrix} \sim \begin{pmatrix} 1 & 0 & -5 & | & -29 \\ 0 & 1 & 6 & | & 53 \end{pmatrix}$$

$R_2 + (-3)R_1 \to R_2 R_1 + (-1)R_2 \to R_1$

Thus $x_1 - 5x_3 = -29$
$x_2 + 6x_3 = 53$

Let $x_3 = t$. Then $x_1 = 5t - 29$ and $x_2 = 53 - 6t$

Thus, $(5t - 29)$ 6000-gallon tank cars, $(53 - 6t)$ 8000-gallon tank cars and (t) 18000-gallon tank cars should be purchased. Also, since t, $5t - 29$ and $53 - 6t$ must each be non-negative integers, it follows that $t = 6, 7$ or 8.

55. Let x_1 = amount of federal income tax (in thousands of dollars),

x_2 = amount of state income tax (in thousands of dollars),

and x_3 = amount of local income tax (in thousands of dollars).

Then,

$x_1 = 0.25(1664 - x_2 - x_3)$

$x_2 = 0.10(1664 - x_1 - x_3)$

$x_3 = 0.05(1664 - x_1 - x_2)$

We rewrite this system in the standard form:

$x_1 + 0.25x_2 + 0.25x_3 = 416$ (1)

$0.10x_1 + x_2 + 0.10x_3 = 166.4$ (2)

$0.05x_1 + 0.05x_2 + x_3 = 83.2$ (3)

Multiply (1) by 100, (2) by 10, and (3) by 100 to obtain:

$100x_1 + 25x_2 + 25x_3 = 41,600$

$x_1 + 10x_2 + x_3 = 1,664$

$5x_1 + 5x_2 + 100x_3 = 8,320$

The augmented matrix corresponding to this system is:

$$\begin{bmatrix} 100 & 25 & 25 & | & 41,600 \\ 1 & 10 & 1 & | & 1,664 \\ 5 & 5 & 100 & | & 8,320 \end{bmatrix} \sim \begin{bmatrix} 1 & 10 & 1 & | & 1,664 \\ 100 & 25 & 25 & | & 41,600 \\ 5 & 5 & 100 & | & 8,320 \end{bmatrix}$$

$$R_1 \leftrightarrow R_2 \qquad\qquad R_2 + (-100)R_1 \to R_2$$
$$R_3 + (-5)R_1 \to R_3$$

$$\sim \begin{bmatrix} 1 & 10 & 1 & | & 1,664 \\ 0 & -975 & -75 & | & -124,800 \\ 0 & -45 & 95 & | & 0 \end{bmatrix} \sim \begin{bmatrix} 1 & 10 & 1 & | & 1,664 \\ 0 & -45 & 95 & | & 0 \\ 0 & -975 & -75 & | & -124,800 \end{bmatrix}$$

$$R_2 \leftrightarrow R_3 \qquad\qquad -\frac{1}{45}R_2 \to R_2$$

$$\sim \begin{bmatrix} 1 & 10 & 1 & | & 1,664 \\ 0 & 1 & -\frac{19}{9} & | & 0 \\ 0 & -975 & -75 & | & -124,800 \end{bmatrix} \sim \begin{bmatrix} 1 & 10 & 1 & | & 1,664 \\ 0 & 1 & -\frac{19}{9} & | & 0 \\ 0 & 0 & -\frac{6400}{3} & | & -124,800 \end{bmatrix}$$

$$R_3 + 975R_2 \to R_3 \qquad\qquad -\frac{3}{6400}R_3 \to R_3$$

$$\sim \begin{bmatrix} 1 & 10 & 1 & | & 1,664 \\ 0 & 1 & -\frac{19}{9} & | & 0 \\ 0 & 0 & 1 & | & 58.5 \end{bmatrix} \sim \begin{bmatrix} 1 & 10 & 0 & | & 1,605.5 \\ 0 & 1 & 0 & | & 123.5 \\ 0 & 0 & 1 & | & 58.5 \end{bmatrix}$$

$$R_2 + \frac{19}{9}R_3 \to R_2 \qquad\qquad R_1 + (-10)R_2 \to R_1$$
$$R_1 + (-1)R_3 \to R_1$$

$$\sim \begin{bmatrix} 1 & 0 & 0 & | & 370.5 \\ 0 & 1 & 0 & | & 123.5 \\ 0 & 0 & 1 & | & 58.5 \end{bmatrix}$$

Thus, $x_1 = \$370,500,$

$x_2 = \$123,500,$

and $x_3 = \$58,500.$

57. Let x_1 = number of ounces of food A,

x_2 = number of ounces of food B,

and x_3 = number of ounces of food C.

We have the following system of equations to solve:

$30x_1 + 10x_2 + 20x_3 = 340$

$10x_1 + 10x_2 + 20x_3 = 180$

$10x_1 + 30x_2 + 20x_3 = 220$

$$\begin{bmatrix} 30 & 10 & 20 & | & 340 \\ 10 & 10 & 20 & | & 180 \\ 10 & 30 & 20 & | & 220 \end{bmatrix} \sim \begin{bmatrix} 10 & 10 & 20 & | & 180 \\ 30 & 10 & 20 & | & 340 \\ 10 & 30 & 20 & | & 220 \end{bmatrix} \sim \begin{bmatrix} 1 & 1 & 2 & | & 18 \\ 3 & 1 & 2 & | & 34 \\ 1 & 3 & 2 & | & 22 \end{bmatrix}$$

$R_1 \leftrightarrow R_2$ 　　　　$\frac{1}{10}R_1 \to R_1$ 　　　　$R_2 + (-3)R_1 \to R_2$

　　　　　　　　　　$\frac{1}{10}R_2 \to R_2$ 　　　　$R_3 + (-1)R_1 \to R_3$

　　　　　　　　　　$\frac{1}{10}R_3 \to R_3$

$$\sim \begin{bmatrix} 1 & 1 & 2 & | & 18 \\ 0 & -2 & -4 & | & -20 \\ 0 & 2 & 0 & | & 4 \end{bmatrix} \sim \begin{bmatrix} 1 & 1 & 2 & | & 18 \\ 0 & 1 & 2 & | & 10 \\ 0 & 2 & 0 & | & 4 \end{bmatrix} \sim \begin{bmatrix} 1 & 1 & 2 & | & 18 \\ 0 & 1 & 2 & | & 10 \\ 0 & 0 & -4 & | & -16 \end{bmatrix}$$

$-\frac{1}{2}R_2 \to R_2$ 　　　$R_3 + (-2)R_2 \to R_3$ 　　　$-\frac{1}{4}R_3 \to R_3$

$$\sim \begin{bmatrix} 1 & 1 & 2 & | & 18 \\ 0 & 1 & 2 & | & 10 \\ 0 & 0 & 1 & | & 4 \end{bmatrix} \sim \begin{bmatrix} 1 & 1 & 0 & | & 10 \\ 0 & 1 & 0 & | & 2 \\ 0 & 0 & 1 & | & 4 \end{bmatrix} \sim \begin{bmatrix} 1 & 0 & 0 & | & 8 \\ 0 & 1 & 0 & | & 2 \\ 0 & 0 & 1 & | & 4 \end{bmatrix}$$

$R_2 + (-2)R_3 \to R_2$ 　　　$R_1 + (-1)R_2 \to R_1$

$R_1 + (-2)R_3 \to R_1$

Thus, $x_1 = 8$, $x_2 = 2$, and $x_3 = 4$ or 8 ounces of food A, 2 ounces of food B, and 4 ounces of food C.

59. Referring to Problem 57, we have:

$30x_1 + 10x_2 = 340$

$10x_1 + 10x_2 = 180$

$10x_1 + 30x_2 = 220$

$$\begin{bmatrix} 30 & 10 & | & 340 \\ 10 & 10 & | & 180 \\ 10 & 30 & | & 220 \end{bmatrix} \sim \begin{bmatrix} 3 & 1 & | & 34 \\ 1 & 1 & | & 18 \\ 1 & 3 & | & 22 \end{bmatrix} \sim \begin{bmatrix} 1 & 1 & | & 18 \\ 3 & 1 & | & 34 \\ 1 & 3 & | & 22 \end{bmatrix} \sim \begin{bmatrix} 1 & 1 & | & 18 \\ 0 & -2 & | & -20 \\ 0 & 2 & | & 4 \end{bmatrix} \sim \begin{bmatrix} 1 & 1 & | & 18 \\ 0 & 1 & | & 10 \\ 0 & 2 & | & 4 \end{bmatrix}$$

$\frac{1}{10}R_1 \to R_1$ 　　$R_1 \leftrightarrow R_2$ 　　$R_2 + (-3)R_1 \to R_2$ 　　$-\frac{1}{2}R_2 \to R_2$ 　　$R_3 + (-2)R_2 \to R_3$

$\frac{1}{10}R_2 \to R_2$ 　　　　　　$R_3 + (-1)R_1 \to R_3$

$\frac{1}{10}R_3 \to R_3$

$$\sim \begin{bmatrix} 1 & 1 & | & 18 \\ 0 & 1 & | & 10 \\ 0 & 0 & | & -16 \end{bmatrix}$$

From this matrix, we conclude that there is no solution.

61. Referring to Problem 57, we have the following system of equations to solve:

$30x_1 + 10x_2 + 20x_3 = 340$

$10x_1 + 10x_2 + 20x_3 = 180$

$$\begin{bmatrix} 30 & 10 & 20 & | & 340 \\ 10 & 10 & 20 & | & 180 \end{bmatrix} \sim \begin{bmatrix} 10 & 10 & 20 & | & 180 \\ 30 & 10 & 20 & | & 340 \end{bmatrix} \sim \begin{bmatrix} 1 & 1 & 2 & | & 18 \\ 3 & 1 & 2 & | & 34 \end{bmatrix} \sim \begin{bmatrix} 1 & 1 & 2 & | & 18 \\ 0 & -2 & -4 & | & -20 \end{bmatrix}$$

$\quad\quad R_1 \leftrightarrow R_2 \quad\quad\quad\quad \frac{1}{10}R_1 \to R_1 \quad\quad\quad R_2 + (-3)R_1 \to R_2 \quad\quad -\frac{1}{2}R_2 \to R_2$

$\quad\quad\quad\quad\quad\quad\quad\quad \frac{1}{10}R_2 \to R_2$

$$\sim \begin{bmatrix} 1 & 1 & 2 & | & 18 \\ 0 & 1 & 2 & | & 10 \end{bmatrix} \sim \begin{bmatrix} 1 & 0 & 0 & | & 8 \\ 0 & 1 & 2 & | & 10 \end{bmatrix}$$
Thus, $x_1 \qquad = 8$

$\qquad\qquad\qquad\qquad\qquad\qquad\qquad\qquad\qquad\qquad\qquad\qquad x_2 + 2x_3 = 10$

$\quad\quad R_1 + (-1)R_2 \to R_1$

Let $x_3 = t$ (t any real number). Then, $x_2 = 10 - 2t$, $0 \le t \le 5$, for x_2 to be positive.

The solution is: $x_1 = 8$ ounces of food A; $x_2 = 10 - 2t$ ounces of food B; $x_3 = t$ ounces of food C, $0 \le t \le 5$.

63. Let x_1 = number of barrels of mix A,

$\qquad x_2$ = number of barrels of mix B,

$\qquad x_3$ = number of barrels of mix C,

and x_4 = number of barrels of mix D,

Then,

$30x_1 + 30x_2 + 30x_3 + 60x_4 = 900 \qquad (1)$

$50x_1 + 75x_2 + 25x_3 + 25x_4 = 750 \qquad (2)$

$30x_1 + 20x_2 + 20x_3 + 50x_4 = 700 \qquad (3)$

Divide each side of equation (1) by 30, each side of equation (2) by 25, and each side of equation (3) by 10. This yields the system of linear equations:

$\quad x_1 + \quad x_2 + \quad x_3 + 2x_4 = 30$

$2x_1 + 3x_2 + \quad x_3 + \quad x_4 = 30$

$3x_1 + 2x_2 + 2x_3 + 5x_4 = 70$

$$\begin{bmatrix} 1 & 1 & 1 & 2 & | & 30 \\ 2 & 3 & 1 & 1 & | & 30 \\ 3 & 2 & 2 & 5 & | & 70 \end{bmatrix} \sim \begin{bmatrix} 1 & 1 & 1 & 2 & | & 30 \\ 0 & 1 & -1 & -3 & | & -30 \\ 0 & -1 & -1 & -1 & | & -20 \end{bmatrix} \sim \begin{bmatrix} 1 & 1 & 1 & 2 & | & 30 \\ 0 & 1 & -1 & -3 & | & -30 \\ 0 & 0 & -2 & -4 & | & -50 \end{bmatrix}$$

$\quad R_2 + (-2)R_1 \to R_2 \qquad\qquad\qquad R_3 + R_2 \to R_3 \qquad\qquad\qquad -\frac{1}{2}R_3 \to R_3$

$\quad R_3 + (-3)R_1 \to R_3$

$$\sim \begin{bmatrix} 1 & 1 & 1 & 2 & | & 30 \\ 0 & 1 & -1 & -3 & | & -30 \\ 0 & 0 & 1 & 2 & | & 25 \end{bmatrix} \sim \begin{bmatrix} 1 & 1 & 0 & 0 & | & 5 \\ 0 & 1 & 0 & -1 & | & -5 \\ 0 & 0 & 1 & 2 & | & 25 \end{bmatrix} \sim \begin{bmatrix} 1 & 0 & 0 & 1 & | & 10 \\ 0 & 1 & 0 & -1 & | & -5 \\ 0 & 0 & 1 & 2 & | & 25 \end{bmatrix}$$

$\quad R_2 + R_3 \to R_2 \qquad\qquad\qquad R_1 + (-1)R_2 \to R_1$

$\quad R_1 + (-1)R_3 \to R_1$

Thus, $x_1 \qquad + \quad x_4 = 10$

$\qquad x_2 \quad - \quad x_4 = -5$

$\qquad\qquad x_3 + 2x_4 = 25$

Let $x_4 = t$ = number of barrels of mix D. Then $x_1 = 10 - t$ = number of barrels of mix A, $x_2 = t - 5$ = number of barrels of mix B, and $x_3 = 25 - 2t$ = number of barrels of mix C. Since the number of barrels of each mix must be nonnegative, $5 \le t \le 10$. Also, t is an integer.

65. Let x_1 = number of hours for Company A,

and x_2 = number of hours for Company B.

Then, $30x_1 + 20x_2 = 600$

$\qquad 10x_1 + 20x_2 = 400$

Divide each side of each equation by 10. This yields the system of linear equations:

$3x_1 + 2x_2 = 60$

$\ x_1 + 2x_2 = 40$

$$\begin{bmatrix} 3 & 2 & | & 60 \\ 1 & 2 & | & 40 \end{bmatrix} \sim \begin{bmatrix} 1 & 2 & | & 40 \\ 3 & 2 & | & 60 \end{bmatrix} \sim \begin{bmatrix} 1 & 2 & | & 40 \\ 0 & -4 & | & -60 \end{bmatrix} \sim \begin{bmatrix} 1 & 2 & | & 40 \\ 0 & 1 & | & 15 \end{bmatrix} \sim \begin{bmatrix} 1 & 0 & | & 10 \\ 0 & 1 & | & 15 \end{bmatrix}$$

$\quad R_1 \leftrightarrow R_2 \qquad R_2 + (-3)R_1 \to R_2 \qquad -\frac{1}{4}R_2 \to R_2 \qquad R_1 + (-2)R_2 \to R_1$

Thus, $x_1 = 10$ and $x_2 = 15$, or 10 hours for Company A and 15 hours for Company B.

67. (A) 6th and Washington Ave.: $x_1 + x_2 = 1200$

$\qquad$ 6th and Lincoln Ave.: $\quad x_2 + x_3 = 1000$

$\qquad$ 5th and Lincoln Ave.: $\quad x_3 + x_4 = 1300$

(B) The system of equations is:

$x_1 \qquad\qquad + \ x_4 = 1500$

$x_1 + x_2 \qquad\qquad = 1200$

$\qquad x_2 + x_3 \qquad = 1000$

$\qquad\qquad x_3 + x_4 = 1300$

$$\begin{pmatrix} 1 & 0 & 0 & 1 & | & 1500 \\ 1 & 1 & 0 & 0 & | & 1200 \\ 0 & 1 & 1 & 0 & | & 1000 \\ 0 & 0 & 1 & 1 & | & 1300 \end{pmatrix} \sim \begin{pmatrix} 1 & 0 & 0 & 1 & | & 1500 \\ 0 & 1 & 0 & -1 & | & -300 \\ 0 & 1 & 1 & 0 & | & 1000 \\ 0 & 0 & 1 & 1 & | & 1300 \end{pmatrix} \sim \begin{pmatrix} 1 & 0 & 0 & 1 & | & 1500 \\ 0 & 1 & 0 & -1 & | & -300 \\ 0 & 0 & 1 & 1 & | & 1300 \\ 0 & 0 & 1 & 1 & | & 1300 \end{pmatrix}$$

$\qquad R_2 + (-1)R_1 \to R_2 \qquad\qquad R_3 + (-1)R_2 \to R_3 \qquad\qquad R_4 + (-1)R_3 \to R_4$

$$\sim \begin{pmatrix} 1 & 0 & 0 & 1 & | & 1500 \\ 0 & 1 & 0 & -1 & | & -300 \\ 0 & 0 & 1 & 1 & | & 1300 \\ 0 & 0 & 0 & 0 & | & 0 \end{pmatrix}$$

$\qquad\qquad$ Thus $x_1 \qquad\qquad + x_4 = 1500$

$\qquad\qquad\qquad\qquad\qquad x_2 \qquad\quad - x_4 = -300$

$\qquad\qquad\qquad\qquad\qquad\qquad x_3 + x_4 = 1300$

Let $x_4 = t$. Then $x_1 = 1500 - t$, $x_2 = t - 300$ and $x_3 = 1300 - t$. Since x_1, x_2, x_3, and x_4 must be nonnegative integers, we have $300 \le t \le 1300$.

(C) The flow from Washington Ave. to Lincoln Ave. on 5th Street is given by $x_4 = t$. As shown in part (B), $300 \le t \le 1300$, that is, the maximum number of vehicles is 1300 and the minimum number is 300.

(D) If $x_4 = t = 1000$, then

Washington Ave.: $x_1 = 1500 - 1000 = 500$
6th St.: $x_2 = 1000 - 300 = 700$
Lincoln Ave.: $x_3 = 1300 - 1000 = 300$

EXERCISE 5-4

Things to remember:

1. A matrix with m rows and n columns is said to have SIZE or DIMENSION $m \times n$. If a matrix has the same number of rows and columns, then it is called a SQUARE MATRIX. A matrix with only one column is a COLUMN MATRIX, and a matrix with only one row is a ROW MATRIX.

2. Two matrices are EQUAL if they have the same dimension and their corresponding elements are equal.

3. The SUM of two matrices of the same dimension, $m \times n$, is an $m \times n$ matrix whose elements are the sum of the corresponding elements of the two given matrices. Addition is not defined for matrices with different dimensions. Matrix addition is commutative: $A + B = B + A$, and assocative: $(A + B) + C = A + (B + C)$.

4. A matrix with all elements equal to zero is called a ZERO MATRIX.

5. The NEGATIVE OF A MATRIX M, denoted by $-M$, is the matrix whose elements are the negatives of the elements of M.

6. If A and B are matrices of the same dimension, then subtraction is defined by $A - B = A + (-B)$. Thus, to subtract B from A, simply subtract corresponding elements.

7. If M is a matrix and k is a number, then kM is the matrix formed by multiplying each element of M by k.

8. DOUBLE SUBSCRIPT NOTATION is used to denote a matrix of arbitrary dimension $m \times n$. For example a general 3×4 matrix would be written

$$A = \begin{bmatrix} a_{11} & a_{12} & a_{13} & a_{14} \\ a_{21} & a_{22} & a_{23} & a_{24} \\ a_{31} & a_{32} & a_{33} & a_{34} \end{bmatrix}$$

The subscript ij denotes the position of the element a_{ij} in the rectangular array—the first subscript indicates the row in which the element appears and the second indicates the column. Thus a_{23} is the element in second row, third column. The PRINCIPAL DIAGONAL of a matrix A consists of the elements a_{11}, a_{22}, a_{33}, ...

---

1. Dimension of B is 2×2 (2 rows $\times$ 2 columns)
    Dimension of E is 1×4 (1 row $\times$ 4 columns)

3. f_{12} is the element in the first row, second column. Thus, $f_{12} = -3$.
    f_{22} is the element in the second row, second column; $f_{22} = 0$
    $f_{42} = 5$

5. $\begin{bmatrix} 0 & 0 \\ 0 & 0 \end{bmatrix}$ 7. C and D are column matrices.

9. A and B are square matrices. [Note: Number of rows = number of columns.]

11. $A + B = \begin{bmatrix} 2 & -1 \\ 3 & 0 \end{bmatrix} + \begin{bmatrix} -3 & 1 \\ 2 & -3 \end{bmatrix} = \begin{bmatrix} 2-3 & -1+1 \\ 3+2 & 0-3 \end{bmatrix} = \begin{bmatrix} -1 & 0 \\ 5 & -3 \end{bmatrix}$

13. $E + F$ is not defined; E and F have different dimensions.

15. $-C = \begin{bmatrix} -2 \\ -(-3) \\ 0 \end{bmatrix} = \begin{bmatrix} -2 \\ 3 \\ 0 \end{bmatrix}$ 17. $D - C = \begin{bmatrix} 1 \\ 3 \\ 5 \end{bmatrix} - \begin{bmatrix} 2 \\ -3 \\ 0 \end{bmatrix} = \begin{bmatrix} 1 \\ 3 \\ 5 \end{bmatrix} + \begin{bmatrix} -2 \\ 3 \\ 0 \end{bmatrix} = \begin{bmatrix} 1-2 \\ 3+3 \\ 5+0 \end{bmatrix} = \begin{bmatrix} -1 \\ 6 \\ 5 \end{bmatrix}$

19. $5B = 5 \begin{bmatrix} -3 & 1 \\ 2 & -3 \end{bmatrix} = \begin{bmatrix} -15 & 5 \\ 10 & -15 \end{bmatrix}$

21. $\begin{bmatrix} 3 & -2 & 0 & 1 \\ 2 & -3 & -1 & 4 \\ 0 & 2 & -1 & 6 \end{bmatrix} + \begin{bmatrix} -2 & 5 & -1 & 0 \\ -3 & -2 & 8 & -2 \\ 4 & 6 & 1 & -8 \end{bmatrix} = \begin{bmatrix} 3-2 & -2+5 & 0-1 & 1+0 \\ 2-3 & -3-2 & -1+8 & 4-2 \\ 0+4 & 2+6 & -1+1 & 6-8 \end{bmatrix}$

$= \begin{bmatrix} 1 & 3 & -1 & 1 \\ -1 & -5 & 7 & 2 \\ 4 & 8 & 0 & -2 \end{bmatrix}$

23. $\begin{bmatrix} 1.3 & 2.5 & -6.1 \\ 8.3 & -1.4 & 6.7 \end{bmatrix} - \begin{bmatrix} -4.1 & 1.8 & -4.3 \\ 0.7 & 2.6 & -1.2 \end{bmatrix} = \begin{bmatrix} 1.3 + 4.1 & 2.5 - 1.8 & -6.1 + 4.3 \\ 8.3 - 0.7 & -1.4 - 2.6 & 6.7 + 1.2 \end{bmatrix}$

$$= \begin{bmatrix} 5.4 & 0.7 & -1.8 \\ 7.6 & -4.0 & 7.9 \end{bmatrix}$$

25. $1000 \begin{bmatrix} 0.25 & 0.36 \\ 0.04 & 0.35 \end{bmatrix} = \begin{bmatrix} 250 & 360 \\ 40 & 350 \end{bmatrix}$

27. $0.08 \begin{bmatrix} 24,000 & 35,000 \\ 12,000 & 24,000 \end{bmatrix} + 0.03 \begin{bmatrix} 12,000 & 22,000 \\ 14,000 & 13,000 \end{bmatrix} = \begin{bmatrix} 1920 & 2800 \\ 960 & 1920 \end{bmatrix} + \begin{bmatrix} 360 & 660 \\ 420 & 390 \end{bmatrix}$

$$= \begin{bmatrix} 1920 + 360 & 2800 + 660 \\ 960 + 420 & 1920 + 390 \end{bmatrix}$$

$$= \begin{bmatrix} 2280 & 3460 \\ 1380 & 2310 \end{bmatrix}$$

29. $\begin{bmatrix} a & b \\ c & d \end{bmatrix} + \begin{bmatrix} 2 & -3 \\ 0 & 1 \end{bmatrix} = \begin{bmatrix} a + 2 & b - 3 \\ c + 0 & d + 1 \end{bmatrix} = \begin{bmatrix} 1 & -2 \\ 3 & -4 \end{bmatrix}$

Thus, $a + 2 = 1$, $a = -1$
$b - 3 = -2$, $b = 1$
$c + 0 = 3$, $c = 3$
$d + 1 = -4$, $d = -5$

31. $\begin{bmatrix} 2x & 4 \\ -3 & 5x \end{bmatrix} + \begin{bmatrix} 3y & -2 \\ -2 & -y \end{bmatrix} = \begin{bmatrix} -5 & 2 \\ -5 & 13 \end{bmatrix}$

$\begin{bmatrix} 2x + 3y & 4 - 2 \\ -3 - 2 & 5x - y \end{bmatrix} = \begin{bmatrix} -5 & 2 \\ -5 & 13 \end{bmatrix}$

$\begin{bmatrix} 2x + 3y & 2 \\ -5 & 5x - y \end{bmatrix} = \begin{bmatrix} -5 & 2 \\ -5 & 13 \end{bmatrix}$ Thus, $2x + 3y = -5$ (1)
$5x - y = 13$ (2)
Solve the above system for x, y.

From (2), $y = 5x - 13$. Substitute $y = 5x - 13$ in (1):

$2x + 3(5x - 13) = -5$
$2x + 15x - 39 = -5$
$17x = -5 + 39$
$17x = 34$
$x = 2$

Substitute $x = 2$ in (1):

$2(2) + 3y = -5$
$4 + 3y = -5$
$3y = -9$
$y = -3$

Thus, the solution is $x = 2$ and $y = -3$.

33. No. $A + B = B + A$ for all matrices A, B of the same dimension. Matrix addition is commutative.

35. No. $k(A + B) = kA + kB$ for all matrices A, B of the same dimension and any real number k. This is a version of the Distributive Law for matrix addition and multiplication by a number.

37. $A + B = \begin{bmatrix} \$30 & \$25 \\ \$60 & \$80 \end{bmatrix} + \begin{bmatrix} \$36 & \$27 \\ \$54 & \$74 \end{bmatrix} = \begin{bmatrix} \$66 & \$52 \\ \$114 & \$154 \end{bmatrix}$

$$\frac{1}{2}(A + B) = \frac{1}{2}\begin{bmatrix} \$66 & \$52 \\ \$114 & \$154 \end{bmatrix} = \begin{array}{c} \\ \\ \end{array}\overset{\text{Guitar} \quad \text{Banjo}}{\begin{bmatrix} \$33 & \$26 \\ \$57 & \$77 \end{bmatrix}} \begin{array}{l} \text{Materials} \\ \text{Labor} \end{array}$$

39. The dealer is increasing the retail prices by 10%. Thus, the new retail price matrix M' (to the nearest dollar) is given by

$$M' = M + 0.1M = 1.1M = 1.1\begin{bmatrix} 10900 & 683 & 253 & 195 \\ 13000 & 738 & 382 & 206 \\ 16300 & 867 & 537 & 225 \end{bmatrix} = \begin{bmatrix} 11990 & 751 & 278 & 214 \\ 14300 & 812 & 420 & 227 \\ 17930 & 954 & 591 & 248 \end{bmatrix}$$

The new dealer invoice matrix N' is given by

$$N' = N + 0.15N = 1.15N = 1.15\begin{bmatrix} 9400 & 582 & 195 & 160 \\ 11500 & 621 & 295 & 171 \\ 14100 & 737 & 420 & 184 \end{bmatrix}$$

$$= \begin{bmatrix} 10810 & 669 & 224 & 184 \\ 13225 & 714 & 339 & 197 \\ 16215 & 848 & 483 & 212 \end{bmatrix}$$

The new markup is:

$$M' - N' = \begin{bmatrix} 11990 & 751 & 278 & 214 \\ 14300 & 812 & 420 & 227 \\ 17930 & 954 & 591 & 248 \end{bmatrix} - \begin{bmatrix} 10810 & 669 & 224 & 184 \\ 13225 & 714 & 339 & 197 \\ 16215 & 848 & 483 & 212 \end{bmatrix}$$

$$= \begin{array}{r} \text{Model A} \\ \text{Model B} \\ \text{Model C} \end{array}\overset{\text{Basic Car} \quad \text{Air/Cond.} \quad \text{AM/FM} \quad \text{Cruise}}{\begin{bmatrix} \$1180 & \$82 & \$54 & \$30 \\ \$1075 & \$98 & \$81 & \$30 \\ \$1715 & \$106 & \$108 & \$36 \end{bmatrix}}$$

41. $M + N = \begin{bmatrix} 319 & 101 \\ 108 & 32 \end{bmatrix} + \begin{bmatrix} 370 & 124 \\ 110 & 36 \end{bmatrix} = \begin{array}{c} \text{Yellow} \\ \text{Green} \end{array}\overset{\text{Round} \quad \text{Wrinkled}}{\begin{bmatrix} 689 & 225 \\ 218 & 68 \end{bmatrix}}$

$$\frac{1}{1200}(M + N) = \frac{1}{1200}\begin{bmatrix} 689 & 225 \\ 218 & 68 \end{bmatrix} = \begin{array}{c} \text{Yellow} \\ \text{Green} \end{array}\overset{\text{Round} \quad \text{Wrinkled}}{\begin{bmatrix} 57\% & 19\% \\ 18\% & 6\% \end{bmatrix}}$$

Things to remember:

<u>1</u>. PRODUCT OF A ROW MATRIX AND A COLUMN MATRIX

The product of a $1 \times n$ row matrix and an $n \times 1$ column matrix is the 1×1 matrix given by

$$[a_1 \quad a_2 \quad \cdots \quad a_n] \underset{1 \times n}{} \begin{bmatrix} b_1 \\ b_2 \\ \vdots \\ b_n \end{bmatrix} \overset{n \times 1}{} = [a_1 b_1 + a_2 b_2 + \cdots + a_n b_n]$$

Note that the number of elements in the row matrix and the number of elements in the column matrix must be the same for the product to be defined.

<u>2</u>. Let A be an $m \times p$ matrix and B be a $p \times n$ matrix. The MATRIX PRODUCT of A and B, denoted AB, is the $m \times n$ matrix whose element in the ith row and the jth column is the real number obtained from the product of the ith row of A and the jth column of B. If the number of columns in A does not equal the number of rows in B, then the matrix product AB is not defined.

<u>3</u>. PROPERTIES OF MATRIX MULTIPLICATION

Assuming all products and sums are defined for the indicated matrices A, B, and C, then we have the following:

1. $A(BC) = (AB)C$ Associative property
2. $A(B + C) = AB + AC$ Left distributive property
3. $(B + C)A = BA + CA$ Right distributive property
4. If $A = B$, then $CA = CB$ Left multiplication property
5. If $A = B$, then $AC = BC$ Right multiplication property
6. $k(AB) = (kA)B = A(kB)$ (k a real number)

NOTE: Matrix multiplication is *not* commutative. That is, AB does not always equal BA, even when both multiplications are defined.

---

1. $[2 \quad 4] \begin{bmatrix} 3 \\ 1 \end{bmatrix} = [2 \cdot 3 + 4 \cdot 1]$ (using <u>1</u>)   **3.** $[-3 \quad 2] \begin{bmatrix} -1 \\ -2 \end{bmatrix} = [(-3)(-1) + 2(-2)]$

$\qquad\qquad\qquad = [10]$ $\qquad\qquad\qquad\qquad\qquad\qquad = [3 - 4] = [-1]$

5. $[2 \quad 5] \begin{bmatrix} 1 & -1 \\ 2 & 3 \end{bmatrix} = \begin{bmatrix} [2 & 5] \begin{bmatrix} 1 \\ 2 \end{bmatrix} & [2 & 5] \begin{bmatrix} -1 \\ 3 \end{bmatrix} \end{bmatrix} = [2 + 10 \quad -2 + 15] = [12 \quad 13]$

7. $\begin{bmatrix} 3 & 4 \\ -1 & -2 \end{bmatrix} \begin{bmatrix} -1 \\ 2 \end{bmatrix} = \begin{bmatrix} [3 \quad 4] \begin{bmatrix} -1 \\ 2 \end{bmatrix} \\ [-1 \quad -2] \begin{bmatrix} -1 \\ 2 \end{bmatrix} \end{bmatrix} = \begin{bmatrix} -3 + 8 \\ 1 - 4 \end{bmatrix} = \begin{bmatrix} 5 \\ -3 \end{bmatrix}$

9. $\begin{bmatrix} 2 & -3 \\ 1 & 2 \end{bmatrix} \begin{bmatrix} 1 & -1 \\ 0 & -2 \end{bmatrix} = \begin{bmatrix} [2 \quad -3] \begin{bmatrix} 1 \\ 0 \end{bmatrix} & [2 \quad -3] \begin{bmatrix} -1 \\ -2 \end{bmatrix} \\ [1 \quad 2] \begin{bmatrix} 1 \\ 0 \end{bmatrix} & [1 \quad 2] \begin{bmatrix} -1 \\ -2 \end{bmatrix} \end{bmatrix} = \begin{bmatrix} 2 + 0 & -2 + 6 \\ 1 + 0 & -1 - 4 \end{bmatrix} = \begin{bmatrix} 2 & 4 \\ 1 & -5 \end{bmatrix}$

11. $\begin{bmatrix} 1 & -1 \\ 0 & -2 \end{bmatrix} \begin{bmatrix} 2 & -3 \\ 1 & 2 \end{bmatrix} = \begin{bmatrix} [1 \quad -1] \begin{bmatrix} 2 \\ 1 \end{bmatrix} & [1 \quad -1] \begin{bmatrix} -3 \\ 2 \end{bmatrix} \\ [0 \quad -2] \begin{bmatrix} 2 \\ 1 \end{bmatrix} & [0 \quad -2] \begin{bmatrix} -3 \\ 2 \end{bmatrix} \end{bmatrix} = \begin{bmatrix} 2 - 1 & -3 - 2 \\ 0 - 2 & 0 - 4 \end{bmatrix} = \begin{bmatrix} 1 & -5 \\ -2 & -4 \end{bmatrix}$

13. $[5 \quad -2] \begin{bmatrix} -3 \\ -4 \end{bmatrix} = [-15 + 8] = [-7]$

15. $\begin{bmatrix} -3 \\ -4 \end{bmatrix} [5 \quad -2] = \begin{bmatrix} [-3][5] & [-3][-2] \\ [-4][5] & [-4][-2] \end{bmatrix} = \begin{bmatrix} -15 & 6 \\ -20 & 8 \end{bmatrix}$

17. $[-1 \quad -2 \quad 2] \begin{bmatrix} 2 \\ -1 \\ 3 \end{bmatrix} = [(-1)2 + (-2)(-1) + 2 \cdot 3] = [6]$

19. $[-1 \quad -3 \quad 0 \quad 5] \begin{bmatrix} 4 \\ -3 \\ -1 \\ 2 \end{bmatrix} = [(-1)4 + (-3)(-3) + 0(-1) + 5 \cdot 2] = [15]$

21. $\begin{bmatrix} 2 & -1 & 1 \\ 1 & 3 & -2 \end{bmatrix} \begin{bmatrix} 1 & 3 \\ 0 & -1 \\ -2 & 2 \end{bmatrix} = \begin{bmatrix} [2 \quad -1 \quad 1] \begin{bmatrix} 1 \\ 0 \\ -2 \end{bmatrix} & [2 \quad -1 \quad 1] \begin{bmatrix} 3 \\ -1 \\ 2 \end{bmatrix} \\ [1 \quad 3 \quad -2] \begin{bmatrix} 1 \\ 0 \\ -2 \end{bmatrix} & [1 \quad 3 \quad -2] \begin{bmatrix} 3 \\ -1 \\ 2 \end{bmatrix} \end{bmatrix}$

$= \begin{bmatrix} 2 + 0 - 2 & 6 + 1 + 2 \\ 1 + 0 + 4 & 3 - 3 - 4 \end{bmatrix} = \begin{bmatrix} 0 & 9 \\ 5 & -4 \end{bmatrix}$

23. $\begin{bmatrix} 1 & 3 \\ 0 & -1 \\ -2 & 2 \end{bmatrix} \begin{bmatrix} 2 & -1 & 1 \\ 1 & 3 & -2 \end{bmatrix} = \begin{bmatrix} [1 \quad 3]\begin{bmatrix} 2 \\ 1 \end{bmatrix} & [1 \quad 3]\begin{bmatrix} -1 \\ 3 \end{bmatrix} & [1 \quad 3]\begin{bmatrix} 1 \\ -2 \end{bmatrix} \\[6pt] [0 \quad -1]\begin{bmatrix} 2 \\ 1 \end{bmatrix} & [0 \quad -1]\begin{bmatrix} -1 \\ 3 \end{bmatrix} & [0 \quad -1]\begin{bmatrix} 1 \\ -2 \end{bmatrix} \\[6pt] [-2 \quad 2]\begin{bmatrix} 2 \\ 1 \end{bmatrix} & [-2 \quad 2]\begin{bmatrix} -1 \\ 3 \end{bmatrix} & [-2 \quad 2]\begin{bmatrix} 1 \\ -2 \end{bmatrix} \end{bmatrix}$

$= \begin{bmatrix} 2+3 & -1+9 & 1-6 \\ 0-1 & 0-3 & 0+2 \\ -4+2 & 2+6 & -2-4 \end{bmatrix} = \begin{bmatrix} 5 & 8 & -5 \\ -1 & -3 & 2 \\ -2 & 8 & -6 \end{bmatrix}$

25. $[3 \quad -2 \quad -4]\begin{bmatrix} 1 \\ 2 \\ -3 \end{bmatrix} = [(3 - 4 + 12)] = [11]$

27. $\begin{bmatrix} 1 \\ 2 \\ -3 \end{bmatrix}[3 \quad -2 \quad -4] = \begin{bmatrix} [1][3] & [1][-2] & [1][-4] \\ [2][3] & [2][-2] & [2][-4] \\ [-3][3] & [-3][-2] & [-3][-4] \end{bmatrix} = \begin{bmatrix} 3 & -2 & -4 \\ 6 & -4 & -8 \\ -9 & 6 & 12 \end{bmatrix}$

29. $\begin{bmatrix} 1 & 2 \\ 2 & -1 \\ -3 & 1 \end{bmatrix}[3 \quad -2 \quad -4]$ This product is not defined since the number of columns of the matrix on the left does not equal the number of rows of the matrix on the right.

31. $\begin{bmatrix} 1 & 2 & -1 \\ 3 & -1 & 4 \\ 2 & -4 & 5 \end{bmatrix}\begin{bmatrix} 4 \\ 5 \\ 7 \end{bmatrix} = \begin{bmatrix} [1 \quad 2 \quad -1]\begin{bmatrix} 4 \\ 5 \\ 7 \end{bmatrix} \\[10pt] [3 \quad -1 \quad 4]\begin{bmatrix} 4 \\ 5 \\ 7 \end{bmatrix} \\[10pt] [2 \quad -4 \quad 5]\begin{bmatrix} 4 \\ 5 \\ 7 \end{bmatrix} \end{bmatrix} = \begin{bmatrix} 1\cdot4 + 2\cdot5 + (-1)\cdot7 \\ 3\cdot4 + (-1)\cdot5 + 4\cdot7 \\ 2\cdot4 + (-4)\cdot5 + 5\cdot7 \end{bmatrix} = \begin{bmatrix} 7 \\ 35 \\ 23 \end{bmatrix}$

33. $\begin{bmatrix} 2 & -1 & 3 & 0 \\ -3 & 4 & 2 & -1 \\ 0 & -2 & 1 & 4 \end{bmatrix}\begin{bmatrix} 2 & -3 & -2 \\ 1 & 0 & 1 \\ -1 & 2 & 0 \\ 2 & -2 & -3 \end{bmatrix}$

$$= \begin{bmatrix} [2 \quad -1 \quad 3 \quad 0]\begin{bmatrix} 2 \\ 1 \\ -1 \\ 2 \end{bmatrix} & [2 \quad -1 \quad 3 \quad 0]\begin{bmatrix} -3 \\ 0 \\ 2 \\ -2 \end{bmatrix} & [2 \quad -1 \quad 3 \quad 0]\begin{bmatrix} -2 \\ 1 \\ 0 \\ -3 \end{bmatrix} \\ [-3 \quad 4 \quad 2 \quad -1]\begin{bmatrix} 2 \\ 1 \\ -1 \\ 2 \end{bmatrix} & [-3 \quad 4 \quad 2 \quad -1]\begin{bmatrix} -3 \\ 0 \\ 2 \\ -2 \end{bmatrix} & [-3 \quad 4 \quad 2 \quad -1]\begin{bmatrix} -2 \\ 1 \\ 0 \\ -3 \end{bmatrix} \\ [0 \quad -2 \quad 1 \quad 4]\begin{bmatrix} 2 \\ 1 \\ -1 \\ 2 \end{bmatrix} & [0 \quad -2 \quad 1 \quad 4]\begin{bmatrix} -3 \\ 0 \\ 2 \\ -2 \end{bmatrix} & [0 \quad -2 \quad 1 \quad 4]\begin{bmatrix} -2 \\ 1 \\ 0 \\ -3 \end{bmatrix} \end{bmatrix}$$

$$= \begin{bmatrix} 4-1-3+0 & -6+0+6+0 & -4-1+0+0 \\ -6+4-2-2 & 9+0+4+2 & 6+4+0+3 \\ 0-2-1+8 & 0+0+2-8 & 0-2+0-12 \end{bmatrix} = \begin{bmatrix} 0 & 0 & -5 \\ -6 & 15 & 13 \\ 5 & -6 & -14 \end{bmatrix}$$

35. Let $A = \begin{bmatrix} 6 & 9 \\ -4 & -6 \end{bmatrix}$. Then

$$A^2 = AA = \begin{bmatrix} 6 & 9 \\ -4 & -6 \end{bmatrix}\begin{bmatrix} 6 & 9 \\ -4 & -6 \end{bmatrix} = \begin{bmatrix} [6 \quad 9]\begin{bmatrix} 6 \\ -4 \end{bmatrix} & [6 \quad 9]\begin{bmatrix} 9 \\ -6 \end{bmatrix} \\ [-4 \quad -6]\begin{bmatrix} 6 \\ -4 \end{bmatrix} & [-4 \quad -6]\begin{bmatrix} 9 \\ -6 \end{bmatrix} \end{bmatrix}$$

$$= \begin{bmatrix} 6\cdot6 + 9(-4) & 6\cdot9 + 9(-6) \\ (-4)6 + (-6)(-4) & (-4)9 + (-6)(-6) \end{bmatrix} = \begin{bmatrix} 0 & 0 \\ 0 & 0 \end{bmatrix} = 0 \quad \text{(the zero matrix)}$$

37. Let $A = \begin{bmatrix} \frac{1}{3} & \frac{1}{3} \\ \frac{2}{3} & \frac{2}{3} \end{bmatrix}$. Then $A^2 = AA = \begin{bmatrix} \frac{1}{3} & \frac{1}{3} \\ \frac{2}{3} & \frac{2}{3} \end{bmatrix}\begin{bmatrix} \frac{1}{3} & \frac{1}{3} \\ \frac{2}{3} & \frac{2}{3} \end{bmatrix} = \begin{bmatrix} [\frac{1}{3} \quad \frac{1}{3}]\begin{bmatrix} \frac{1}{3} \\ \frac{2}{3} \end{bmatrix} & [\frac{1}{3} \quad \frac{1}{3}]\begin{bmatrix} \frac{1}{3} \\ \frac{2}{3} \end{bmatrix} \\ [\frac{2}{3} \quad \frac{2}{3}]\begin{bmatrix} \frac{1}{3} \\ \frac{2}{3} \end{bmatrix} & [\frac{2}{3} \quad \frac{2}{3}]\begin{bmatrix} \frac{1}{3} \\ \frac{2}{3} \end{bmatrix} \end{bmatrix}$

$$\begin{bmatrix} \frac{1}{3}\cdot\frac{1}{3} + \frac{1}{3}\cdot\frac{2}{3} & \frac{1}{3}\cdot\frac{1}{3} + \frac{1}{3}\cdot\frac{2}{3} \\ \frac{2}{3}\cdot\frac{1}{3} + \frac{2}{3}\cdot\frac{2}{3} & \frac{2}{3}\cdot\frac{1}{3} + \frac{2}{3}\cdot\frac{2}{3} \end{bmatrix} = \begin{bmatrix} \frac{1}{3} & \frac{1}{3} \\ \frac{2}{3} & \frac{2}{3} \end{bmatrix}$$

39. $AB = \begin{bmatrix} 1 & 2 \\ 0 & 1 \end{bmatrix}\begin{bmatrix} 1 & 1 \\ 2 & 3 \end{bmatrix} = \begin{bmatrix} [1 \quad 2]\begin{bmatrix} 1 \\ 2 \end{bmatrix} & [1 \quad 2]\begin{bmatrix} 1 \\ 3 \end{bmatrix} \\ [0 \quad 1]\begin{bmatrix} 1 \\ 2 \end{bmatrix} & [0 \quad 1]\begin{bmatrix} 1 \\ 3 \end{bmatrix} \end{bmatrix} = \begin{bmatrix} 5 & 7 \\ 2 & 3 \end{bmatrix}$

$BA = \begin{bmatrix} 1 & 1 \\ 2 & 3 \end{bmatrix}\begin{bmatrix} 1 & 2 \\ 0 & 1 \end{bmatrix} = \begin{bmatrix} [1 \quad 1]\begin{bmatrix} 1 \\ 0 \end{bmatrix} & [1 \quad 1]\begin{bmatrix} 2 \\ 1 \end{bmatrix} \\ [2 \quad 3]\begin{bmatrix} 1 \\ 0 \end{bmatrix} & [2 \quad 3]\begin{bmatrix} 2 \\ 1 \end{bmatrix} \end{bmatrix} = \begin{bmatrix} 1 & 3 \\ 2 & 7 \end{bmatrix}$

41. $A(B + C) = \begin{bmatrix} 1 & 2 \\ 0 & 1 \end{bmatrix} \left(\begin{bmatrix} 1 & 1 \\ 2 & 3 \end{bmatrix} + \begin{bmatrix} -3 & 1 \\ -1 & 2 \end{bmatrix} \right) = \begin{bmatrix} 1 & 2 \\ 0 & 1 \end{bmatrix} \begin{bmatrix} -2 & 2 \\ 1 & 5 \end{bmatrix}$

$$= \begin{bmatrix} [1 \quad 2]\begin{bmatrix} -2 \\ 1 \end{bmatrix} & [1 \quad 2]\begin{bmatrix} 2 \\ 5 \end{bmatrix} \\ [0 \quad 1]\begin{bmatrix} -2 \\ 1 \end{bmatrix} & [0 \quad 1]\begin{bmatrix} 2 \\ 5 \end{bmatrix} \end{bmatrix} = \begin{bmatrix} 0 & 12 \\ 1 & 5 \end{bmatrix}$$

$AB + AC = \begin{bmatrix} 1 & 2 \\ 0 & 1 \end{bmatrix}\begin{bmatrix} 1 & 1 \\ 2 & 3 \end{bmatrix} + \begin{bmatrix} 1 & 2 \\ 0 & 1 \end{bmatrix}\begin{bmatrix} -3 & 1 \\ -1 & 2 \end{bmatrix}$

$$= \begin{bmatrix} [1 \quad 2]\begin{bmatrix} 1 \\ 2 \end{bmatrix} & [1 \quad 2]\begin{bmatrix} 1 \\ 3 \end{bmatrix} \\ [0 \quad 1]\begin{bmatrix} 1 \\ 2 \end{bmatrix} & [0 \quad 1]\begin{bmatrix} 1 \\ 3 \end{bmatrix} \end{bmatrix} + \begin{bmatrix} [1 \quad 2]\begin{bmatrix} -3 \\ -1 \end{bmatrix} & [1 \quad 2]\begin{bmatrix} 1 \\ 2 \end{bmatrix} \\ [0 \quad 1]\begin{bmatrix} -3 \\ -1 \end{bmatrix} & [0 \quad 1]\begin{bmatrix} 1 \\ 2 \end{bmatrix} \end{bmatrix}$$

$$= \begin{bmatrix} 5 & 7 \\ 2 & 3 \end{bmatrix} + \begin{bmatrix} -5 & 5 \\ -1 & 2 \end{bmatrix} = \begin{bmatrix} 0 & 12 \\ 1 & 5 \end{bmatrix} \quad \text{Thus,} \quad A(B + C) = AB + AC.$$

43. $A^2 = AA = \begin{bmatrix} 1 & 2 \\ 0 & 1 \end{bmatrix}\begin{bmatrix} 1 & 2 \\ 0 & 1 \end{bmatrix} = \begin{bmatrix} [1 \quad 2]\begin{bmatrix} 1 \\ 0 \end{bmatrix} & [1 \quad 2]\begin{bmatrix} 2 \\ 1 \end{bmatrix} \\ [0 \quad 1]\begin{bmatrix} 1 \\ 0 \end{bmatrix} & [0 \quad 1]\begin{bmatrix} 2 \\ 1 \end{bmatrix} \end{bmatrix} = \begin{bmatrix} 1 & 4 \\ 0 & 1 \end{bmatrix}$

$B^2 = BB = \begin{bmatrix} 1 & 1 \\ 2 & 3 \end{bmatrix}\begin{bmatrix} 1 & 1 \\ 2 & 3 \end{bmatrix} = \begin{bmatrix} [1 \quad 1]\begin{bmatrix} 1 \\ 2 \end{bmatrix} & [1 \quad 1]\begin{bmatrix} 1 \\ 3 \end{bmatrix} \\ [2 \quad 3]\begin{bmatrix} 1 \\ 2 \end{bmatrix} & [2 \quad 3]\begin{bmatrix} 1 \\ 3 \end{bmatrix} \end{bmatrix} = \begin{bmatrix} 3 & 4 \\ 8 & 11 \end{bmatrix}$

Now $A^2 - B^2 = \begin{bmatrix} 1 & 4 \\ 0 & 1 \end{bmatrix} - \begin{bmatrix} 3 & 4 \\ 8 & 11 \end{bmatrix} = \begin{bmatrix} -2 & 0 \\ -8 & -10 \end{bmatrix}$

$A - B = \begin{bmatrix} 1 & 2 \\ 0 & 1 \end{bmatrix} - \begin{bmatrix} 1 & 1 \\ 2 & 3 \end{bmatrix} = \begin{bmatrix} 0 & 1 \\ -2 & -2 \end{bmatrix}$

$A + B = \begin{bmatrix} 1 & 2 \\ 0 & 1 \end{bmatrix} + \begin{bmatrix} 1 & 1 \\ 2 & 3 \end{bmatrix} = \begin{bmatrix} 2 & 3 \\ 2 & 4 \end{bmatrix}$

and

$$(A - B)(A + B) = \begin{bmatrix} 0 & 1 \\ -2 & -2 \end{bmatrix}\begin{bmatrix} 2 & 3 \\ 2 & 4 \end{bmatrix} = \begin{bmatrix} [0 \quad 1]\begin{bmatrix} 2 \\ 2 \end{bmatrix} & [0 \quad 1]\begin{bmatrix} 3 \\ 4 \end{bmatrix} \\ [-2 \quad -2]\begin{bmatrix} 2 \\ 2 \end{bmatrix} & [-2 \quad -2]\begin{bmatrix} 3 \\ 4 \end{bmatrix} \end{bmatrix} = \begin{bmatrix} 2 & 4 \\ -8 & -14 \end{bmatrix}$$

Thus, $A^2 - B^2 \neq (A - B)(A + B)$.

45. (A) $[0.6 \quad 0.6 \quad 0.2] \begin{bmatrix} 8 \\ 10 \\ 5 \end{bmatrix} = [4.8 + 6.0 + 1.0] = [11.8]$

Thus, the labor cost per boat for one-person boats at plant I is $11.80.

(B) $[1.5 \quad 1.2 \quad 0.4] \begin{bmatrix} 9 \\ 12 \\ 6 \end{bmatrix} = [13.5 + 14.4 + 2.4] = [30.3]$

Thus, the labor cost per boat for four-person boats at plant II is $30.30.

(C) Dimension of $M = 3 \times 3$.

Dimension of $N = 3 \times 2$. Thus, the dimension of MN is 3×2.

(D) $MN = \begin{bmatrix} 0.6 & 0.6 & 0.2 \\ 1.0 & 0.9 & 0.3 \\ 1.5 & 1.2 & 0.4 \end{bmatrix} \begin{bmatrix} 8 & 9 \\ 10 & 12 \\ 5 & 6 \end{bmatrix}$

$$= \begin{bmatrix} [0.6 \quad 0.6 \quad 0.2]\begin{bmatrix}8\\10\\5\end{bmatrix} & [0.6 \quad 0.6 \quad 0.2]\begin{bmatrix}9\\12\\6\end{bmatrix} \\ [1.0 \quad 0.9 \quad 0.3]\begin{bmatrix}8\\10\\5\end{bmatrix} & [1.0 \quad 0.9 \quad 0.3]\begin{bmatrix}9\\12\\6\end{bmatrix} \\ [1.5 \quad 1.2 \quad 0.4]\begin{bmatrix}8\\10\\5\end{bmatrix} & [1.5 \quad 1.2 \quad 0.4]\begin{bmatrix}9\\12\\6\end{bmatrix} \end{bmatrix}$$

	Plant I	Plant II	
$=$	$11.80	$13.80	One-person boat
	$18.50	$21.60	Two-person boat
	$26.00	$30.30	Four-person boat

This matrix represents the labor cost per boat for each kind of boat at each plant. For example, $21.60 represents labor costs per boat for two-person boats at plant II.

47. $A = \begin{bmatrix} 0 & 1 & 0 & 1 & 0 \\ 0 & 0 & 1 & 0 & 0 \\ 1 & 0 & 0 & 0 & 1 \\ 0 & 0 & 1 & 0 & 0 \\ 0 & 0 & 0 & 1 & 0 \end{bmatrix}$

(A) $A^2 = \begin{bmatrix} 0 & 1 & 0 & 1 & 0 \\ 0 & 0 & 1 & 0 & 0 \\ 1 & 0 & 0 & 0 & 1 \\ 0 & 0 & 1 & 0 & 0 \\ 0 & 0 & 0 & 1 & 0 \end{bmatrix} \begin{bmatrix} 0 & 1 & 0 & 1 & 0 \\ 0 & 0 & 1 & 0 & 0 \\ 1 & 0 & 0 & 0 & 1 \\ 0 & 0 & 1 & 0 & 0 \\ 0 & 0 & 0 & 1 & 0 \end{bmatrix} = \begin{bmatrix} 0 & 0 & 2 & 0 & 0 \\ 1 & 0 & 0 & 0 & 1 \\ 0 & 1 & 0 & 2 & 0 \\ 1 & 0 & 0 & 0 & 1 \\ 0 & 0 & 1 & 0 & 0 \end{bmatrix}$

The 1 in row two, column one indicates that there is one way to travel from Baltimore to Atlanta with one intermediate connection, namely Baltimore-to-Chicago-to-Atlanta.

The 2 in row one, column three indicates that there are two ways to travel from Atlanta to Chicago with one intermediate connection, namely Atlanta-to-Baltimore-to-Chicago, and Atlanta-to-Denver-to-Chicago.

In general, the element b_{ij}, $i \neq j$, in A^2 indicates the number of different ways to travel from the ith city to the jth city with one intermediate connection.

(B) $A^3 = AA^2 = \begin{bmatrix} 0 & 1 & 0 & 1 & 0 \\ 0 & 0 & 1 & 0 & 0 \\ 1 & 0 & 0 & 0 & 1 \\ 0 & 0 & 1 & 0 & 0 \\ 0 & 0 & 0 & 1 & 0 \end{bmatrix} \begin{bmatrix} 0 & 0 & 2 & 0 & 0 \\ 1 & 0 & 0 & 0 & 1 \\ 0 & 1 & 0 & 2 & 0 \\ 1 & 0 & 0 & 0 & 1 \\ 0 & 0 & 1 & 0 & 0 \end{bmatrix} = \begin{bmatrix} 2 & 0 & 0 & 0 & 2 \\ 0 & 1 & 0 & 2 & 0 \\ 0 & 0 & 3 & 0 & 0 \\ 0 & 1 & 0 & 2 & 0 \\ 1 & 0 & 0 & 0 & 1 \end{bmatrix}$

The 1 in row four, column 2 indicates that there is one way to travel from Denver to Baltimore with two intermediate connections.

The 2 in row one, column five indicates that there are two ways to travel from Atlanta to El Paso with two intermediate connections.

In general, the element c_{ij}, $i \neq j$, in A^3 indicates the number of ways to travel from the ith city to the jth city with two intermediate connections.

49. The incidence matrix is:

$A = \begin{bmatrix} 0 & 0 & 0 & 1 & 1 \\ 1 & 0 & 0 & 1 & 0 \\ 0 & 1 & 0 & 0 & 0 \\ 1 & 0 & 0 & 0 & 1 \\ 0 & 1 & 1 & 0 & 0 \end{bmatrix}$

$A^2 = \begin{bmatrix} 0 & 0 & 0 & 1 & 1 \\ 1 & 0 & 0 & 1 & 0 \\ 0 & 1 & 0 & 0 & 0 \\ 1 & 0 & 0 & 0 & 1 \\ 0 & 1 & 1 & 0 & 0 \end{bmatrix} \begin{bmatrix} 0 & 0 & 0 & 1 & 1 \\ 1 & 0 & 0 & 1 & 0 \\ 0 & 1 & 0 & 0 & 0 \\ 1 & 0 & 0 & 0 & 1 \\ 0 & 1 & 1 & 0 & 0 \end{bmatrix} = \begin{bmatrix} 1 & 1 & 1 & 0 & 1 \\ 1 & 0 & 0 & 1 & 2 \\ 1 & 0 & 0 & 1 & 0 \\ 0 & 1 & 1 & 1 & 1 \\ 1 & 1 & 0 & 1 & 0 \end{bmatrix}$

$$A^3 = AA^2 = \begin{bmatrix} 0 & 0 & 0 & 1 & 1 \\ 1 & 0 & 0 & 1 & 0 \\ 0 & 1 & 0 & 0 & 0 \\ 1 & 0 & 0 & 0 & 1 \\ 0 & 1 & 1 & 0 & 0 \end{bmatrix} \begin{bmatrix} 1 & 1 & 1 & 0 & 1 \\ 1 & 0 & 0 & 1 & 2 \\ 1 & 0 & 0 & 1 & 0 \\ 0 & 1 & 1 & 1 & 1 \\ 1 & 1 & 0 & 1 & 0 \end{bmatrix} = \begin{bmatrix} 1 & 2 & 1 & 2 & 1 \\ 1 & 2 & 2 & 1 & 2 \\ 1 & 0 & 0 & 1 & 2 \\ 2 & 2 & 1 & 1 & 1 \\ 2 & 0 & 0 & 2 & 2 \end{bmatrix}$$

$$A + A^2 + A^3 = \begin{bmatrix} 0 & 0 & 0 & 1 & 1 \\ 1 & 0 & 0 & 1 & 0 \\ 0 & 1 & 0 & 0 & 0 \\ 1 & 0 & 0 & 0 & 1 \\ 0 & 1 & 1 & 0 & 0 \end{bmatrix} + \begin{bmatrix} 1 & 1 & 1 & 0 & 1 \\ 1 & 0 & 0 & 1 & 2 \\ 1 & 0 & 0 & 1 & 0 \\ 0 & 1 & 1 & 1 & 1 \\ 1 & 1 & 0 & 1 & 0 \end{bmatrix} + \begin{bmatrix} 1 & 2 & 1 & 2 & 1 \\ 1 & 2 & 2 & 1 & 2 \\ 1 & 0 & 0 & 1 & 2 \\ 2 & 2 & 1 & 1 & 1 \\ 2 & 0 & 0 & 2 & 2 \end{bmatrix}$$

$$= \begin{bmatrix} 2 & 3 & 2 & 3 & 3 \\ 3 & 2 & 2 & 3 & 4 \\ 2 & 1 & 0 & 2 & 2 \\ 3 & 3 & 2 & 2 & 3 \\ 3 & 2 & 1 & 3 & 2 \end{bmatrix}$$

Interpretation: it is possible to travel from any origin to any destination with at most two intermediate connections.

51. (A) $[1000 \quad 500 \quad 5000] \begin{bmatrix} 0.4 \\ 0.75 \\ 0.25 \end{bmatrix} = [1000(0.4) + 500(0.75) + 5000(0.25)] = [2025]$

Thus, the total amount spent in Berkeley = $2025.

(B) $[2000 \quad 800 \quad 8000] \begin{bmatrix} 0.40 \\ 0.75 \\ 0.25 \end{bmatrix} = [2000(0.4) + 800(0.75) + 8000(0.25)] = [3400]$

Thus, the total amount spent in Oakland = $3400.

(C) $NM = \begin{bmatrix} 1000 & 500 & 5000 \\ 2000 & 800 & 8000 \end{bmatrix} \begin{bmatrix} 0.4 \\ 0.75 \\ 0.25 \end{bmatrix} = \begin{bmatrix} [1000 \quad 500 \quad 5000] \begin{bmatrix} 0.4 \\ 0.75 \\ 0.25 \end{bmatrix} \\ [2000 \quad 800 \quad 8000] \begin{bmatrix} 0.4 \\ 0.75 \\ 0.25 \end{bmatrix} \end{bmatrix}$

This matrix represents cost per town. $= \begin{bmatrix} \$2025 \\ \$3400 \end{bmatrix} \begin{matrix} \text{Berkeley} \\ \text{Oakland} \end{matrix}$

(D) $[1 \quad 1] \begin{bmatrix} 1000 & 500 & 5000 \\ 2000 & 800 & 8000 \end{bmatrix} = \begin{bmatrix} [1 \quad 1] \begin{bmatrix} 1000 \\ 2000 \end{bmatrix} & [1 \quad 1] \begin{bmatrix} 500 \\ 800 \end{bmatrix} & [1 \quad 1] \begin{bmatrix} 5000 \\ 8000 \end{bmatrix} \end{bmatrix}$

$$\begin{matrix} \text{Phone} & \text{House} & \text{Letter} \\ = [3000 & 1300 & 13,000] \end{matrix}$$

This matrix indicates the number of each type of contact made in both towns.

Things to remember:

1. The IDENTITY element for multiplication for the set of square
    matrices of order n (dimension $n \times n$) is the square matrix I of
    order n which has 1's on the principal diagonal (upper left
    corner to lower right corner) and 0's elsewhere. The identity
    matrices of order 2 and 3, respectively, are

    $$I = \begin{bmatrix} 1 & 0 \\ 0 & 1 \end{bmatrix} \quad \text{and} \quad I = \begin{bmatrix} 1 & 0 & 0 \\ 0 & 1 & 0 \\ 0 & 0 & 1 \end{bmatrix}.$$

2. If M is any square matrix of order n and I is the identity
    matrix of order n, then

    $$IM = MI = M.$$

3. INVERSE OF A SQUARE MATRIX

    Let M be a square matrix of order n and I be the identity
    matrix of order n. If there exists a matrix M^{-1} such that
    $$MM^{-1} = M^{-1}M = I$$
    then M^{-1} is called the MULTIPLICATIVE INVERSE OF M or, more
    simply, the INVERSE OF M. M^{-1} is read "M inverse."

4. If the augmented matrix $[M \mid I]$ is transformed by row operations
    into $[I \mid B]$, then the resulting matrix B is M^{-1}. However, if
    all zeros are obtained in one or more rows to the left of the
    vertical line during the row transformation procedure, then M^{-1}
    does not exist.

---

1. $\begin{bmatrix} 1 & 0 \\ 0 & 1 \end{bmatrix} \begin{bmatrix} 2 & -3 \\ 4 & 5 \end{bmatrix} = \begin{bmatrix} 1 \cdot 2 + 0 \cdot 4 & 1(-3) + 0 \cdot 5 \\ 0 \cdot 2 + 1 \cdot 4 & 0(-3) + 1 \cdot 5 \end{bmatrix} = \begin{bmatrix} 2 & -3 \\ 4 & 5 \end{bmatrix}$

3. $\begin{bmatrix} 2 & -3 \\ 4 & 5 \end{bmatrix} \begin{bmatrix} 1 & 0 \\ 0 & 1 \end{bmatrix} = \begin{bmatrix} 2 \cdot 1 + (-3)0 & 2 \cdot 0 + (-3)1 \\ 4 \cdot 1 + 5 \cdot 0 & 4 \cdot 0 + 5 \cdot 1 \end{bmatrix} = \begin{bmatrix} 2 & -3 \\ 4 & 5 \end{bmatrix}$

5. $\begin{bmatrix} 1 & 0 & 0 \\ 0 & 1 & 0 \\ 0 & 0 & 1 \end{bmatrix} \begin{bmatrix} -2 & 1 & 3 \\ 2 & 4 & -2 \\ 5 & 1 & 0 \end{bmatrix}$

$= \begin{bmatrix} 1(-2) + 0 \cdot 2 + 0 \cdot 5 & 1 \cdot 1 + 0 \cdot 4 + 0 \cdot 1 & 1 \cdot 3 + 0(-2) + 0 \cdot 0 \\ 0(-2) + 1 \cdot 2 + 0 \cdot 5 & 0 \cdot 1 + 1 \cdot 4 + 0 \cdot 1 & 0 \cdot 3 + 1(-2) + 0 \cdot 0 \\ 0(-2) + 0 \cdot 2 + 1 \cdot 5 & 0 \cdot 1 + 0 \cdot 4 + 1 \cdot 1 & 0 \cdot 3 + 0(-2) + 1 \cdot 0 \end{bmatrix} = \begin{bmatrix} -2 & 1 & 3 \\ 2 & 4 & -2 \\ 5 & 1 & 0 \end{bmatrix}$

7. $\begin{bmatrix} -2 & 1 & 3 \\ 2 & 4 & -2 \\ 5 & 1 & 0 \end{bmatrix} \begin{bmatrix} 1 & 0 & 0 \\ 0 & 1 & 0 \\ 0 & 0 & 1 \end{bmatrix}$

$$= \begin{bmatrix} (-2)\cdot 1 + 1\cdot 0 + 3\cdot 0 & (-2)0 + 1\cdot 1 + 3\cdot 0 & (-2)0 + 1\cdot 0 + 3\cdot 1 \\ 2\cdot 1 + 4\cdot 0 + (-2)0 & 2\cdot 0 + 4\cdot 1 + (-2)0 & 2\cdot 0 + 4\cdot 0 + (-2)1 \\ 5\cdot 1 + 1\cdot 0 + 0\cdot 0 & 5\cdot 0 + 1\cdot 1 + 0\cdot 0 & 5\cdot 0 + 1\cdot 0 + 0\cdot 1 \end{bmatrix}$$

$$= \begin{bmatrix} -2 & 1 & 3 \\ 2 & 4 & -2 \\ 5 & 1 & 0 \end{bmatrix}$$

9. $\begin{bmatrix} 3 & -4 \\ -2 & 3 \end{bmatrix} \begin{bmatrix} 3 & 4 \\ 2 & 3 \end{bmatrix} = \begin{bmatrix} 3\cdot 3 + (-4)2 & 3\cdot 4 + (-4)3 \\ (-2)3 + 3\cdot 2 & (-2)4 + 3\cdot 3 \end{bmatrix} = \begin{bmatrix} 1 & 0 \\ 0 & 1 \end{bmatrix}$

11. $\begin{bmatrix} -5 & 2 \\ -8 & 3 \end{bmatrix} \begin{bmatrix} 3 & -2 \\ 8 & -5 \end{bmatrix} = \begin{bmatrix} (-5)3 + 2\cdot 8 & (-5)(-2) + 2(-5) \\ (-8)3 + 3\cdot 8 & (-8)(-2) + 3(-5) \end{bmatrix} = \begin{bmatrix} 1 & 0 \\ 0 & 1 \end{bmatrix}$

13. $\begin{bmatrix} 1 & -1 & 1 \\ 0 & 2 & -1 \\ 2 & 3 & 0 \end{bmatrix} \begin{bmatrix} 3 & 3 & -1 \\ -2 & -2 & 1 \\ -4 & -5 & 2 \end{bmatrix}$

$$= \begin{bmatrix} 1\cdot 3 + (-1)(-2) + 1(-4) & 1\cdot 3 + (-1)(-2) + 1(-5) & 1(-1) + (-1)1 + 1\cdot 2 \\ 0\cdot 3 + 2(-2) + (-1)(-4) & 0\cdot 3 + 2(-2) + (-1)(-5) & 0(-1) + 2\cdot 1 + (-1)2 \\ 2\cdot 3 + 3(-2) + 0(-4) & 2\cdot 3 + 3(-2) + 0(-5) & 2(-1) + 3\cdot 1 + 0\cdot 2 \end{bmatrix}$$

$$= \begin{bmatrix} 1 & 0 & 0 \\ 0 & 1 & 0 \\ 0 & 0 & 1 \end{bmatrix}$$

15. $\left[\begin{array}{cc|cc} -1 & 0 & 1 & 0 \\ -3 & 1 & 0 & 1 \end{array}\right] \sim \left[\begin{array}{cc|cc} 1 & 0 & -1 & 0 \\ -3 & 1 & 0 & 1 \end{array}\right] \sim \left[\begin{array}{cc|cc} 1 & 0 & -1 & 0 \\ 0 & 1 & -3 & 1 \end{array}\right]$

$\qquad (-1)R_1 \rightarrow R_1 \qquad\qquad R_2 + 3R_1 \rightarrow R_2$

Thus, $M^{-1} = \begin{bmatrix} -1 & 0 \\ -3 & 1 \end{bmatrix}$

Check:

$M \cdot M^{-1} = \begin{bmatrix} -1 & 0 \\ -3 & 1 \end{bmatrix} \begin{bmatrix} -1 & 0 \\ -3 & 1 \end{bmatrix} = \begin{bmatrix} (-1)(-1) + 0(-3) & (-1)0 + 0\cdot 1 \\ (-3)(-1) + 1(-3) & (-3)0 + 1\cdot 1 \end{bmatrix} = \begin{bmatrix} 1 & 0 \\ 0 & 1 \end{bmatrix}$

17. $\left[\begin{array}{cc|cc} 1 & 2 & 1 & 0 \\ 1 & 3 & 0 & 1 \end{array}\right] \sim \left[\begin{array}{cc|cc} 1 & 2 & 1 & 0 \\ 0 & 1 & -1 & 1 \end{array}\right] \sim \left[\begin{array}{cc|cc} 1 & 0 & 3 & -2 \\ 0 & 1 & -1 & 1 \end{array}\right]$

$R_2 + (-1)R_1 \rightarrow R_2 \qquad R_1 + (-2)R_2 \rightarrow R_1$

Thus, $M^{-1} = \begin{bmatrix} 3 & -2 \\ -1 & 1 \end{bmatrix}$.

Check:

$$M \cdot M^{-1} = \begin{bmatrix} 1 & 2 \\ 1 & 3 \end{bmatrix}\begin{bmatrix} 3 & -2 \\ -1 & 1 \end{bmatrix} = \begin{bmatrix} 1 \cdot 3 + 2(-1) & 1(-2) + 2 \cdot 1 \\ 1 \cdot 3 + 3(-1) & 1(-2) + 3 \cdot 1 \end{bmatrix} = \begin{bmatrix} 1 & 0 \\ 0 & 1 \end{bmatrix}$$

19. $\left[\begin{array}{cc|cc} 1 & 3 & 1 & 0 \\ 2 & 7 & 0 & 1 \end{array}\right] \sim \left[\begin{array}{cc|cc} 1 & 3 & 1 & 0 \\ 0 & 1 & -2 & 1 \end{array}\right] \sim \left[\begin{array}{cc|cc} 1 & 0 & 7 & -3 \\ 0 & 1 & -2 & 1 \end{array}\right]$

$\quad R_2 + (-2)R_1 \to R_2 \qquad R_1 + (-3)R_2 \to R_1$

Thus, $M^{-1} = \begin{bmatrix} 7 & -3 \\ -2 & 1 \end{bmatrix}$.

Check:

$$\begin{bmatrix} 1 & 3 \\ 2 & 7 \end{bmatrix}\begin{bmatrix} 7 & -3 \\ -2 & 1 \end{bmatrix} = \begin{bmatrix} 1 \cdot 7 + 3(-2) & 1(-3) + 3 \cdot 1 \\ 2 \cdot 7 + 7(-2) & 2(-3) + 7 \cdot 1 \end{bmatrix} = \begin{bmatrix} 1 & 0 \\ 0 & 1 \end{bmatrix}$$

21. $\left[\begin{array}{ccc|ccc} 1 & -3 & 0 & 1 & 0 & 0 \\ 0 & 3 & 1 & 0 & 1 & 0 \\ 2 & -1 & 2 & 0 & 0 & 1 \end{array}\right] \sim \left[\begin{array}{ccc|ccc} 1 & -3 & 0 & 1 & 0 & 0 \\ 0 & 3 & 1 & 0 & 1 & 0 \\ 0 & 5 & 2 & -2 & 0 & 1 \end{array}\right]$

$\qquad R_3 + (-2)R_1 \to R_3 \qquad\qquad \frac{1}{3}R_2 \to R_2$

$\sim \left[\begin{array}{ccc|ccc} 1 & -3 & 0 & 1 & 0 & 0 \\ 0 & 1 & \frac{1}{3} & 0 & \frac{1}{3} & 0 \\ 0 & 5 & 2 & -2 & 0 & 1 \end{array}\right] \sim \left[\begin{array}{ccc|ccc} 1 & -3 & 0 & 1 & 0 & 0 \\ 0 & 1 & \frac{1}{3} & 0 & \frac{1}{3} & 0 \\ 0 & 0 & \frac{1}{3} & -2 & -\frac{5}{3} & 1 \end{array}\right]$

$\qquad R_3 + (-5)R_2 \to R_3 \qquad\qquad 3R_3 \to R_3$

$\sim \left[\begin{array}{ccc|ccc} 1 & -3 & 0 & 1 & 0 & 0 \\ 0 & 1 & \frac{1}{3} & 0 & \frac{1}{3} & 0 \\ 0 & 0 & 1 & -6 & -5 & 3 \end{array}\right] \sim \left[\begin{array}{ccc|ccc} 1 & -3 & 0 & 1 & 0 & 0 \\ 0 & 1 & 0 & 2 & 2 & -1 \\ 0 & 0 & 1 & -6 & -5 & 3 \end{array}\right]$

$\qquad R_2 + \left(-\frac{1}{3}\right)R_3 \to R_2 \qquad\qquad R_1 + 3R_2 \to R_1$

$\sim \left[\begin{array}{ccc|ccc} 1 & 0 & 0 & 7 & 6 & -3 \\ 0 & 1 & 0 & 2 & 2 & -1 \\ 0 & 0 & 1 & -6 & -5 & 3 \end{array}\right]$ Thus, $M^{-1} = \begin{bmatrix} 7 & 6 & -3 \\ 2 & 2 & -1 \\ -6 & -5 & 3 \end{bmatrix}$.

Check:

$$\begin{bmatrix} 1 & -3 & 0 \\ 0 & 3 & 1 \\ 2 & -1 & 2 \end{bmatrix}\begin{bmatrix} 7 & 6 & -3 \\ 2 & 2 & -1 \\ -6 & -5 & 3 \end{bmatrix}$$

$$= \begin{bmatrix} 1\cdot 7 + (-3)2 + 0\cdot(-6) & 1\cdot 6 + (-3)2 + 0(-5) & 1(-3) + (-3)(-1) + 0\cdot 3 \\ 0\cdot 7 + 3\cdot 2 + 1(-6) & 0\cdot 6 + 3\cdot 2 + 1(-5) & 0(-3) + 3(-1) + 1\cdot 3 \\ 2\cdot 7 + (-1)2 + 2(-6) & 2\cdot 6 + (-1)2 + 2(-5) & 2(-3) + (-1)(-1) + 2\cdot 3 \end{bmatrix}$$

$$= \begin{bmatrix} 1 & 0 & 0 \\ 0 & 1 & 0 \\ 0 & 0 & 1 \end{bmatrix}$$

23. $\begin{bmatrix} 1 & 1 & 0 & | & 1 & 0 & 0 \\ 0 & 3 & -1 & | & 0 & 1 & 0 \\ 1 & 0 & 1 & | & 0 & 0 & 1 \end{bmatrix} \sim \begin{bmatrix} 1 & 1 & 0 & | & 1 & 0 & 0 \\ 0 & 3 & -1 & | & 0 & 1 & 0 \\ 0 & -1 & 1 & | & -1 & 0 & 1 \end{bmatrix}$

$\qquad\qquad R_3 + (-1)R_1 \rightarrow R_3 \qquad\qquad\qquad R_2 \leftrightarrow R_3$

$\sim \begin{bmatrix} 1 & 1 & 0 & | & 1 & 0 & 0 \\ 0 & -1 & 1 & | & -1 & 0 & 1 \\ 0 & 3 & -1 & | & 0 & 1 & 0 \end{bmatrix} \sim \begin{bmatrix} 1 & 1 & 0 & | & 1 & 0 & 0 \\ 0 & 1 & -1 & | & 1 & 0 & -1 \\ 0 & 3 & -1 & | & 0 & 1 & 0 \end{bmatrix}$

$\qquad\qquad (-1)R_2 \rightarrow R_2 \qquad\qquad\qquad R_3 + (-3)R_2 \rightarrow R_3$

$\sim \begin{bmatrix} 1 & 1 & 0 & | & 1 & 0 & 0 \\ 0 & 1 & -1 & | & 1 & 0 & -1 \\ 0 & 0 & 2 & | & -3 & 1 & 3 \end{bmatrix} \sim \begin{bmatrix} 1 & 1 & 0 & | & 1 & 0 & 0 \\ 0 & 1 & -1 & | & 1 & 0 & -1 \\ 0 & 0 & 1 & | & -\frac{3}{2} & \frac{1}{2} & \frac{3}{2} \end{bmatrix}$

$\qquad\qquad \frac{1}{2}R_3 \rightarrow R_3 \qquad\qquad\qquad R_2 + R_3 \rightarrow R_2$

$\sim \begin{bmatrix} 1 & 1 & 0 & | & 1 & 0 & 0 \\ 0 & 1 & 0 & | & -\frac{1}{2} & \frac{1}{2} & \frac{1}{2} \\ 0 & 0 & 1 & | & -\frac{3}{2} & \frac{1}{2} & \frac{3}{2} \end{bmatrix} \sim \begin{bmatrix} 1 & 0 & 0 & | & \frac{3}{2} & -\frac{1}{2} & -\frac{1}{2} \\ 0 & 1 & 0 & | & -\frac{1}{2} & \frac{1}{2} & \frac{1}{2} \\ 0 & 0 & 1 & | & -\frac{3}{2} & \frac{1}{2} & \frac{3}{2} \end{bmatrix}$

$\qquad\qquad R_1 + (-1)R_2 \rightarrow R_1$

Thus, $M^{-1} = \begin{bmatrix} \frac{3}{2} & -\frac{1}{2} & -\frac{1}{2} \\ -\frac{1}{2} & \frac{1}{2} & \frac{1}{2} \\ -\frac{3}{2} & \frac{1}{2} & \frac{3}{2} \end{bmatrix} = \frac{1}{2}\begin{bmatrix} 3 & -1 & -1 \\ -1 & 1 & 1 \\ -3 & 1 & 3 \end{bmatrix}.$

<u>Check</u>:

$\begin{bmatrix} 1 & 1 & 0 \\ 0 & 3 & -1 \\ 1 & 0 & 1 \end{bmatrix}\begin{bmatrix} \frac{3}{2} & -\frac{1}{2} & -\frac{1}{2} \\ -\frac{1}{2} & \frac{1}{2} & \frac{1}{2} \\ -\frac{3}{2} & \frac{1}{2} & \frac{3}{2} \end{bmatrix} = \begin{bmatrix} \frac{3}{2} - \frac{1}{2} & -\frac{1}{2} + \frac{1}{2} & -\frac{1}{2} + \frac{1}{2} \\ -\frac{3}{2} + \frac{3}{2} & \frac{3}{2} - \frac{1}{2} & \frac{3}{2} - \frac{3}{2} \\ \frac{3}{2} - \frac{3}{2} & -\frac{1}{2} + \frac{1}{2} & -\frac{1}{2} + \frac{3}{2} \end{bmatrix} = \begin{bmatrix} 1 & 0 & 0 \\ 0 & 1 & 0 \\ 0 & 0 & 1 \end{bmatrix}$

25. $\begin{bmatrix} 3 & 2 & | & 1 & 0 \\ -4 & -3 & | & 0 & 1 \end{bmatrix} \sim \begin{bmatrix} 1 & \frac{2}{3} & | & \frac{1}{3} & 0 \\ -4 & -3 & | & 0 & 1 \end{bmatrix} \sim \begin{bmatrix} 1 & \frac{2}{3} & | & \frac{1}{3} & 0 \\ 0 & -\frac{1}{3} & | & \frac{4}{3} & 1 \end{bmatrix}$

$\qquad\quad \frac{1}{3}R_1 \rightarrow R_1 \qquad\qquad R_2 + 4R_1 \rightarrow R_2 \qquad\qquad -3R_2 \rightarrow R_2$

$$\sim \begin{bmatrix} 1 & \frac{2}{3} & \Big| & \frac{1}{3} & 0 \\ 0 & 1 & \Big| & -4 & -3 \end{bmatrix} \sim \begin{bmatrix} 1 & 0 & \Big| & 3 & 2 \\ 0 & 1 & \Big| & -4 & -3 \end{bmatrix}$$

$$R_1 + \left(-\frac{2}{3}\right) R_2 \rightarrow R_1$$

Thus, $M^{-1} = \begin{bmatrix} 3 & 2 \\ -4 & -3 \end{bmatrix}$.

27. $\begin{bmatrix} 3 & 9 & \Big| & 1 & 0 \\ 2 & 6 & \Big| & 0 & 1 \end{bmatrix} \sim \begin{bmatrix} 1 & 3 & \Big| & \frac{1}{3} & 0 \\ 2 & 6 & \Big| & 0 & 1 \end{bmatrix} \sim \begin{bmatrix} 1 & 3 & \Big| & \frac{1}{3} & 0 \\ 0 & 0 & \Big| & -\frac{2}{3} & 1 \end{bmatrix}$

$\quad\quad \frac{1}{3} R_1 \rightarrow R_1 \quad\quad\quad R_2 + (-2) R_1 \rightarrow R_2$

From this matrix, we conclude that the inverse does not exist. See 4.
All entries to the left of the vertical line of row 2 (R_2) are zero.

29. $\begin{bmatrix} 3 & 1 & \Big| & 1 & 0 \\ 4 & 2 & \Big| & 0 & 1 \end{bmatrix} \sim \begin{bmatrix} 1 & \frac{1}{3} & \Big| & \frac{1}{3} & 0 \\ 4 & 2 & \Big| & 0 & 1 \end{bmatrix} \sim \begin{bmatrix} 1 & \frac{1}{3} & \Big| & \frac{1}{3} & 0 \\ 0 & \frac{2}{3} & \Big| & -\frac{4}{3} & 1 \end{bmatrix} \sim \begin{bmatrix} 1 & \frac{1}{3} & \Big| & \frac{1}{3} & 0 \\ 0 & 1 & \Big| & -2 & \frac{3}{2} \end{bmatrix}$

$\quad\quad \frac{1}{3} R_1 \rightarrow R_1 \quad\quad\quad R_2 + (-4) R_1 \rightarrow R_2 \quad\quad\quad \frac{3}{2} R_2 \rightarrow R_2 \quad\quad\quad R_1 + \left(-\frac{1}{3}\right) R_2 \rightarrow R_1$

$\sim \begin{bmatrix} 1 & 0 & \Big| & 1 & -\frac{1}{2} \\ 0 & 1 & \Big| & -2 & \frac{3}{2} \end{bmatrix}$ Thus, the inverse is $\begin{bmatrix} 1 & -\frac{1}{2} \\ -2 & \frac{3}{2} \end{bmatrix}$.

31. $\begin{bmatrix} -5 & -2 & -2 & \Big| & 1 & 0 & 0 \\ 2 & 1 & 0 & \Big| & 0 & 1 & 0 \\ 1 & 0 & 1 & \Big| & 0 & 0 & 1 \end{bmatrix} \sim \begin{bmatrix} 1 & 0 & 1 & \Big| & 0 & 0 & 1 \\ 2 & 1 & 0 & \Big| & 0 & 1 & 0 \\ -5 & -2 & -2 & \Big| & 1 & 0 & 0 \end{bmatrix}$

$\quad\quad\quad\quad R_1 \leftrightarrow R_3 \quad\quad\quad\quad\quad\quad\quad R_2 + (-2) R_1 \rightarrow R_2$
$\quad\quad\quad\quad\quad\quad\quad\quad\quad\quad\quad\quad\quad\quad R_3 + 5 R_1 \rightarrow R_3$

$\sim \begin{bmatrix} 1 & 0 & 1 & \Big| & 0 & 0 & 1 \\ 0 & 1 & -2 & \Big| & 0 & 1 & -2 \\ 0 & -2 & 3 & \Big| & 1 & 0 & 5 \end{bmatrix} \sim \begin{bmatrix} 1 & 0 & 1 & \Big| & 0 & 0 & 1 \\ 0 & 1 & -2 & \Big| & 0 & 1 & -2 \\ 0 & 0 & -1 & \Big| & 1 & 2 & 1 \end{bmatrix}$

$\quad\quad\quad R_3 + 2 R_2 \rightarrow R_3 \quad\quad\quad\quad\quad\quad (-1) R_3 \rightarrow R_3$

$\sim \begin{bmatrix} 1 & 0 & 1 & \Big| & 0 & 0 & 1 \\ 0 & 1 & -2 & \Big| & 0 & 1 & -2 \\ 0 & 0 & 1 & \Big| & -1 & -2 & -1 \end{bmatrix} \sim \begin{bmatrix} 1 & 0 & 0 & \Big| & 1 & 2 & 2 \\ 0 & 1 & 0 & \Big| & -2 & -3 & -4 \\ 0 & 0 & 1 & \Big| & -1 & -2 & -1 \end{bmatrix}$

$\quad R_2 + 2 R_3 \rightarrow R_2$
$\quad R_1 + (-1) R_3 \rightarrow R_1$

Thus, the inverse is $\begin{bmatrix} 1 & 2 & 2 \\ -2 & -3 & -4 \\ -1 & -2 & -1 \end{bmatrix}$.

33. $\begin{bmatrix} 2 & 1 & 1 \\ 1 & 1 & 0 \\ -1 & -1 & 0 \end{bmatrix} \begin{matrix} 1 & 0 & 0 \\ 0 & 1 & 0 \\ 0 & 0 & 1 \end{matrix}$ ~ $\begin{bmatrix} 1 & 1 & 0 \\ 2 & 1 & 1 \\ -1 & -1 & 0 \end{bmatrix} \begin{matrix} 0 & 1 & 0 \\ 1 & 0 & 0 \\ 0 & 0 & 1 \end{matrix}$ ~ $\begin{bmatrix} 1 & 1 & 0 \\ 0 & -1 & 1 \\ 0 & 0 & 0 \end{bmatrix} \begin{matrix} 0 & 1 & 0 \\ 1 & -2 & 0 \\ 0 & 1 & 1 \end{matrix}$

$\quad\quad R_1 \leftrightarrow R_2 \quad\quad\quad\quad\quad\quad R_2 + (-2)R_1 \rightarrow R_2$

$\quad\quad\quad\quad\quad\quad\quad\quad\quad\quad\quad\quad R_3 + R_1 \rightarrow R_3$

From this matrix, we conclude that the inverse does not exist.

35. $\begin{bmatrix} -1 & -2 & 2 \\ 4 & 2 & 0 \\ 4 & 0 & 4 \end{bmatrix} \begin{matrix} 1 & 0 & 0 \\ 0 & 1 & 0 \\ 0 & 0 & 1 \end{matrix}$ ~ $\begin{bmatrix} 1 & 2 & -2 \\ 4 & 2 & 0 \\ 4 & 0 & 4 \end{bmatrix} \begin{matrix} -1 & 0 & 0 \\ 0 & 1 & 0 \\ 0 & 0 & 1 \end{matrix}$

$\quad\quad (-1)R_1 \rightarrow R_1 \quad\quad\quad\quad\quad\quad R_2 + (-4)R_1 \rightarrow R_2$

$\quad\quad\quad\quad\quad\quad\quad\quad\quad\quad\quad\quad R_3 + (-4)R_1 \rightarrow R_3$

~ $\begin{bmatrix} 1 & 2 & -2 \\ 0 & -6 & 8 \\ 0 & -8 & 12 \end{bmatrix} \begin{matrix} -1 & 0 & 0 \\ 4 & 1 & 0 \\ 4 & 0 & 1 \end{matrix}$ ~ $\begin{bmatrix} 1 & 2 & -2 \\ 0 & 1 & -\frac{4}{3} \\ 0 & -8 & 12 \end{bmatrix} \begin{matrix} -1 & 0 & 0 \\ -\frac{2}{3} & -\frac{1}{6} & 0 \\ 4 & 0 & 1 \end{matrix}$

$\quad\quad -\dfrac{1}{6}R_2 \rightarrow R_2 \quad\quad\quad\quad\quad\quad R_3 + 8R_2 \rightarrow R_3$

~ $\begin{bmatrix} 1 & 2 & -2 \\ 0 & 1 & -\frac{4}{3} \\ 0 & 0 & \frac{4}{3} \end{bmatrix} \begin{matrix} -1 & 0 & 0 \\ -\frac{2}{3} & -\frac{1}{6} & 0 \\ -\frac{4}{3} & -\frac{4}{3} & 1 \end{matrix}$ ~ $\begin{bmatrix} 1 & 2 & -2 \\ 0 & 1 & -\frac{4}{3} \\ 0 & 0 & 1 \end{bmatrix} \begin{matrix} -1 & 0 & 0 \\ -\frac{2}{3} & -\frac{1}{6} & 0 \\ -1 & -1 & \frac{3}{4} \end{matrix}$

$\quad\quad \dfrac{3}{4}R_3 \rightarrow R_3 \quad\quad\quad\quad\quad\quad R_2 + \dfrac{4}{3}R_3 \rightarrow R_2$

$\quad\quad\quad\quad\quad\quad\quad\quad\quad\quad\quad\quad R_1 + 2R_3 \rightarrow R_1$

$\begin{bmatrix} 1 & 2 & 0 \\ 0 & 1 & 0 \\ 0 & 0 & 1 \end{bmatrix} \begin{matrix} -3 & -2 & \frac{3}{2} \\ -2 & -\frac{3}{2} & 1 \\ -1 & -1 & \frac{3}{4} \end{matrix}$ ~ $\begin{bmatrix} 1 & 0 & 0 \\ 0 & 1 & 0 \\ 0 & 0 & 1 \end{bmatrix} \begin{matrix} 1 & 1 & -\frac{1}{2} \\ -2 & -\frac{3}{2} & 1 \\ -1 & -1 & \frac{3}{4} \end{matrix}$

$\quad\quad R_1 + (-2)R_2 \rightarrow R_1$

Thus, the inverse is $\begin{bmatrix} 1 & 1 & -\frac{1}{2} \\ -2 & -\frac{3}{2} & 1 \\ -1 & -1 & \frac{3}{4} \end{bmatrix}$.

37. $A = \begin{bmatrix} 3 & 4 \\ 2 & 3 \end{bmatrix}$

$\begin{bmatrix} 3 & 4 \\ 2 & 3 \end{bmatrix} \begin{matrix} 1 & 0 \\ 0 & 1 \end{matrix}$ ~ $\begin{bmatrix} 1 & 1 \\ 2 & 3 \end{bmatrix} \begin{matrix} 1 & -1 \\ 0 & 1 \end{matrix}$ ~ $\begin{bmatrix} 1 & 1 \\ 0 & 1 \end{bmatrix} \begin{matrix} 1 & -1 \\ -2 & 3 \end{matrix}$ ~ $\begin{bmatrix} 1 & 0 \\ 0 & 1 \end{bmatrix} \begin{matrix} 3 & -4 \\ -2 & 3 \end{matrix}$

$\quad R_1 + (-1)R_2 \rightarrow R_1 \quad R_2 + (-2)R_1 \rightarrow R_2 \quad R_1 + (-1)R_2 \rightarrow R_1$

Therefore $A^{-1} = \begin{bmatrix} 3 & -4 \\ -2 & 3 \end{bmatrix}$

Now

$$\begin{bmatrix} 3 & -4 & | & 1 & 0 \\ -2 & 3 & | & 0 & 1 \end{bmatrix} \sim \begin{bmatrix} 1 & -1 & | & 1 & 1 \\ -2 & 3 & | & 0 & 1 \end{bmatrix} \sim \begin{bmatrix} 1 & -1 & | & 1 & 1 \\ 0 & 1 & | & 2 & 3 \end{bmatrix} \sim \begin{bmatrix} 1 & 0 & | & 3 & 4 \\ 0 & 1 & | & 2 & 3 \end{bmatrix}$$

$$R_1 + R_2 \to R_1 \qquad\qquad R_2 + 2R_1 \to R_2 \qquad\qquad R_1 + R_2 \to R_1$$

and $(A^{-1})^{-1} = \begin{bmatrix} 3 & 4 \\ 2 & 3 \end{bmatrix} = A$

39. $A = \begin{bmatrix} 6 & 2 & 0 & 4 \\ 5 & 3 & 2 & 1 \\ 0 & -1 & 1 & -2 \\ 2 & -3 & 1 & 0 \end{bmatrix}$; $A^{-1} = \begin{bmatrix} 0.5 & -0.3 & 0.85 & -0.25 \\ 0 & 0.1 & 0.05 & -0.25 \\ -1 & 0.9 & -1.55 & 0.75 \\ -0.5 & 0.4 & -1.3 & 0.5 \end{bmatrix}$

41. $A = \begin{bmatrix} 3 & 2 & 3 & 4 & 4 \\ 5 & 4 & 3 & 2 & 1 \\ -1 & -1 & 2 & -2 & 3 \\ 3 & -3 & 1 & 0 & 1 \\ 1 & 1 & 2 & 0 & 2 \end{bmatrix}$;

$$A^{-1} = \begin{bmatrix} 1.75 & 5.25 & 8.75 & -1 & -18.75 \\ 1.25 & 3.75 & 6.25 & -1 & -13.25 \\ -4.75 & -13.25 & -22.75 & 3 & 48.75 \\ -1.375 & -4.625 & -7.875 & 1 & 16.375 \\ 3.25 & 8.75 & 15.25 & -2 & -32.25 \end{bmatrix}$$

43. $A = \begin{bmatrix} 1 & 2 \\ 1 & 3 \end{bmatrix}$

Assign the numbers 1—26 to the letters of the alphabet, in order, and let 27 correspond to a blank space. Then the message "THE SUN ALSO RISES" corresponds to the sequence

  20   8   5 27 19 21 14 27   1 12 19 15 27 18   9 19   5 19

To encode this message, divide the numbers into groups of two and use the groups as columns of a matrix B with two rows

$$B = \begin{bmatrix} 20 & 5 & 19 & 14 & 1 & 19 & 27 & 9 & 5 \\ 8 & 27 & 21 & 27 & 12 & 15 & 18 & 19 & 19 \end{bmatrix}$$

Now

$$AB = \begin{bmatrix} 1 & 2 \\ 1 & 3 \end{bmatrix} \begin{bmatrix} 20 & 5 & 19 & 14 & 1 & 19 & 27 & 9 & 5 \\ 8 & 27 & 21 & 27 & 12 & 15 & 18 & 19 & 19 \end{bmatrix}$$

$$= \begin{bmatrix} 36 & 59 & 61 & 68 & 25 & 49 & 63 & 47 & 43 \\ 44 & 86 & 82 & 95 & 37 & 64 & 81 & 66 & 62 \end{bmatrix}$$

The coded message is:
36 44 59 86 61 82 68 95 25 37 49 64 63 81 47 66 43 62.

45. First, we must find the inverse of $A = \begin{bmatrix} 1 & 2 \\ 1 & 3 \end{bmatrix}$

$$\left[\begin{array}{cc|cc} 1 & 2 & 1 & 0 \\ 1 & 3 & 0 & 1 \end{array}\right] \sim \left[\begin{array}{cc|cc} 1 & 2 & 1 & 0 \\ 0 & 1 & -1 & 1 \end{array}\right] \sim \left[\begin{array}{cc|cc} 1 & 0 & 3 & -2 \\ 0 & 1 & -1 & 1 \end{array}\right]$$

$R_2 + (-1)R_1 \rightarrow R_2 \qquad R_1 + (-2)R_2 \rightarrow R_1$

Thus, $A^{-1} = \begin{bmatrix} 3 & -2 \\ -1 & 1 \end{bmatrix}$

Now $\begin{bmatrix} 3 & -2 \\ -1 & 1 \end{bmatrix} \begin{bmatrix} 37 & 24 & 73 & 49 & 62 & 36 & 59 & 41 & 22 \\ 52 & 29 & 96 & 69 & 89 & 44 & 86 & 50 & 26 \end{bmatrix}$

$= \begin{bmatrix} 7 & 14 & 27 & 9 & 8 & 20 & 5 & 23 & 14 \\ 15 & 5 & 23 & 20 & 27 & 8 & 27 & 9 & 4 \end{bmatrix}$

Thus, the decoded message is

   7   15   14   5   27   23   9   20   8   27   20   8   5   27   23   9   14   4

which corresponds to

  GONE WITH THE WIND

47. "THE BEST YEARS OF OUR LIVES" corresponds to the sequence

  20 8 5 27 2 5 19 20 27 25 5 1 18 19 27 15 6 27 15
  21 18 27 12 9 22 5 19

We divide the numbers in the sequence into groups of 5 and use these groups as the columns of a matrix with 5 rows, adding 3 blanks at the end to make the columns come out even. Then we multiply this matrix on the left by the given matrix B.

$$\begin{bmatrix} 1 & 0 & 1 & 0 & 1 \\ 0 & 1 & 1 & 0 & 3 \\ 2 & 1 & 1 & 1 & 1 \\ 0 & 0 & 1 & 0 & 2 \\ 1 & 1 & 1 & 2 & 1 \end{bmatrix} \begin{bmatrix} 20 & 5 & 5 & 15 & 18 & 5 \\ 8 & 19 & 1 & 6 & 27 & 19 \\ 5 & 20 & 18 & 27 & 12 & 27 \\ 27 & 27 & 19 & 15 & 9 & 27 \\ 2 & 25 & 27 & 21 & 22 & 27 \end{bmatrix}$$

$$= \begin{bmatrix} 27 & 50 & 50 & 63 & 52 & 59 \\ 19 & 114 & 100 & 96 & 105 & 127 \\ 82 & 101 & 75 & 99 & 106 & 110 \\ 9 & 70 & 72 & 69 & 56 & 81 \\ 89 & 123 & 89 & 99 & 97 & 132 \end{bmatrix}$$

The encoded message is:
27 19 82 9 89 50 114 101 70 123 50 100 75 72 89 63 96
99 69 99 52 105 106 56 97 59 127 110 81 132.

49. First, we must find the inverse of B:

$$B^{-1} = \begin{bmatrix} -2 & -1 & 2 & 2 & -1 \\ 3 & 2 & -2 & -4 & 1 \\ 6 & 2 & -4 & -5 & 2 \\ -2 & -1 & 1 & 2 & -7 \\ 3 & -1 & 2 & 3 & -1 \end{bmatrix}$$

Now $B^{-1}\begin{bmatrix} 32 & 24 & 54 & 56 & 48 & 62 \\ 34 & 21 & 71 & 92 & 66 & 135 \\ 87 & 67 & 112 & 109 & 98 & 124 \\ 19 & 11 & 43 & 55 & 41 & 81 \\ 94 & 69 & 112 & 109 & 89 & 143 \end{bmatrix} = \begin{bmatrix} 20 & 18 & 19 & 15 & 27 & 8 \\ 8 & 5 & 20 & 23 & 5 & 27 \\ 5 & 1 & 27 & 27 & 1 & 27 \\ 27 & 20 & 19 & 15 & 18 & 27 \\ 7 & 5 & 8 & 14 & 20 & 27 \end{bmatrix}$

Thus, the decoded message is

20 8 5 27 7 18 5 1 20 5 19 20 27 19 8 15 23 27 15
14 27 5 1 18 20 8

which corresponds to

THE GREATEST SHOW ON EARTH

EXERCISE 5-7

Things to remember:

<u>1.</u>   PROPERTIES OF IDENTITY AND INVERSE MATRICES

If A is a square matrix of order n and I is the identity matrix of order n, then
$$IA = AI = A$$

If, in addition, A^{-1} exists, then
$$A^{-1}A = AA^{-1} = I$$

<u>2.</u>   A system of n linear equations in n unknowns has the matrix representation $AX = B$, where A is a square matrix of order n, B is an $n \times 1$ column matrix and X is the $n \times 1$ column matrix of unknowns. If A^{-1} exists, then the (unique) solution to the system is
$$X = A^{-1}B.$$

1. $\begin{bmatrix} 3 & 1 \\ 2 & -1 \end{bmatrix} \begin{bmatrix} x_1 \\ x_2 \end{bmatrix} = \begin{bmatrix} 5 \\ -4 \end{bmatrix}$

$\begin{bmatrix} 3x_1 + x_2 \\ 2x_1 - x_2 \end{bmatrix} = \begin{bmatrix} 5 \\ -4 \end{bmatrix}$

Thus, $3x_1 + x_2 = 5$
$2x_1 - x_2 = -4$

3. $\begin{bmatrix} -3 & 1 & 0 \\ 2 & 0 & 1 \\ -1 & 3 & -2 \end{bmatrix} \begin{bmatrix} x_1 \\ x_2 \\ x_3 \end{bmatrix} = \begin{bmatrix} 3 \\ -4 \\ 2 \end{bmatrix}$

$\begin{bmatrix} -3x_1 + x_2 \\ 2x_1 + x_3 \\ -x_1 + 3x_2 - 2x_3 \end{bmatrix} = \begin{bmatrix} 3 \\ -4 \\ 2 \end{bmatrix}$

Thus, $-3x_1 + x_2 = 3$
$2x_1 + x_3 = -4$
$-x_1 + 3x_2 - 2x_3 = 2$

5. $3x_1 - 4x_2 = 1$
$2x_1 + x_2 = 5$

$\begin{bmatrix} 3x_1 - 4x_2 \\ 2x_1 + x_2 \end{bmatrix} = \begin{bmatrix} 1 \\ 5 \end{bmatrix}$ and $\begin{bmatrix} 3 & -4 \\ 2 & 1 \end{bmatrix} \begin{bmatrix} x_1 \\ x_2 \end{bmatrix} = \begin{bmatrix} 1 \\ 5 \end{bmatrix}$

7. $x_1 - 3x_2 + 2x_3 = -3$
$-2x_1 + 3x_2 = 1$
$x_1 + x_2 + 4x_3 = -2$

$\begin{bmatrix} x_1 - 3x_2 + 2x_3 \\ -2x_1 + 3x_2 \\ x_1 + x_2 + 4x_3 \end{bmatrix} = \begin{bmatrix} -3 \\ 1 \\ -2 \end{bmatrix}$ and $\begin{bmatrix} 1 & -3 & 2 \\ -2 & 3 & 0 \\ 1 & 1 & 4 \end{bmatrix} \begin{bmatrix} x_1 \\ x_2 \\ x_3 \end{bmatrix} = \begin{bmatrix} -3 \\ 1 \\ -2 \end{bmatrix}$

9. $\begin{bmatrix} x_1 \\ x_2 \end{bmatrix} = \begin{bmatrix} 3 & -2 \\ 1 & 4 \end{bmatrix} \begin{bmatrix} -2 \\ 1 \end{bmatrix} = \begin{bmatrix} 3(-2) + (-2)1 \\ 1(-2) + 4 \cdot 1 \end{bmatrix} = \begin{bmatrix} -8 \\ 2 \end{bmatrix}$ Thus, $x_1 = -8$
and $x_2 = 2$

11. $\begin{bmatrix} x_1 \\ x_2 \end{bmatrix} = \begin{bmatrix} -2 & 3 \\ 2 & -1 \end{bmatrix} \begin{bmatrix} 3 \\ 2 \end{bmatrix} = \begin{bmatrix} (-2)3 + 3 \cdot 2 \\ 2 \cdot 3 + (-1)2 \end{bmatrix} = \begin{bmatrix} 0 \\ 4 \end{bmatrix}$ Thus, $x_1 = 0$
and $x_2 = 4$

13. The matrix equation for the given system is:

$\begin{bmatrix} 1 & 2 \\ 1 & 3 \end{bmatrix} \begin{bmatrix} x_1 \\ x_2 \end{bmatrix} = \begin{bmatrix} k_1 \\ k_2 \end{bmatrix}$

From Exercise 5-6, Problem 17, $\begin{bmatrix} 1 & 2 \\ 1 & 3 \end{bmatrix}^{-1} = \begin{bmatrix} 3 & -2 \\ -1 & 1 \end{bmatrix}$

Thus, $\begin{bmatrix} x_1 \\ x_2 \end{bmatrix} = \begin{bmatrix} 3 & -2 \\ -1 & 1 \end{bmatrix} \begin{bmatrix} k_1 \\ k_2 \end{bmatrix}$

(A) $\begin{bmatrix} x_1 \\ x_2 \end{bmatrix} = \begin{bmatrix} 3 & -2 \\ -1 & 1 \end{bmatrix} \begin{bmatrix} 1 \\ 3 \end{bmatrix} = \begin{bmatrix} -3 \\ 2 \end{bmatrix}$ Thus, $x_1 = -3$ and $x_2 = 2$

(B) $\begin{bmatrix} x_1 \\ x_2 \end{bmatrix} = \begin{bmatrix} 3 & -2 \\ -1 & 1 \end{bmatrix} \begin{bmatrix} 3 \\ 5 \end{bmatrix} = \begin{bmatrix} -1 \\ 2 \end{bmatrix}$ Thus, $x_1 = -1$ and $x_2 = 2$

(C) $\begin{bmatrix} x_1 \\ x_2 \end{bmatrix} = \begin{bmatrix} 3 & -2 \\ -1 & 1 \end{bmatrix} \begin{bmatrix} -2 \\ 1 \end{bmatrix} = \begin{bmatrix} -8 \\ 3 \end{bmatrix}$ Thus, $x_1 = -8$ and $x_2 = 3$

15. The matrix equation for the given system is:

$$\begin{bmatrix} 1 & 3 \\ 2 & 7 \end{bmatrix} \begin{bmatrix} x_1 \\ x_2 \end{bmatrix} = \begin{bmatrix} k_1 \\ k_2 \end{bmatrix}$$

From Exercise 5-6, Problem 19, $\begin{bmatrix} 1 & 3 \\ 2 & 7 \end{bmatrix}^{-1} = \begin{bmatrix} 7 & -3 \\ -2 & 1 \end{bmatrix}$

Thus, $\begin{bmatrix} x_1 \\ x_2 \end{bmatrix} = \begin{bmatrix} 7 & -3 \\ -2 & 1 \end{bmatrix} \begin{bmatrix} k_1 \\ k_2 \end{bmatrix}$

(A) $\begin{bmatrix} x_1 \\ x_2 \end{bmatrix} = \begin{bmatrix} 7 & -3 \\ -2 & 1 \end{bmatrix} \begin{bmatrix} 2 \\ -1 \end{bmatrix} = \begin{bmatrix} 17 \\ -5 \end{bmatrix}$ Thus, $x_1 = 17$ and $x_2 = -5$

(B) $\begin{bmatrix} x_1 \\ x_2 \end{bmatrix} = \begin{bmatrix} 7 & -3 \\ -2 & 1 \end{bmatrix} \begin{bmatrix} 1 \\ 0 \end{bmatrix} = \begin{bmatrix} 7 \\ -2 \end{bmatrix}$ Thus, $x_1 = 7$ and $x_2 = -2$

(C) $\begin{bmatrix} x_1 \\ x_2 \end{bmatrix} = \begin{bmatrix} 7 & -3 \\ -2 & 1 \end{bmatrix} \begin{bmatrix} 3 \\ -1 \end{bmatrix} = \begin{bmatrix} 24 \\ -7 \end{bmatrix}$ Thus, $x_1 = 24$ and $x_2 = -7$

17. The matrix equation for the given system is:

$$\begin{bmatrix} 1 & -3 & 0 \\ 0 & 3 & 1 \\ 2 & -1 & 2 \end{bmatrix} \begin{bmatrix} x_1 \\ x_2 \\ x_3 \end{bmatrix} = \begin{bmatrix} k_1 \\ k_2 \\ k_3 \end{bmatrix}$$

From Exercise 5-6, Problem 21, $\begin{bmatrix} 1 & -3 & 0 \\ 0 & 3 & 1 \\ 2 & -1 & 2 \end{bmatrix}^{-1} = \begin{bmatrix} 7 & 6 & -3 \\ 2 & 2 & -1 \\ -6 & -5 & 3 \end{bmatrix}$

Thus,

$$\begin{bmatrix} x_1 \\ x_2 \\ x_3 \end{bmatrix} = \begin{bmatrix} 7 & 6 & -3 \\ 2 & 2 & -1 \\ -6 & -5 & 3 \end{bmatrix} \begin{bmatrix} k_1 \\ k_2 \\ k_3 \end{bmatrix}$$

(A) $\begin{bmatrix} x_1 \\ x_2 \\ x_3 \end{bmatrix} = \begin{bmatrix} 7 & 6 & -3 \\ 2 & 2 & -1 \\ -6 & -5 & 3 \end{bmatrix} \begin{bmatrix} 1 \\ 0 \\ 2 \end{bmatrix} = \begin{bmatrix} 1 \\ 0 \\ 0 \end{bmatrix}$ Thus, $x_1 = 1$, $x_2 = 0$, $x_3 = 0$.

(B) $\begin{bmatrix} x_1 \\ x_2 \\ x_3 \end{bmatrix} = \begin{bmatrix} 7 & 6 & -3 \\ 2 & 2 & -1 \\ -6 & -5 & 3 \end{bmatrix} \begin{bmatrix} -1 \\ 1 \\ 0 \end{bmatrix} = \begin{bmatrix} -1 \\ 0 \\ 1 \end{bmatrix}$ Thus, $x_1 = -1$, $x_2 = 0$, $x_3 = 1$.

(C) $\begin{bmatrix} x_1 \\ x_2 \\ x_3 \end{bmatrix} = \begin{bmatrix} 7 & 6 & -3 \\ 2 & 2 & -1 \\ -6 & -5 & 3 \end{bmatrix} \begin{bmatrix} 2 \\ -2 \\ 1 \end{bmatrix} = \begin{bmatrix} -1 \\ -1 \\ 1 \end{bmatrix}$ Thus, $x_1 = -1$, $x_2 = -1$, $x_3 = 1$.

19. The matrix equation for the given system is:

$$\begin{bmatrix} 1 & 1 & 0 \\ 0 & 3 & -1 \\ 1 & 0 & 1 \end{bmatrix} \begin{bmatrix} x_1 \\ x_2 \\ x_3 \end{bmatrix} = \begin{bmatrix} k_1 \\ k_2 \\ k_3 \end{bmatrix}$$

From Exercise 5-6, Problem 23, the inverse of the coefficient matrix is

$$\frac{1}{2} \begin{bmatrix} 3 & -1 & -1 \\ -1 & 1 & 1 \\ -3 & 1 & 3 \end{bmatrix}.$$

Thus,

$$\begin{bmatrix} x_1 \\ x_2 \\ x_3 \end{bmatrix} = \frac{1}{2} \begin{bmatrix} 3 & -1 & -1 \\ -1 & 1 & 1 \\ -3 & 1 & 3 \end{bmatrix} \begin{bmatrix} k_1 \\ k_2 \\ k_3 \end{bmatrix}$$

(A) $\begin{bmatrix} x_1 \\ x_2 \\ x_3 \end{bmatrix} = \frac{1}{2} \begin{bmatrix} 3 & -1 & -1 \\ -1 & 1 & 1 \\ -3 & 1 & 3 \end{bmatrix} \begin{bmatrix} 2 \\ 0 \\ 4 \end{bmatrix} = \frac{1}{2} \begin{bmatrix} 2 \\ 2 \\ 6 \end{bmatrix} = \begin{bmatrix} 1 \\ 1 \\ 3 \end{bmatrix}$ Thus, $x_1 = 1$, $x_2 = 1$, and $x_3 = 3$.

(B) $\begin{bmatrix} x_1 \\ x_2 \\ x_3 \end{bmatrix} = \frac{1}{2} \begin{bmatrix} 3 & -1 & -1 \\ -1 & 1 & 1 \\ -3 & 1 & 3 \end{bmatrix} \begin{bmatrix} 0 \\ 4 \\ -2 \end{bmatrix} = \frac{1}{2} \begin{bmatrix} -2 \\ 2 \\ -2 \end{bmatrix} = \begin{bmatrix} -1 \\ 1 \\ -1 \end{bmatrix}$ Thus, $x_1 = -1$, $x_2 = 1$, and $x_3 = -1$.

(C) $\begin{bmatrix} x_1 \\ x_2 \\ x_3 \end{bmatrix} = \frac{1}{2} \begin{bmatrix} 3 & -1 & -1 \\ -1 & 1 & 1 \\ -3 & 1 & 3 \end{bmatrix} \begin{bmatrix} 4 \\ 2 \\ 0 \end{bmatrix} = \frac{1}{2} \begin{bmatrix} 10 \\ -2 \\ -10 \end{bmatrix} = \begin{bmatrix} 5 \\ -1 \\ -5 \end{bmatrix}$ Thus, $x_1 = 5$, $x_2 = -1$, and $x_3 = -5$.

21. $AX - BX = C$

$\quad (A - B)X = C$

$\quad X = (A - B)^{-1}C$

23. $AX + X = C$

$\quad (A + I)X = C$, where I is the identity matrix of order n

$$X = (A + I)^{-1}C$$

25. $AX - C = D - BX$

$\quad\quad AX + BX = C + D$

$\quad (A + B)X = C + D$

$$X = (A + B)^{-1}(C + D)$$

27. The matrix equation for the given system is:

$$\begin{bmatrix} 1 & 8 & 7 \\ 6 & 6 & 8 \\ 3 & 4 & 6 \end{bmatrix} \begin{bmatrix} x_1 \\ x_2 \\ x_3 \end{bmatrix} = \begin{bmatrix} 135 \\ 155 \\ 75 \end{bmatrix}$$

Thus, $\begin{bmatrix} x_1 \\ x_2 \\ x_3 \end{bmatrix} = \begin{bmatrix} 1 & 8 & 7 \\ 6 & 6 & 8 \\ 3 & 4 & 6 \end{bmatrix}^{-1} \begin{bmatrix} 135 \\ 155 \\ 75 \end{bmatrix} = \begin{bmatrix} -0.08 & 0.4 & -0.44 \\ 0.24 & 0.3 & -0.68 \\ -0.12 & -0.4 & 0.84 \end{bmatrix} \begin{bmatrix} 135 \\ 155 \\ 75 \end{bmatrix}$

$$= \begin{bmatrix} 18.2 \\ 27.9 \\ -15.2 \end{bmatrix} \text{ and } \begin{array}{l} x_1 = 18.2 \\ x_2 = 27.9 \\ x_3 = -15.2 \end{array}$$

29. The matrix equation for the given system is:

$$\begin{bmatrix} 6 & 9 & 7 & 5 \\ 6 & 4 & 7 & 3 \\ 4 & 5 & 3 & 2 \\ 4 & 3 & 8 & 2 \end{bmatrix} \begin{bmatrix} x_1 \\ x_2 \\ x_3 \\ x_4 \end{bmatrix} = \begin{bmatrix} 250 \\ 195 \\ 145 \\ 125 \end{bmatrix}$$

Thus

$\begin{bmatrix} x_1 \\ x_2 \\ x_3 \\ x_4 \end{bmatrix} = \begin{bmatrix} 6 & 9 & 7 & 5 \\ 6 & 4 & 7 & 3 \\ 4 & 5 & 3 & 2 \\ 4 & 3 & 8 & 2 \end{bmatrix}^{-1} \begin{bmatrix} 250 \\ 195 \\ 145 \\ 125 \end{bmatrix} = \begin{bmatrix} -0.25 & 0.37 & 0.28 & -0.21 \\ 0 & -0.4 & 0.4 & 0.2 \\ 0 & -0.16 & -0.04 & 0.28 \\ 0.5 & 0.5 & -1 & -0.5 \end{bmatrix} \begin{bmatrix} 250 \\ 195 \\ 145 \\ 125 \end{bmatrix}$

$$= \begin{bmatrix} 24 \\ 5 \\ -2 \\ 15 \end{bmatrix} \text{ and } \begin{array}{l} x_1 = 24 \\ x_2 = 5 \\ x_3 = -2 \\ x_4 = 15 \end{array}$$

31. Let x_1 = number of \$4 tickets sold
and x_2 = number of \$8 tickets sold.

For the first return of \$56,000 we have the following system to solve:
$$x_1 + x_2 = 10,000$$
$$4x_1 + 8x_2 = 56,000$$

The corresponding matrix equation is: $\begin{bmatrix} 1 & 1 \\ 4 & 8 \end{bmatrix} \begin{bmatrix} x_1 \\ x_2 \end{bmatrix} = \begin{bmatrix} 10,000 \\ 56,000 \end{bmatrix}$.

First, we compute the inverse of $\begin{bmatrix} 1 & 1 \\ 4 & 8 \end{bmatrix}$.

$$\left[\begin{array}{cc|cc} 1 & 1 & 1 & 0 \\ 4 & 8 & 0 & 1 \end{array}\right] \sim \left[\begin{array}{cc|cc} 1 & 1 & 1 & 0 \\ 0 & 4 & -4 & 1 \end{array}\right] \sim \left[\begin{array}{cc|cc} 1 & 1 & 1 & 0 \\ 0 & 1 & -1 & \frac{1}{4} \end{array}\right] \sim \left[\begin{array}{cc|cc} 1 & 0 & 2 & -\frac{1}{4} \\ 0 & 1 & -1 & \frac{1}{4} \end{array}\right]$$

$$R_2 + (-4)R_1 \to R_2 \qquad \tfrac{1}{4}R_2 \to R_2 \qquad R_1 + (-1)R_2 \to R_1$$

Thus, $\begin{bmatrix} 1 & 1 \\ 4 & 8 \end{bmatrix}^{-1} = \begin{bmatrix} 2 & -\frac{1}{4} \\ -1 & \frac{1}{4} \end{bmatrix}$ and $\begin{bmatrix} x_1 \\ x_2 \end{bmatrix} = \begin{bmatrix} 2 & -\frac{1}{4} \\ -1 & \frac{1}{4} \end{bmatrix} \begin{bmatrix} 10,000 \\ 56,000 \end{bmatrix}$

$$= \begin{bmatrix} 20,000 - 14,000 \\ -10,000 + 14,000 \end{bmatrix} = \begin{bmatrix} 6000 \\ 4000 \end{bmatrix}.$$

So, for Concert 1, x_1 = 6000 \$4 tickets
x_2 = 4000 \$8 tickets.

For a return of \$60,000:
$$\begin{bmatrix} x_1 \\ x_2 \end{bmatrix} = \begin{bmatrix} 2 & -\frac{1}{4} \\ -1 & \frac{1}{4} \end{bmatrix} \begin{bmatrix} 10,000 \\ 60,000 \end{bmatrix} = \begin{bmatrix} 20,000 - 15,000 \\ -10,000 + 15,000 \end{bmatrix} = \begin{bmatrix} 5000 \\ 5000 \end{bmatrix}.$$

For Concert 2, x_1 = 5000 \$4 tickets
x_2 = 5000 \$8 tickets.

Finally, for a return of \$68,000:
$$\begin{bmatrix} x_1 \\ x_2 \end{bmatrix} = \begin{bmatrix} 2 & -\frac{1}{4} \\ -1 & \frac{1}{4} \end{bmatrix} \begin{bmatrix} 10,000 \\ 68,000 \end{bmatrix} = \begin{bmatrix} 20,000 - 17,000 \\ -10,000 + 17,000 \end{bmatrix} = \begin{bmatrix} 3000 \\ 7000 \end{bmatrix}.$$

Thus, for Concert 3, x_1 = 3000 \$4 tickets
x_2 = 7000 \$8 tickets.

33. Let x_1 = number of hours at Plant A
and x_2 = number of hours at Plant B

Then $10x_1 + 8x_2 = k_1$ (number of car frames)
$5x_1 + 8x_2 = k_2$ (number of truck frames)

The corresponding matrix equation is:
$$\begin{bmatrix} 10 & 8 \\ 5 & 8 \end{bmatrix} \begin{bmatrix} x_1 \\ x_2 \end{bmatrix} = \begin{bmatrix} k_1 \\ k_2 \end{bmatrix}$$

First we compute the inverse of $\begin{bmatrix} 10 & 8 \\ 5 & 8 \end{bmatrix}$

$$\begin{bmatrix} 10 & 8 & | & 1 & 0 \\ 5 & 8 & | & 0 & 1 \end{bmatrix} \sim \begin{bmatrix} 1 & \frac{4}{5} & | & \frac{1}{10} & 0 \\ 5 & 8 & | & 0 & 1 \end{bmatrix} \sim \begin{bmatrix} 1 & \frac{4}{5} & | & \frac{1}{10} & 0 \\ 0 & 4 & | & -\frac{1}{2} & 1 \end{bmatrix}$$

$$\frac{1}{10}R_1 \to R_1 \qquad\qquad R_2 + (-5)R_1 \to R_2 \qquad\qquad \frac{1}{4}R_2 \to R_2$$

$$\sim \begin{bmatrix} 1 & \frac{4}{5} & | & \frac{1}{10} & 0 \\ 0 & 1 & | & -\frac{1}{8} & \frac{1}{4} \end{bmatrix} \sim \begin{bmatrix} 1 & 0 & | & \frac{1}{5} & -\frac{1}{5} \\ 0 & 1 & | & -\frac{1}{8} & \frac{1}{4} \end{bmatrix}$$

$$R_1 + \left(-\frac{4}{5}\right)R_2 \to R_1$$

Thus $\begin{bmatrix} 10 & 8 \\ 5 & 8 \end{bmatrix}^{-1} = \begin{bmatrix} \frac{1}{5} & -\frac{1}{5} \\ -\frac{1}{8} & \frac{1}{4} \end{bmatrix}$ and $\begin{bmatrix} x_1 \\ x_2 \end{bmatrix} = \begin{bmatrix} \frac{1}{5} & -\frac{1}{5} \\ -\frac{1}{8} & \frac{1}{4} \end{bmatrix} \begin{bmatrix} k_1 \\ k_2 \end{bmatrix}$

Now, for order 1:

$\begin{bmatrix} x_1 \\ x_2 \end{bmatrix} = \begin{bmatrix} \frac{1}{5} & -\frac{1}{5} \\ -\frac{1}{8} & \frac{1}{4} \end{bmatrix} \begin{bmatrix} 3000 \\ 1600 \end{bmatrix} = \begin{bmatrix} 280 \\ 25 \end{bmatrix}$ and $\begin{matrix} x_1 = 280 \text{ hours at Plant A} \\ x_2 = 25 \text{ hours at Plant B} \end{matrix}$

For order 2:

$\begin{bmatrix} x_1 \\ x_2 \end{bmatrix} = \begin{bmatrix} \frac{1}{5} & -\frac{1}{5} \\ -\frac{1}{8} & \frac{1}{4} \end{bmatrix} \begin{bmatrix} 2800 \\ 2000 \end{bmatrix} = \begin{bmatrix} 160 \\ 150 \end{bmatrix}$ and $\begin{matrix} x_1 = 160 \text{ hours at Plant A} \\ x_2 = 150 \text{ hours at Plant B} \end{matrix}$

For order 3:

$\begin{bmatrix} x_1 \\ x_2 \end{bmatrix} = \begin{bmatrix} \frac{1}{5} & -\frac{1}{5} \\ -\frac{1}{8} & \frac{1}{4} \end{bmatrix} \begin{bmatrix} 2600 \\ 2200 \end{bmatrix} = \begin{bmatrix} 80 \\ 225 \end{bmatrix}$ and $\begin{matrix} x_1 = 80 \text{ hours at Plant A} \\ x_2 = 225 \text{ hours at Plant B} \end{matrix}$

35. Let x_1 = number of ounces of mix A
and x_2 = number of ounces of mix B.

For Diet 1, we have the following system to solve:
$0.2x_1 + 0.1x_2 = 20$
$0.02x_1 + 0.06x_2 = 6$
or

$$\begin{array}{l} \text{Diet 1} \quad \left\{\begin{array}{l}\text{Diet 2}\end{array}\right. \quad \left\{\begin{array}{l}\text{Diet 3}\end{array}\right. \\ 2x_1 + x_2 = 200 \left.\begin{array}{l} = 100 \end{array}\right\} \left.\begin{array}{l} = 100 \end{array}\right\} \\ 2x_1 + 6x_2 = 600 \left.\begin{array}{l} = 400 \end{array}\right. \quad \left.\begin{array}{l} = 600 \end{array}\right. \end{array}$$

First, compute the inverse matrix of $\begin{bmatrix} 2 & 1 \\ 2 & 6 \end{bmatrix}$.

$$\begin{bmatrix} 2 & 1 & | & 1 & 0 \\ 2 & 6 & | & 0 & 1 \end{bmatrix} \sim \begin{bmatrix} 1 & \frac{1}{2} & | & \frac{1}{2} & 0 \\ 2 & 6 & | & 0 & 1 \end{bmatrix} \sim \begin{bmatrix} 1 & \frac{1}{2} & | & \frac{1}{2} & 0 \\ 0 & 5 & | & -1 & 1 \end{bmatrix} \sim \begin{bmatrix} 1 & \frac{1}{2} & | & \frac{1}{2} & 0 \\ 0 & 1 & | & -\frac{1}{5} & \frac{1}{5} \end{bmatrix}$$

$$\frac{1}{2} R_1 \to R_1 \qquad R_2 + (-2) R_1 \to R_2 \qquad \frac{1}{5} R_2 \to R_2 \qquad R_1 + \left(-\frac{1}{2}\right) R_2 \to R_1$$

$$\sim \begin{bmatrix} 1 & 0 & | & \frac{6}{10} & -\frac{1}{10} \\ 0 & 1 & | & -\frac{1}{5} & \frac{1}{5} \end{bmatrix} \quad \text{Thus,} \quad \begin{bmatrix} 2 & 1 \\ 2 & 6 \end{bmatrix}^{-1} = \begin{bmatrix} \frac{3}{5} & -\frac{1}{10} \\ -\frac{1}{5} & \frac{1}{5} \end{bmatrix}$$

and $\begin{bmatrix} x_1 \\ x_2 \end{bmatrix} = \begin{bmatrix} \frac{3}{5} & -\frac{1}{10} \\ -\frac{1}{5} & \frac{1}{5} \end{bmatrix} \begin{bmatrix} 200 \\ 600 \end{bmatrix} = \begin{bmatrix} 120 - 60 \\ -40 + 120 \end{bmatrix} = \begin{bmatrix} 60 \\ 80 \end{bmatrix}$.

So, for Diet 1, $x_1 = 60$ ounces of mix A
$\qquad\qquad\qquad x_2 = 80$ ounces of mix B.

For Diet 2, the solution is:

$\begin{bmatrix} x_1 \\ x_2 \end{bmatrix} = \begin{bmatrix} \frac{3}{5} & -\frac{1}{10} \\ -\frac{1}{5} & \frac{1}{5} \end{bmatrix} \begin{bmatrix} 100 \\ 400 \end{bmatrix} = \begin{bmatrix} 60 - 40 \\ -20 + 80 \end{bmatrix} = \begin{bmatrix} 20 \\ 60 \end{bmatrix}$

So for Diet 2, $x_1 = 20$ ounces of mix A
$\qquad\qquad\qquad x_2 = 60$ ounces of mix B.

For Diet 3, we have:

$\begin{bmatrix} x_1 \\ x_2 \end{bmatrix} = \begin{bmatrix} \frac{3}{5} & -\frac{1}{10} \\ -\frac{1}{5} & \frac{1}{5} \end{bmatrix} \begin{bmatrix} 100 \\ 600 \end{bmatrix} = \begin{bmatrix} 60 - 60 \\ -20 + 120 \end{bmatrix} = \begin{bmatrix} 0 \\ 100 \end{bmatrix}$

Thus, for Diet 3, $x_1 = 0$ ounces of mix A
$\qquad\qquad\qquad x_2 = 100$ ounces of mix B.

37. Let x_1 = President's bonus
$\qquad x_2$ = Executive Vice President's bonus
$\qquad x_3$ = Associate Vice President's bonus
$\qquad x_4$ = Assistant Vice President's bonus

Then

$x_1 = 0.03(2,000,000 - x_2 - x_3 - x_4)$
$x_2 = 0.025(2,000,000 - x_1 - x_3 - x_4)$
$x_3 = 0.02(2,000,000 - x_1 - x_2 - x_4)$
$x_4 = 0.015(2,000,000 - x_1 - x_2 - x_3)$

or

$$\begin{aligned}
x_1 + 0.03x_2 + 0.03x_3 + 0.03x_4 &= 60,000 \\
0.025x_1 + x_2 + 0.025x_3 + 0.025x_4 &= 50,000 \\
0.02x_1 + 0.02x_2 + x_3 + 0.02x_4 &= 40,000 \\
0.015x_1 + 0.015x_2 + 0.015x_3 + x_4 &= 30,000
\end{aligned}$$

and $\begin{pmatrix} 1 & 0.03 & 0.03 & 0.03 \\ 0.025 & 1 & 0.025 & 0.025 \\ 0.02 & 0.02 & 1 & 0.02 \\ 0.015 & 0.015 & 0.015 & 1 \end{pmatrix} \begin{pmatrix} x_1 \\ x_2 \\ x_3 \\ x_4 \end{pmatrix} = \begin{pmatrix} 60,000 \\ 50,000 \\ 40,000 \\ 30,000 \end{pmatrix}$

Thus $\begin{pmatrix} x_1 \\ x_2 \\ x_3 \\ x_4 \end{pmatrix} = \begin{pmatrix} 1 & 0.03 & 0.03 & 0.03 \\ 0.025 & 1 & 0.025 & 0.025 \\ 0.02 & 0.02 & 1 & 0.02 \\ 0.015 & 0.015 & 0.015 & 1 \end{pmatrix}^{-1} \begin{pmatrix} 60,000 \\ 50,000 \\ 40,000 \\ 30,000 \end{pmatrix} \approx \begin{pmatrix} 56,600 \\ 47,000 \\ 37,400 \\ 27,900 \end{pmatrix}$

or $x_1 = \$56,600$, $x_2 = \$47,000$, $x_3 = \$37,400$, $x_4 = \$27,900$ to the nearest hundred dollars.

39. To find the taxable income of each company, we calculate
$$\begin{bmatrix} 0.82 & 0.08 & 0.03 & 0.07 \\ 0.12 & 0.64 & 0.11 & 0.13 \\ 0.11 & 0.09 & 0.72 & 0.08 \\ 0.06 & 0.02 & 0.14 & 0.78 \end{bmatrix} \begin{bmatrix} 3.2 \\ 2.6 \\ 3.8 \\ 4.4 \end{bmatrix} = \begin{bmatrix} 3.254 \\ 3.038 \\ 3.674 \\ 4.208 \end{bmatrix}$$

Thus, to the nearest hundred thousand dollars, the taxable income for Company A = $3,300,000; for Company B = $3,000,000; for Company C = $3,700,000; and for Company D = $4,200,000.

Now, (Total Taxable Income) $-$ (Total Net Income) =
$(3.3 + 3.0 + 3.7 + 4.2) - (3.2 + 2.6 + 3.8 + 4.4) = 0.2$
or $200,000

41. Let F = total cost for the Freezer Department
R = total cost for the Refrigerator Department
A = total cost for the Accounting Department
M = total cost for the Maintenance Department

(A) $F = 260,000 + 0.38A + 0.42M$
$R = 190,000 + 0.34A + 0.28M$
$A = 55,000 + 0.16A + 0.13M$
$M = 95,000 + 0.12A + 0.17M$

From the last two equations, we have
$0.84A - 0.13M = 55,000$
$-0.12A + 0.83M = 95,000$

or $\begin{pmatrix} 0.84 & -0.13 \\ -0.12 & 0.83 \end{pmatrix} \begin{pmatrix} A \\ M \end{pmatrix} = \begin{pmatrix} 55,000 \\ 95,000 \end{pmatrix}$ and $\begin{pmatrix} A \\ M \end{pmatrix} = \begin{pmatrix} 0.84 & -0.13 \\ -0.12 & 0.83 \end{pmatrix}^{-1} \begin{pmatrix} 55,000 \\ 95,000 \end{pmatrix}$

$\approx \begin{pmatrix} 1.22 & 0.19 \\ 0.18 & 1.23 \end{pmatrix} \begin{pmatrix} 55,000 \\ 95,000 \end{pmatrix} \approx \begin{pmatrix} 85,000 \\ 127,000 \end{pmatrix}$

Thus $A = \$85,000$ and $M = \$127,000$ (to the nearest thousand)

Now, from the first two equations

$$\begin{pmatrix} F \\ R \end{pmatrix} = \begin{pmatrix} 260,000 \\ 190,000 \end{pmatrix} + \begin{pmatrix} 0.38 & 0.42 \\ 0.34 & 0.28 \end{pmatrix} \begin{pmatrix} A \\ M \end{pmatrix} = \begin{pmatrix} 260,000 \\ 190,000 \end{pmatrix} + \begin{pmatrix} 0.38 & 0.42 \\ 0.34 & 0.28 \end{pmatrix} \begin{pmatrix} 85,000 \\ 127,000 \end{pmatrix}$$

$$\begin{pmatrix} F \\ R \end{pmatrix} = \begin{pmatrix} 260,000 \\ 190,000 \end{pmatrix} + \begin{pmatrix} 85,640 \\ 64,460 \end{pmatrix} = \begin{pmatrix} 345,640 \\ 254,460 \end{pmatrix}$$

Therefore, $F = \$346,000$ and $M = \$254,000$ (to the nearest thousand).

(B) Sum of Direct costs—all four departments: $600,000

   $F + R = 346,000 + 254,000 = \$600,000$

   Interpretation: The direct costs of the service departments are distributed among the production departments.

EXERCISE 5-8 CHAPTER REVIEW

1. $y = 2x - 4$ (1)

   $y = \frac{1}{2}x + 2$ (2)

   The point of intersection is the solution. This is $x = 4$, $y = 4$.

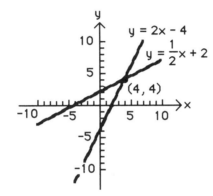

2. Substitute equation (1) into (2):

   $2x - 4 = \frac{1}{2}x + 2$

   $\frac{3}{2}x = 6$

   $x = 4$

   Substitute $x = 4$ into (1):

   $y = 2 \cdot 4 - 4 = 4$

   Solution:

   $x = 4$, $y = 4$

3. $A + B = \begin{bmatrix} 1+2 & 2+1 \\ 3+1 & 1+1 \end{bmatrix} = \begin{bmatrix} 3 & 3 \\ 4 & 2 \end{bmatrix}$

4. $B + D = \begin{bmatrix} 2 & 1 \\ 1 & 1 \end{bmatrix} + \begin{bmatrix} 1 \\ 2 \end{bmatrix}$

   The matrices B and D cannot be added because their dimensions are different.

5. $A - 2B = \begin{bmatrix} 1 & 2 \\ 3 & 1 \end{bmatrix} - 2 \begin{bmatrix} 2 & 1 \\ 1 & 1 \end{bmatrix} = \begin{bmatrix} 1 & 2 \\ 3 & 1 \end{bmatrix} + \begin{bmatrix} -4 & -2 \\ -2 & -2 \end{bmatrix} = \begin{bmatrix} -3 & 0 \\ 1 & -1 \end{bmatrix}$

6. $AB = \begin{bmatrix} 1 & 2 \\ 3 & 1 \end{bmatrix} \begin{bmatrix} 2 & 1 \\ 1 & 1 \end{bmatrix} = \begin{bmatrix} [1 \quad 2]\begin{bmatrix} 2 \\ 1 \end{bmatrix} & [1 \quad 2]\begin{bmatrix} 1 \\ 1 \end{bmatrix} \\ [3 \quad 1]\begin{bmatrix} 2 \\ 1 \end{bmatrix} & [3 \quad 1]\begin{bmatrix} 1 \\ 1 \end{bmatrix} \end{bmatrix} = \begin{bmatrix} 4 & 3 \\ 7 & 4 \end{bmatrix}$

7. AC is *not defined* because the dimension of A is 2 × 2 and the dimension of C is 1 × 2. So, the number of columns in A is not equal to the number of rows in C.

8. $AD = \begin{bmatrix} 1 & 2 \\ 3 & 1 \end{bmatrix}\begin{bmatrix} 1 \\ 2 \end{bmatrix} = \begin{bmatrix} [1 & 2]\begin{bmatrix} 1 \\ 2 \end{bmatrix} \\ [3 & 1]\begin{bmatrix} 1 \\ 2 \end{bmatrix} \end{bmatrix} = \begin{bmatrix} 5 \\ 5 \end{bmatrix}$

9. $DC = \begin{bmatrix} 1 \\ 2 \end{bmatrix}[2 \quad 3] = \begin{bmatrix} [1]\cdot[2] & [1]\cdot[3] \\ [2]\cdot[2] & [2]\cdot[3] \end{bmatrix} = \begin{bmatrix} 2 & 3 \\ 4 & 6 \end{bmatrix}$

10. $CD = [2 \quad 3]\begin{bmatrix} 1 \\ 2 \end{bmatrix} = [2+6] = [8]$ **11.** $C + D = [2 \quad 3] + \begin{bmatrix} 1 \\ 2 \end{bmatrix}$

Not defined because the dimensions of C and D are different.

12. $\begin{bmatrix} 3 & 2 & | & 1 & 0 \\ 4 & 3 & | & 0 & 1 \end{bmatrix} \sim \begin{bmatrix} 1 & \frac{2}{3} & | & \frac{1}{3} & 0 \\ 4 & 3 & | & 0 & 1 \end{bmatrix} \sim \begin{bmatrix} 1 & \frac{2}{3} & | & \frac{1}{3} & 0 \\ 0 & \frac{1}{3} & | & -\frac{4}{3} & 1 \end{bmatrix} \sim \begin{bmatrix} 1 & \frac{2}{3} & | & \frac{1}{3} & 0 \\ 0 & 1 & | & -4 & 3 \end{bmatrix}$

$\quad\quad \frac{1}{3}R_1 \to R_1 \quad\quad\quad R_2 + (-4)R_1 \to R_2 \quad\quad 3R_2 \to R_2 \quad\quad R_1 + \left(-\frac{2}{3}\right)R_2 \to R_1$

$\sim \begin{bmatrix} 1 & 0 & | & 3 & -2 \\ 0 & 1 & | & -4 & 3 \end{bmatrix}$ Thus, $A^{-1} = \begin{bmatrix} 3 & -2 \\ -4 & 3 \end{bmatrix}$ and

$A^{-1}A = \begin{bmatrix} 3 & -2 \\ -4 & 3 \end{bmatrix}\begin{bmatrix} 3 & 2 \\ 4 & 3 \end{bmatrix} = \begin{bmatrix} [3 & -2]\begin{bmatrix} 3 \\ 4 \end{bmatrix} & [3 & -2]\begin{bmatrix} 2 \\ 3 \end{bmatrix} \\ [-4 & 3]\begin{bmatrix} 3 \\ 4 \end{bmatrix} & [-4 & 3]\begin{bmatrix} 2 \\ 3 \end{bmatrix} \end{bmatrix} = \begin{bmatrix} 1 & 0 \\ 0 & 1 \end{bmatrix}$

13. $3x_1 + 2x_2 = 3 \quad (1)$
$4x_1 + 3x_2 = 5 \quad (2)$

Multiply (1) by 4 and (2) by −3, then add:

$0 - x_2 = -3$
$\quad\quad x_2 = 3$

Substitute $x_2 = 3$ into (1) to get:

$3x_1 + 2(3) = 3$
$\quad 3x_1 + 6 = 3$
$\quad\quad 3x_1 = -3$
$\quad\quad\quad x_1 = -1$

Solution: $x_1 = -1$, $x_2 = 3$.

14. The augmented matrix of the system is:

$\begin{bmatrix} 3 & 2 & | & 3 \\ 4 & 3 & | & 5 \end{bmatrix} \sim \begin{bmatrix} 1 & \frac{2}{3} & | & 1 \\ 4 & 3 & | & 5 \end{bmatrix} \sim \begin{bmatrix} 1 & \frac{2}{3} & | & 1 \\ 0 & \frac{1}{3} & | & 1 \end{bmatrix} \sim \begin{bmatrix} 1 & \frac{2}{3} & | & 1 \\ 0 & 1 & | & 3 \end{bmatrix} \sim \begin{bmatrix} 1 & 0 & | & -1 \\ 0 & 1 & | & 3 \end{bmatrix}$

$\frac{1}{3}R_1 \to R_1 \quad R_2 + (-4)R_1 \to R_2 \quad 3R_2 \to R_2 \quad R_1 + \left(-\frac{2}{3}\right)R_2 \to R_1$

Thus, the solution is: $x_1 = -1$, $x_2 = 3$.

15. (A) The matrix equation of the given system is: $\begin{bmatrix} 3 & 2 \\ 4 & 3 \end{bmatrix} \begin{bmatrix} x_1 \\ x_2 \end{bmatrix} = \begin{bmatrix} 3 \\ 5 \end{bmatrix}$.

The inverse of $\begin{bmatrix} 3 & 2 \\ 4 & 3 \end{bmatrix}$, by Problem 12, is: $\begin{bmatrix} 3 & -2 \\ -4 & 3 \end{bmatrix}$.

Thus, $\begin{bmatrix} x_1 \\ x_2 \end{bmatrix} = \begin{bmatrix} 3 & -2 \\ -4 & 3 \end{bmatrix} \begin{bmatrix} 3 \\ 5 \end{bmatrix} = \begin{bmatrix} -1 \\ 3 \end{bmatrix}$. Solution: $x_1 = -1$, $x_2 = 3$.

(B) $\begin{bmatrix} 3 & 2 \\ 4 & 3 \end{bmatrix} \begin{bmatrix} x_1 \\ x_2 \end{bmatrix} = \begin{bmatrix} 7 \\ 10 \end{bmatrix}$

$\begin{bmatrix} x_1 \\ x_2 \end{bmatrix} = \begin{bmatrix} 3 & -2 \\ -4 & 3 \end{bmatrix} \begin{bmatrix} 7 \\ 10 \end{bmatrix} = \begin{bmatrix} 1 \\ 2 \end{bmatrix}$. Thus, $x_1 = 1$ and $x_2 = 2$.

(C) $\begin{bmatrix} 3 & 2 \\ 4 & 3 \end{bmatrix} \begin{bmatrix} x_1 \\ x_2 \end{bmatrix} = \begin{bmatrix} 4 \\ 2 \end{bmatrix}$

$\begin{bmatrix} x_1 \\ x_2 \end{bmatrix} = \begin{bmatrix} 3 & -2 \\ -4 & 3 \end{bmatrix} \begin{bmatrix} 4 \\ 2 \end{bmatrix} = \begin{bmatrix} 8 \\ -10 \end{bmatrix}$. Thus, $x_1 = 8$ and $x_2 = -10$.

16. $A + D = \begin{bmatrix} 2 & -2 \\ 1 & 0 \\ 3 & 2 \end{bmatrix} + \begin{bmatrix} 3 & -2 & 1 \\ -1 & 1 & 2 \end{bmatrix}$ Not defined, because the dimensions of A and D are different.

17. $E + DA = \begin{bmatrix} 3 & -4 \\ -1 & 0 \end{bmatrix} + \begin{bmatrix} 3 & -2 & 1 \\ -1 & 1 & 2 \end{bmatrix} \begin{bmatrix} 2 & -2 \\ 1 & 0 \\ 3 & 2 \end{bmatrix} = \begin{bmatrix} 3 & -4 \\ -1 & 0 \end{bmatrix} + \begin{bmatrix} 7 & -4 \\ 5 & 6 \end{bmatrix} = \begin{bmatrix} 10 & -8 \\ 4 & 6 \end{bmatrix}$

18. From Problem 17, $DA = \begin{bmatrix} 7 & -4 \\ 5 & 6 \end{bmatrix}$. Thus,

$DA - 3E = \begin{bmatrix} 7 & -4 \\ 5 & 6 \end{bmatrix} - 3 \begin{bmatrix} 3 & -4 \\ -1 & 0 \end{bmatrix} = \begin{bmatrix} 7 & -4 \\ 5 & 6 \end{bmatrix} + \begin{bmatrix} -9 & 12 \\ 3 & 0 \end{bmatrix} = \begin{bmatrix} -2 & 8 \\ 8 & 6 \end{bmatrix}$

19. $BC = \begin{bmatrix} -1 \\ 2 \\ 3 \end{bmatrix} [2 \quad 1 \quad 3] = \begin{bmatrix} -2 & -1 & -3 \\ 4 & 2 & 6 \\ 6 & 3 & 9 \end{bmatrix}$

20. $CB = [2 \quad 1 \quad 3] \begin{bmatrix} -1 \\ 2 \\ 3 \end{bmatrix} = [-2 + 2 + 9] = [9]$ (a 1×1 matrix)

21. $AD - BC$

$$AD = \begin{bmatrix} 2 & -2 \\ 1 & 0 \\ 3 & 2 \end{bmatrix} \begin{bmatrix} 3 & -2 & 1 \\ -1 & 1 & 2 \end{bmatrix} = \begin{bmatrix} [2 \;\; -2]\begin{bmatrix} 3 \\ -1 \end{bmatrix} & [2 \;\; -2]\begin{bmatrix} -2 \\ 1 \end{bmatrix} & [2 \;\; -2]\begin{bmatrix} 1 \\ 2 \end{bmatrix} \\[2mm] [1 \;\; 0]\begin{bmatrix} 3 \\ -1 \end{bmatrix} & [1 \;\; 0]\begin{bmatrix} -2 \\ 1 \end{bmatrix} & [1 \;\; 0]\begin{bmatrix} 1 \\ 2 \end{bmatrix} \\[2mm] [3 \;\; 2]\begin{bmatrix} 3 \\ -1 \end{bmatrix} & [3 \;\; 2]\begin{bmatrix} -2 \\ 1 \end{bmatrix} & [3 \;\; 2]\begin{bmatrix} 1 \\ 2 \end{bmatrix} \end{bmatrix}$$

$$= \begin{bmatrix} 8 & -6 & -2 \\ 3 & -2 & 1 \\ 7 & -4 & 7 \end{bmatrix}$$

$$BC = \begin{bmatrix} -1 \\ 2 \\ 3 \end{bmatrix} [2 \;\; 1 \;\; 3] = \begin{bmatrix} -2 & -1 & -3 \\ 4 & 2 & 6 \\ 6 & 3 & 9 \end{bmatrix}$$

$$AD - BC = \begin{bmatrix} 8 & -6 & -2 \\ 3 & -2 & 1 \\ 7 & -4 & 7 \end{bmatrix} - \begin{bmatrix} -2 & -1 & -3 \\ 4 & 2 & 6 \\ 6 & 3 & 9 \end{bmatrix} = \begin{bmatrix} 8 - (-2) & -6 - (-1) & -2 - (-3) \\ 3 - 4 & -2 - 2 & 1 - 6 \\ 7 - 6 & -4 - 3 & 7 - 9 \end{bmatrix} = \begin{bmatrix} 10 & -5 & 1 \\ -1 & -4 & -5 \\ 1 & -7 & -2 \end{bmatrix}$$

22.
$$\begin{bmatrix} 1 & 2 & 3 & | & 1 & 0 & 0 \\ 2 & 3 & 4 & | & 0 & 1 & 0 \\ 1 & 2 & 1 & | & 0 & 0 & 1 \end{bmatrix} \sim \begin{bmatrix} 1 & 2 & 3 & | & 1 & 0 & 0 \\ 0 & -1 & -2 & | & -2 & 1 & 0 \\ 0 & 0 & -2 & | & -1 & 0 & 1 \end{bmatrix} \sim \begin{bmatrix} 1 & 2 & 3 & | & 1 & 0 & 0 \\ 0 & 1 & 2 & | & 2 & -1 & 0 \\ 0 & 0 & 1 & | & \frac{1}{2} & 0 & -\frac{1}{2} \end{bmatrix}$$

$$\begin{array}{ccc} R_2 + (-2)R_1 \rightarrow R_2 & -R_2 \rightarrow R_2 & R_2 + (-2)R_3 \rightarrow R_2 \\ R_3 + (-1)R_1 \rightarrow R_3 & -\frac{1}{2}R_3 \rightarrow R_3 & R_1 + (-3)R_3 \rightarrow R_1 \end{array}$$

$$\sim \begin{bmatrix} 1 & 2 & 0 & | & -\frac{1}{2} & 0 & \frac{3}{2} \\ 0 & 1 & 0 & | & 1 & -1 & 1 \\ 0 & 0 & 1 & | & \frac{1}{2} & 0 & -\frac{1}{2} \end{bmatrix} \sim \begin{bmatrix} 1 & 0 & 0 & | & -\frac{5}{2} & 2 & -\frac{1}{2} \\ 0 & 1 & 0 & | & 1 & -1 & 1 \\ 0 & 0 & 1 & | & \frac{1}{2} & 0 & -\frac{1}{2} \end{bmatrix}, \; A^{-1} = \begin{bmatrix} -\frac{5}{2} & 2 & -\frac{1}{2} \\ 1 & -1 & 1 \\ \frac{1}{2} & 0 & -\frac{1}{2} \end{bmatrix}.$$

$$R_1 + (-2)R_2 \rightarrow R_1$$

Check:

$$A^{-1}A = \begin{bmatrix} -\frac{5}{2} & 2 & -\frac{1}{2} \\ 1 & -1 & 1 \\ \frac{1}{2} & 0 & -\frac{1}{2} \end{bmatrix} \begin{bmatrix} 1 & 2 & 3 \\ 2 & 3 & 4 \\ 1 & 2 & 1 \end{bmatrix} = \begin{bmatrix} -\frac{5}{2} + 4 - \frac{1}{2} & -5 + 6 - 1 & -\frac{15}{2} + 8 - \frac{1}{2} \\ 1 - 2 + 1 & 2 - 3 + 2 & 3 - 4 + 1 \\ \frac{1}{2} + 0 - \frac{1}{2} & 1 + 0 - 1 & \frac{3}{2} + 0 - \frac{1}{2} \end{bmatrix}$$

$$= \begin{bmatrix} 1 & 0 & 0 \\ 0 & 1 & 0 \\ 0 & 0 & 1 \end{bmatrix}$$

23. (A) The augmented matrix corresponding to the given system is:

$$\begin{bmatrix} 1 & 2 & 3 & | & 1 \\ 2 & 3 & 4 & | & 3 \\ 1 & 2 & 1 & | & 3 \end{bmatrix} \sim \begin{bmatrix} 1 & 2 & 3 & | & 1 \\ 0 & -1 & -2 & | & 1 \\ 0 & 0 & -2 & | & 2 \end{bmatrix} \sim \begin{bmatrix} 1 & 2 & 3 & | & 1 \\ 0 & 1 & 2 & | & -1 \\ 0 & 0 & 1 & | & -1 \end{bmatrix} \sim \begin{bmatrix} 1 & 2 & 0 & | & 4 \\ 0 & 1 & 0 & | & 1 \\ 0 & 0 & 1 & | & -1 \end{bmatrix}$$

$R_2 + (-2)R_1 \rightarrow R_2$ $-R_2 \rightarrow R_2$ $R_2 + (-2)R_3 \rightarrow R_2$ $R_1 + (-2)R_2 \rightarrow R_1$

$R_3 + (-1)R_1 \rightarrow R_3$ $-\frac{1}{2}R_3 \rightarrow R_3$ $R_1 + (-3)R_3 \rightarrow R_1$

$$\sim \begin{bmatrix} 1 & 0 & 0 & | & 2 \\ 0 & 1 & 0 & | & 1 \\ 0 & 0 & 1 & | & -1 \end{bmatrix}$$

Thus, the solution is: $x_1 = 2$

$x_2 = 1$

$x_3 = -1$.

(B) The augmented matrix corresponding to the given system is:

$$\begin{bmatrix} 1 & 2 & -1 & | & 2 \\ 2 & 3 & 1 & | & -3 \\ 3 & 5 & 0 & | & -1 \end{bmatrix} \sim \begin{bmatrix} 1 & 2 & -1 & | & 2 \\ 0 & -1 & 3 & | & -7 \\ 0 & -1 & 3 & | & -7 \end{bmatrix} \sim \begin{bmatrix} 1 & 2 & -1 & | & 2 \\ 0 & 1 & -3 & | & 7 \\ 0 & -1 & 3 & | & -7 \end{bmatrix} \sim \begin{bmatrix} 1 & 2 & -1 & | & 2 \\ 0 & 1 & -3 & | & 7 \\ 0 & 0 & 0 & | & 0 \end{bmatrix}$$

$R_2 + (-2)R_1 \rightarrow R_2$ $(-1)R_2 \rightarrow R_2$ $R_3 + R_2 \rightarrow R_3$ $R_1 + (-2)R_2 \rightarrow R_1$

$R_3 + (-3)R_1 \rightarrow R_3$

$$\sim \begin{bmatrix} 1 & 0 & 5 & | & -12 \\ 0 & 1 & -3 & | & 7 \\ 0 & 0 & 0 & | & 0 \end{bmatrix}$$

Thus, $x_1 \quad + 5x_3 = -12$ (1)

$x_2 - 3x_3 = \quad 7$ (2)

Let $x_3 = t$ (t any real number). Then, from (1),

$x_1 = -5t - 12$

and, from (2),

$x_2 = 3t + 7$.

Thus, the solution is $x_1 = -5t - 12$, $x_2 = 3t + 7$, $x_3 = t$.

24. (A) The matrix equation for the given system is:

$$\begin{bmatrix} 1 & 2 & 3 \\ 2 & 3 & 4 \\ 1 & 2 & 1 \end{bmatrix} \begin{bmatrix} x_1 \\ x_2 \\ x_3 \end{bmatrix} = \begin{bmatrix} 1 \\ 3 \\ 3 \end{bmatrix}$$

The inverse matrix of the coefficient matrix of the system, from Problem 22, is:

$$\begin{bmatrix} -\frac{5}{2} & 2 & -\frac{1}{2} \\ 1 & -1 & 1 \\ \frac{1}{2} & 0 & -\frac{1}{2} \end{bmatrix}$$ Thus,

$$\begin{bmatrix} x_1 \\ x_2 \\ x_3 \end{bmatrix} = \begin{bmatrix} -\frac{5}{2} & 2 & -\frac{1}{2} \\ 1 & -1 & 1 \\ \frac{1}{2} & 0 & -\frac{1}{2} \end{bmatrix} \begin{bmatrix} 1 \\ 3 \\ 3 \end{bmatrix} = \begin{bmatrix} \frac{-5+12-3}{2} \\ 1 - 3 + 3 \\ \frac{1+0-3}{2} \end{bmatrix} = \begin{bmatrix} 2 \\ 1 \\ -1 \end{bmatrix} \qquad \text{Solution: } x_1 = 2, \\ x_2 = 1, \\ x_3 = -1.$$

(B) $\begin{bmatrix} x_1 \\ x_2 \\ x_3 \end{bmatrix} = \begin{bmatrix} -\frac{5}{2} & 2 & -\frac{1}{2} \\ 1 & -1 & 1 \\ \frac{1}{2} & 0 & -\frac{1}{2} \end{bmatrix} \begin{bmatrix} 0 \\ 0 \\ -2 \end{bmatrix} = \begin{bmatrix} 1 \\ -2 \\ 1 \end{bmatrix}$ Solution: $x_1 = 1,$ $x_2 = -2,$ $x_3 = 1.$

(C) $\begin{bmatrix} x_1 \\ x_2 \\ x_3 \end{bmatrix} = \begin{bmatrix} -\frac{5}{2} & 2 & -\frac{1}{2} \\ 1 & -1 & 1 \\ \frac{1}{2} & 0 & -\frac{1}{2} \end{bmatrix} \begin{bmatrix} -3 \\ -4 \\ 1 \end{bmatrix} = \begin{bmatrix} \frac{15-16-1}{2} \\ -3 + 4 + 1 \\ \frac{-3+0-1}{2} \end{bmatrix} \begin{bmatrix} -1 \\ 2 \\ -2 \end{bmatrix}$ Solution: $x_1 = -1,$ $x_2 = 2,$ $x_3 = -2.$

25. $\begin{bmatrix} 4 & 5 & 6 & | & 1 & 0 & 0 \\ 4 & 5 & -6 & | & 0 & 1 & 0 \\ 1 & 1 & 1 & | & 0 & 0 & 1 \end{bmatrix} \sim \begin{bmatrix} 1 & 1 & 1 & | & 0 & 0 & 1 \\ 4 & 5 & -6 & | & 0 & 1 & 0 \\ 4 & 5 & 6 & | & 1 & 0 & 0 \end{bmatrix} \sim \begin{bmatrix} 1 & 1 & 1 & | & 0 & 0 & 1 \\ 0 & 1 & -10 & | & 0 & 1 & -4 \\ 0 & 1 & 2 & | & 1 & 0 & -4 \end{bmatrix}$

$R_1 \leftrightarrow R_3$ $R_2 + (-4)R_1 \rightarrow R_2$ $R_3 + (-1)R_2 \rightarrow R_3$

                       $R_3 + (-4)R_1 \rightarrow R_3$

$\sim \begin{bmatrix} 1 & 1 & 1 & | & 0 & 0 & 1 \\ 0 & 1 & -10 & | & 0 & 1 & -4 \\ 0 & 0 & 12 & | & 1 & -1 & 0 \end{bmatrix} \sim \begin{bmatrix} 1 & 1 & 1 & | & 0 & 0 & 1 \\ 0 & 1 & -10 & | & 0 & 1 & -4 \\ 0 & 0 & 1 & | & \frac{1}{12} & -\frac{1}{12} & 0 \end{bmatrix}$

       $\frac{1}{12}R_3 \rightarrow R_3$ $R_2 + 10R_3 \rightarrow R_2$

                   $R_1 + (-1)R_3 \rightarrow R_1$

$\sim \begin{bmatrix} 1 & 1 & 0 & | & -\frac{1}{12} & \frac{1}{12} & 1 \\ 0 & 1 & 0 & | & \frac{10}{12} & \frac{2}{12} & -4 \\ 0 & 0 & 1 & | & \frac{1}{12} & -\frac{1}{12} & 0 \end{bmatrix} \sim \begin{bmatrix} 1 & 0 & 0 & | & -\frac{11}{12} & -\frac{1}{12} & 5 \\ 0 & 1 & 0 & | & \frac{10}{12} & \frac{2}{12} & -4 \\ 0 & 0 & 1 & | & \frac{1}{12} & -\frac{1}{12} & 0 \end{bmatrix}$

       $R_1 + (-1)R_2 \rightarrow R_1$

Thus, $A^{-1} = \begin{bmatrix} -\frac{11}{12} & -\frac{1}{12} & 5 \\ \frac{10}{12} & \frac{2}{12} & -4 \\ \frac{1}{12} & -\frac{1}{12} & 0 \end{bmatrix}$;

$$A^{-1}A \begin{bmatrix} -\frac{11}{12} & -\frac{1}{12} & 5 \\ \frac{10}{12} & \frac{2}{12} & -4 \\ \frac{1}{12} & -\frac{1}{12} & 0 \end{bmatrix} \begin{bmatrix} 4 & 5 & 6 \\ 4 & 5 & -6 \\ 1 & 1 & 1 \end{bmatrix}$$

$$= \begin{bmatrix} \frac{-44-4+60}{12} & \frac{-55-5+60}{12} & \frac{-66+6+60}{12} \\ \frac{40+8-48}{12} & \frac{50+10-48}{12} & \frac{60-12-48}{12} \\ \frac{4-4+0}{12} & \frac{5-5+0}{12} & \frac{6+6+0}{12} \end{bmatrix} = \begin{bmatrix} 1 & 0 & 0 \\ 0 & 1 & 0 \\ 0 & 0 & 1 \end{bmatrix}$$

26. Multiply by 100 to eliminate the decimals from the first two equations. We get the following system:

$$4x_1 + 5x_2 + 6x_3 = 36,000$$
$$4x_1 + 5x_2 - 6x_3 = 12,000$$
$$x_1 + x_2 + x_3 = 7,000$$

The matrix equation of the above system is: $\begin{bmatrix} 4 & 5 & 6 \\ 4 & 5 & -6 \\ 1 & 1 & 1 \end{bmatrix} \begin{bmatrix} x_1 \\ x_2 \\ x_3 \end{bmatrix} = \begin{bmatrix} 36,000 \\ 12,000 \\ 7,000 \end{bmatrix}$

It follows that:

$$\begin{bmatrix} x_1 \\ x_2 \\ x_3 \end{bmatrix} = \begin{bmatrix} -\frac{11}{12} & -\frac{1}{12} & 5 \\ \frac{10}{12} & \frac{2}{12} & -4 \\ \frac{1}{12} & -\frac{1}{12} & 0 \end{bmatrix} \begin{bmatrix} 36,000 \\ 12,000 \\ 7,000 \end{bmatrix} = \begin{bmatrix} -33,000 - 1,000 + 35,000 \\ 30,000 + 2,000 - 28,000 \\ 3,000 - 1,000 + 0 \end{bmatrix} = \begin{bmatrix} 1,000 \\ 4,000 \\ 2,000 \end{bmatrix}$$

Solution: $x_1 = 1,000$, $x_2 = 4,000$, $x_3 = 2,000$.

27. The augmented matrix corresponding to the given system is:

$$\begin{bmatrix} 0.04 & 0.05 & 0.06 & | & 360 \\ 0.04 & 0.05 & -0.06 & | & 120 \\ 1 & 1 & 1 & | & 7,000 \end{bmatrix} \sim \begin{bmatrix} 4 & 5 & 6 & | & 36,000 \\ 4 & 5 & -6 & | & 12,000 \\ 1 & 1 & 1 & | & 7,000 \end{bmatrix} \sim \begin{bmatrix} 1 & 1 & 1 & | & 7,000 \\ 4 & 5 & -6 & | & 12,000 \\ 4 & 5 & 6 & | & 36,000 \end{bmatrix}$$

$$100R_1 \rightarrow R_1 \qquad\qquad R_1 \leftrightarrow R_3 \qquad\qquad R_2 + (-4)R_1 \rightarrow R_2$$
$$100R_2 \rightarrow R_2 \qquad\qquad\qquad\qquad\qquad\qquad R_3 + (-4)R_1 \rightarrow R_3$$

$$\sim \begin{bmatrix} 1 & 1 & 1 & | & 7,000 \\ 0 & 1 & -10 & | & -16,000 \\ 0 & 1 & 2 & | & 8,000 \end{bmatrix} \sim \begin{bmatrix} 1 & 1 & 1 & | & 7,000 \\ 0 & 1 & -10 & | & -16,000 \\ 0 & 0 & 12 & | & 24,000 \end{bmatrix} \sim \begin{bmatrix} 1 & 1 & 1 & | & 7,000 \\ 0 & 1 & -10 & | & -16,000 \\ 0 & 0 & 1 & | & 2,000 \end{bmatrix}$$

$$R_3 + (-1)R_2 \rightarrow R_3 \qquad\qquad \frac{1}{12}R_3 \rightarrow R_3 \qquad\qquad R_2 + 10R_3 \rightarrow R_2$$
$$\qquad\qquad\qquad\qquad\qquad\qquad\qquad\qquad\qquad R_1 + (-1)R_3 \rightarrow R_1$$

$$\sim \begin{bmatrix} 1 & 1 & 0 & | & 5,000 \\ 0 & 1 & 0 & | & 4,000 \\ 0 & 0 & 1 & | & 2,000 \end{bmatrix} \sim \begin{bmatrix} 1 & 0 & 0 & | & 1,000 \\ 0 & 1 & 0 & | & 4,000 \\ 0 & 0 & 1 & | & 2,000 \end{bmatrix}$$

Thus, $x_1 = 1,000$,
$x_2 = 4,000$,
$x_3 = 2,000$.

$$R_1 + (-1)R_2 \rightarrow R_1$$

28. Let x_1 = number of tons of ore A
and x_2 = number of tons of ore B.

Then, we have the following system of equations:
$$0.01x_1 + 0.02x_2 = 4.5$$
$$0.02x_1 + 0.05x_2 = 10$$

Multiply each equation by 100. This yields
$$x_1 + 2x_2 = 450$$
$$2x_1 + 5x_2 = 1000$$

The augmented matrix corresponding to this system is:
$$\begin{bmatrix} 1 & 2 & | & 450 \\ 2 & 5 & | & 1000 \end{bmatrix} \sim \begin{bmatrix} 1 & 2 & | & 450 \\ 0 & 1 & | & 100 \end{bmatrix} \sim \begin{bmatrix} 1 & 0 & | & 250 \\ 0 & 1 & | & 100 \end{bmatrix}$$
$$R_2 + (-2)R_1 \rightarrow R_2 \qquad R_1 + (-2)R_2 \rightarrow R_1$$

Thus, the solution is: $x_1 = 250$ tons of ore A, $x_2 = 100$ tons of ore B.

29. (A) The matrix equation for Problem 28 is:
$$\begin{bmatrix} 0.01 & 0.02 \\ 0.02 & 0.05 \end{bmatrix}\begin{bmatrix} x_1 \\ x_2 \end{bmatrix} = \begin{bmatrix} 4.5 \\ 10 \end{bmatrix}$$

First, compute the inverse of $\begin{bmatrix} 0.01 & 0.02 \\ 0.02 & 0.05 \end{bmatrix}$;

$$\begin{bmatrix} 0.01 & 0.02 & | & 1 & 0 \\ 0.02 & 0.05 & | & 0 & 1 \end{bmatrix} \sim \begin{bmatrix} 1 & 2 & | & 100 & 0 \\ 0.02 & 0.05 & | & 0 & 1 \end{bmatrix} \sim \begin{bmatrix} 1 & 2 & | & 100 & 0 \\ 0 & 0.01 & | & -2 & 1 \end{bmatrix}$$
$$100R_1 \rightarrow R_1 \qquad\qquad R_2 + (-0.02)R_1 \rightarrow R_2 \qquad\qquad 100R_2 \rightarrow R_2$$

$$\sim \begin{bmatrix} 1 & 2 & | & 100 & 0 \\ 0 & 1 & | & -200 & 100 \end{bmatrix} \sim \begin{bmatrix} 1 & 0 & | & 500 & -200 \\ 0 & 1 & | & -200 & 100 \end{bmatrix}$$
$$R_1 + (-2)R_2 \rightarrow R_1$$

Thus, the inverse matrix is $\begin{bmatrix} 500 & -200 \\ -200 & 100 \end{bmatrix}$.

Hence, $\begin{bmatrix} x_1 \\ x_2 \end{bmatrix} = \begin{bmatrix} 500 & -200 \\ -200 & 100 \end{bmatrix}\begin{bmatrix} 4.5 \\ 10 \end{bmatrix} = \begin{bmatrix} 2250 - 2000 \\ -900 + 1000 \end{bmatrix} = \begin{bmatrix} 250 \\ 100 \end{bmatrix}$

Again the solution is: $x_1 = 250$ tons of ore A.
$x_2 = 100$ tons of ore B.

(B) $\begin{bmatrix} x_1 \\ x_2 \end{bmatrix} = \begin{bmatrix} 500 & -200 \\ -200 & 100 \end{bmatrix}\begin{bmatrix} 2.3 \\ 5 \end{bmatrix} = \begin{bmatrix} 1150 - 1000 \\ -460 + 500 \end{bmatrix} = \begin{bmatrix} 150 \\ 40 \end{bmatrix}$

Now the solution is: $x_1 = 150$ tons of ore A.
$x_2 = 40$ tons of ore B.

30. Let x_1 = number of model A trucks
$\quad\quad x_2$ = number of model B trucks
$\quad\quad x_3$ = number of model C trucks

Then $x_1 + x_2 + x_3 = 12$

and $18{,}000x_1 + 22{,}000x_2 + 30{,}000x_3 = 300{,}000$

or $\quad x_1 + \quad x_2 + \quad x_3 = 12$
$\quad\quad 9x_1 + 11x_2 + 15x_3 = 150$

The augmented matrix corresponding to this system is $\begin{pmatrix} 1 & 1 & 1 & 12 \\ 9 & 11 & 15 & 150 \end{pmatrix}$

Now

$\begin{pmatrix} 1 & 1 & 1 & 12 \\ 9 & 11 & 15 & 150 \end{pmatrix} \sim \begin{pmatrix} 1 & 1 & 1 & 12 \\ 0 & 2 & 6 & 42 \end{pmatrix} \sim \begin{pmatrix} 1 & 1 & 1 & 12 \\ 0 & 1 & 3 & 21 \end{pmatrix} \sim \begin{pmatrix} 1 & 0 & -2 & -9 \\ 0 & 1 & 3 & 21 \end{pmatrix}$

$\quad R_2 + (-9)R_1 \to R_2 \quad\quad \frac{1}{2}R_2 \to R_2 \quad\quad R_1 + (-1)R_2 \to R_1$

The corresponding system of equations is

$x_1 \quad\quad - 2x_3 = -9$
$\quad\quad x_2 + 3x_3 = 21$ and the solutions are

$x_1 = 2t - 9$
$x_2 = 21 - 3t$
$x_3 = t$

Now, since x_1, x_2, and x_3 are nonnegative integers, we must have

$\frac{9}{2} \le t \le 7$ or $t = 5$, 6, or 7.

For $t = 5$: 1 model A truck, 6 model B trucks, 5 model C trucks
$\quad\quad t = 6$: 3 model A trucks, 3 model B trucks, 6 model C trucks
$\quad\quad t = 7$: 5 model A trucks, 0 model B trucks, 7 model C trucks

31. (A) $MN = \begin{bmatrix} 4800 & 600 & 300 \\ 6000 & 1400 & 700 \end{bmatrix} \begin{bmatrix} 0.75 & 0.70 \\ 6.50 & 6.70 \\ 0.40 & 0.50 \end{bmatrix}$

$= \begin{bmatrix} 4800(0.75) + 600(6.50) + 300(0.40) & 4800(0.70) + 600(6.70) + 300(0.50) \\ 6000(0.75) + 1400(6.50) + 700(0.40) & 6000(0.70) + 1400(6.70) + 700(0.50) \end{bmatrix}$

$\quad\quad$ Supplier A $\quad$ Supplier B

$= \begin{bmatrix} \$\ 7{,}620 & \$\ 7{,}530 \\ \$13{,}880 & \$13{,}930 \end{bmatrix}$ Alloy 1
$\quad\quad\quad\quad\quad\quad\quad\quad\quad\quad$ Alloy 2

This matrix represents the cost of each alloy from each supplier.

(B) $[1 \quad 1]MN = [1 \quad 1]\begin{bmatrix} 7{,}620 & 7{,}530 \\ 13{,}880 & 13{,}930 \end{bmatrix}$

$\quad\quad\quad\quad = [7{,}620 + 13{,}880 \quad 7{,}530 + 13{,}930]$

$\quad\quad\quad\quad = [\$21{,}500 \quad \$21{,}460]$

This matrix represents the total cost of both alloys from each supplier.

32. (A) The labor cost for producing one Model B calculator in California is:

$$[0.25 \text{ hr} \quad 0.20 \text{ hr} \quad 0.05 \text{ hr}] \begin{bmatrix} \$15 \\ \$12 \\ \$ 4 \end{bmatrix}$$

$$= [(0.25)(15) + (0.20)(12) + (0.05)(4)]$$
$$= [3.75 + 2.40 + 0.20]$$
$$= [\$6.35]$$

(B) $MN = \begin{bmatrix} 0.15 \text{ hr} & 0.10 \text{ hr} & 0.05 \text{ hr} \\ 0.25 \text{ hr} & 0.20 \text{ hr} & 0.05 \text{ hr} \end{bmatrix} \begin{bmatrix} \$15 & \$12 \\ \$12 & \$10 \\ \$ 4 & \$ 4 \end{bmatrix}$

$$= \begin{bmatrix} (0.15)15 + (0.10)12 + (0.05)4 & (0.15)12 + (0.10)10 + (0.05)4 \\ (0.25)15 + (0.20)12 + (0.05)4 & (0.25)12 + (0.20)10 + (0.05)4 \end{bmatrix}$$

$$= \begin{matrix} & \text{Calif.} & \text{Texas} \\ & \begin{bmatrix} \$3.65 & \$3.00 \\ \$6.35 & \$5.20 \end{bmatrix} & \begin{matrix} \text{Model A} \\ \text{Model B} \end{matrix} \end{matrix}$$

This matrix represents the total labor costs for each model at each plant.

33. Let x_1 = amount invested at 5%
and x_2 = amount invested at 10%.

Then, $\quad x_1 + \quad x_2 = 5000$
$\quad\quad 0.05x_1 + 0.1x_2 = 400$

The augmented matrix for the system given above is:

$$\begin{bmatrix} 1 & 1 & | & 5000 \\ 0.05 & 0.1 & | & 400 \end{bmatrix} \sim \begin{bmatrix} 1 & 1 & | & 5000 \\ 0 & 0.05 & | & 150 \end{bmatrix} \sim \begin{bmatrix} 1 & 1 & | & 5000 \\ 0 & 1 & | & 3000 \end{bmatrix} \sim \begin{bmatrix} 1 & 0 & | & 2000 \\ 0 & 1 & | & 3000 \end{bmatrix}$$

$R_2 + (-0.05)R_1 \to R_2 \quad\quad \frac{1}{0.05}R_2 \to R_2 \quad\quad R_1 + (-1)R_2 \to R_1$

Hence, x_1 = $2000 at 5%, x_2 = $3000 at 10%.

34. The matrix equation corresponding to the system in Problem 33 is:

$$\begin{bmatrix} 1 & 1 \\ 0.05 & 0.1 \end{bmatrix} \begin{bmatrix} x_1 \\ x_2 \end{bmatrix} = \begin{bmatrix} 5000 \\ 400 \end{bmatrix}$$

Now we compute the inverse matrix of $\begin{bmatrix} 1 & 1 \\ 0.05 & 0.1 \end{bmatrix}$.

$$\begin{bmatrix} 1 & 1 & | & 1 & 0 \\ 0.05 & 0.1 & | & 0 & 1 \end{bmatrix} \sim \begin{bmatrix} 1 & 1 & | & 1 & 0 \\ 0 & 0.05 & | & -0.05 & 1 \end{bmatrix} \sim \begin{bmatrix} 1 & 1 & | & 1 & 0 \\ 0 & 1 & | & -1 & 20 \end{bmatrix} \sim \begin{bmatrix} 1 & 0 & | & 2 & -20 \\ 0 & 1 & | & -1 & 20 \end{bmatrix}$$

$R_2 + (-0.05)R_1 \to R_2 \quad\quad \frac{1}{0.05}R_2 \to R_2 \quad\quad R_1 + (-1)R_2 \to R_1$

Thus, the inverse of the coefficient matrix is $\begin{bmatrix} 2 & -20 \\ -1 & 20 \end{bmatrix}$, and

$$\begin{bmatrix} x_1 \\ x_2 \end{bmatrix} = \begin{bmatrix} 2 & -20 \\ -1 & 20 \end{bmatrix}\begin{bmatrix} 5000 \\ 400 \end{bmatrix} = \begin{bmatrix} 10{,}000 - 8{,}000 \\ -5{,}000 + 8{,}000 \end{bmatrix} = \begin{bmatrix} 2000 \\ 3000 \end{bmatrix}.$$

So, $x_1 = \$2000$ at 5%, $x_2 = \$3000$ at 10%.

35. Let x_1 = number of \$8 tickets
x_2 = number of \$12 tickets
x_3 = number of \$20 tickets

Since the number of \$8 tickets must equal the number of \$20 tickets, we have
$$x_1 = x_3 \quad \text{or} \quad x_1 - x_3 = 0$$
Also, since all seats are sold
$$x_1 + x_2 + x_3 = 25{,}000$$
Finally, the return is
$$8x_1 + 12x_2 + 20x_3 = R \text{ (where } R \text{ is the return required)}.$$
Thus, the system of equations is:

$$\begin{array}{rrrcl} x_1 & & - & x_3 & = 0 \\ x_1 & + & x_2 + & x_3 & = 25{,}000 \\ 8x_1 & + & 12x_2 + & 20x_3 & = R \end{array} \quad \text{or} \quad \begin{bmatrix} 1 & 0 & -1 \\ 1 & 1 & 1 \\ 8 & 12 & 20 \end{bmatrix}\begin{bmatrix} x_1 \\ x_2 \\ x_3 \end{bmatrix} = \begin{bmatrix} 0 \\ 25{,}000 \\ R \end{bmatrix}$$

First, we compute the inverse of the coefficient matrix

$$\left[\begin{array}{ccc|ccc} 1 & 0 & -1 & 1 & 0 & 0 \\ 1 & 1 & 1 & 0 & 1 & 0 \\ 8 & 12 & 20 & 0 & 0 & 1 \end{array}\right] \sim \left[\begin{array}{ccc|ccc} 1 & 0 & -1 & 1 & 0 & 0 \\ 0 & 1 & 2 & -1 & 1 & 0 \\ 0 & 12 & 28 & -8 & 0 & 1 \end{array}\right]$$

$R_2 + (-1)R_1 \rightarrow R_2$ $R_3 + (-12)R_2 \rightarrow R_3$
$R_3 + (-8)R_1 \rightarrow R_3$

$$\sim \left[\begin{array}{ccc|ccc} 1 & 0 & -1 & 1 & 0 & 0 \\ 0 & 1 & 2 & -1 & 1 & 0 \\ 0 & 0 & 4 & 4 & -12 & 1 \end{array}\right] \sim \left[\begin{array}{ccc|ccc} 1 & 0 & -1 & 1 & 0 & 0 \\ 0 & 1 & 2 & -1 & 1 & 0 \\ 0 & 0 & 1 & 1 & -3 & \frac{1}{4} \end{array}\right]$$

$\frac{1}{4}R_3 \rightarrow R_3$ $R_2 + (-2)R_3 \rightarrow R_2$
                         $R_1 + R_3 \rightarrow R_1$

$$\sim \left[\begin{array}{ccc|ccc} 1 & 0 & 0 & 2 & -3 & \frac{1}{4} \\ 0 & 1 & 0 & -3 & 7 & -\frac{1}{2} \\ 0 & 0 & 1 & 1 & -3 & \frac{1}{4} \end{array}\right]. \quad \text{Thus, the inverse is} \quad \begin{bmatrix} 2 & -3 & \frac{1}{4} \\ -3 & 7 & -\frac{1}{2} \\ 1 & -3 & \frac{1}{4} \end{bmatrix}$$

Concert 1:

$$\begin{array}{rrrcl} x_1 & & - & x_3 & = 0 \\ x_1 & + & x_2 + & x_3 & = 25{,}000 \\ 8x_1 & + & 12x_2 + & 20x_3 & = 320{,}000 \end{array} \quad \text{or} \quad \begin{bmatrix} 1 & 0 & -1 \\ 1 & 1 & 1 \\ 8 & 12 & 20 \end{bmatrix}\begin{bmatrix} x_1 \\ x_2 \\ x_3 \end{bmatrix} = \begin{bmatrix} 0 \\ 25{,}000 \\ 320{,}000 \end{bmatrix}$$

$$\text{Thus } \begin{bmatrix} x_1 \\ x_2 \\ x_3 \end{bmatrix} = \begin{bmatrix} 2 & -3 & \frac{1}{4} \\ -3 & 7 & -\frac{1}{2} \\ 1 & -3 & \frac{1}{4} \end{bmatrix} \begin{bmatrix} 0 \\ 25{,}000 \\ 320{,}000 \end{bmatrix} = \begin{bmatrix} 5{,}000 \\ 15{,}000 \\ 5{,}000 \end{bmatrix}$$

and $x_1 = 5{,}000$ \$8 tickets
$x_2 = 15{,}000$ \$12 tickets
$x_3 = 5{,}000$ \$20 tickets

Concert 2:

$$\begin{array}{rcr} x_1 \quad\quad\; - \quad x_3 &=& 0 \\ x_1 + \quad x_2 + \quad x_3 &=& 25{,}000 \\ 8x_1 + 12x_2 + 20x_3 &=& 330{,}000 \end{array} \quad \text{or} \quad \begin{bmatrix} 1 & 0 & -1 \\ 1 & 1 & 1 \\ 8 & 12 & 20 \end{bmatrix} \begin{bmatrix} x_1 \\ x_2 \\ x_3 \end{bmatrix} = \begin{bmatrix} 0 \\ 25{,}000 \\ 330{,}000 \end{bmatrix}$$

$$\text{Thus } \begin{bmatrix} x_1 \\ x_2 \\ x_3 \end{bmatrix} = \begin{bmatrix} 2 & -3 & \frac{1}{4} \\ -3 & 7 & -\frac{1}{2} \\ 1 & -3 & \frac{1}{4} \end{bmatrix} \begin{bmatrix} 0 \\ 25{,}000 \\ 330{,}000 \end{bmatrix} = \begin{bmatrix} 7{,}500 \\ 10{,}000 \\ 7{,}500 \end{bmatrix}$$

and $x_1 = 7{,}500$ \$8 tickets
$x_2 = 10{,}000$ \$12 tickets
$x_3 = 7{,}500$ \$20 tickets

Concert 3:

$$\begin{array}{rcr} x_1 \quad\quad\; - \quad x_3 &=& 0 \\ x_1 + \quad x_2 + \quad x_3 &=& 25{,}000 \\ 8x_1 + 12x_2 + 20x_3 &=& 340{,}000 \end{array} \quad \text{or} \quad \begin{bmatrix} 1 & 0 & -1 \\ 1 & 1 & 1 \\ 8 & 12 & 20 \end{bmatrix} \begin{bmatrix} x_1 \\ x_2 \\ x_3 \end{bmatrix} = \begin{bmatrix} 0 \\ 25{,}000 \\ 340{,}000 \end{bmatrix}$$

$$\text{Thus } \begin{bmatrix} x_1 \\ x_2 \\ x_3 \end{bmatrix} = \begin{bmatrix} 2 & -3 & \frac{1}{4} \\ -3 & 7 & -\frac{1}{2} \\ 1 & -3 & \frac{1}{4} \end{bmatrix} \begin{bmatrix} 0 \\ 25{,}000 \\ 340{,}000 \end{bmatrix} = \begin{bmatrix} 10{,}000 \\ 5{,}000 \\ 10{,}000 \end{bmatrix}$$

and $x_1 = 10{,}000$ \$8 tickets
$x_2 = 5{,}000$ \$12 tickets
$x_3 = 10{,}000$ \$20 tickets

EXERCISE 6-1

Things to remember:

1. The graph of the linear inequality

    $Ax + By < C$ or $Ax + By > C$

    with $B \neq 0$ is either the upper half-plane or the lower half-plane (but not both) determined by the line $Ax + By = C$.
    If $B = 0$, the graph of

    $Ax < C$ or $Ax > C$

    is either the right half-plane or the left half-plane (but not both) determined by the vertical line $Ax = C$.

2. For strict inequalities ("<" or ">"), the line is not included in the graph. For weak inequalities ("≤" or "≥"), the line is included in the graph.

3. PROCEDURE FOR GRAPHING LINEAR INEQUALITIES

    (a) First graph $Ax + By = C$ as a broken line if equality is not included in the original statement or as a solid line if equality is included.

    (b) Choose a test point anywhere in the plane not on the line [the origin (0, 0) often requires the least computation] and substitute the coordinates into the inequality.

    (c) The graph of the original inequality includes the half-plane containing the test point if the inequality is satisfied by that point or the half-plane not containing the test point if the inequality is not satisfied by that point.

4. To solve a system of linear inequalities graphically, graph each inequality in the system and then take the intersection of all the graphs. The resulting graph is called the SOLUTION REGION, or FEASIBLE REGION.

5. A CORNER POINT of a solution region is a point in the solution region which is the intersection of two boundary lines.

6. The solution region of a system of linear inequalities is BOUNDED if it can be enclosed within a circle; if it cannot be enclosed within a circle, then it is UNBOUNDED.

1. $y \le x - 1$

Graph $y = x - 1$ as a solid line.

Test point $(0, 0)$:

$0 \le 0 - 1$

$0 \le -1$

The inequality is false. Thus, the graph is below the line $y = x - 1$, including the line.

x	y
0	-1
1	0

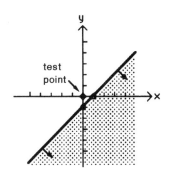

3. $3x - 2y > 6$

Graph $3x - 2y = 6$ as a broken line.

Test point $(0, 0)$:

$3 \cdot 0 - 2 \cdot 0 > 6$

$ 0 > 6$

The inequality is false. Thus, the graph is below the line $3x - 2y = 6$, not including the line.

x	y
0	-3
2	0

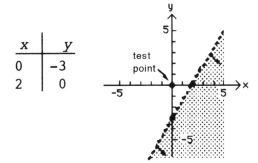

5. $x \ge -4$

Graph $x = -4$ [the vertical line through $(-4, 0)$] as a solid line.

Test point $(0, 0)$:

$0 \ge -4$

The inequality is true. Thus, the graph is to the right of the line $x = -4$, including the line.

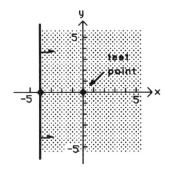

7. $-4 \le y < 4$

Graph $y = -4$ as a solid line and $y = 4$ as a broken line [horizontal lines through $(0, -4)$ and $(0, 4)$, respectively].

Test point $(0, 0)$:

$-4 \le 0$ and $0 < 4$

i.e., $-4 \le 0 < 4$

Both inequalities are true. Thus, the graph is between the lines $y = -4$ and $y = 4$, including the line $y = -4$ but not including the line $y = 4$.

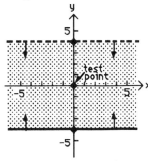

9. $6x + 4y \geq 24$

Graph the line $6x + 4y = 24$ as a solid line.

Test point $(0, 0)$:

$6 \cdot 0 + 4 \cdot 0 \geq 24$

$0 \geq 24$

The inequality is false. Thus, the graph is the region above the line, including the line.

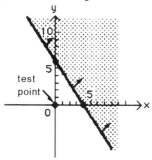

11. $5x \leq -2y$ or $5x + 2y \leq 0$

Graph the line $5x + 2y = 0$ as a solid line. Since the line passes through the origin $(0, 0)$, we use $(1, 0)$ as a test point:

$5 \cdot 1 + 2 \cdot 0 \leq 0$

$5 \leq 0$

This inequality is false. Thus, the graph is below the line $5x + 2y = 0$, including the line.

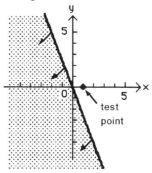

13. The graph of $x + 2y \leq 8$ is the region below the line $x + 2y = 8$ [e.g., $(0, 0)$ satisfies the inequality]. The graph of $3x - 2y \geq 0$ is the region below the line $3x - 2y = 0$ [e.g., $(1, 0)$ satisfies the inequality]. The intersection of these two regions is region IV.

15. The graph of $x + 2y \geq 8$ is the region above the line $x + 2y = 8$ [e.g., $(0, 0)$ does not satisfy the inequality]. The graph of $3x - 2y \geq 0$ is the region below the line $3x - 2y = 0$ [e.g., $(1, 0)$ satisfies the inequality]. The intersection of these two regions is region I.

17. The graphs of the inequalities $3x + y \geq 6$ and $x \leq 4$ are:

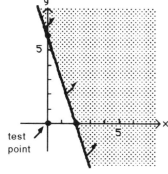
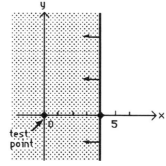

The intersection of these regions (drawn on the same coordinate plane) is shown in the graph at the right.

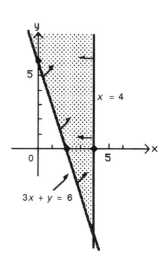

19. The graphs of the inequalities $x - 2y \le 12$ and $2x + y \ge 4$ are:

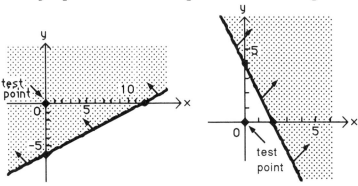

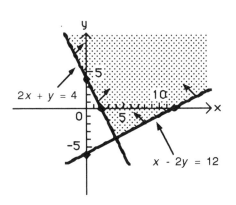

The intersection of these regions (drawn on the same coordinate plane) is shown in the graph at the right.

21. The graph of $x + 3y \le 18$ is the region below the line $x + 3y = 18$ and the graph of $2x + y \ge 16$ is the region above the line $2x + y = 16$. The graph of $x \ge 0$, $y \ge 0$ is the first quadrant. The intersection of these regions is region IV. The corner points are $(8, 0)$, $(18, 0)$, and $(6, 4)$.

23. The graph of $x + 3y \ge 18$ is the region above the line $x + 3y = 18$ and the graph of $2x + y \ge 16$ is the region above the line $2x + y = 16$. The graph of $x \ge 0$, $y \ge 0$ is the first quadrant. The intersection of these regions is region I. The corner points are $(0, 16)$, $(6, 4)$, and $(18, 0)$.

25. The graphs of the inequalities are shown at the right. The solution region is indicated by the shaded region. The solution region is *bounded*.

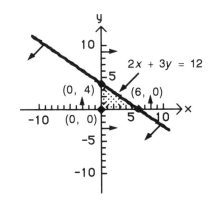

The corner points of the solution region are:

$(0, 0)$, the intersection of $x = 0$, $y = 0$;
$(0, 4)$, the intersection of $x = 0$,
$\qquad 2x + 3y = 12$;
$(6, 0)$, the intersection of $y = 0$,
$\qquad 2x + 3y = 12$.

27. The graphs of the inequalities are shown at the right. The solution region is shaded. The solution region is *bounded*.

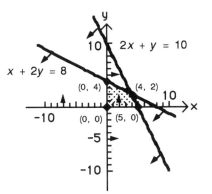

The corner points of the solution region are:

$(0, 0)$, the intersection of $x = 0$, $y = 0$;
$(0, 4)$, the intersection of $x = 0$, $x + 2y = 8$;
$(4, 2)$, the intersection of $x + 2y = 8$,
$\qquad 2x + y = 10$;
$(5, 0)$, the intersection of $y = 0$,
$\qquad 2x + y = 10$.

29. The graphs of the inequalities are shown at the right. The solution region is shaded. The solution region is *unbounded*.

The corner points of the solution region are:

(0, 10), the intersection of $x = 0$,
      $2x + y = 10$;

(4, 2), the intersection of $x + 2y = 8$,
      $2x + y = 10$;

(8, 0), the intersection of $y = 0$, $x + 2y = 8$.

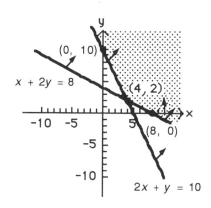

31. The graphs of the inequalities are shown at the right. The solution is indicated by the shaded region. The solution region is *bounded*.

The corner points of the solution region are:

(0, 0), the intersection of $x = 0$, $y = 0$,
(0, 6), the intersection of $x = 0$,
      $x + 2y = 12$;

(2, 5), the intersection of $x + 2y = 12$,
      $x + y = 7$;

(3, 4), the intersection of $x + y = 7$,
      $2x + y = 10$;

(5, 0), the intersection of $y = 0$,
      $2x + y = 10$.

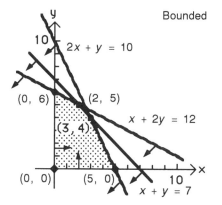

Note that the point of intersection of the lines $2x + y = 10$, $x + 2y = 12$ is not a corner point because it is not in the solution region.

33. The graphs of the inequalities are shown at the right. The solution is indicated by the shaded region, which is *unbounded*.

The corner points are:

(0, 16), the intersection of $x = 0$,
      $2x + y = 16$;

(4, 8), the intersection of $2x + y = 16$,
      $x + y = 12$;

(10, 2), the intersection of $x + y = 12$,
      $x + 2y = 14$;

(14, 0), the intersection of $y = 0$,
      $x + 2y = 14$.

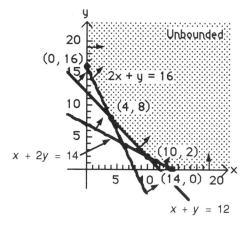

The intersection of $x + 2y = 14$, $2x + y = 16$ is not a corner point because it is not in the solution region.

35. The graphs of the inequalities are shown at the right. The solution is indicated by the shaded region, which is *bounded*.

The corner points are (8, 6), (4, 7), and (9, 3).

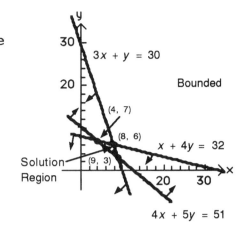

37. The graphs of the inequalities are shown at the right. The system of inequalities does not have a solution because the intersection of the graphs is empty.

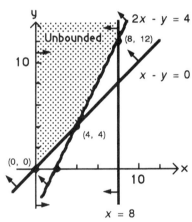

39. The graphs of the inequalities are shown at the right. The solution is indicated by the shaded region, which is *unbounded*.

The corner points are (0, 0), (4, 4), and (8, 12).

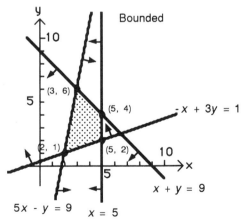

41. The graphs of the inequalities are shown at the right. The solution is indicated by the shaded region, which is *bounded*.

The corner points are (2, 1), (3, 6), (5, 4), and (5, 2).

43. Let x = the number of trick skis and y = the number of slalom skis produced per day. The information is summarized in the following table.

	Hours per ski		Maximum labor-hours per day available
	Trick ski	Slalom ski	
Fabrication	6 hrs	4 hrs	108 hrs
Finishing	1 hr	1 hr	24 hrs

We have the following inequalities:

$6x + 4y \leq 108$ for fabrication
$x + y \leq 24$ for finishing

Also, $x \geq 0$ and $y \geq 0$.

The graphs of these inequalities are shown at the right. The shaded region indicates the set of feasible solutions.

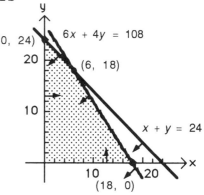

45. Let x = the number of cubic yards of mix A and y = the number of cubic yards of mix B. The information is summarized in the following table:

	Amount of substance per cubic yard		Minimum monthly requirement
	Mix A	Mix B	
Phosphoric acid	20 lbs	10 lbs	460 lbs
Nitrogen	30 lbs	30 lbs	960 lbs
Potash	5 lbs	10 lbs	220 lbs

We have the following inequalities:

$20x + 10y \geq 460$
$30x + 30y \geq 960$
$5x + 10y \geq 220$

Also, $x \geq 0$ and $y \geq 0$.

The graphs of these inequalities are shown at the right. The shaded region indicates the set of feasible solutions.

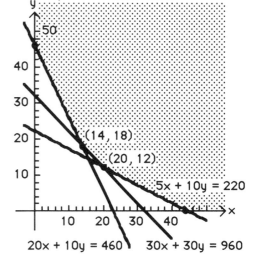

47. Let x = the number of mice used and y = the number of rats used. The information is summarized in the following table.

	Mice	Rats	Maximum time available per day
Box A	10 min	20 min	800 min
Box B	20 min	10 min	640 min

We have the following inequalities:

$10x + 20y \leq 800$ for box A
$20x + 10y \leq 640$ for box B

Also, $x \geq 0$ and $y \geq 0$.

The graphs of these inequalities are shown at the right. The shaded region indicates the set of feasible solutions.

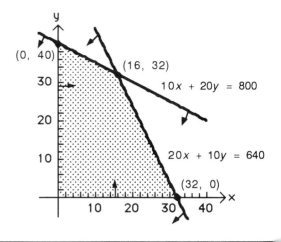

EXERCISE 6-2

Things to remember:

1. A LINEAR PROGRAMMING PROBLEM is a problem concerned with finding the maximum or minimum value of a linear OBJECTIVE FUNCTION of the form

   $$z = c_1 x_1 + c_2 x_2 + \cdots + c_n x_n,$$

   where the DECISION VARIABLES x_1, x_2, ..., x_n are subject to PROBLEM CONSTRAINTS in the form of linear inequalities and equations. In addition, the decision variables must satisfy the NONNEGATIVE CONSTRAINTS $x_i \geq 0$, for $i = 1, 2, ..., n$. The set of points satisfying both the problem constraints and the nonnegative constraints is called the FEASIBLE REGION for the problem. Any point in the feasible region that produces the optimal value of the objective function over the feasible region is called an OPTIMAL SOLUTION.

2. FUNDAMENTAL THEOREM OF LINEAR PROGRAMMING

   If a linear programming problem has an optimal solution, then this solution must occur at one (or more) of the corner points of the feasible region.

3. EXISTENCE OF SOLUTIONS

   (A) If the feasible region for a linear programming problem is bounded, then both the maximum value and the minimum value of the objective function always exist.

   (B) If the feasible region is unbounded, and the coefficients of the objective function are positive, then the minimum value of the objective function exists, but the maximum value does not.

   (C) If the feasible region is empty (that is, there are no points that satisfy all the constraints), then both the maximum value and the minimum value of the objective function do not exist.

4. GEOMETRIC SOLUTION OF A LINEAR PROGRAMMING PROBLEM WITH TWO DECISION VARIABLES.

(1) Summarize relevant material in table form if solving an application.

(2) Form a mathematical model for the problem:

    (a) Introduce decision variables and write a linear objective function.

    (b) Write problem constraints using linear inequalities and/or equations.

    (c) Write nonnegative constraints.

(3) Graph the feasible region, then (if according to 3 an optimal solution exists) find the coordinates of each corner point.

(4) Make a table listing the value of the objective function at each corner point.

(5) Determine the optimal solution(s) from the table in Step (4).

(6) Interpret the optimal solution(s) in terms of the original problem, if solving an application.

---

1. Steps (1)—(3) in 4 do not apply. Thus, we begin with Step (4).

Step (4): Evaluate the objective function at each corner point.

Corner Point	$z = x + y$
(0, 0)	0
(0, 12)	12
(7, 9)	16
(10, 0)	10

Step (5): Determine the optimal solution from Step (4).
The maximum value of z is 16 at (7, 9).

3. Steps (1)—(3) in 4 do not apply. Thus, we begin with Step (4).

Step (4): Evaluate the objective function at each corner point.

Corner Point	$z = 3x + 7y$
(0, 0)	0
(0, 12)	84
(7, 9)	84
(10, 0)	30

Step (5): Determine the optimal solution from Step (4).
The maximum value of z is 84 at (0, 12) *and* (7, 9). This is a multiple optimal solution.

5. Steps (1)–(3) in **4** do not apply. Thus, we begin with Step (4).

Step (4): Evaluate the objective function at each corner point.

Corner Point	$z = 7x + 4y$
(0, 12)	48
(0, 8)	32
(4, 3)	40
(12, 0)	84

Step (5): Determine the optimal solution from Step (4).
The minimum value of z is 32 at (0, 8).

7. Steps (1)–(3) in **4** do not apply. Thus, we begin with Step (4).

Step (4): Evaluate the objective function at each corner point.

Corner Point	$z = 3x + 8y$
(0, 12)	96
(0, 8)	64
(4, 3)	36
(12, 0)	36

Step (5): Determine the optimal solution from Step (4).
The minimum value of z is 36 at (4, 3) and (12, 0). This is a multiple optimal solution.

9. Step (3): Graph the feasible region and find the corner points.

The feasible region S is the solution set of the given inequalities. This region is indicated by the shading in the graph at the right.

The corner points are (0, 0), (0, 4), (4, 2), and (5, 0).

Since S is bounded, it follows from **3**(a) that P has a maximum value.

Step (4): Evaluate the objective function at each corner point.

The value of P at each corner point is given in the following table.

Corner Point	$P = 5x_1 + 5x_2$
(0, 0)	$P = 5(0) + 5(0) = 0$
(0, 4)	$P = 5(0) + 5(4) = 20$
(4, 2)	$P = 5(4) + 5(2) = 30$
(5, 0)	$P = 5(5) + 5(0) = 25$

Step (5): Determine the optimal solution.

The maximum value of P is 30 at $x_1 = 4$, $x_2 = 2$.

11. Step (3): Graph the feasible region and find the corner points.

The feasible region S is the solution set of the given inequalities. This region is indicated by the shading in the graph at the right.

The corner points are $(0, 10)$, $(4, 2)$, and $(8, 0)$.

Since S is unbounded and $a = 2 > 0$, $b = 3 > 0$, it follows from 3(b) that P has a minimum value but not a maximum value.

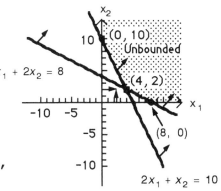

Step (4): Evaluate the objective function at each corner point.

The value of P at each corner point is given in the following table:

Corner Point	$z = 2x_1 + 3x_2$
$(0, 10)$	$z = 2(0) + 3(10) = 30$
$(4, 2)$	$z = 2(4) + 3(2) = 14$
$(8, 0)$	$z = 2(8) + 3(0) = 16$

Step (5): Determine the optimal solutions.

The minimum occurs at $x_1 = 4$, $x_2 = 2$, and the minimum value is $z = 14$; z does not have a maximum value.

13. Step (3): Graph the feasible region and find the corner points.

The feasible region S is the solution set of the given inequalities. This region is indicated by the shading in the graph at the right.

The corner points are $(0, 0)$, $(0, 6)$, $(2, 5)$, $(3, 4)$, and $(5, 0)$.

Since S is bounded, it follows from 3(a) that P has a maximum value.

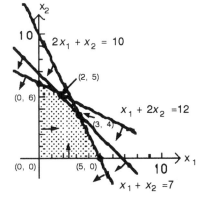

Step (4): Evaluate the objective function at each corner point.

The value of P at each corner point is:

Corner Point	$P = 30x_1 + 40x_2$
$(0, 0)$	$P = 30(0) + 40(0) = 0$
$(0, 6)$	$P = 30(0) + 40(6) = 240$
$(2, 5)$	$P = 30(2) + 40(5) = 260$
$(3, 4)$	$P = 30(3) + 40(4) = 250$
$(5, 0)$	$P = 30(5) + 40(0) = 150$

Step (5): Determine the optimal solution.

The maximum occurs at $x_1 = 2$, $x_2 = 5$, and the maximum value is $P = 260$.

15. Step (3): Graph the feasible region and find the corner points.

The feasible region S is the solution set of the given inequalities. This region is indicated by the shading in the graph at the right.

The corner points are $(0, 16)$, $(4, 8)$, $(10, 2)$, and $(14, 0)$.

Since S is unbounded and $a = 10 > 0$, $b = 30 > 0$, it follows from $\underline{3}$(b) that z has a minimum value but not a maximum value.

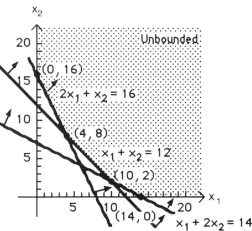

Step (4): Evaluate the objective function at each corner point.

The value of z at each corner point is:

Corner Point	$z = 10x_1 + 30x_2$
$(0, 16)$	$z = 10(0) + 30(16) = 480$
$(4, 8)$	$z = 10(4) + 30(8) = 280$
$(10, 2)$	$z = 10(10) + 30(2) = 160$
$(14, 0)$	$z = 10(14) + 30(0) = 140$

Step (5): Determine the optimal solution.

The minimum occurs at $x_1 = 14$, $x_2 = 0$, and the minimum value is $z = 140$; z does not have a maximum value.

17. Step (3): Graph the feasible region and find the corner points.

The feasible region S is the solution set of the given inequalities, and is indicated by the shading in the graph at the right.

The corner points are $(0, 2)$, $(0, 9)$, $(2, 6)$, $(5, 0)$, and $(2, 0)$.

Since S is bounded, it follows from $\underline{3}$(a) that P has a maximum value and a minimum value.

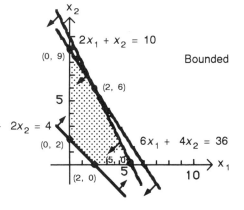

Step (4): Evaluate the objective function at each corner point.

The value of P at each corner point is given in the following table:

Corner Point	$P = 30x_1 + 10x_2$
$(0, 2)$	$P = 30(0) + 10(2) = 20$
$(0, 9)$	$P = 30(0) + 10(9) = 90$
$(2, 6)$	$P = 30(2) + 10(6) = 120$
$(5, 0)$	$P = 30(5) + 10(0) = 150$
$(2, 0)$	$P = 30(2) + 10(0) = 60$

Step (5): Determine the optimal solutions.
The maximum occurs at $x_1 = 5$, $x_2 = 0$, and the maximum value
is $P = 150$; the minimum occurs at
$x_1 = 0$, $x_2 = 2$, and the minimum
value is $P = 20$.

19. Step (3): Graph the feasible region and
find the corner points.

The feasible region S is the solution set
of the given inequalities. As indicated,
the feasible region is empty. Thus, by
3(c), there are no optimal solutions.

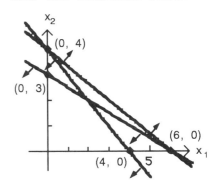

21. Step (3): Graph the feasible region and
find the corner points.

The feasible region S is the solution set
of the given inequalities, and is
indicated by the shading in the graph at
the right.

The corner points are (3, 8), (8, 10),
and (12, 2).

Since S is bounded, it follows from 3(a)
that P has a maximum value and a minimum
value.

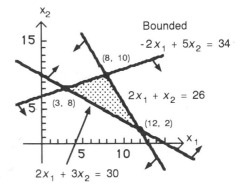

Step (4): Evaluate the objective function at each corner point.
The value of P at each corner point is:

Corner Point	$P = 20x_1 + 10x_2$
(3, 8)	$P = 20(3) + 10(8) = 140$
(8, 10)	$P = 20(8) + 10(10) = 260$
(12, 2)	$P = 20(12) + 10(2) = 260$

Step (5): Determine the optimal solutions.
The minimum occurs at $x_1 = 3$, $x_2 = 8$, and the minimum value
is $P = 140$; the maximum occurs at $x_1 = 8$, $x_2 = 10$, at $x_1 = 12$,
$x_2 = 2$, and at any point along the line segment joining
(8, 10) and (12, 2). The maximum value is $P = 260$.

23. Step (3): Graph the feasible region and
find the corner points.
The feasible region S is the set of
solutions of the given inequalities,
and is indicated by the shading in the
graph at the right.
The corner points are (0, 0), (0, 800),
(400, 600), (600, 450), and (900, 0).
Since S is bounded, it follows from
3(a) that P has a maximum value.

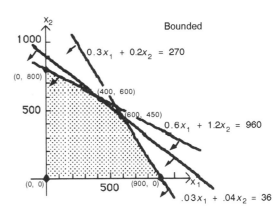

Step (4): Evaluate the objective function at each corner point.
The value of P at each corner point is:

Corner Point	$P = 20x_1 + 30x_2$
(0, 0)	$P = 20(0) + 30(0) = 0$
(0, 800)	$P = 20(0) + 30(800) = 24,000$
(400, 600)	$P = 20(400) + 30(600) = 26,000$
(600, 450)	$P = 20(600) + 30(450) = 25,500$
(900, 0)	$P = 20(900) + 30(0) = 18,000$

Step (5): Determine the optimal solution.
The maximum occurs at $x_1 = 400$, $x_2 = 600$, and the maximum value is $P = 26,000$.

25. The value of $P = ax_1 + bx_2$, $a > 0$, $b > 0$, at each corner point is:

Corner Point	P
O: (0, 0)	$P = a(0) + b(0) = 0$
A: (0, 5)	$P = a(0) + b(5) = 5b$
B: (4, 3)	$P = a(4) + b(3) = 4a + 3b$
C: (5, 0)	$P = a(5) + b(0) = 5a$

(A) For the maximum value of P to occur at A only, we must have $5b > 4a + 3b$ and $5b > 5a$. solving the first inequality, we get $2b > 4a$ or $b > 2a$; from the second inequality, we get $b > a$. Therefore, we must have $b > 2a$ or $2a < b$ in order for P to have its maximum value at A only.

(B) For the maximum value of P to occur at B only, we must have $4a + 3b > 5b$ and $4a + 3b > 5a$. Solving this pair of inequalities, we get $4a > 2b$ and $3b > a$, which is the same as $\frac{a}{3} < b < 2a$.

(C) For the maximum value of P to occur at C only, we must have $5a > 4a + 3b$ and $5a > 5b$. This pair of inequalities inplies that $a > 3b$ or $b < \frac{a}{3}$.

(D) For the maximum value of P to occur at both A and B, we must have $5b = 4a + 3b$ or $b = 2a$.

(E) For the maximum value of P to occur at both B and C, we must have $4a + 3b = 5a$ or $b = \frac{a}{3}$.

27. Step (1): Has been done.

Step (2): Form a mathematical model for the problem.

Let x_1 = the number of trick skis
and x_2 = the number of slalom skis produced per day. The mathematical model for this problem is: Maximize $P = 40x_1 + 30x_2$

$$\text{Subject to: } 6x_1 + 4x_2 \le 108$$
$$x_1 + x_2 \le 24$$
$$x_1 \ge 0, \; x_2 \ge 0$$

Step (3): Graph the feasible region and find the corner points.

The feasible region S is the solution set of the given system of inequalities, and is indicated by the shading in the graph at the right.

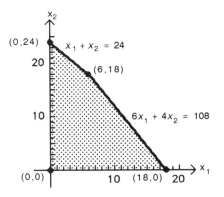

The corner points are (0, 0), (0, 24), (6, 18), and (18, 0).

Since S is bounded, P has a maximum value by 3(a).

Step (4): Evaluate the objective function at each corner point.
The value of P at each corner point is:

Corner Point	$P = 40x_1 + 30x_2$
(0, 0)	$P = 40(0) + 30(0) = 0$
(0, 24)	$P = 40(0) + 30(24) = 720$
(6, 18)	$P = 40(6) + 30(18) = 780$
(18, 0)	$P = 40(18) + 30(0) = 720$

Step (5): Determine the optimal solution.
The maximum occurs when $x_1 = 6$ (trick skis) and $x_2 = 18$ (slalom skis) are produced. The maximum profit is $P = \$780$.

29. (A) Step (1): Summarize relevant material in table form.

	Plant A	Plant B	Amount required
Tables	20	25	200
Chairs	60	50	500
Cost per day	$1000	$900	

Step (2): Form a mathematical model for the problem.

Let x_1 = the number of days to operate Plant A and x_2 = the number of days to operate Plant B.

The mathematical model for this problem is:

Minimize $C = 1000x_1 + 900x_2$
Subject to: $20x_1 + 25x_2 \geq 200$
$60x_1 + 50x_2 \geq 500$
$x_1 \geq 0, \ x_2 \geq 0$

Step (3): Graph the feasible region and find the corner points.

The feasible region S is the solution set of the system of inequalities, and is indicated by the shading in the graph shown on the following page.

The corner points are (0, 10), (5, 4), and (10, 0).

Since S is unbounded and $a = 1000 > 0$, $b = 900 > 0$, C has a minimum value by 3(b).

Step (4): Evaluate the objective function at each corner point.

The value of C at each corner point is:

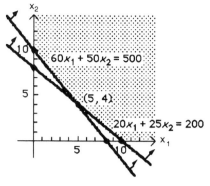

Corner Point	$C = 1000x_1 + 900x_2$
(0, 10)	$C = 1000(0) + 900(10) = 9,000$
(5, 4)	$C = 1000(5) + 900(4) = 8,600$
(10, 0)	$C = 1000(10) + 900(0) = 10,000$

Step (5): Determine the optimal solution.

The minimum occurs when $x_1 = 5$ and $x_2 = 4$.

That is, Plant A should be operated five days and Plant B should be operated four days. The minimum cost is $C = \$8600$.

(B) The mathematical model for this problem is: Minimize $C = 600x_1 + 900x_2$
Subject to: $20x_1 + 25x_2 \geq 200$
$60x_1 + 50x_2 \geq 500$
$x_1 \geq 0, x_2 \geq 0$

The feasible region S and the corner points are the same as in part (A), and C has a minimum value.

The value of C at each corner point is:

Corner Point	$C = 600x_1 + 900x_2$
(0, 10)	$C = 600(0) + 900(10) = 9000$
(5, 4)	$C = 600(5) + 900(4) = 6600$
(10, 0)	$C = 600(10) + 900(0) = 6000$

Thus, the minimum occurs when $x_1 = 10$ and $x_2 = 0$. That is, Plant A should be operated 10 days and Plant B should not be operated at all. The minimum cost is $C = \$6000$.

(C) The mathematical model for this problem is:
Minimize $C = 1000x_1 + 800x_2$
Subject to: $20x_1 + 25x_2 \geq 200$
$60x_1 + 50x_2 \geq 500$
$x_1 \geq 0, x_2 \geq 0$

The feasible region S and the corner points are the same as in Part (A) and C has a minimum value.

The value of C at each corner point is:

Corner Point	$C = 1000x_1 + 800x_2$
(0, 10)	$C = 1000(0) + 800(10) = 8,000$
(5, 4)	$C = 1000(5) + 800(4) = 8,200$
(10, 0)	$C = 1000(10) + 800(0) = 10,000$

Thus, the minimum occurs when $x_1 = 0$ and $x_2 = 10$. That is, Plant A should not be operated and Plant B should be operated 10 days. The minimum cost is $C = \$8000$.

31. (A) <u>Step (1)</u>: Summarize relevant material.

	Buses	Vans	Number to accommodate
Students	40	8	400
Chaperones	3	1	36

Rental cost $1200 per bus $100 per van

<u>Step (2)</u>: Form a mathematical model for the problem.

Let x_1 = the number of buses
and x_2 = the number of vans.

The mathematical model for this problem is:
Minimize $C = 1200x_1 + 100x_2$
Subject to: $40x_1 + 8x_2 \geq 400$
$\qquad\qquad 3x_1 + x_2 \leq 36$
$\qquad\qquad x_1 \geq 0,\ x_2 \geq 0$

<u>Step (3)</u>: Graph the feasible region and find the corner points.

The feasible region S is the solution set of the system of inequalities, and is indicated by the shading in the graph at the right.

The corner points are (10, 0), (7, 15), and (12, 0).

Since S is bounded, C has a minimum value by <u>3</u>(a).

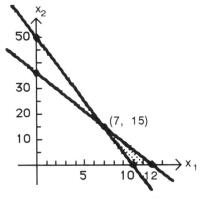

<u>Step (4)</u>: Evaluate the objective function at each corner point.

The value of C at each corner point is:

Corner Point	$C = 1200x_1 + 100x_2$
(10, 0)	$C = 1200(10) + 100(0)\ = 12{,}000$
(7, 15)	$C = 1200(7)\ + 100(15) =\ 9{,}900$
(12, 0)	$C = 1200(12) + 100(0)\ = 14{,}400$

<u>Step (5)</u>: Determine the optimal solution.

The minimum occurs when $x_1 = 7$ and $x_2 = 15$. That is, the officers should rent 7 buses and 15 vans at the minimum cost of $9900.

33. <u>Step (1)</u>: Summarize relevant material.

	Grams per Gallon Emitted by:		Maximum allowed
	Old process	New Process	
Sulfur dioxide	15	5	10,500
Particulate	40	20	30,000
Profit/gallon	30¢	20¢	

<u>Step (2)</u>: Form a mathematical model for the problem.

Let x_1 = the number of gallons produced by the old process and x_2 = the number of gallons produced by the new process.

The mathematical model for this problem is:

Maximize $P = 30x_1 + 20x_2$

Subject to: $15x_1 + 5x_2 \leq 10,500$

$\qquad\qquad 40x_1 + 20x_2 \leq 30,000$

$\qquad\qquad x_1 \geq 0, \; x_2 \geq 0$

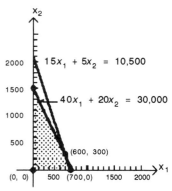

<u>Step (3)</u>: Graph the feasible region and find the corner points.

The feasible region S is the solution set of the given inequalities, and is indicated by the shading in the graph at the right.

The corner points are $(0, 0)$, $(0, 1500)$, $(600, 300)$, and $(700, 0)$.

Since S is bounded, P has a maximum value by $\underline{3}$(a).

<u>Step (4)</u>: Evaluate the objective function at each corner point.

The value of P at each corner point is:

Corner Point	$P = 30x_1 + 20x_2$
$(0, 0)$	$P = 30(0) + 20(0) = 0$
$(0, 1500)$	$P = 30(0) + 20(1500) = 30,000$
$(600, 300)$	$P = 30(600) + 20(300) = 24,000$
$(700, 0)$	$P = 30(700) + 20(0) = 21,000$

<u>Step (5)</u>: Determine the optimal solution.

The maximum occurs when the amount of chemical produced by the old process is zero and that produced by the new process is 1500 gallons. The maximum profit is $P = 30,000¢$, i.e., $P = \$300$.

35. Let x_1 = the number of bags of Brand A and x_2 = the number of bags of Brand B.

(A) The mathematical model for this problem is:

Maximize $N = 8x_1 + 3x_2$

Subject to: $4x_1 + 4x_2 \geq 1000$

$\qquad\qquad 2x_1 + x_2 \leq 400$

$\qquad\qquad x_1 \geq 0, \; x_2 \geq 0$

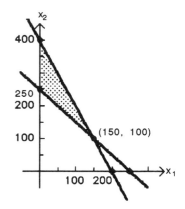

The feasible region S is the solution set of the system of inequalities, and is indicated by the shading in the graph at the right.

The corner points are $(0, 250)$, $(0, 400)$, and $(150, 100)$.

Since S is bounded, N has a maximum value by $\underline{3}$(a).

The value of N at each corner point is given in the table below:

Corner Point	$N = 8x_1 + 3x_2$
(0, 250)	$N = 8(0) + 3(250) = 750$
(150, 100)	$N = 8(150) + 3(100) = 1500$
(0, 400)	$N = 8(0) + 3(400) = 1200$

Thus, the maximum occurs when $x_1 = 150$ and $x_2 = 100$. That is, the grower should use 150 bags of Brand A and 100 bags of Brand B. The maximum number of pounds of nitrogen is 1500.

(B) The mathematical model for this problem is:

Minimize $N = 8x_1 + 3x_2$
Subject to: $4x_1 + 4x_2 \geq 1000$
$\qquad\qquad 2x_1 + x_2 \leq 400$
$\qquad\qquad x_1 \geq 0, \ x_2 \geq 0$

The feasible region S and the corner points are the same as in part (A). Thus, the minimum occurs when $x_1 = 0$ and $x_2 = 250$. That is, the grower should use 0 bags of Brand A and 250 bags of Brand B. The minimum number of pounds of nitrogen is 750.

37.

	Amount per Cubic Yard (in pounds)		Minimum monthly requirement
	Mix A	Mix B	
Phosphoric acid	20	10	460
Nitrogen	30	30	960
Potash	5	10	220
Cost/cubic yd.	$30	$35	

Let x_1 = the number of cubic yards of mix A
and x_2 = the number of cubic yards of mix B.

The mathematical model for this problem is:
Minimize $C = 30x_1 + 35x_2$
Subject to: $20x_1 + 10x_2 \geq 460$
$\qquad\qquad 30x_1 + 30x_2 \geq 960$
$\qquad\qquad 5x_1 + 10x_2 \geq 220$
$\qquad\qquad x_1 \geq 0, \ x_2 \geq 0$

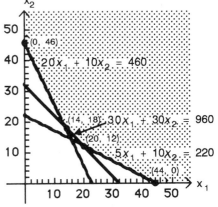

The feasible region S is the solution set of the given inequalities and is indicated by the shading in the graph at the right.

The corner points are (0, 46), (14, 18), (20, 12), and (44, 0).

Since S is unbounded and $a = 30 > 0$, $b = 35 > 0$, C has a minimum value by 3(b).

The value of C at each corner point is:

Corner Point	$C = 30x_1 + 35x_2$
(0, 46)	$C = 30(0) + 35(46) = 1610$
(14, 18)	$C = 30(14) + 35(18) = 1050$
(20, 12)	$C = 30(20) + 35(12) = 1020$
(44, 0)	$C = 30(44) + 35(0) = 1320$

Thus, the minimum occurs when the amount of mix A used is 20 cubic yards and the amount of mix B used is 12 cubic yards. The minimum cost is $C = \$1020$.

39. Let x_1 = the number of mice used
and x_2 = the number of rats used.

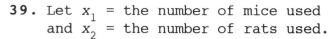

The mathematical model for this problem is:

Maximize $P = x_1 + x_2$
Subject to: $10x_1 + 20x_2 \le 800$
$20x_1 + 10x_2 \le 640$
$x_1 \ge 0, \ x_2 \ge 0$

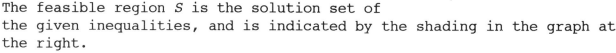

The feasible region S is the solution set of the given inequalities, and is indicated by the shading in the graph at the right.

The corner points are (0, 0), (0, 40), (16, 32), and (32, 0).
Since S is bounded, P has a maximum value by $\underline{3}$(a).

The value of P at each corner point is:

Corner Point	$P = x_1 + x_2$
(0, 0)	$P = 0 + 0 = 0$
(0, 40)	$P = 0 + 40 = 40$
(16, 32)	$P = 16 + 32 = 48$
(32, 0)	$P = 32 + 0 = 32$

Thus, the maximum occurs when the number of mice used is 16 and the number of rats used is 32. The maximum number of mice and rats that can be used is 48.

EXERCISE 6-3

Things to remember:

$\underline{1}$. STANDARD MAXIMIZATION PROBLEM IN STANDARD FORM

A linear programming problem is said to be a STANDARD MAXIMIZATION PROBLEM IN STANDARD FORM if its mathematical model is of the form:

Maximize $P = c_1x_1 + c_2x_2 + \cdots + c_nx_n$

Subject to problem constraints of the form:

$a_1x_1 + a_2x_2 + \cdots + a_nx_n \le b, \quad b \ge 0$

with nonnegative constraints:

$x_1, \ x_2, \ \ldots, \ x_n \ge 0$.

[<u>Note</u>: The coefficients of the objective function can be any real numbers.]

2. SLACK VARIABLES

Given a linear programming problem. SLACK VARIABLES are
nonnegative quantities that are introduced to convert problem
constraint inequalities into equations.

3. BASIC VARIABLES AND NONBASIC VARIABLES; BASIC SOLUTIONS AND
BASIC FEASIBLE SOLUTIONS

Given a system of linear equations associated with a linear
programming problem. (Such a system will always have more
variables than equations.)

The variables are divided into two (mutually exclusive) groups,
called BASIC VARIABLES and NONBASIC VARIABLES, as follows:
Basic variables are selected arbitrarily with the one
restriction that there be as many basic variables as there are
equations. The remaining variables are called nonbasic
variables.

A solution found by setting the nonbasic variables equal to
zero and solving for the basic variables is called a BASIC
SOLUTION. If a basic solution has no negative values, it is a
BASIC FEASIBLE SOLUTION.

4. FUNDAMENTAL THEOREM OF LINEAR PROGRAMMING

If the optimal value of the objective function in a linear
programming problem exists, then that value must occur at one
(or more) of the basic feasible solutions.

---

1. (A) There are 5 constraint equations; the number of equations is the
same as the number of slack variables.

(B) There are 4 decision variables since there are 9 variables
altogether, and 5 of them are slack variables.

(C) There are 5 basic variables and 4 nonbasic variables; the number of
basic variables equals the number of equations.

(D) Five linear equations with 5 variables.

3.

	Nonbasic	Basic	Feasible?
(A)	x_1, x_2	s_1, s_2	Yes, all values are nonnegative.
(B)	x_1, s_1	x_2, s_2	Yes, all values are nonnegative.
(C)	x_1, s_2	x_2, s_1	No, $s_1 = -12 < 0$.
(D)	x_2, s_1	x_1, s_2	No, $s_2 = -12 < 0$.
(E)	x_2, s_2	x_1, s_1	Yes, all values are nonnegative.
(F)	s_1, s_2	x_1, x_2	Yes, all values are nonnegative.

5.

	x_1	x_2	s_1	s_2	Feasible?
(A)	0	0	50	40	Yes, all values are nonnegative.
(B)	0	50	0	-60	No, $s_2 = -60 < 0$.
(C)	0	20	30	0	Yes, all values are nonnegative.
(D)	25	0	0	15	Yes, all values are nonnegative.
(E)	40	0	-30	0	No, $s_1 = -30 < 0$.
(F)	20	10	0	0	Yes, all values are nonnegative.

7.

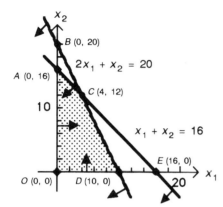

Introduce slack variables s_1 and s_2 to obtain the system of equations:

$$x_1 + x_2 + s_1 \qquad = 16$$
$$2x_1 + x_2 \qquad + s_2 = 20$$

x_1	x_2	s_1	s_2	Intersection Point	Feasible?
0	0	16	20	O	Yes
0	16	0	4	A	Yes
0	20	-4	0	B	No, $s_1 = -4 < 0$
16	0	0	-12	E	No, $s_2 = -12 < 0$
10	0	6	0	D	Yes
4	12	0	0	C	Yes

9.

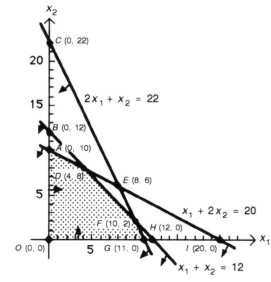

Introduce slack variables s_1, s_2, and s_3 to obtain the system of equations:

$$2x_1 + x_2 + s_1 \qquad = 22$$
$$x_1 + x_2 \qquad + s_2 \qquad = 12$$
$$x_1 + 2x_2 \qquad + s_3 = 20$$

x_1	x_2	s_1	s_2	s_3	Intersection Point	Feasible?
0	0	22	12	20	O	Yes
0	22	0	-10	-24	C	No
0	12	10	0	-4	B	No
0	10	12	2	0	A	Yes
11	0	0	1	9	G	Yes
12	0	-2	0	8	H	No
20	0	-18	-8	0	I	No
10	2	0	0	6	F	Yes
8	6	0	-2	0	E	No
4	8	6	0	0	D	Yes

EXERCISE 6-4

Things to remember:

1. **SELECTING BASIC AND NONBASIC VARIABLES FOR THE SIMPLEX PROCESS**

   Given a simplex tableau.

   (a) Determine the number of basic and the number of nonbasic variables. These numbers do not change during the simplex process.

   (b) SELECTING BASIC VARIABLES: A variable can be selected as a basic variable only if it corresponds to a column in the tableau that has exactly one nonzero element (usually 1) and the nonzero element in the column is not in the same row as the nonzero element of another basic variable column. (This procedure always selects P as a basic variable, since the P column never changes during the simplex process.)

   (c) SELECTING NONBASIC VARIABLES: After the basic variables are selected in Step (b), the remaining variables are selected as the nonbasic variables. (The tableau columns under the nonbasic variables will usually contain more than one nonzero element.)

2. **SELECTING THE PIVOT ELEMENT**

   (a) Locate the most negative indicator in the bottom row of the tableau to the left of the P column (the number with the largest absolute value). The column containing this element is the PIVOT COLUMN. If there is a tie for the most negative, choose either.

   (b) Divide each POSITIVE element in the pivot column above the dashed line into the corresponding element in the last column. The PIVOT ROW is the row corresponding to the smallest quotient. If there is a tie for the smallest quotient, choose either. If the pivot column above the dashed line has no positive elements, then there is no solution and we stop.

(c) The PIVOT (or PIVOT ELEMENT) is the element in the intersection of the pivot column and pivot row. [<u>Note</u>: The pivot element is always positive and is never in the bottom row.]

[<u>Remember</u>: The entering variable is at the top of the pivot column and the exiting variable is at the left of the pivot row.]

<u>3</u>. PERFORMING THE PIVOT OPERATION

A PIVOT OPERATION or PIVOTING consists of performing row operations as follows:

(a) Multiply the pivot row by the reciprocal of the pivot element to transform the pivot element into a 1. (If the pivot element is already a 1, omit this step.)

(b) Add multiples of the pivot row to other rows in the tableau to transform all other nonzero elements in the pivot column into 0's.

[<u>Note</u>: Rows are not to be interchanged while performing a pivot operation. The only way the (positive) pivot element can be transformed into 1 (if it is not a 1 already) is for the pivot row to be multiplied by the reciprocal of the pivot element.]

<u>4</u>. THE SIMPLEX METHOD

(a) Start with a standard maximization problem written in standard form.

(b) Introduce slack variables and write the initial system.

(c) Write the simplex tableau associated with the initial system.

(d) Determine the pivot element (if it exists) and the entering and exiting variables.

(e) Perform the pivot operation.

(f) Repeat steps (d) and (e) until all indicators in the bottom row are nonnegative. When this occurs, we stop the process and read the optimal solution.

<u>REMARKS</u>

There are two different reasons for stopping the simplex process:

(i) If we cannot select a new pivot column, we stop because the optimal solution has been found [see step (f) above].

(ii) If we select a new pivot column and then are unable to select a new pivot row, we stop because the problem has no solution [see step (d) above and step <u>2</u>(b) above].

1. Given the simplex tableau:

$$
\begin{array}{ccccc}
x_1 & x_2 & s_1 & s_2 & P \\
\end{array}
$$

$$
\left[
\begin{array}{ccccc|c}
2 & 1 & 0 & 3 & 0 & 12 \\
3 & 0 & 1 & -2 & 0 & 15 \\
\hline
-4 & 0 & 0 & 4 & 1 & 20
\end{array}
\right]
$$

which corresponds to the system of equations:

$$
\text{(I)} \quad
\begin{cases}
2x_1 + x_2 \quad\quad + 3s_2 \quad\quad = 12 \\
3x_1 \quad\quad + s_1 - 2s_2 \quad\quad = 15 \\
-4x_1 \quad\quad\quad\quad + 4s_2 + P = 20
\end{cases}
$$

(A) The basic variables are x_2, s_1, and P, and the nonbasic variables are x_1 and s_2.

(B) The corresponding basic feasible solution is found by setting the nonbasic variables equal to 0 in system (I). This yields:
$$x_1 = 0, \ x_2 = 12, \ s_1 = 15, \ s_2 = 0, \ P = 20$$

(C) An additional pivot is required, since the last row of the tableau has a negative indicator, the -4 in the first column.

3. Given the simplex tableau:

$$
\begin{array}{ccccccc}
x_1 & x_2 & x_3 & s_1 & s_2 & s_3 & P \\
\end{array}
$$

$$
\left[
\begin{array}{ccccccc|c}
-2 & 0 & 1 & 3 & 1 & 0 & 0 & 5 \\
0 & 1 & 0 & -2 & 0 & 0 & 0 & 15 \\
-1 & 0 & 0 & 4 & 1 & 1 & 0 & 12 \\
\hline
-4 & 0 & 0 & 2 & 4 & 0 & 1 & 45
\end{array}
\right]
$$

which corresponds to the system of equations:

$$
\text{(I)} \quad
\begin{cases}
-2x_1 \quad\quad + x_3 + 3s_1 + s_2 \quad\quad = 5 \\
\quad\quad x_2 \quad\quad\quad - 2s_1 \quad\quad\quad = 15 \\
-x_1 \quad\quad\quad + 4s_1 + s_2 + s_3 \quad = 12 \\
-4x_1 \quad\quad\quad + 2s_1 + 4s_2 \quad\quad + P = 45
\end{cases}
$$

(A) The basic variables are x_2, x_3, s_3, and P, and the nonbasic variables are x_1, s_1, and s_2.

(B) The corresponding basic feasible solution is found by setting the nonbasic variables equal to 0 in system (I). This yields:
$$x_1 = 0, \ x_2 = 15, \ x_3 = 5, \ s_1 = 0, \ s_2 = 0, \ s_3 = 12, \ P = 45$$

(C) Since the last row of the tableau has a negative indicator, the -4 in the first column, an additional pivot should be required. However, since there are no positive elements in the pivot column (the first column), the problem has *no solution*.

5. Given the simplex tableau:

$$\begin{array}{ccccc} x_1 & x_2 & s_1 & s_2 & P \end{array}$$

$$\left[\begin{array}{ccccc|c} 1 & 4 & 1 & 0 & 0 & 4 \\ 3 & 5 & 0 & 1 & 0 & 24 \\ \hline -8 & -5 & 0 & 0 & 1 & 0 \end{array}\right]$$

The most negative indicator is -8 in the first column. Thus, the first column is the pivot column. Now, $\frac{4}{1} = 4$ and $\frac{24}{3} = 8$. Thus, the first row is the pivot row and the pivot element is the element in the first row, first column. These are indicated in the following tableau.

$$\begin{array}{c} \text{Enter} \\ \begin{array}{ccccc} x_1 & x_2 & s_1 & s_2 & P \end{array} \end{array}$$

Exit s_1
$$\left[\begin{array}{ccccc|c} \textcircled{1} & 4 & 1 & 0 & 0 & 4 \\ 3 & 5 & 0 & 1 & 0 & 24 \\ \hline -8 & -5 & 0 & 0 & 1 & 0 \end{array}\right] \begin{array}{l} \frac{4}{1} = 4 \ (\text{minimum}) \\[6pt] \frac{24}{3} = 8 \end{array}$$

with s_2 and P labels on the second and third rows.

$$\left[\begin{array}{ccccc|c} \textcircled{1} & 4 & 1 & 0 & 0 & 4 \\ 3 & 5 & 0 & 1 & 0 & 24 \\ \hline -8 & -5 & 0 & 0 & 1 & 0 \end{array}\right] \sim \left[\begin{array}{ccccc|c} 1 & 4 & 1 & 0 & 0 & 4 \\ 0 & -7 & -3 & 1 & 0 & 12 \\ \hline 0 & 27 & 8 & 0 & 1 & 32 \end{array}\right]$$

$$R_2 + (-3)R_1 \to R_2$$
$$R_3 + 8R_1 \to R_3$$

7. Given the simplex tableau:

$$\begin{array}{cccccc} x_1 & x_2 & s_1 & s_2 & s_3 & P \end{array}$$

$$\left[\begin{array}{cccccc|c} 2 & 1 & 1 & 0 & 0 & 0 & 4 \\ 3 & 0 & 1 & 1 & 0 & 0 & 8 \\ 0 & 0 & 2 & 0 & 1 & 0 & 2 \\ \hline -4 & 0 & -3 & 0 & 0 & 1 & 5 \end{array}\right]$$

The most negative indicator is -4. Thus, the first column is the pivot column. Now, $\frac{4}{2} = 2$, $\frac{8}{3} = 2\frac{2}{3}$. Thus, the first row is the pivot row, and the pivot element is the element in the first row, first column. These are indicated in the tableau.

$$\begin{array}{c} \text{Enter} \\ \begin{array}{cccccc} x_1 & x_2 & s_1 & s_2 & s_3 & P \end{array} \end{array}$$

Exit x_2
$$\left[\begin{array}{cccccc|c} \textcircled{2} & 1 & 1 & 0 & 0 & 0 & 4 \\ 3 & 0 & 1 & 1 & 0 & 0 & 8 \\ 0 & 0 & 2 & 0 & 1 & 0 & 2 \\ \hline -4 & 0 & -3 & 0 & 0 & 1 & 5 \end{array}\right] \begin{array}{l} \frac{4}{2} = 2 \ (\text{minimum}) \\[6pt] \frac{8}{3} = 2\frac{2}{3} \end{array}$$

with s_2, s_3 and P labels on the second, third and fourth rows.

$$\begin{bmatrix} \textcircled{2} & 1 & 1 & 0 & 0 & 0 & | & 4 \\ 3 & 0 & 1 & 1 & 0 & 0 & | & 8 \\ 0 & 0 & 2 & 0 & 1 & 0 & | & 2 \\ \hline -4 & 0 & -3 & 0 & 0 & 1 & | & 5 \end{bmatrix}$$

$$\begin{array}{ccccccc} x_1 & x_2 & s_1 & s_2 & s_3 & P & \\ \end{array}$$

$$\sim \begin{bmatrix} \textcircled{1} & \tfrac{1}{2} & \tfrac{1}{2} & 0 & 0 & 0 & | & 2 \\ 3 & 0 & 1 & 1 & 0 & 0 & | & 8 \\ 0 & 0 & 2 & 0 & 1 & 0 & | & 2 \\ \hline -4 & 0 & -3 & 0 & 0 & 1 & | & 5 \end{bmatrix}$$

$$\tfrac{1}{2} R_1 \to R_1 \qquad\qquad R_2 + (-3)R_1 \to R_2, \quad R_4 + 4R_1 \to R_4$$

$$\sim \begin{bmatrix} 1 & \tfrac{1}{2} & \tfrac{1}{2} & 0 & 0 & 0 & | & 2 \\ 0 & -\tfrac{3}{2} & -\tfrac{1}{2} & 1 & 0 & 0 & | & 2 \\ 0 & 0 & 2 & 0 & 1 & 0 & | & 2 \\ \hline 0 & 2 & -1 & 0 & 0 & 1 & | & 13 \end{bmatrix}$$

9. (A) Introduce slack variables s_1 and s_2 to obtain:

Maximize $P = 15x_1 + 10x_2$
Subject to: $2x_1 + x_2 + s_1 \qquad\quad = 10$
$\qquad\qquad\quad x_1 + 2x_2 \qquad + s_2 = 8$
$\qquad\qquad\qquad x_1, \ x_2, \ s_1, \ s_2 \ \geq \ 0$

This system can be written in initial form:

$\quad 2x_1 + \quad x_2 + s_1 \qquad\qquad\quad = 10$
$\qquad x_1 + \ 2x_2 \qquad + s_2 \qquad\quad = 8$
$\ -15x_1 - 10x_2 \qquad\qquad + P = 0$
$\qquad\qquad x_1, \ x_2, \ s_1, \ s_2 \ \geq \ 0$

(B) The simplex tableau for this problem is:

$$\begin{array}{c} \text{Enter} \\ \begin{array}{ccccc} x_1 & x_2 & s_1 & s_2 & P \end{array} \end{array}$$

$$\begin{array}{c} \text{Exit} \ s_1 \\ s_2 \\ P \end{array} \begin{bmatrix} \textcircled{2} & 1 & 1 & 0 & 0 & | & 10 \\ 1 & 2 & 0 & 1 & 0 & | & 8 \\ \hline -15 & -10 & 0 & 0 & 1 & | & 0 \end{bmatrix} \begin{array}{l} \tfrac{10}{2} = 5 \ (\text{minimum}) \\ \tfrac{8}{1} = 8 \end{array}$$

Column 1 is the pivot column (-15 is the most negative indicator). Row 1 is the pivot row (5 is the smallest positive quotient). Thus, the pivot element is the circled 2.

(C) We use the simplex method as outlined above. The pivot elements are circled.

$$\begin{array}{ccccc} x_1 & x_2 & s_1 & s_2 & P \end{array}$$

$$\begin{bmatrix} \textcircled{2} & 1 & 1 & 0 & 0 & | & 10 \\ 1 & 2 & 0 & 1 & 0 & | & 8 \\ \hline -15 & -10 & 0 & 0 & 1 & | & 0 \end{bmatrix} \sim \begin{bmatrix} \textcircled{1} & \tfrac{1}{2} & \tfrac{1}{2} & 0 & 0 & | & 5 \\ 1 & 2 & 0 & 1 & 0 & | & 8 \\ \hline -15 & -10 & 0 & 0 & 1 & | & 0 \end{bmatrix} \sim$$

$$\tfrac{1}{2} R_1 \to R_1 \qquad\qquad R_2 + (-1)R_1 \to R_2, \quad R_3 + 15R_1 \to R_3$$

Enter

$$\begin{array}{c} \\ x_1 \\ \text{Exit } s_2 \\ P \end{array}
\begin{array}{ccccc} x_1 & x_2 & s_1 & s_2 & P \\ \end{array}
\left[\begin{array}{ccccc|c} 1 & \frac{1}{2} & \frac{1}{2} & 0 & 0 & 5 \\ 0 & \boxed{\frac{3}{2}} & -\frac{1}{2} & 1 & 0 & 3 \\ \hline 0 & -\frac{5}{2} & \frac{15}{2} & 0 & 1 & 75 \end{array}\right]
\begin{array}{l} \frac{5}{1/2} = 10 \\ \frac{3}{3/2} = 2 \leftarrow \text{pivot row} \\ \end{array}$$

$$\underset{\substack{\uparrow \\ \text{pivot} \\ \text{column}}}{} \qquad \frac{2}{3}R_2 \rightarrow R_2$$

$$\begin{array}{ccccc} & & & & & x_1 & x_2 & s_1 & s_2 & P \end{array}$$

$$\sim \left[\begin{array}{ccccc|c} 1 & \frac{1}{2} & \frac{1}{2} & 0 & 0 & 5 \\ 0 & \boxed{1} & -\frac{1}{3} & \frac{2}{3} & 0 & 2 \\ \hline 0 & -\frac{5}{2} & \frac{15}{2} & 0 & 1 & 75 \end{array}\right]
\sim \left[\begin{array}{ccccc|c} 1 & 0 & \frac{2}{3} & -\frac{1}{3} & 0 & 4 \\ 0 & 1 & -\frac{1}{3} & \frac{2}{3} & 0 & 2 \\ \hline 0 & 0 & \frac{40}{6} & \frac{5}{3} & 1 & 80 \end{array}\right]$$

$$R_1 + \left(-\frac{1}{2}\right)R_2 \rightarrow R_1 \text{ and}$$
$$R_3 + \frac{5}{2}R_2 \rightarrow R_3$$

All the elements in the last row are nonnegative. Thus, max $P = 80$ at $x_1 = 4$, $x_2 = 2$, $s_1 = 0$, $s_2 = 0$.

11. (A) Introduce slack variables s_1 and s_2 to obtain:

Maximize $P = 30x_1 + x_2$
Subject to: $2x_1 + x_2 + s_1 \qquad = 10$
$ x_1 + 2x_2 \qquad + s_2 = 8$
$ x_1, \ x_2, \ s_1, \ s_2 \geq 0$

This system can be written in the initial form:
$$2x_1 + x_2 + s_1 \qquad\qquad = 10$$
$$x_1 + 2x_2 \qquad + s_2 \qquad = 8$$
$$-30x_1 - x_2 \qquad\qquad + P = 0$$

(B) The simplex tableau for this problem is:

Enter

$$\begin{array}{c} \\ \text{Exit } s_1 \\ s_2 \\ P \end{array}
\begin{array}{ccccc} x_1 & x_2 & s_1 & s_2 & P \\ \end{array}
\left[\begin{array}{ccccc|c} \boxed{2} & 1 & 1 & 0 & 0 & 10 \\ 1 & 2 & 0 & 1 & 0 & 8 \\ \hline -30 & -1 & 0 & 0 & 1 & 0 \end{array}\right]
\begin{array}{l} \frac{10}{2} = 5 \text{ (minimum)} \\ \frac{8}{1} = 8 \\ \end{array}$$

$$\underset{\substack{\uparrow \\ \text{pivot} \\ \text{column}}}{}$$

(C)

$$\begin{array}{c} \quad x_1 \quad x_2 \quad s_1 \quad s_2 \quad P \\ \begin{bmatrix} ②& 1 & 1 & 0 & 0 & | & 10 \\ 1 & 2 & 0 & 1 & 0 & | & 8 \\ \hline -30 & -1 & 0 & 0 & 1 & | & 0 \end{bmatrix} \end{array} \sim \begin{array}{c} \begin{bmatrix} ① & \frac{1}{2} & \frac{1}{2} & 0 & 0 & | & 5 \\ 1 & 2 & 0 & 1 & 0 & | & 8 \\ \hline -30 & -1 & 0 & 0 & 1 & | & 0 \end{bmatrix} \end{array}$$

$$\frac{1}{2}R_1 \to R_1 \qquad\qquad R_2 + (-1)R_1 \to R_2 \text{ and } R_3 + 30R_1 \to R_3$$

$$\begin{array}{c} \quad\; x_1 \quad x_2 \quad\; s_1 \quad s_2 \quad P \\ \sim \begin{bmatrix} 1 & \frac{1}{2} & \frac{1}{2} & 0 & 0 & | & 5 \\ 0 & \frac{3}{2} & -\frac{1}{2} & 1 & 0 & | & 3 \\ \hline 0 & 14 & 15 & 0 & 1 & | & 150 \end{bmatrix} \end{array}$$

All elements in the last row are nonnegative. Thus, max $P = 150$ at $x_1 = 5$, $x_2 = 0$, $s_1 = 0$, $s_2 = 3$.

13. The simplex tableau for this problem is:

$$\begin{array}{cc} & \qquad\qquad \text{Enter} \\ & \qquad x_1 \quad x_2 \quad s_1 \quad s_2 \quad s_3 \quad P \\ \begin{array}{c} s_1 \\ s_2 \\ \text{pivot}\!\to\! s_3 \\ \text{row} \\ \text{Exit} \end{array} & \begin{bmatrix} 2 & 1 & 1 & 0 & 0 & 0 & | & 10 \\ 1 & 1 & 0 & 1 & 0 & 0 & | & 7 \\ 1 & ② & 0 & 0 & 1 & 0 & | & 12 \\ \hline -30 & -40 & 0 & 0 & 0 & 1 & | & 0 \end{bmatrix} \end{array} \begin{array}{c} 10 \\ 7 \\ \frac{12}{2} = 6 \text{ (minimum)} \end{array}$$

[Note: The pivot elements have been circled.]

$$\uparrow\ \text{pivot}\quad \frac{1}{2}R_3 \to R_3$$
$$\text{column}$$

$$\sim \begin{bmatrix} 2 & 1 & 1 & 0 & 0 & 0 & | & 10 \\ 1 & 1 & 0 & 1 & 0 & 0 & | & 7 \\ \frac{1}{2} & ① & 0 & 0 & \frac{1}{2} & 0 & | & 6 \\ \hline -30 & -40 & 0 & 0 & 0 & 1 & | & 0 \end{bmatrix}$$

$$R_1 + (-1)R_3 \to R_1, \quad R_2 + (-1)R_3 \to R_2, \text{ and } R_4 + 40R_3 \to R_4$$

$$\begin{array}{c} \\ \text{pivot}\to \\ \text{row} \end{array} \sim \begin{bmatrix} \frac{3}{2} & 0 & 1 & 0 & -\frac{1}{2} & 0 & | & 4 \\ ①/② & 0 & 0 & 1 & -\frac{1}{2} & 0 & | & 1 \\ \frac{1}{2} & 1 & 0 & 0 & \frac{1}{2} & 0 & | & 6 \\ \hline -10 & 0 & 0 & 0 & 20 & 1 & | & 240 \end{bmatrix} \begin{array}{c} \frac{4}{3/2} = \frac{8}{3} \\ \frac{1}{1/2} = 2 \text{ (minimum)} \\ \frac{6}{1/2} = 12 \end{array}$$

$$\uparrow\ \text{pivot}\quad 2R_2 \to R_2$$
$$\text{column}$$

$$\sim \begin{bmatrix} \frac{3}{2} & 0 & 1 & 0 & -\frac{1}{2} & 0 & | & 4 \\ \textcircled{1} & 0 & 0 & 2 & -1 & 0 & | & 2 \\ \frac{1}{2} & 1 & 0 & 0 & \frac{1}{2} & 0 & | & 6 \\ \hdashline -10 & 0 & 0 & 0 & 20 & 1 & | & 240 \end{bmatrix}$$

$$R_1 + \left(-\frac{3}{2}\right)R_2 \rightarrow R_1, \; R_3 + \left(-\frac{1}{2}\right)R_2 \rightarrow R_3, \text{ and } R_4 + 10R_2 \rightarrow R_4$$

$$\begin{array}{cccccc} x_1 & x_2 & s_1 & s_2 & s_3 & P \end{array}$$

$$\sim \begin{bmatrix} 0 & 0 & 1 & -3 & 1 & 0 & | & 1 \\ 1 & 0 & 0 & 2 & -1 & 0 & | & 2 \\ 0 & 1 & 0 & -1 & 1 & 0 & | & 5 \\ \hdashline 0 & 0 & 0 & 20 & 10 & 1 & | & 260 \end{bmatrix}$$

Optimal solution: max $P = 260$ at
$x_1 = 2$, $x_2 = 5$, $s_1 = 1$, $s_2 = 0$, $s_3 = 0$.

15. The simplex tableau for this problem is:

$$\begin{array}{c} \text{Enter} \\ \text{Exit} \quad \begin{array}{cccccc} x_1 & x_2 & s_1 & s_2 & s_3 & P \end{array} \end{array}$$

$$\begin{array}{c} \text{pivot} \\ \text{row} \end{array} \rightarrow \begin{array}{c} s_1 \\ s_2 \\ s_3 \\ \\ P \end{array} \begin{bmatrix} -2 & \textcircled{1} & 1 & 0 & 0 & 0 & | & 2 \\ -1 & 1 & 0 & 1 & 0 & 0 & | & 5 \\ 0 & 1 & 0 & 0 & 1 & 0 & | & 6 \\ \hdashline -2 & -3 & 0 & 0 & 0 & 1 & | & 0 \end{bmatrix} \begin{array}{l} \frac{2}{1} = 2 \text{ (minimum)} \\ \frac{5}{1} = 5 \\ \frac{6}{1} = 6 \end{array}$$

$$\begin{array}{c} \uparrow \\ \text{pivot} \\ \text{column} \end{array} \quad R_2 + (-1)R_1 \rightarrow R_2, \; R_3 + (-1)R_1 \rightarrow R_3, \text{ and } R_4 + 3R_1 \rightarrow R_4$$

$$\begin{array}{c} \\ \\ \text{pivot} \\ \text{row} \end{array} \xrightarrow{\sim} \begin{bmatrix} -2 & 1 & 1 & 0 & 0 & 0 & | & 2 \\ 1 & 0 & -1 & 1 & 0 & 0 & | & 3 \\ \textcircled{2} & 0 & -1 & 0 & 1 & 0 & | & 4 \\ \hdashline -8 & 0 & 3 & 0 & 0 & 1 & | & 6 \end{bmatrix} \begin{array}{l} \\ \frac{3}{1} = 3 \\ \frac{4}{2} = 2 \text{ (minimum)} \end{array}$$

$$\begin{array}{c} \uparrow \\ \text{pivot} \\ \text{column} \end{array} \quad \frac{1}{2}R_3 \rightarrow R_3$$

$$\sim \begin{bmatrix} -2 & 1 & 1 & 0 & 0 & 0 & | & 2 \\ 1 & 0 & -1 & 1 & 0 & 0 & | & 3 \\ \textcircled{1} & 0 & -\frac{1}{2} & 0 & \frac{1}{2} & 0 & | & 2 \\ \hdashline -8 & 0 & 3 & 0 & 0 & 1 & | & 6 \end{bmatrix} \sim \begin{array}{cccccc} x_1 & x_2 & s_1 & s_2 & s_3 & P \end{array}$$
$$\begin{bmatrix} 0 & 1 & 0 & 0 & 1 & 0 & | & 6 \\ 0 & 0 & -\frac{1}{2} & 1 & -\frac{1}{2} & 0 & | & 1 \\ 1 & 0 & -\frac{1}{2} & 0 & \frac{1}{2} & 0 & | & 2 \\ \hdashline 0 & 0 & -1 & 0 & 4 & 1 & | & 22 \end{bmatrix}$$

$$R_1 + 2R_3 \rightarrow R_1, \; R_2 + (-1)R_3 \rightarrow R_2,$$
$$\text{and } R_4 + 8R_3 \rightarrow R_4$$

$$\begin{array}{c} \uparrow \\ \text{pivot} \\ \text{column} \end{array}$$

Since there are no positive elements in the pivot column (above the dashed line), we conclude that there is no solution.

CHAPTER 6 LINEAR INEQUALITIES AND LINEAR PROGRAMMING

17. The simplex tableau for this problem is:

$$
\begin{array}{cccccc}
 & x_1 & x_2 & s_1 & s_2 & s_3 & P
\end{array}
$$

pivot row →
$$
\left[\begin{array}{cccccc|c}
-1 & ① & 1 & 0 & 0 & 0 & 2 \\
-1 & 3 & 0 & 1 & 0 & 0 & 12 \\
1 & -4 & 0 & 0 & 1 & 0 & 4 \\
\hline
1 & -2 & 0 & 0 & 0 & 1 & 0
\end{array}\right]
\begin{array}{l}
\frac{2}{1} = 2 \ (\text{minimum}) \\[4pt]
\frac{12}{3} = 4
\end{array}
$$

↑ pivot column $R_2 + (-3)R_1 \rightarrow R_2$, $R_3 + 4R_1 \rightarrow R_3$, and $R_4 + 2R_1 \rightarrow R_4$

pivot row →
$$
\left[\begin{array}{cccccc|c}
-1 & 1 & 1 & 0 & 0 & 0 & 2 \\
② & 0 & -3 & 1 & 0 & 0 & 6 \\
-3 & 0 & 4 & 0 & 1 & 0 & 12 \\
\hline
-1 & 0 & 2 & 0 & 0 & 1 & 4
\end{array}\right]
\begin{array}{l}
\frac{6}{2} = 3 \leftarrow \text{pivot row} \\[6pt]
[\underline{\text{Note}}: \text{We only use the} \\
\textit{positive} \text{ elements above} \\
\text{the dashed line in the} \\
\text{pivot column.}]
\end{array}
$$

↑ pivot column $\frac{1}{2}R_2 \rightarrow R_2$

$$
\left[\begin{array}{cccccc|c}
-1 & 1 & 1 & 0 & 0 & 0 & 2 \\
① & 0 & -\frac{3}{2} & \frac{1}{2} & 0 & 0 & 3 \\
-3 & 0 & 4 & 0 & 1 & 0 & 12 \\
\hline
-1 & 0 & 2 & 0 & 0 & 1 & 4
\end{array}\right]
\sim
\begin{array}{cccccc}
x_1 & x_2 & s_1 & s_2 & s_3 & P
\end{array}
$$

$$
\left[\begin{array}{cccccc|c}
0 & 1 & -\frac{1}{2} & \frac{1}{2} & 0 & 0 & 5 \\
1 & 0 & -\frac{3}{2} & \frac{1}{2} & 0 & 0 & 3 \\
0 & 0 & -\frac{1}{2} & \frac{3}{2} & 1 & 0 & 21 \\
\hline
0 & 0 & \frac{1}{2} & \frac{1}{2} & 0 & 1 & 7
\end{array}\right]
$$

$R_1 + R_2 \rightarrow R_1$, $R_3 + 3R_2 \rightarrow R_3$, and $R_4 + R_2 \rightarrow R_4$

Optimal solution: max $P = 7$ at $x_1 = 3$, $x_2 = 5$, $s_1 = 0$, $s_2 = 0$, $s_3 = 21$.

19. The simplex tableau for this problem is:

$$
\begin{array}{cccccc}
 & x_1 & x_2 & x_3 & s_1 & s_2 & P
\end{array}
$$

pivot row →
$$
\left[\begin{array}{cccccc|c}
① & 1 & -1 & 1 & 0 & 0 & 10 \\
2 & 4 & 1 & 0 & 1 & 0 & 30 \\
\hline
-5 & -2 & 1 & 0 & 0 & 1 & 0
\end{array}\right]
\begin{array}{l}
\frac{10}{1} = 10 \ (\text{minimum}) \\[4pt]
\frac{30}{2} = 15
\end{array}
$$

↑ pivot column $R_2 + (-2)R_1 \rightarrow R_2$ and $R_3 + 5R_1 \rightarrow R_3$

$$
\left[\begin{array}{cccccc|c}
1 & 1 & -1 & 1 & 0 & 0 & 10 \\
0 & 2 & ③ & -2 & 1 & 0 & 10 \\
\hline
0 & 3 & -4 & 5 & 0 & 1 & 50
\end{array}\right]
\sim
\left[\begin{array}{cccccc|c}
1 & 1 & -1 & 1 & 0 & 0 & 10 \\
0 & \frac{2}{3} & ① & -\frac{2}{3} & \frac{1}{3} & 0 & \frac{10}{3} \\
\hline
0 & 3 & -4 & 5 & 0 & 1 & 50
\end{array}\right]
$$

$\frac{1}{3}R_2 \rightarrow R_2$ $R_1 + R_2 \rightarrow R_1$ and $R_3 + 4R_2 \rightarrow R_3$

$$\begin{array}{ccccccc} & x_1 & x_2 & x_3 & s_1 & s_2 & P \\ \sim \left[\begin{array}{cccccc|c} 1 & \frac{5}{3} & 0 & \frac{1}{3} & \frac{1}{3} & 0 & \frac{40}{3} \\ 0 & \frac{2}{3} & 1 & -\frac{2}{3} & \frac{1}{3} & 0 & \frac{10}{3} \\ \hline 0 & \frac{17}{3} & 0 & \frac{7}{3} & \frac{4}{3} & 1 & \frac{190}{3} \end{array}\right] \end{array}$$

Optimal solution: max $P = \dfrac{190}{3}$ at $x_1 = \dfrac{40}{3}$, $x_2 = 0$, $x_3 = \dfrac{10}{3}$, $s_1 = 0$, $s_2 = 0$.

21. The simplex tableau for this problem is:

$$\begin{array}{ccccccc} & x_1 & x_2 & x_3 & s_1 & s_2 & P \\ & \left[\begin{array}{cccccc|c} 1 & 0 & 1 & 1 & 0 & 0 & 4 \\ 0 & 1 & \textcircled{1} & 0 & 1 & 0 & 3 \\ \hline -2 & -3 & -4 & 0 & 0 & 1 & 0 \end{array}\right] \end{array}$$

pivot row → (second row)

$\dfrac{4}{1} = 4$

$\dfrac{3}{1} = 3$ (minimum)

pivot column (↑ under x_3)

$R_1 + (-1)R_2 \rightarrow R_1$
$R_3 + 4R_2 \rightarrow R_3$

$$\sim \left[\begin{array}{cccccc|c} \textcircled{1} & -1 & 0 & 1 & -1 & 0 & 1 \\ 0 & 1 & 1 & 0 & 1 & 0 & 3 \\ \hline -2 & 1 & 0 & 0 & 4 & 1 & 12 \end{array}\right] \sim \left[\begin{array}{cccccc|c} 1 & -1 & 0 & 1 & -1 & 0 & 1 \\ 0 & \textcircled{1} & 1 & 0 & 1 & 0 & 3 \\ \hline 0 & -1 & 0 & 2 & 2 & 1 & 14 \end{array}\right]$$

$R_3 + 2R_1 \rightarrow R_3$ \qquad\qquad $R_1 + R_2 \rightarrow R_1$ and $R_3 + R_2 \rightarrow R_3$

$$\begin{array}{ccccccc} & x_1 & x_2 & x_3 & s_1 & s_2 & P \\ \sim & \left[\begin{array}{cccccc|c} 1 & 0 & 1 & 1 & 0 & 0 & 4 \\ 0 & 1 & 1 & 0 & 1 & 0 & 3 \\ \hline 0 & 0 & 1 & 2 & 3 & 1 & 17 \end{array}\right] \end{array}$$

Optimal solution: max $P = 17$ at $x_1 = 4$, $x_2 = 3$, $x_3 = 0$, $s_1 = 0$, $s_2 = 0$.

23. The simplex tableau for this problem is:

$$\begin{array}{cccccccc} & x_1 & x_2 & x_3 & s_1 & s_2 & s_3 & P \\ & \left[\begin{array}{ccccccc|c} 3 & 2 & 5 & 1 & 0 & 0 & 0 & 23 \\ \textcircled{2} & 1 & 1 & 0 & 1 & 0 & 0 & 8 \\ 1 & 1 & 2 & 0 & 0 & 1 & 0 & 7 \\ \hline -4 & -3 & -2 & 0 & 0 & 0 & 1 & 0 \end{array}\right] \end{array}$$

pivot row → (second row)

$\dfrac{23}{3} = 7\dfrac{2}{3}$

$\dfrac{8}{2} = 4$ (minimum)

$\dfrac{7}{1} = 7$

pivot column (↑ under x_1) \qquad $\dfrac{1}{2}R_2 \rightarrow R_2$

$$\sim \left[\begin{array}{ccccccc|c} 3 & 2 & 5 & 1 & 0 & 0 & 0 & 23 \\ \textcircled{1} & \frac{1}{2} & \frac{1}{2} & 0 & \frac{1}{2} & 0 & 0 & 4 \\ 1 & 1 & 2 & 0 & 0 & 1 & 0 & 7 \\ \hline -4 & -3 & -2 & 0 & 0 & 0 & 1 & 0 \end{array}\right] \sim \left[\begin{array}{ccccccc|c} 0 & \frac{1}{2} & \frac{7}{2} & 1 & -\frac{3}{2} & 0 & 0 & 11 \\ 1 & \frac{1}{2} & \frac{1}{2} & 0 & \frac{1}{2} & 0 & 0 & 4 \\ 0 & \textcircled{$\frac{1}{2}$} & \frac{3}{2} & 0 & -\frac{1}{2} & 1 & 0 & 3 \\ \hline 0 & -1 & 0 & 0 & 2 & 0 & 1 & 16 \end{array}\right]$$

$R_1 + (-3)R_2 \rightarrow R_1$, $R_3 + (-1)R_2 \rightarrow R_3$, and \qquad $2R_3 \rightarrow R_3$
$R_4 + 4R_2 \rightarrow R_4$

$$\sim \begin{bmatrix} 0 & \frac{1}{2} & \frac{7}{2} & 1 & -\frac{3}{2} & 0 & 0 & 11 \\ 1 & \frac{1}{2} & \frac{1}{2} & 0 & \frac{1}{2} & 0 & 0 & 4 \\ 0 & \textcircled{1} & 3 & 0 & -1 & 2 & 0 & 6 \\ \hline 0 & -1 & 0 & 0 & 2 & 0 & 1 & 16 \end{bmatrix}$$

$$\begin{array}{ccccccc} x_1 & x_2 & x_3 & s_1 & s_2 & s_3 & P \end{array}$$
$$\sim \begin{bmatrix} 0 & 0 & 2 & 1 & -1 & -1 & 0 & 8 \\ 1 & 0 & -1 & 0 & 1 & -1 & 0 & 1 \\ 0 & 1 & 3 & 0 & -1 & 2 & 0 & 6 \\ \hline 0 & 0 & 3 & 0 & 1 & 2 & 1 & 22 \end{bmatrix}$$

$R_1 + \left(-\frac{1}{2}\right)R_3 \to R_1$, $R_2 + \left(-\frac{1}{2}\right)R_3 \to R_2$, and

$R_4 + R_3 \to R_4$

Optimal solution: max $P = 22$ at $x_1 = 1$, $x_2 = 6$, $x_3 = 0$, $s_1 = 8$, $s_2 = 0$, $s_3 = 0$.

25. Multiply the first problem constraint by $\frac{10}{6}$, the second by 100, and the third by 10 to clear the fractions. Then, the simplex tableau for this problem is:

$$\begin{array}{cccccc} x_1 & x_2 & s_1 & s_2 & s_3 & P \end{array}$$
$$\begin{bmatrix} 1 & \textcircled{2} & 1 & 0 & 0 & 0 & 1{,}600 \\ 3 & 4 & 0 & 1 & 0 & 0 & 3{,}600 \\ 3 & 2 & 0 & 0 & 1 & 0 & 2{,}700 \\ \hline -20 & -30 & 0 & 0 & 0 & 1 & 0 \end{bmatrix} \begin{array}{l} \frac{1{,}600}{2} = 800 \\[4pt] \frac{3{,}600}{4} = 900 \\[4pt] \frac{2{,}700}{2} = 1{,}350 \end{array}$$

$\frac{1}{2}R_1 \to R_1$

$$\sim \begin{bmatrix} \frac{1}{2} & \textcircled{1} & \frac{1}{2} & 0 & 0 & 0 & 800 \\ 3 & 4 & 0 & 1 & 0 & 0 & 3{,}600 \\ 3 & 2 & 0 & 0 & 1 & 0 & 2{,}700 \\ \hline -20 & -30 & 0 & 0 & 0 & 1 & 0 \end{bmatrix}$$

$R_2 + (-4)R_1 \to R_2$, $R_3 + (-2)R_1 \to R_3$, and $R_4 + 30R_1 \to R_4$

$$\sim \begin{bmatrix} \frac{1}{2} & 1 & \frac{1}{2} & 0 & 0 & 0 & 800 \\ \textcircled{1} & 0 & -2 & 1 & 0 & 0 & 400 \\ 2 & 0 & -1 & 0 & 1 & 0 & 1{,}100 \\ \hline -5 & 0 & 15 & 0 & 0 & 1 & 24{,}000 \end{bmatrix} \begin{array}{l} \frac{800}{1/2} = 1{,}600 \\[4pt] \frac{400}{1} = 400 \\[4pt] \frac{1{,}100}{2} = 550 \end{array}$$

$R_1 + \left(-\frac{1}{2}\right)R_2 \to R_1$, $R_3 + (-2)R_2 \to R_3$, and $R_4 + 5R_2 \to R_4$

$$
\sim
\begin{array}{c}
\begin{array}{cccccc} x_1 & x_2 & s_1 & s_2 & s_3 & P \end{array} \\
\left[
\begin{array}{cccccc|c}
0 & 1 & \frac{3}{2} & -\frac{1}{2} & 0 & 0 & 600 \\
1 & 0 & -2 & 1 & 0 & 0 & 400 \\
0 & 0 & 3 & -2 & 1 & 0 & 300 \\
\hdashline
0 & 0 & 5 & 5 & 0 & 1 & 26{,}000
\end{array}
\right]
\end{array}
$$

Optimal solution: max $P = 26{,}000$ at $x_1 = 400$, $x_2 = 600$, $s_1 = 0$, $s_2 = 0$, $s_3 = 300$.

27. The simplex tableau for this problem is:

$$
\begin{array}{c}
\begin{array}{ccccccc} x_1 & x_2 & x_3 & s_1 & s_2 & s_3 & P \end{array} \\
\left[
\begin{array}{ccccccc|c}
2 & 2 & \circled{8} & 1 & 0 & 0 & 0 & 600 \\
1 & 3 & 2 & 0 & 1 & 0 & 0 & 600 \\
3 & 2 & 1 & 0 & 0 & 1 & 0 & 400 \\
\hdashline
-1 & -2 & -3 & 0 & 0 & 0 & 1 & 0
\end{array}
\right]
\end{array}
\begin{array}{l}
\frac{600}{8} = 75 \\[6pt]
\frac{600}{2} = 300 \\[6pt]
\frac{400}{1} = 400
\end{array}
$$

$\frac{1}{8} R_1 \rightarrow R_1$

$$
\sim
\left[
\begin{array}{ccccccc|c}
\frac{1}{4} & \frac{1}{4} & \circled{1} & \frac{1}{8} & 0 & 0 & 0 & 75 \\
1 & 3 & 2 & 0 & 1 & 0 & 0 & 600 \\
3 & 2 & 1 & 0 & 0 & 1 & 0 & 400 \\
\hdashline
-1 & -2 & -3 & 0 & 0 & 0 & 1 & 0
\end{array}
\right]
$$

$R_2 + (-2)R_1 \rightarrow R_2$, $R_3 + (-1)R_1 \rightarrow R_3$, and $R_4 + 3R_1 \rightarrow R_4$

$$
\sim
\left[
\begin{array}{ccccccc|c}
\frac{1}{4} & \frac{1}{4} & 1 & \frac{1}{8} & 0 & 0 & 0 & 75 \\
\frac{1}{2} & \circled{\frac{5}{2}} & 0 & -\frac{1}{4} & 1 & 0 & 0 & 450 \\
\frac{11}{4} & \frac{7}{4} & 0 & -\frac{1}{8} & 0 & 1 & 0 & 325 \\
\hdashline
-\frac{1}{4} & -\frac{5}{4} & 0 & \frac{3}{8} & 0 & 0 & 1 & 225
\end{array}
\right]
\begin{array}{l}
\frac{75}{1/4} = 300 \\[6pt]
\frac{450}{5/2} = 180 \\[6pt]
\frac{325}{7/4} = 185.71
\end{array}
$$

$\frac{2}{5} R_2 \rightarrow R_2$

$$
\sim
\left[
\begin{array}{ccccccc|c}
\frac{1}{4} & \frac{1}{4} & 1 & \frac{1}{8} & 0 & 0 & 0 & 75 \\
\frac{1}{5} & \circled{1} & 0 & -\frac{1}{10} & \frac{2}{5} & 0 & 0 & 180 \\
\frac{11}{4} & \frac{7}{4} & 0 & -\frac{1}{8} & 0 & 1 & 0 & 325 \\
\hdashline
-\frac{1}{4} & -\frac{5}{4} & 0 & \frac{3}{8} & 0 & 0 & 1 & 225
\end{array}
\right]
$$

$R_1 + \left(-\frac{1}{4}\right)R_2 \rightarrow R_1$, $R_3 + \left(-\frac{7}{4}\right)R_2 \rightarrow R_3$,

and $R_4 + \frac{5}{4}R_2 \rightarrow R_4$

$$
\sim
\begin{array}{c}
\begin{array}{ccccccc} x_1 & x_2 & x_3 & s_1 & s_2 & s_3 & P \end{array} \\
\left[
\begin{array}{ccccccc|c}
\frac{1}{5} & 0 & 1 & \frac{3}{20} & -\frac{1}{10} & 0 & 0 & 30 \\
\frac{1}{5} & 1 & 0 & -\frac{1}{10} & \frac{2}{5} & 0 & 0 & 180 \\
\frac{12}{5} & 0 & 0 & \frac{1}{20} & -\frac{7}{10} & 1 & 0 & 10 \\
\hdashline
0 & 0 & 0 & \frac{1}{4} & \frac{1}{2} & 0 & 1 & 450
\end{array}
\right]
\end{array}
$$

Optimal solution: max $P = 450$ at $x_1 = 0$, $x_2 = 180$, $x_3 = 30$, $s_1 = 0$, $s_2 = 0$, $s_3 = 10$.

29. The simplex tableau for this problem is:

$$
\begin{array}{ccccccc}
x_1 & x_2 & s_1 & s_2 & s_3 & s_4 & P \\
\end{array}
$$

$$
\left[
\begin{array}{ccccccc|c}
1 & 2 & 1 & 0 & 0 & 0 & 0 & 40 \\
1 & 3 & 0 & 1 & 0 & 0 & 0 & 48 \\
1 & 4 & 0 & 0 & 1 & 0 & 0 & 60 \\
0 & \boxed{1} & 0 & 0 & 0 & 1 & 0 & 14 \\
\hline
-2 & -5 & 0 & 0 & 0 & 0 & 1 & 0 \\
\end{array}
\right]
\quad
\begin{array}{l}
\frac{40}{2} = 20 \\[4pt]
\frac{48}{3} = 16 \\[4pt]
\frac{60}{4} = 15 \\[4pt]
\frac{14}{1} = 14 \\
\end{array}
$$

$R_1 + (-2)R_4 \rightarrow R_1$, $R_2 + (-3)R_4 \rightarrow R_2$, $R_3 + (-4)R_4 \rightarrow R_3$,
and $R_5 + 5R_4 \rightarrow R_5$

$$
\sim
\left[
\begin{array}{ccccccc|c}
1 & 0 & 1 & 0 & 0 & -2 & 0 & 12 \\
1 & 0 & 0 & 1 & 0 & -3 & 0 & 6 \\
\boxed{1} & 0 & 0 & 0 & 1 & -4 & 0 & 4 \\
0 & 1 & 0 & 0 & 0 & 1 & 0 & 14 \\
\hline
-2 & 0 & 0 & 0 & 0 & 5 & 1 & 70 \\
\end{array}
\right]
\quad
\begin{array}{l}
\frac{12}{1} = 12 \\[4pt]
\frac{6}{1} = 6 \\[4pt]
\frac{4}{1} = 4 \\
\end{array}
$$

$R_1 + (-1)R_3 \rightarrow R_1$, $R_2 + (-1)R_3 \rightarrow R_2$, and $R_5 + 2R_3 \rightarrow R_5$

$$
\sim
\left[
\begin{array}{ccccccc|c}
0 & 0 & 1 & 0 & -1 & 2 & 0 & 8 \\
0 & 0 & 0 & 1 & -1 & \boxed{1} & 0 & 2 \\
1 & 0 & 0 & 0 & 1 & -4 & 0 & 4 \\
0 & 1 & 0 & 0 & 0 & 1 & 0 & 14 \\
\hline
0 & 0 & 0 & 0 & 2 & -3 & 1 & 78 \\
\end{array}
\right]
\quad
\begin{array}{l}
\frac{8}{2} = 4 \\[4pt]
\frac{2}{1} = 2 \\[4pt]
\\
\frac{14}{1} = 14 \\
\end{array}
$$

$R_1 + (-2)R_2 \rightarrow R_1$, $R_3 + 4R_2 \rightarrow R_3$, $R_4 + (-1)R_2 \rightarrow R_4$,
and $R_5 + 3R_2 \rightarrow R_5$

$$
\sim
\left[
\begin{array}{ccccccc|c}
0 & 0 & 1 & -2 & \boxed{1} & 0 & 0 & 4 \\
0 & 0 & 0 & 1 & -1 & 1 & 0 & 2 \\
1 & 0 & 0 & 4 & -3 & 0 & 0 & 12 \\
0 & 1 & 0 & -1 & 1 & 0 & 0 & 12 \\
\hline
0 & 0 & 0 & 3 & -1 & 0 & 1 & 84 \\
\end{array}
\right]
\quad
\begin{array}{l}
\frac{4}{1} = 4 \\[4pt]
\\
\\
\frac{12}{1} = 12 \\
\end{array}
$$

$R_2 + R_1 \rightarrow R_2$, $R_3 + 3R_1 \rightarrow R_3$, $R_4 + (-1)R_1 \rightarrow R_4$, and $R_5 + R_1 \rightarrow R_5$

$$\sim \begin{bmatrix} x_1 & x_2 & s_1 & s_2 & s_3 & s_4 & P & \\ 0 & 0 & 1 & -2 & 1 & 0 & 0 & 4 \\ 0 & 0 & 1 & -1 & 0 & 1 & 0 & 6 \\ 1 & 0 & 3 & -2 & 0 & 0 & 0 & 24 \\ 0 & 1 & -1 & 1 & 0 & 0 & 0 & 8 \\ \hdashline 0 & 0 & 1 & 1 & 0 & 0 & 1 & 88 \end{bmatrix}$$

Optimal solution: max $P = 88$ at $x_1 = 24$, $x_2 = 8$, $s_1 = 0$, $s_2 = 0$, $s_3 = 4$, $s_4 = 6$.

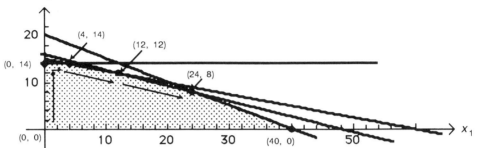

31. Let x_1 = the number of A components,
x_2 = the number of B components,
and x_3 = the number of C components.

The mathematical model for this problem is:

Maximize $P = 7x_1 + 9x_2 + 10x_3$
Subject to:
$$2x_1 + x_2 + 2x_3 \leq 1000$$
$$x_1 + 2x_2 + 2x_3 \leq 800$$
$$x_1, \ x_2, \ x_3 \geq 0$$

We introduce slack variables s_1 and s_2 to obtain the equivalent form:

$$2x_1 + x_2 + 2x_3 + s_1 \qquad\quad = 1000$$
$$x_1 + 2x_2 + 2x_3 \qquad + s_2 \quad = 800$$
$$-7x_1 - 9x_2 - 10x_3 \qquad\qquad + P = 0$$

The simplex tableau for this problem is:

$$\begin{bmatrix} x_1 & x_2 & x_3 & s_1 & s_2 & P & \\ 2 & 1 & 2 & 1 & 0 & 0 & 1000 \\ 1 & 2 & ② & 0 & 1 & 0 & 800 \\ \hdashline -7 & -9 & -10 & 0 & 0 & 1 & 0 \end{bmatrix} \quad \begin{array}{l} \frac{1000}{2} = 500 \\[4pt] \frac{800}{2} = 400 \end{array}$$

$\frac{1}{2}R_2 \rightarrow R_2$

$$\sim \begin{bmatrix} 2 & 1 & 2 & 1 & 0 & 0 & 1000 \\ \frac{1}{2} & 1 & ① & 0 & \frac{1}{2} & 0 & 400 \\ \hdashline -7 & -9 & -10 & 0 & 0 & 1 & 0 \end{bmatrix}$$

$R_1 + (-2)R_2 \rightarrow R_1$ and $R_3 + 10R_2 \rightarrow R_3$

$$\sim \begin{bmatrix} \boxed{1} & -1 & 0 & 1 & -1 & 0 & 200 \\ \frac{1}{2} & 1 & 1 & 0 & \frac{1}{2} & 0 & 400 \\ \hline -2 & 1 & 0 & 0 & 5 & 1 & 4000 \end{bmatrix} \begin{array}{l} \frac{200}{1} = 200 \\[4pt] \frac{400}{\frac{1}{2}} = 800 \end{array}$$

$$R_2 + \left(-\frac{1}{2}\right)R_1 \;\rightarrow\; R_2 \text{ and } R_3 + 2R_1 \;\rightarrow\; R_3$$

$$\sim \begin{bmatrix} 1 & -1 & 0 & 1 & -1 & 0 & 200 \\ 0 & \boxed{\tfrac{3}{2}} & 1 & -\frac{1}{2} & 1 & 0 & 300 \\ \hline 0 & -1 & 0 & 2 & 3 & 1 & 4400 \end{bmatrix} \sim \begin{bmatrix} 1 & -1 & 0 & 1 & -1 & 0 & 200 \\ 0 & \boxed{1} & \frac{2}{3} & -\frac{1}{3} & \frac{2}{3} & 0 & 200 \\ \hline 0 & -1 & 0 & 2 & 3 & 1 & 4400 \end{bmatrix}$$

$$\frac{2}{3}R_2 \;\rightarrow\; R_2 \qquad\qquad\qquad R_1 + R_2 \;\rightarrow\; R_1 \text{ and } R_3 + R_2 \;\rightarrow\; R_3$$

$$\begin{array}{ccccccc} x_1 & x_2 & x_3 & s_1 & s_2 & P & \end{array}$$

$$\sim \begin{bmatrix} 1 & 0 & \frac{2}{3} & \frac{2}{3} & -\frac{1}{3} & 0 & 400 \\ 0 & 1 & \frac{2}{3} & -\frac{1}{3} & \frac{2}{3} & 0 & 200 \\ \hline 0 & 0 & \frac{2}{3} & \frac{5}{3} & \frac{11}{3} & 1 & 4600 \end{bmatrix}$$

Optimal solution: the maximum profit is $4600 when 400 A components, 200 B components, and 0 C components are manufactured.

33. Let x_1 = the amount invested in government bonds,

$\quad\quad\; x_2$ = the amount invested in mutual funds,

and x_3 = the amount invested in money market funds.

The mathematical model for this problem is:

Maximize $P = .08x_1 + .13x_2 + .15x_3$

Subject to: $x_1 + x_2 + x_3 \le 100,000$

$$x_2 + x_3 \le x_1$$
$$x_1,\; x_2,\; x_3 \ge 0$$

We introduce slack variables s_1 and s_2 to obtain the equivalent form:

$$\begin{array}{rcr} x_1 + x_2 + x_3 + s_1 \qquad\qquad & = & 100,000 \\ -x_1 + x_2 + x_3 \qquad + s_2 \qquad & = & 0 \\ -.08x_1 - .13x_2 - .15x_3 \qquad\qquad + P & = & 0 \end{array}$$

The simplex tableau for this problem is:

$$\begin{array}{cccccc} x_1 & x_2 & x_3 & s_1 & s_2 & P \end{array}$$

$$\begin{bmatrix} 1 & 1 & 1 & 1 & 0 & 0 & 100,000 \\ -1 & 1 & \boxed{1} & 0 & 1 & 0 & 0 \\ \hline -.08 & -.13 & -.15 & 0 & 0 & 1 & 0 \end{bmatrix} \begin{array}{l} \frac{100,000}{1} = 100,000 \end{array}$$

$$R_1 + (-1)R_2 \;\rightarrow\; R_1 \text{ and } R_3 + .15R_2 \;\rightarrow\; R_3$$

$$\sim \begin{bmatrix} ②\ & 0 & 0 & 1 & -1 & 0 & | & 100{,}000 \\ -1 & 1 & 1 & 0 & 1 & 0 & | & 0 \\ \hline -.23 & .02 & 0 & 0 & .15 & 1 & | & 0 \end{bmatrix} \sim \begin{bmatrix} ①\ & 0 & 0 & \tfrac{1}{2} & -\tfrac{1}{2} & 0 & | & 50{,}000 \\ -1 & 1 & 1 & 0 & 1 & 0 & | & 0 \\ \hline -.23 & .02 & 0 & 0 & .15 & 1 & | & 0 \end{bmatrix}$$

$\dfrac{1}{2}R_1 \to R_1$ $R_2 + R_1 \to R_2$ and $R_3 + .23R_1 \to R_3$

$$\begin{array}{cccccc} x_1 & x_2 & x_3 & s_1 & s_2 & P \end{array}$$

$$\sim \begin{bmatrix} 1 & 0 & 0 & \tfrac{1}{2} & -\tfrac{1}{2} & 0 & | & 50{,}000 \\ 0 & 1 & 1 & \tfrac{1}{2} & \tfrac{1}{2} & 0 & | & 50{,}000 \\ \hline 0 & .02 & 0 & .115 & .035 & 1 & | & 11{,}500 \end{bmatrix}$$

Optimal solution: the maximum return is \$11,500 when $x_1 = $ \$50,000 is invested in government bonds, $x_2 = \$0$ is invested in mutual funds, and $x_3 = \$50{,}000$ is invested in money market funds.

35. Let x_1 = the number of daytime ads,
     x_2 = the number of prime-time ads,
and x_3 = the number of late-night ads.

The mathematical model for this problem is:

Maximize $P = 14{,}000x_1 + 24{,}000x_2 + 18{,}000x_3$
Subject to: $1000x_1 + 2000x_2 + 1500x_3 \le 20{,}000$
$$x_1 + x_2 + x_3 \le 15$$
$$x_1,\ x_2,\ x_3 \ge 0$$

We introduce slack variables to obtain the following initial form:

$$1000x_1 + 2000x_2 + 1500x_3 + s_1 \qquad\qquad = 20{,}000$$
$$x_1 + x_2 + x_3 \qquad + s_2 \qquad = 15$$
$$-14{,}000x_1 - 24{,}000x_2 - 18{,}000x_3 \qquad\qquad + P = 0$$

The simplex tableau for this problem is:

$$\begin{array}{cccccc} x_1 & x_2 & x_3 & s_1 & s_2 & P \end{array}$$

$$\begin{bmatrix} 1000 & ⓪\!2000 & 1500 & 1 & 0 & 0 & | & 20{,}000 \\ 1 & 1 & 1 & 0 & 1 & 0 & | & 15 \\ \hline -14{,}000 & -24{,}000 & -18{,}000 & 0 & 0 & 1 & | & 0 \end{bmatrix} \quad \begin{array}{l} \dfrac{20{,}000}{2000} = 10 \\[2mm] \dfrac{15}{1} = 15 \end{array}$$

$\dfrac{1}{2000}R_1 \to R_1$

$$\sim \begin{bmatrix} \tfrac{1}{2} & ① & \tfrac{3}{4} & \tfrac{1}{2000} & 0 & 0 & | & 10 \\ 1 & 1 & 1 & 0 & 1 & 0 & | & 15 \\ \hline -14{,}000 & -24{,}000 & -18{,}000 & 0 & 0 & 1 & | & 0 \end{bmatrix}$$

$R_2 + (-1)R_1 \to R_2,\ \ R_3 + 24{,}000R_1 \to R_3$

$$\sim \begin{bmatrix} \tfrac{1}{2} & 1 & \tfrac{3}{4} & \tfrac{1}{2000} & 0 & 0 & | & 10 \\ ⓪\tfrac{1}{2} & 0 & \tfrac{1}{4} & -\tfrac{1}{2000} & 1 & 0 & | & 5 \\ \hline -2000 & 0 & 0 & 12 & 0 & 1 & | & 240{,}000 \end{bmatrix}$$

$2R_2 \to R_2$

$$\sim \begin{bmatrix} \frac{1}{2} & 1 & \frac{3}{4} & \frac{1}{2000} & 0 & 0 & 10 \\ \boxed{1} & 0 & \frac{1}{2} & -\frac{1}{1000} & 2 & 0 & 10 \\ \hline -2000 & 0 & 0 & 12 & 0 & 1 & 240{,}000 \end{bmatrix}$$

$$R_1 + \left(-\frac{1}{2}\right)R_2 \to R_1, \quad R_3 + 2000R_2 \to R_3$$

$$
\begin{array}{cccccc}
x_1 & x_2 & x_3 & s_1 & s_2 & P
\end{array}
$$

$$\sim \begin{bmatrix} 0 & 1 & \frac{1}{2} & \frac{1}{1000} & -1 & 0 & 5 \\ 1 & 0 & \frac{1}{2} & -\frac{1}{1000} & 2 & 0 & 10 \\ \hline 0 & 0 & 1000 & 10 & 4000 & 1 & 260{,}000 \end{bmatrix}$$

Optimal solution: maximum number of potential customers is 260,000 when $x_1 = 10$ daytime ads, $x_2 = 5$ prime-time ads, and $x_3 = 0$ late-night ads are placed.

37. Let x_1 = the number of colonial houses,
  x_2 = the number of split-level houses,
and x_3 = the number of ranch-style houses.

The mathematical model for this problem is:

Maximize $P = 20{,}000x_1 + 18{,}000x_2 + 24{,}000x_3$

Subject to:
$$\frac{1}{2}x_1 + \frac{1}{2}x_2 + x_3 \le 30$$
$$60{,}000x_1 + 60{,}000x_2 + 80{,}000x_3 \le 3{,}200{,}000$$
$$4{,}000x_1 + 3{,}000x_2 + 4{,}000x_3 \le 180{,}000$$
$$x_1,\ x_2,\ x_3 \ge 0$$

We simplify the inequalities and then introduce slack variables to obtain the initial form:

$$
\begin{aligned}
\frac{1}{2}x_1 + \frac{1}{2}x_2 + x_3 + s_1 &= 30 \\
6x_1 + 6x_2 + 8x_3 + s_2 &= 320 \\
4x_1 + 3x_2 + 4x_3 + s_3 &= 180 \\
-20{,}000x_1 - 18{,}000x_2 - 24{,}000x_3 + P &= 0
\end{aligned}
$$

[Note: This simplification will change the interpretation of the slack variables.]

The simplex tableau for this problem is:

$$
\begin{array}{ccccccc}
x_1 & x_2 & x_3 & s_1 & s_2 & s_3 & P
\end{array}
$$

$$\begin{bmatrix} \frac{1}{2} & \frac{1}{2} & \boxed{1} & 1 & 0 & 0 & 0 & 30 \\ 6 & 6 & 8 & 0 & 1 & 0 & 0 & 320 \\ 4 & 3 & 4 & 0 & 0 & 1 & 0 & 180 \\ \hline -20{,}000 & -18{,}000 & -24{,}000 & 0 & 0 & 0 & 1 & 0 \end{bmatrix} \quad \begin{array}{l} \frac{30}{1} = 30 \\[4pt] \frac{320}{8} = 40 \\[4pt] \frac{180}{4} = 45 \end{array}$$

$$R_2 + (-8)R_1 \to R_2, \quad R_3 + (-4)R_1 \to R_3, \quad R_4 + 24{,}000R_1 \to R_4$$

$$\sim \begin{bmatrix} \frac{1}{2} & \frac{1}{2} & 1 & 1 & 0 & 0 & 0 & | & 30 \\ 2 & 2 & 0 & -8 & 1 & 0 & 0 & | & 80 \\ ② & 1 & 0 & -4 & 0 & 1 & 0 & | & 60 \\ \hdashline -8000 & -6000 & 0 & 24{,}000 & 0 & 0 & 1 & | & 720{,}000 \end{bmatrix}$$

$$\frac{1}{2}R_3 \rightarrow R_3$$

$$\sim \begin{bmatrix} \frac{1}{2} & \frac{1}{2} & 1 & 1 & 0 & 0 & 0 & | & 30 \\ 2 & 2 & 0 & -8 & 1 & 0 & 0 & | & 80 \\ ① & \frac{1}{2} & 0 & -2 & 0 & \frac{1}{2} & 0 & | & 30 \\ \hdashline -8000 & -6000 & 0 & 24{,}000 & 0 & 0 & 1 & | & 720{,}000 \end{bmatrix}$$

$$R_1 + \left(-\frac{1}{2}\right)R_3 \rightarrow R_1, \quad R_2 + (-2)R_3 \rightarrow R_2, \quad R_4 + 8000R_3 \rightarrow R_4$$

$$\sim \begin{bmatrix} 0 & \frac{1}{4} & 1 & 2 & 0 & -\frac{1}{4} & 0 & | & 15 \\ 0 & ① & 0 & -4 & 1 & -1 & 0 & | & 20 \\ 1 & \frac{1}{2} & 0 & -2 & 0 & \frac{1}{2} & 0 & | & 30 \\ \hdashline 0 & -2000 & 0 & 8000 & 0 & 4000 & 1 & | & 960{,}000 \end{bmatrix}$$

$$R_1 + \left(-\frac{1}{4}\right)R_2 \rightarrow R_1, \quad R_3 + \left(-\frac{1}{2}\right)R_2 \rightarrow R_3, \quad R_4 + 2000R_2 \rightarrow R_4$$

$$\begin{array}{ccccccc} x_1 & x_2 & x_3 & s_1 & s_2 & s_3 & P \end{array}$$

$$\sim \begin{bmatrix} 0 & 0 & 1 & 3 & -\frac{1}{4} & 0 & 0 & | & 10 \\ 0 & 1 & 0 & -4 & 1 & -1 & 0 & | & 20 \\ 1 & 0 & 0 & 0 & -\frac{1}{2} & 1 & 0 & | & 20 \\ \hdashline 0 & 0 & 0 & 0 & 2000 & 2000 & 1 & | & 1{,}000{,}000 \end{bmatrix}$$

Optimal solution: maximum profit is $1,000,000 when x_1 = 20 colonial houses, x_2 = 20 split-level houses, and x_3 = 10 ranch-style houses are built.

39. Let x_1 = the number of boxes of Assortment I,
x_2 = the number of boxes of Assortment II,
and x_3 = the number of boxes of Assortment III.
The profit per box of Assortment I is:
$$9.40 - [4(0.20) + 4(0.25) + 12(0.30)] = \$4.00$$
The profit per box of Assortment II is:
$$7.60 - [12(0.20) + 4(0.25) + 4(0.30)] = \$3.00$$
The profit per box of Assortment III is:
$$11.00 - [8(0.20) + 8(0.25) + 8(0.30)] = \$5.00$$

The mathematical model for this problem is:

Maximize $P = 4x_1 + 3x_2 + 5x_3$

Subject to:
$$4x_1 + 12x_2 + 8x_3 \leq 4800$$
$$4x_1 + 4x_2 + 8x_3 \leq 4000$$
$$12x_1 + 4x_2 + 8x_3 \leq 5600$$
$$x_1,\ x_2,\ x_3 \geq 0$$

We introduce slack variables to obtain the initial form:

$$4x_1 + 12x_2 + 8x_3 + s_1 \qquad\qquad = 4800$$
$$4x_1 + 4x_2 + 8x_3 \qquad + s_2 \qquad = 4000$$
$$12x_1 + 4x_2 + 8x_3 \qquad\qquad + s_3 \qquad = 5600$$
$$-4x_1 - 3x_2 - 5x_3 \qquad\qquad\qquad + P = 0$$

$$
\begin{array}{ccccccc}
x_1 & x_2 & x_3 & s_1 & s_2 & s_3 & P
\end{array}
$$

$$
\left[
\begin{array}{ccccccc|c}
4 & 12 & 8 & 1 & 0 & 0 & 0 & 4800 \\
4 & 4 & ⑧ & 0 & 1 & 0 & 0 & 4000 \\
12 & 4 & 8 & 0 & 0 & 1 & 0 & 5600 \\
\hline
-4 & -3 & -5 & 0 & 0 & 0 & 1 & 0
\end{array}
\right]
\qquad
\begin{array}{l}
\dfrac{4800}{8} = 600 \\[2mm]
\dfrac{4000}{8} = 500 \\[2mm]
\dfrac{5600}{8} = 700
\end{array}
$$

$\dfrac{1}{8}R_2 \rightarrow R_2$

$$
\sim
\left[
\begin{array}{ccccccc|c}
4 & 12 & 8 & 1 & 0 & 0 & 0 & 4800 \\
\frac{1}{2} & \frac{1}{2} & ① & 0 & \frac{1}{8} & 0 & 0 & 500 \\
12 & 4 & 8 & 0 & 0 & 1 & 0 & 5600 \\
\hline
-4 & -3 & -5 & 0 & 0 & 0 & 1 & 0
\end{array}
\right]
\sim
\left[
\begin{array}{ccccccc|c}
0 & 8 & 0 & 1 & -1 & 0 & 0 & 800 \\
\frac{1}{2} & \frac{1}{2} & 1 & 0 & \frac{1}{8} & 0 & 0 & 500 \\
⑧ & 0 & 0 & 0 & -1 & 1 & 0 & 1600 \\
\hline
-\frac{3}{2} & -\frac{1}{2} & 0 & 0 & \frac{5}{8} & 0 & 1 & 2500
\end{array}
\right]
$$

$R_1 + (-8)R_2 \rightarrow R_1,\ R_3 + (-8)R_2 \rightarrow R_3,$ $\qquad \dfrac{1}{8}R_3 \rightarrow R_3$

$R_4 + 5R_2 \rightarrow R_4$

$$
\sim
\left[
\begin{array}{ccccccc|c}
0 & 8 & 0 & 1 & -1 & 0 & 0 & 800 \\
\frac{1}{2} & \frac{1}{2} & 1 & 0 & \frac{1}{8} & 0 & 0 & 500 \\
① & 0 & 0 & 0 & -\frac{1}{8} & \frac{1}{8} & 0 & 200 \\
\hline
-\frac{3}{2} & -\frac{1}{2} & 0 & 0 & \frac{5}{8} & 0 & 1 & 2500
\end{array}
\right]
\sim
\left[
\begin{array}{ccccccc|c}
0 & ⑧ & 0 & 1 & -1 & 0 & 0 & 800 \\
0 & \frac{1}{2} & 1 & 0 & \frac{3}{16} & -\frac{1}{16} & 0 & 400 \\
1 & 0 & 0 & 0 & -\frac{1}{8} & \frac{1}{8} & 0 & 200 \\
\hline
0 & -\frac{1}{2} & 0 & 0 & \frac{7}{16} & \frac{3}{16} & 1 & 2800
\end{array}
\right]
$$

$R_2 + \left(-\dfrac{1}{2}\right)R_3 \rightarrow R_2,\ R_4 + \dfrac{3}{2}R_3 \rightarrow R_4$ $\qquad \dfrac{1}{8}R_1 \rightarrow R_1$

$$
\begin{array}{ccccccc}
 & & & & & & \\
x_1 & x_2 & x_3 & s_1 & s_2 & s_3 & P
\end{array}
$$

$$
\sim
\left[
\begin{array}{ccccccc|c}
0 & ① & 0 & \frac{1}{8} & -\frac{1}{8} & 0 & 0 & 100 \\
0 & \frac{1}{2} & 1 & 0 & \frac{3}{16} & -\frac{1}{16} & 0 & 400 \\
1 & 0 & 0 & 0 & -\frac{1}{8} & \frac{1}{8} & 0 & 200 \\
\hline
0 & -\frac{1}{2} & 0 & 0 & \frac{7}{16} & \frac{3}{16} & 1 & 2800
\end{array}
\right]
\sim
\left[
\begin{array}{ccccccc|c}
0 & 1 & 0 & \frac{1}{8} & -\frac{1}{8} & 0 & 0 & 100 \\
0 & 0 & 1 & -\frac{1}{16} & \frac{1}{4} & -\frac{1}{16} & 0 & 350 \\
1 & 0 & 0 & 0 & -\frac{1}{8} & \frac{1}{8} & 0 & 200 \\
\hline
0 & 0 & 0 & \frac{1}{16} & \frac{3}{8} & \frac{3}{16} & 1 & 2850
\end{array}
\right]
$$

$R_2 + \left(-\dfrac{1}{2}\right)R_1 \rightarrow R_2,\ R_4 + \dfrac{1}{2}R_1 \rightarrow R_4$

Optimal solution: maximum profit is \$2850 when 200 boxes of Assortment I, 100 boxes of Assortment II, and 350 boxes of Assortment III are made.

41. Let x_1 = the number of grams of food A,
x_2 = the number of grams of food B,
and x_3 = the number of grams of food C.
The mathematical model for this problem is: Maximize $P = 3x_1 + 3x_2 + 5x_3$
Subject to: $x_1 + 3x_2 + 2x_3 \leq 30$
$2x_1 + x_2 + x_3 \leq 24$
$x_1, x_2, x_3 \geq 0$

We introduce slack variables s_1 and s_2 to obtain the initial form:
$$x_1 + 3x_2 + 2x_3 + s_1 = 30$$
$$2x_1 + x_2 + x_3 + s_2 = 24$$
$$-3x_1 - 3x_2 - 5x_3 + P = 0$$

The simplex tableau for this problem is:

$$
\begin{array}{cccccc}
x_1 & x_2 & x_3 & s_1 & s_2 & P
\end{array}
$$

$$
\left[
\begin{array}{cccccc|c}
1 & 3 & ② & 1 & 0 & 0 & 30 \\
2 & 1 & 1 & 0 & 1 & 0 & 24 \\
\hline
-3 & -3 & -5 & 0 & 0 & 1 & 0
\end{array}
\right]
\quad
\begin{array}{l}
\frac{30}{2} = 15 \\[6pt]
\frac{24}{1} = 24
\end{array}
$$

$\frac{1}{2}R_1 \rightarrow R_1$

$$
\sim
\left[
\begin{array}{cccccc|c}
\frac{1}{2} & \frac{3}{2} & ① & \frac{1}{2} & 0 & 0 & 15 \\
2 & 1 & 1 & 0 & 1 & 0 & 24 \\
\hline
-3 & -3 & -5 & 0 & 0 & 1 & 0
\end{array}
\right]
\sim
\left[
\begin{array}{cccccc|c}
\frac{1}{2} & \frac{3}{2} & 1 & \frac{1}{2} & 0 & 0 & 15 \\
③\!\!/\!\!② & -\frac{1}{2} & 0 & -\frac{1}{2} & 1 & 0 & 9 \\
\hline
-\frac{1}{2} & \frac{9}{2} & 0 & \frac{5}{2} & 0 & 1 & 75
\end{array}
\right]
\begin{array}{l}
\frac{15}{1/2} = 30 \\[6pt]
\frac{9}{3/2} = 6
\end{array}
$$

$R_2 + (-1)R_1 \rightarrow R_2$ and $R_3 + 5R_1 \rightarrow R_3$ $\frac{2}{3}R_2 \rightarrow R_2$

$$
\begin{array}{cccccc}
 & & & & & \\
x_1 & x_2 & x_3 & s_1 & s_2 & P
\end{array}
$$

$$
\sim
\left[
\begin{array}{cccccc|c}
\frac{1}{2} & \frac{3}{2} & 1 & \frac{1}{2} & 0 & 0 & 15 \\
① & -\frac{1}{3} & 0 & -\frac{1}{3} & \frac{2}{3} & 0 & 6 \\
\hline
-\frac{1}{2} & \frac{9}{2} & 0 & \frac{5}{2} & 0 & 1 & 75
\end{array}
\right]
\sim
\left[
\begin{array}{cccccc|c}
0 & \frac{5}{3} & 1 & \frac{2}{3} & -\frac{1}{3} & 0 & 12 \\
1 & -\frac{1}{3} & 0 & -\frac{1}{3} & \frac{2}{3} & 0 & 6 \\
\hline
0 & \frac{13}{3} & 0 & \frac{7}{3} & \frac{1}{3} & 1 & 78
\end{array}
\right]
$$

$R_1 + \left(-\frac{1}{2}\right)R_2 \rightarrow R_1$ and $R_3 + \frac{1}{2}R_2 \rightarrow R_3$

Optimal solution: the maximum amount of protein is 78 units when $x_1 = 6$ grams of food A, $x_2 = 0$ grams of food B, and $x_3 = 12$ grams of food C.

43. Let x_1 = the number of undergraduate students,
x_2 = the number of graduate students,
and x_3 = the number of faculty members.

The mathematical model for this problem is:
Maximize $P = 18x_1 + 25x_2 + 30x_3$
Subject to: $x_1 + x_2 + x_3 \leq 20$
$60x_1 + 90x_2 + 120x_3 \leq 1620$
$x_1, x_2, x_3 \geq 0$

We introduce slack variables s_1 and s_2 to obtain the initial form:

$$x_1 + x_2 + x_3 + s_1 \qquad\quad = 20$$
$$60x_1 + 90x_2 + 120x_3 \qquad + s_2 \qquad = 1620$$
$$-18x_1 - 25x_2 - 30x_3 \qquad\qquad + P = 0$$

The simplex tableau for this problem is:

$$
\begin{array}{cccccc}
x_1 & x_2 & x_3 & s_1 & s_2 & P
\end{array}
$$

$$
\left[\begin{array}{cccccc|c}
1 & 1 & 1 & 1 & 0 & 0 & 20 \\
60 & 90 & \boxed{120} & 0 & 1 & 0 & 1620 \\
\hdashline
-18 & -25 & -30 & 0 & 0 & 1 & 0
\end{array}\right]
\begin{array}{l}
\frac{20}{1} = 20 \\[4pt]
\frac{1620}{120} = \frac{27}{2}
\end{array}
\sim
\left[\begin{array}{cccccc|c}
1 & 1 & 1 & 1 & 0 & 0 & 20 \\
\frac{1}{2} & \frac{3}{4} & \boxed{1} & 0 & \frac{1}{120} & 0 & \frac{27}{2} \\
\hdashline
-18 & -25 & -30 & 0 & 0 & 1 & 0
\end{array}\right]
$$

$$\frac{1}{120}R_2 \to R_2 \qquad\qquad\qquad\qquad R_1 + (-1)R_2 \to R_1 \text{ and } R_3 + 30R_2 \to R_3$$

$$
\sim
\left[\begin{array}{cccccc|c}
\boxed{\frac{1}{2}} & \frac{1}{4} & 0 & 1 & -\frac{1}{120} & 0 & \frac{13}{2} \\
\frac{1}{2} & \frac{3}{4} & 1 & 0 & \frac{1}{120} & 0 & \frac{27}{2} \\
\hdashline
-3 & -\frac{5}{2} & 0 & 0 & \frac{1}{4} & 1 & 405
\end{array}\right]
\begin{array}{l}
\frac{13/2}{1/2} = 13 \\[4pt]
\frac{27/2}{1/2} = 27
\end{array}
\sim
\left[\begin{array}{cccccc|c}
\boxed{1} & \frac{1}{2} & 0 & 2 & -\frac{1}{60} & 0 & 13 \\
\frac{1}{2} & \frac{3}{4} & 1 & 0 & \frac{1}{120} & 0 & \frac{27}{2} \\
\hdashline
-3 & -\frac{5}{2} & 0 & 0 & \frac{1}{4} & 1 & 405
\end{array}\right]
$$

$$2R_1 \to R_1 \qquad\qquad\qquad\qquad R_2 + \left(-\frac{1}{2}\right)R_1 \to R_2 \text{ and } R_3 + 3R_1 \to R_3$$

$$
\sim
\left[\begin{array}{cccccc|c}
1 & \frac{1}{2} & 0 & 2 & -\frac{1}{60} & 0 & 13 \\
0 & \boxed{\frac{1}{2}} & 1 & -1 & \frac{1}{60} & 0 & 7 \\
\hdashline
0 & -1 & 0 & 6 & \frac{1}{5} & 1 & 444
\end{array}\right]
\begin{array}{l}
\frac{13}{1/2} = 26 \\[4pt]
\frac{7}{1/2} = 14
\end{array}
\sim
\left[\begin{array}{cccccc|c}
1 & \frac{1}{2} & 0 & 2 & -\frac{1}{60} & 0 & 13 \\
0 & \boxed{1} & 2 & -2 & \frac{1}{30} & 0 & 14 \\
\hdashline
0 & -1 & 0 & 6 & \frac{1}{5} & 1 & 444
\end{array}\right]
$$

$$2R_2 \to R_2 \qquad\qquad\qquad\qquad R_1 + \left(-\frac{1}{2}\right)R_2 \to R_1 \text{ and } R_3 + R_2 \to R_3$$

$$
\begin{array}{cccccc}
x_1 & x_2 & x_3 & s_1 & s_2 & P
\end{array}
$$

$$
\sim
\left[\begin{array}{cccccc|c}
1 & 0 & -1 & 3 & -\frac{1}{30} & 0 & 6 \\
0 & 1 & 2 & -2 & \frac{1}{30} & 0 & 14 \\
\hdashline
0 & 0 & 2 & 4 & \frac{7}{30} & 1 & 458
\end{array}\right]
$$

Optimal solution: the maximum number of interviews is 458 when $x_1 = 6$ undergraduate students, $x_2 = 14$ graduate students, and $x_3 = 0$ faculty members are hired.

EXERCISE 6-5   CHAPTER REVIEW

1. $2x_1 + x_2 \leq 8$
$3x_1 + 9x_2 \leq 27$
$\quad x_1, \ x_2 \geq 0$

The graphs of the inequalities are shown at the right. The solution region is shaded; it is *bounded*.

The corner points are:
(0, 0), (0, 3), (3, 2), (4, 0)

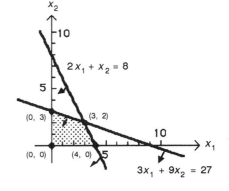

2. $3x_1 + x_2 \geq 9$
$2x_1 + 4x_2 \geq 16$
$x_1, \; x_2 \geq 0$

The graphs of the inequalities are shown at the right. The solution region is shaded; it is *unbounded*.

The corner points are:
$(0, 9)$, $(2, 3)$, $(8, 0)$

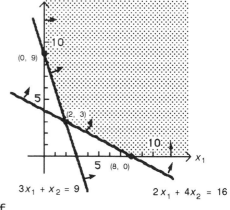

3. The feasible region is the solution set of the given inequalities, and is indicated by the shaded region in the graph at the right.

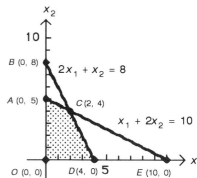

The corner points are $(0, 0)$, $(0, 5)$, $(2, 4)$, and $(4, 0)$.

The value of P at each corner point is:

Corner Point	$P = 6x_1 + 2x_2$
$(0, 0)$	$P = 6(0) + 2(0) = 0$
$(0, 5)$	$P = 6(0) + 2(5) = 10$
$(2, 4)$	$P = 6(2) + 2(4) = 20$
$(4, 0)$	$P = 6(4) + 2(0) = 24$

Thus, the maximum occurs at $x_1 = 4$, $x_2 = 0$, and the maximum value is $P = 24$.

4. We introduce the slack variables s_1 and s_2 to obtain the system of equations:
$$2x_1 + x_2 + s_1 \qquad\;\; = 8$$
$$x_1 + 2x_2 \qquad + s_2 = 10$$

5. There are 2 basic and 2 nonbasic variables.

6. The basic solutions are given in the following table.

x_1	x_2	s_1	s_2	Intersection Point	Feasible?
0	0	8	10	O	Yes
0	8	0	−6	B	No
0	5	3	0	A	Yes
4	0	0	6	D	Yes
10	0	−12	0	E	No
2	4	0	0	C	Yes

7. The simplex tableau for Problem 3 is:

$$
\begin{array}{c}
\text{Enter} \\
\begin{array}{cccccc}
 & x_1 & x_2 & s_1 & s_2 & P \\
\text{Exit } s_1 & \left[\begin{array}{ccccc|c} ② & 1 & 1 & 0 & 0 & 8 \\ \end{array}\right. & & & & & \\
\end{array}
\end{array}
$$

$$
\text{Exit } s_1 \quad \begin{array}{c} s_1 \\ s_2 \\ P \end{array}
\left[\begin{array}{ccccc|c}
② & 1 & 1 & 0 & 0 & 8 \\
1 & 2 & 0 & 1 & 0 & 10 \\
\hline
-6 & -2 & 0 & 0 & 1 & 0
\end{array}\right]
\begin{array}{l}
\frac{8}{2} = 4 \\[4pt]
\frac{10}{1} = 10
\end{array}
$$

8.

$$
\sim \left[\begin{array}{ccccc|c}
② & 1 & 1 & 0 & 0 & 8 \\
1 & 2 & 0 & 1 & 0 & 10 \\
\hline
-6 & -2 & 0 & 0 & 1 & 0
\end{array}\right]
\sim
\left[\begin{array}{ccccc|c}
① & \frac{1}{2} & \frac{1}{2} & 0 & 0 & 4 \\
1 & 2 & 0 & 1 & 0 & 10 \\
\hline
-6 & -2 & 0 & 0 & 1 & 0
\end{array}\right]
$$

$$\tfrac{1}{2}R_1 \rightarrow R_1 \qquad\qquad\qquad R_2 + (-1)R_1 \rightarrow R_2 \text{ and } R_3 + 6R_1 \rightarrow R_3$$

$$
\begin{array}{ccccc}
x_1 & x_2 & s_1 & s_2 & P
\end{array}
$$

$$
\sim \left[\begin{array}{ccccc|c}
1 & \frac{1}{2} & \frac{1}{2} & 0 & 0 & 4 \\
0 & \frac{3}{2} & -\frac{1}{2} & 1 & 0 & 6 \\
\hline
0 & 1 & 3 & 0 & 1 & 24
\end{array}\right]
$$

Optimal solution: max $P = 24$ at $x_1 = 4$, $x_2 = 0$.

9.

$$
\begin{array}{c}
\text{Enter} \\
\begin{array}{ccccccc}
x_1 & x_2 & x_3 & s_1 & s_2 & s_3 & P
\end{array}
\end{array}
$$

$$
\begin{array}{c}
x_2 \\
s_2 \\
\text{Exit } s_3 \\
P
\end{array}
\left[\begin{array}{ccccccc|c}
2 & 1 & 3 & -1 & 0 & 0 & 0 & 20 \\
3 & 0 & 4 & 1 & 1 & 0 & 0 & 30 \\
② & 0 & 5 & 2 & 0 & 1 & 0 & 10 \\
\hline
-8 & 0 & -5 & 3 & 0 & 0 & 1 & 50
\end{array}\right]
\begin{array}{l}
\frac{20}{2} = 10 \\[4pt]
\frac{30}{3} = 10 \\[4pt]
\frac{10}{2} = 5
\end{array}
$$

The basic variables are x_2, s_2, s_3, and P, and the nonbasic variables are x_1, x_3, and s_1.

The first column is the pivot column and the third row is the pivot row. The pivot element is circled.

$$
\sim \left[\begin{array}{ccccccc|c}
2 & 1 & 3 & -1 & 0 & 0 & 0 & 20 \\
3 & 0 & 4 & 1 & 1 & 0 & 0 & 30 \\
① & 0 & \frac{5}{2} & 1 & 0 & \frac{1}{2} & 0 & 5 \\
\hline
-8 & 0 & -5 & 3 & 0 & 0 & 1 & 50
\end{array}\right]
$$

$$R_1 + (-2)R_3 \rightarrow R_1, \; R_2 + (-3)R_3 \rightarrow R_2, \; R_4 + 8R_3 \rightarrow R_4$$

$$
\begin{array}{ccccccc}
x_1 & x_2 & x_3 & s_1 & s_2 & s_3 & P
\end{array}
$$

$$
\sim \begin{array}{c}
x_2 \\
s_2 \\
x_1 \\
P
\end{array}
\left[\begin{array}{ccccccc|c}
0 & 1 & -2 & -3 & 0 & -1 & 0 & 10 \\
0 & 0 & -\frac{7}{2} & -2 & 1 & -\frac{3}{2} & 0 & 15 \\
1 & 0 & \frac{5}{2} & 1 & 0 & \frac{1}{2} & 0 & 5 \\
\hline
0 & 0 & 15 & 11 & 0 & 4 & 1 & 90
\end{array}\right]
$$

10. (A) The basic feasible solution is: $x_1 = 0$, $x_2 = 2$, $s_1 = 0$, $s_2 = 5$, $P = 12$. Additional pivoting is required because the last row contains a negative indicator.

  (B) The basic feasible solution is: $x_1 = 0$, $x_2 = 0$, $s_1 = 0$, $s_2 = 7$, $P = 22$. There is no optimal solution because there are no positive elements above the dashed line in the pivot column, column 1.

  (C) The basic feasible solution is: $x_1 = 6$, $x_2 = 0$, $s_1 = 15$, $s_2 = 0$, $P = 10$. This is the optimal solution.

11. The feasible region is the solution set of the given inequalities and is indicated by the shaded region in the graph at the right.

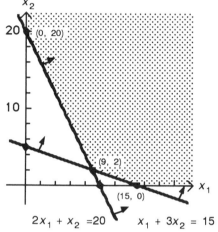

The corner points are $(0, 20)$, $(9, 2)$, and $(15, 0)$.

The value of C at each corner point is:

Corner Point	$C = 5x_1 + 2x_2$
$(0, 20)$	$C = 5(0) + 2(20) = 40$
$(9, 2)$	$C = 5(9) + 2(2) = 49$
$(15, 0)$	$C = 5(15) + 2(0) = 75$

The minimum occurs at $x_1 = 0$, $x_2 = 20$, and the minimum value is $C = 40$.

12. The feasible region is the solution set of the given inequalities and is indicated by the shading in the graph at the right.

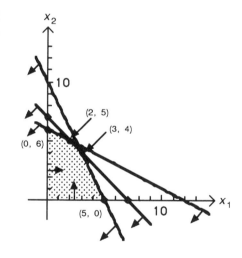

The corner points are $(0, 0)$, $(0, 6)$, $(2, 5)$, $(3, 4)$, and $(5, 0)$.

The value of P at each corner point is:

Corner Point	$P = 3x_1 + 4x_2$
$(0, 0)$	$P = 3(0) + 4(0) = 0$
$(0, 6)$	$P = 3(0) + 4(6) = 24$
$(2, 5)$	$P = 3(2) + 4(5) = 26$
$(3, 4)$	$P = 3(3) + 4(4) = 25$
$(5, 0)$	$P = 3(5) + 4(0) = 15$

Thus, the maximum occurs at $x_1 = 2$, $x_2 = 5$, and the maximum value is $P = 26$.

13. We simplify the inequalities and introduce the slack variables s_1, s_2, and s_3 to obtain the equivalent form:

$$
\begin{aligned}
x_1 + 2x_2 + s_1 &= 12 \\
x_1 + x_2 + s_2 &= 7 \\
2x_1 + x_2 + s_3 &= 10 \\
-3x_1 - 4x_2 + P &= 0
\end{aligned}
$$

The simplex tableau for this problem is:

$$
\begin{array}{c}
\text{Enter}\\
\begin{array}{ccccccc}
& x_1 & x_2 & s_1 & s_2 & s_3 & P
\end{array}\\
\begin{array}{c}
\text{Exit } s_1\\ s_2\\ s_3\\[2pt] P
\end{array}
\left[
\begin{array}{cccccc|c}
1 & \boxed{2} & 1 & 0 & 0 & 0 & 12\\
1 & 1 & 0 & 1 & 0 & 0 & 7\\
2 & 1 & 0 & 0 & 1 & 0 & 10\\
\hline
-3 & -4 & 0 & 0 & 0 & 1 & 0
\end{array}
\right]
\begin{array}{l}
\frac{12}{2} = 6\\[4pt]
\frac{7}{1} = 7\\[4pt]
\frac{10}{1} = 10
\end{array}
\end{array}
$$

$$\tfrac{1}{2}R_1 \rightarrow R_1$$

$$
\sim
\left[
\begin{array}{cccccc|c}
\frac{1}{2} & \boxed{1} & \frac{1}{2} & 0 & 0 & 0 & 6\\
1 & 1 & 0 & 1 & 0 & 0 & 7\\
2 & 1 & 0 & 0 & 1 & 0 & 10\\
\hline
-3 & -4 & 0 & 0 & 0 & 1 & 0
\end{array}
\right]
\sim
\left[
\begin{array}{cccccc|c}
\frac{1}{2} & 1 & \frac{1}{2} & 0 & 0 & 0 & 6\\
\boxed{\frac{1}{2}} & 0 & -\frac{1}{2} & 1 & 0 & 0 & 1\\
\frac{3}{2} & 0 & -\frac{1}{2} & 0 & 1 & 0 & 4\\
\hline
-1 & 0 & 2 & 0 & 0 & 1 & 24
\end{array}
\right]
\begin{array}{l}
\frac{6}{1/2} = 12\\[4pt]
\frac{1}{1/2} = 2\\[4pt]
\frac{4}{3/2} \approx 2.67
\end{array}
$$

$$R_2 + (-1)R_1 \rightarrow R_2,\; R_3 + (-1)R_1 \rightarrow R_3, \qquad 2R_2 \rightarrow R_2$$

$$\text{and } R_4 + 4R_1 \rightarrow R_4$$

$$
\begin{array}{c}
\begin{array}{ccccccc}
& & & & & & \\
\end{array}\\
\sim
\left[
\begin{array}{cccccc|c}
\frac{1}{2} & 1 & \frac{1}{2} & 0 & 0 & 0 & 6\\
\boxed{1} & 0 & -1 & 2 & 0 & 0 & 2\\
\frac{3}{2} & 0 & -\frac{1}{2} & 0 & 1 & 0 & 4\\
\hline
-1 & 0 & 2 & 0 & 0 & 1 & 24
\end{array}
\right]
\end{array}
$$

$$
\begin{array}{c}
\begin{array}{cccccc}
x_1 & x_2 & s_1 & s_2 & s_3 & P
\end{array}\\
\sim
\left[
\begin{array}{cccccc|c}
0 & 1 & 1 & -1 & 0 & 0 & 5\\
1 & 0 & -1 & 2 & 0 & 0 & 2\\
0 & 0 & 1 & -3 & 1 & 0 & 1\\
\hline
0 & 0 & 1 & 2 & 0 & 1 & 26
\end{array}
\right]
\end{array}
$$

$$R_1 + \left(-\tfrac{1}{2}\right)R_2 \rightarrow R_1,\; R_3 + \left(-\tfrac{3}{2}\right)R_2 \rightarrow R_3,$$

$$\text{and } R_4 + R_2 \rightarrow R_4$$

Optimal solution: max $P = 26$ at $x_1 = 2,\; x_2 = 5.$

14. The feasible region is the solution set of the given inequalities and is indicated by the shaded region in the graph at the right.

The corner points are $(0, 10)$, $(5, 5)$, and $(9, 3)$.

The value of C at each corner point is:

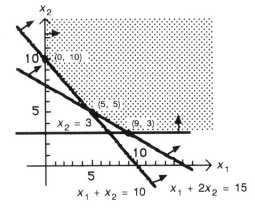

Corner Point	$C = 3x_1 + 8x_2$
$(0, 10)$	$C = 3(0) + 8(10) = 80$
$(5, 5)$	$C = 3(5) + 8(5) = 55$
$(9, 3)$	$C = 3(9) + 8(3) = 51$

Thus, the minimum occurs at $x_1 = 9$, $x_2 = 3$, and the minimum value is $C = 51$.

15. Introduce slack variables s_1 and s_2 to obtain the equivalent form:

$$x_1 - x_2 - 2x_3 + s_1 \qquad\qquad = 3$$
$$2x_1 + 2x_2 - 5x_3 \qquad + s_2 \qquad = 10$$
$$-5x_1 - 3x_2 + 3x_3 \qquad\qquad + P = 0$$

The simplex tableau for this problem is:

Enter

$$
\begin{array}{c}
\text{Exit } s_1 \\
s_2 \\
P
\end{array}
\left[
\begin{array}{cccccc|c}
x_1 & x_2 & x_3 & s_1 & s_2 & P & \\
① & -1 & -2 & 1 & 0 & 0 & 3 \\
2 & 2 & -5 & 0 & 1 & 0 & 10 \\
\hline
-5 & -3 & 3 & 0 & 0 & 1 & 0
\end{array}
\right]
\begin{array}{l}
\frac{3}{1} = 3 \\[4pt]
\frac{10}{2} = 5
\end{array}
\sim
\left[
\begin{array}{cccccc|c}
1 & -1 & -2 & 1 & 0 & 0 & 3 \\
0 & ④ & -1 & -2 & 1 & 0 & 4 \\
\hline
0 & -8 & -7 & 5 & 0 & 1 & 15
\end{array}
\right]
$$

$$R_2 + (-2)R_1 \rightarrow R_2 \text{ and } R_3 + 5R_1 \rightarrow R_3 \qquad\qquad \tfrac{1}{4}R_2 \rightarrow R_2$$

$$
\sim
\left[
\begin{array}{cccccc|c}
1 & -1 & -2 & 1 & 0 & 0 & 3 \\
0 & ① & -\frac{1}{4} & -\frac{1}{2} & \frac{1}{4} & 0 & 1 \\
\hline
0 & -8 & -7 & 5 & 0 & 1 & 15
\end{array}
\right]
\sim
\left[
\begin{array}{cccccc|c}
x_1 & x_2 & x_3 & s_1 & s_2 & P & \\
1 & 0 & -\frac{9}{4} & \frac{1}{2} & \frac{1}{4} & 0 & 4 \\
0 & 1 & -\frac{1}{4} & -\frac{1}{2} & \frac{1}{4} & 0 & 1 \\
\hline
0 & 0 & -9 & 1 & 2 & 1 & 23
\end{array}
\right]
$$

$$R_1 + R_2 \rightarrow R_1 \text{ and } R_3 + 8R_2 \rightarrow R_3$$

No optimal solution exists; the elements in the pivot column (the x_3 column) above the dashed line are negative.

16. Introduce slack variables s_1 and s_2 to obtain the equivalent form:

$$x_1 - x_2 - 2x_3 + s_1 \qquad\qquad = 3$$
$$x_1 + x_2 \qquad\qquad + s_2 \qquad = 5$$
$$-5x_1 - 3x_2 + 3x_3 \qquad\qquad + P = 0$$

The simplex tableau for this problem is:

Enter

$$
\begin{array}{c}
\text{Exit } s_1 \\
s_2 \\
P
\end{array}
\left[
\begin{array}{cccccc|c}
x_1 & x_2 & x_3 & s_1 & s_2 & P & \\
① & -1 & -2 & 1 & 0 & 0 & 3 \\
1 & 1 & 0 & 0 & 1 & 0 & 5 \\
\hline
-5 & -3 & 3 & 0 & 0 & 1 & 0
\end{array}
\right]
\begin{array}{l}
\frac{3}{1} = 3 \\[4pt]
\frac{5}{1} = 5
\end{array}
\sim
\left[
\begin{array}{cccccc|c}
1 & -1 & -2 & 1 & 0 & 0 & 3 \\
0 & ② & 2 & -1 & 1 & 0 & 2 \\
\hline
0 & -8 & -7 & 5 & 0 & 1 & 15
\end{array}
\right]
$$

$$R_2 + (-1)R_1 \rightarrow R_2 \text{ and } R_3 + 5R_1 \rightarrow R_3 \qquad\qquad \tfrac{1}{2}R_2 \rightarrow R_2$$

$$
\sim
\left[
\begin{array}{cccccc|c}
1 & -1 & -2 & 1 & 0 & 0 & 3 \\
0 & ① & 1 & -\frac{1}{2} & \frac{1}{2} & 0 & 1 \\
\hline
0 & -8 & -7 & 5 & 0 & 1 & 15
\end{array}
\right]
\sim
\left[
\begin{array}{cccccc|c}
x_1 & x_2 & x_3 & s_1 & s_2 & P & \\
1 & 0 & -1 & \frac{1}{2} & \frac{1}{2} & 0 & 4 \\
0 & 1 & 1 & -\frac{1}{2} & \frac{1}{2} & 0 & 1 \\
\hline
0 & 0 & 1 & 1 & 4 & 1 & 23
\end{array}
\right]
$$

$$R_1 + R_2 \rightarrow R_1 \text{ and } R_3 + 8R_2 \rightarrow R_3 \text{ Optimal solution: max } P = 23 \text{ at}$$
$$x_1 = 4,\ x_2 = 1,\ x_3 = 0.$$

7. Introduce slack variables s_1, s_2, s_3, and s_4 to obtain the equivalent form:

$$\begin{aligned}
x_1 + 2x_2 + s_1 \qquad\qquad\qquad &= 22 \\
2x_1 + x_2 \qquad + s_2 \qquad\qquad &= 20 \\
x_1 \qquad\qquad\qquad + s_3 \qquad &= 8 \\
x_2 \qquad\qquad\qquad\quad + s_4 \quad &= 10 \\
-2x_1 - 3x_2 \qquad\qquad\qquad\qquad + P &= 0
\end{aligned}$$

The simplex tableau for this problem is:

$$
\begin{array}{ccccccc}
x_1 & x_2 & s_1 & s_2 & s_3 & s_4 & P \\
\end{array}
$$

$$
\left[\begin{array}{ccccccc|c}
1 & 2 & 1 & 0 & 0 & 0 & 0 & 22 \\
2 & 1 & 0 & 1 & 0 & 0 & 0 & 20 \\
1 & 0 & 0 & 0 & 1 & 0 & 0 & 8 \\
0 & \textcircled{1} & 0 & 0 & 0 & 1 & 0 & 10 \\
\hline
-2 & -3 & 0 & 0 & 0 & 0 & 1 & 0
\end{array}\right]
\quad
\begin{array}{l}
\frac{22}{2} = 11 \\[4pt]
\frac{20}{1} = 20 \\[4pt]
\\
\frac{10}{1} = 10
\end{array}
$$

$$R_1 + (-2)R_4 \rightarrow R_1,\quad R_2 + (-1)R_4 \rightarrow R_2,\quad R_5 + 3R_4 \rightarrow R_5$$

$$
\sim
\left[\begin{array}{ccccccc|c}
\textcircled{1} & 0 & 1 & 0 & 0 & -2 & 0 & 2 \\
2 & 0 & 0 & 1 & 0 & -1 & 0 & 10 \\
1 & 0 & 0 & 0 & 1 & 0 & 0 & 8 \\
0 & 1 & 0 & 0 & 0 & 1 & 0 & 10 \\
\hline
-2 & 0 & 0 & 0 & 0 & 3 & 1 & 30
\end{array}\right]
\quad
\begin{array}{l}
\frac{2}{1} = 2 \\[4pt]
\frac{10}{2} = 5 \\[4pt]
\frac{8}{1} = 8 \\
\end{array}
$$

$$R_2 + (-2)R_1 \rightarrow R_2,\quad R_3 + (-1)R_1 \rightarrow R_3,\quad R_5 + 2R_1 \rightarrow R_5$$

$$
\sim
\left[\begin{array}{ccccccc|c}
1 & 0 & 1 & 0 & 0 & -2 & 0 & 2 \\
0 & 0 & -2 & 1 & 0 & \textcircled{3} & 0 & 6 \\
0 & 0 & -1 & 0 & 1 & 2 & 0 & 6 \\
0 & 1 & 0 & 0 & 0 & 1 & 0 & 10 \\
\hline
0 & 0 & 2 & 0 & 0 & -1 & 1 & 34
\end{array}\right]
\quad
\begin{array}{l}
\\
\frac{6}{3} = 2 \\[4pt]
\frac{6}{2} = 3 \\[4pt]
\frac{10}{1} = 10
\end{array}
$$

$$\tfrac{1}{3}R_2 \rightarrow R_2$$

$$
\sim
\left[\begin{array}{ccccccc|c}
1 & 0 & 1 & 0 & 0 & -2 & 0 & 2 \\
0 & 0 & -\frac{2}{3} & \frac{1}{3} & 0 & \textcircled{1} & 0 & 2 \\
0 & 0 & -1 & 0 & 1 & 2 & 0 & 6 \\
0 & 1 & 0 & 0 & 0 & 1 & 0 & 10 \\
\hline
0 & 0 & 2 & 0 & 0 & -1 & 1 & 34
\end{array}\right]
$$

$$R_1 + 2R_2 \rightarrow R_1,\quad R_3 + (-2)R_2 \rightarrow R_3,\quad R_4 + (-1)R_2 \rightarrow R_4,\quad R_5 + R_2 \rightarrow R_5$$

$$\sim \begin{bmatrix} 1 & 0 & -\frac{1}{3} & \frac{2}{3} & 0 & 0 & 0 & 6 \\ 0 & 0 & -\frac{2}{3} & \frac{1}{3} & 0 & 1 & 0 & 2 \\ 0 & 0 & \frac{1}{3} & -\frac{2}{3} & 1 & 0 & 0 & 2 \\ 0 & 1 & \frac{2}{3} & -\frac{1}{3} & 0 & 0 & 0 & 8 \\ \hdashline 0 & 0 & \frac{4}{3} & \frac{1}{3} & 0 & 0 & 1 & 36 \end{bmatrix}$$

with column headers x_1 x_2 s_1 s_2 s_3 s_4 P

Optimal solution:
       max $P = 36$ at $x_1 = 6$, $x_2 = 8$.

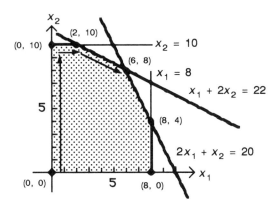

18. Let x_1 = the number of regular sails
and x_2 = the number of competition sails.
The mathematical model for this problem is: Maximize $P = 100x_1 + 200x_2$

Subject to: $2x_1 + 3x_2 \le 150$
$4x_1 + 9x_2 \le 360$
$x_1,\ x_2 \ge \quad 0$

19. Let x_1 = amount invested in oil stock
x_2 = amount invested in steel stock
and x_3 = amount invested in government bonds.

Maximize $P = 0.12x_1 + 0.09x_2 + 0.05x_3$
Subject to: $x_1 + x_2 + x_3 \le 150{,}000$
$x_1 + x_2 \le x_3$
$x_2 \ge 2x_1$

Putting the problem in the standard format, we have:

Maximize $P = 0.12x_1 + 0.09x_2 + 0.05x_3$
Subject to: $x_1 + x_2 + x_3 \le 150{,}000$
$x_1 + x_2 - x_3 \le 0$
$2x_1 - x_2 \quad\ \le 0$
$x_1,\ x_2,\ x_3 \ge 0$

20. Let x_1 = the number of grams of mix A
and x_2 = the number of grams of mix B.
The mathematical model for this problem is: Minimize $C = 0.02x_1 + 0.04x_2$

Subject to: $3x_1 + \quad 4x_2 \ge 300$
$2x_1 + \quad 5x_2 \ge 200$
$6x_1 + 10x_2 \ge 900$
$x_1,\ x_2 \ge \quad 0$

EXERCISE 7-1

Things to remember:

1. Let A be a set with finitely many elements. Then $n(A)$ denotes the number of elements in A.

2. ADDITION PRINCIPLE (for counting)
   For any two sets A and B,
   $$n(A \cup B) = n(A) + n(B) - n(A \cap B)$$
   If A and B are disjoint, i.e., if $A \cap B = \varnothing$, then
   $$n(A \cup B) = n(A) + n(B)$$

3. MULTIPLICATION PRINCIPLE (for counting)
   (a) If two operations O_1 and O_2 are performed in order, with N_1 possible outcomes for the first operation and N_2 possible outcomes for the second operation, then there are
   $$N_1 \cdot N_2$$
   possible combined outcomes of the first operation followed by the second.
   (b) In general, if n operations O_1, O_2, ..., O_n are performed in order with possible number of outcomes N_1, N_2, ..., N_n, respectively, then there are
   $$N_1 \cdot N_2 \cdot \ldots \cdot N_n$$
   possible combined outcomes of the operations performed in the given order.

---

1. $n(A) = 75 + 40 = 115$

3. $n(A \cup B) = 75 + 40 + 95 = 210$
$n[(A \cup B)'] = 90$
Thus, $n(U) = n(A \cup B) + n[(A \cup B)'] = 210 + 90 = 300$
[Note: $U = (A \cup B) \cup (A \cup B)'$ and $(A \cup B) \cap (A \cup B)' = \varnothing$]

5. $B' = A \cap B' \cup (A \cup B)'$ and $(A \cap B') \cap (A \cup B)' = \varnothing$.
Thus, $n(B') = n(A \cap B') + n[(A \cup B)']$
$= 75 + 90 = 165$

7. $n(A \cup B) = n(A) + n(B) - n(A \cap B)$
$\qquad = 115 + 135 - 40 = 210$

Note, also, that $A \cup B = (A \cap B') \cup (A \cap B) \cup (A' \cap B)$, and
$(A \cap B') \cap (A \cap B) = \varnothing$, $(A \cap B') \cap (A' \cap B) = \varnothing$,
$(A \cap B) \cap (A' \cap B) = \varnothing$.
So, $n(A \cup B) = n(A \cap B') + n(A \cap B) + n(A' \cap B)$
$\qquad\qquad = 75 + 40 + 95 = 210$

9. $n(A' \cap B) = 95$

11. $(A \cap B) \cup (A \cap B)' = U$ and $(A \cap B) \cap (A \cap B)' = \varnothing$
Thus, $n(U) = n(A \cap B) + n[(A \cap B)']$
or $n[(A \cap B)'] = n(U) - n(A \cap B) = 300 - 40 = 260$

13. (A) Tree Diagram

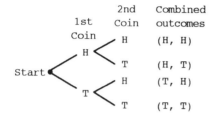

	2nd	Combined
1st	Coin	outcomes
Coin		

H → H (H, H)
H → T (H, T)
T → H (T, H)
T → T (T, T)

Thus, there are 4 ways.

(B) Multiplication Principle
O_1: 1st coin
N_1: 2 ways
O_2: 2nd coin
N_2: 2 ways

Thus, there are
$N_1 \cdot N_2 = 2 \cdot 2 = 4$ ways.

15. (A) Tree Diagram

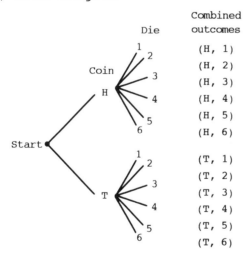

Combined outcomes

(H, 1)
(H, 2)
(H, 3)
(H, 4)
(H, 5)
(H, 6)

(T, 1)
(T, 2)
(T, 3)
(T, 4)
(T, 5)
(T, 6)

Thus, there are 12 combined outcomes.

(B) Multiplication Principle
O_1: Coin
N_1: 2 outcomes
O_2: Die
N_2: 6 outcomes

Thus, there are
$N_1 \cdot N_2 = 2 \cdot 6 = 12$ combined
outcomes

17. $A = (A \cap B') \cup (A \cap B)$ and $B = (A' \cap B) \cup (A \cap B)$
Thus, $\qquad n(A) = n(A \cap B') + n(A \cap B)$, so
$\qquad n(A \cap B') = n(A) - n(A \cap B) = 80 - 20 = 60$
$\qquad n(A \cap B) = 20$.
$\qquad\qquad n(B) = n(A' \cap B) + n(A \cap B)$ so
$\qquad n(A' \cap B) = n(B) - n(A \cap B) = 50 - 20 = 30$

Also, $A \cup B = (A' \cap B) \cup (A \cap B) \cup (A \cap B')$ and
$U = (A \cup B) \cup (A' \cap B')$ where $(A \cup B) \cap (A' \cap B') = \varnothing$

Thus, $n(U) = n(A \cup B) + n(A' \cap B')$ so

$$n(A' \cap B') = n(U) - n(A \cup B)$$
$$= 200 - (60 + 20 + 30) = 200 - 110 = 90$$

19. $n(A \cup B) = n(A) + n(B) - n(A \cap B)$

So $n(A \cap B) = n(A) + n(B) - n(A \cup B)$
$$= 25 + 55 - 60 = 20$$

$n(A \cap B') = n(A) - n(A \cap B) = 25 - 20 = 5$
$n(A' \cap B) = n(B) - n(A \cap B) = 55 - 20 = 35$
and $n(A' \cap B') = n(U) - n(A \cup B)$
$$= 100 - 60 = 40$$

21. $n(A \cap B) = 30$

$n(A \cap B') = n(A) - n(A \cap B) = 70 - 30 = 40$
$n(A' \cap B) = n(B) - n(A \cap B) = 90 - 30 = 60$
$n(A' \cap B') = n(U) - [n(A \cap B') + n(A \cap B) + n(A' \cap B)]$
$$= 200 - [40 + 30 + 60] = 200 - 130 = 70$$

Therefore,

	A	A'	Totals
B	30	60	90
B'	40	70	110
Totals	70	130	200

23. $n(A \cap B) = n(A) + n(B) - n(A \cup B) = 45 + 55 - 80 = 20$

$n(A \cap B') = n(A) - n(A \cap B) = 45 - 20 = 25$
$n(A' \cap B) = n(B) - n(A \cap B) = 55 - 20 = 35$
$n(A' \cap B') = n(U) - n(A \cup B) = 100 - 80 = 20$

Therefore,

	A	A'	Totals
B	20	35	55
B'	25	20	45
Totals	45	55	100

25. Using the Multiplication Principle:

O_1: Choose the color O_3: Choose the interior
N_1: 5 ways N_3: 4 ways

O_2: Choose the transmission O_4: Choose the engine
N_2: 3 ways N_4: 2 ways

Thus, there are

$N_1 \cdot N_2 \cdot N_3 \cdot N_4 = 5 \cdot 3 \cdot 4 \cdot 2 = 120$ different variations of this model car.

27. (A) Number of four-letter code words, no letter repeated.

O_1: Selecting the first letter
N_1: 6 ways

O_2: Selecting the second letter
N_2: 5 ways

O_3: Selecting the third letter
N_3: 4 ways

O_4: Selecting the fourth letter
N_4: 3 ways

Thus, there are

$N_1 \cdot N_2 \cdot N_3 \cdot N_4 = 6 \cdot 5 \cdot 4 \cdot 3 = 360$

possible code words. Note that this is the number of permutations of 6 objects taken 4 at a time:

$$P_{6,4} = \frac{6!}{(6-4)!} = \frac{6 \cdot 5 \cdot 4 \cdot 3 \cdot 2!}{2!} = 360$$

(B) Number of four-letter code words, allowing repetition.

O_1: Selecting the first letter
N_1: 6 ways

O_2: Selecting the second letter
N_2: 6 ways

O_3: Selecting the third letter
N_3: 6 ways

O_4: Selecting the fourth letter
N_4: 6 ways

Thus, there are

$N_1 \cdot N_2 \cdot N_3 \cdot N_4 = 6 \cdot 6 \cdot 6 \cdot 6 = 6^4 = 1296$

possible code words.

(C) Number of four-letter code words, adjacent letters different.

O_1: Selecting the first letter
N_1: 6 ways

O_2: Selecting the second letter
N_2: 5 ways

O_3: Selecting the third letter
N_3: 5 ways

O_4: Selecting the fourth letter
N_4: 5 ways

Thus, there are

$N_1 \cdot N_2 \cdot N_3 \cdot N_4 = 6 \cdot 5 \cdot 5 \cdot 5 = 6 \cdot 5^3 = 750$

possible code words.

29. (A) Number of five-digit combinations, no digit repeated.

O_1: Selecting the first digit
N_1: 10 ways

O_2: Selecting the second digit
N_2: 9 ways

O_3: Selecting the third digit
N_3: 8 ways

O_4: Selecting the fourth digit
N_4: 7 ways

O_5: Selecting the fifth digit
N_5: 6 ways

Thus, there are

$N_1 \cdot N_2 \cdot N_3 \cdot N_4 \cdot N_5 = 10 \cdot 9 \cdot 8 \cdot 7 \cdot 6 = 30,240$

possible combinations

(B) Number of five-digit combinations, allowing repetition.

O_1: Selecting the first digit
N_1: 10 ways

O_4: Selecting the fourth digit
N_4: 10 ways

O_2: Selecting the second digit
N_2: 10 ways

O_5: Selecting the fifth digit
N_5: 10 ways

O_3: Selecting the third digit
N_3: 10 ways

Thus, there are

$N_1 \cdot N_2 \cdot N_3 \cdot N_4 \cdot N_5 = 10 \cdot 10 \cdot 10 \cdot 10 \cdot 10 = 10^5 = 100,000$
possible combinations

31. (A) Letters and/or digits may be repeated.

O_1: Selecting the first letter
N_1: 26 ways

O_4: Selecting the first digit
N_4: 10 ways

O_2: Selecting the second letter
N_2: 26 ways

O_5: Selecting the second digit
N_5: 10 ways

O_3: Selecting the third letter
N_3: 26 ways

O_6: Selecting the third digit
N_6: 10 ways

Thus, there are

$N_1 \cdot N_2 \cdot N_3 \cdot N_4 \cdot N_5 \cdot N_6 = 26 \cdot 26 \cdot 26 \cdot 10 \cdot 10 \cdot 10 = 17,576,000$
different license plates.

(B) No repeated letters and no repeated digits are allowed.

O_1: Select the three letters, no letter repeated
N_1: $26 \cdot 25 \cdot 24 = 15,600$ ways

O_2: Select the three numbers, no number repeated
N_2: $10 \cdot 9 \cdot 8 = 720$ ways

Thus, there are

$N_1 \cdot N_2 = 15,600 \cdot 720 = 11,232,000$
different license plates with no letter or digit repeated.

33. O_1: Select the left-hand glove
N_1: 12 ways

O_2: Select the right-hand glove, different brand from the left-hand glove
N_2: 11 ways

Thus, there are

$N_1 \cdot N_2 = 12 \cdot 11 = 132$
pairs of gloves that do not match.

35. Let T = the people who play tennis, and
G = the people who play golf.

Then $n(T) = 32$, $n(G) = 37$, $n(T \cap G) = 8$ and $n(U) = 75$.

Thus, $n(T \cup G) = n(T) + n(G) - n(T \cap G) = 32 + 37 - 8 = 61$

The set of people who play neither tennis nor golf is represented by $T' \cap G'$. Since $U = (T \cup G) \cup (T' \cap G')$ and $(T \cup G) \cap (T' \cap G') = \varnothing$, it follows that $n(T' \cap G') = n(U) - n(T \cup G) = 75 - 61 = 14$.

There are 14 people who play neither tennis nor golf.

37. Let F = the people who speak French, and
G = the people who speak German.

Then $n(F) = 42$, $n(G) = 55$, $n(F' \cap G') = 17$ and $n(U) = 100$. Since $U = (F \cup G) \cup (F' \cap G')$ and $(F \cup G) \cap (F' \cap G') = \varnothing$, it follows that
$$n(F \cup G) = n(U) - n(F' \cap G') = 100 - 17 = 83$$
Now $n(F \cup G) = n(F) + n(G) - n(F \cap G)$, so
$$\begin{aligned} n(F \cap G) &= n(F) + n(G) - n(F \cup G) \\ &= 42 + 55 - 83 = 14 \end{aligned}$$
There are 14 people who speak both French and German.

39. (A)

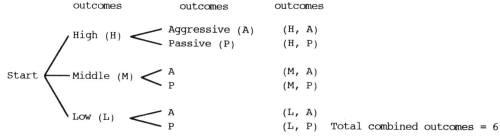

(B) Operation 1: Test scores can be classified into three groups, high, middle, or low:
$$N_1 = 3$$

Operation 2: Interviews can be classified into two groups, aggressive or passive:
$$N_2 = 2$$

The total possible combined classifications is:
$$N_1 \cdot N_2 = 3 \cdot 2 = 6$$

41. O_1: Travel from home to airport and back $\qquad$ O_3: Fly to second city
N_1: 2 ways $\qquad$ N_3: 2 ways

O_2: Fly to first city $\qquad$ O_4: Fly to third city
N_2: 3 ways $\qquad$ N_4: 1 way

Thus, there are
$$N_1 \cdot N_2 \cdot N_3 \cdot N_4 = 2 \cdot 3 \cdot 2 \cdot 1 = 12$$
different travel plans.

43. Let U = the group of people surveyed
   M = people who own a microwave oven, and
   V = people who own a VCR.

Then $n(U) = 1200$, $n(M) = 850$, $n(V) = 740$ and $n(M \cap V) = 580$. Now draw a Venn diagram.

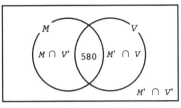

From this diagram, we see that
$$n(M \cap V') = n(M) - n(M \cap V) = 850 - 580 = 270$$
$$n(M' \cap V) = n(V) - n(M \cap V) = 740 - 580 = 160$$
$$n(M \cup V) = n(M \cap V') + n(M \cap V) + n(M' \cap V)$$
$$= 580 + 270 + 160 = 1010$$
and $n(M' \cap V') = n(U) - n(M \cup V) = 1200 - 1010 = 190$

Thus,
(A) $n(M \cup V) = 1010$ (B) $n(M' \cap V') = 190$ (C) $n(M \cap V') = 270$

45. Let U = group of people surveyed
   H = group of people who receive HBO
   S = group of people who receive Showtime.

Then, $n(U) = 8,000$, $n(H) = 2,450$, $n(S) = 1,940$ and $n(H' \cap S') = 5,180$

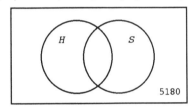

Now, $n(H \cup S) = n(U) - n(H' \cap S') = 8,000 - 5,180 = 2,820$
Since $n(H \cup S) = n(H) + n(S) - n(H \cap S)$, we have
$$n(H \cap S) = n(H) + n(S) - n(H \cup S) = 2,450 + 1,940 - 2,820 = 1,570$$
Thus, 1,570 subscribers receive both channels.

47. From the table:
(A) The number of males aged 20-24 *and* below minimum wage is: 102 (the element in the (2, 2) position in the body of the table.)
(B) The number of females aged 20 or older *and* at minimum wage is: 186 + 503 = 689 (the sum of the elements in the (3, 2) and (3, 3) positions.)
(C) The number of workers who are *either* aged 16-19 *or* are males at minimum wage is:
$$343 + 118 + 367 + 251 + 154 + 237 = 1,470$$
(D) The number of workers below minimum wage is: 379 + 993 = 1,372.

49. (A)

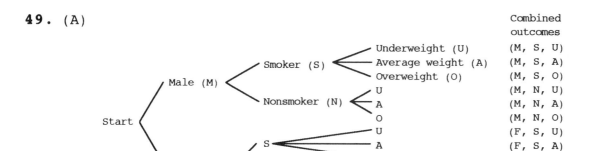

Total combined outcomes = 12

(B) Operation 1: Two classifications, male and female; $N_1 = 2$.

Operation 2: Two classifications, smoker and nonsmoker; $N_2 = 2$.

Operation 3: Three classifications, underweight, average weight, and overweight; $N_3 = 3$.

Thus the total possible combined classifications
$= N_1 \cdot N_2 \cdot N_3 = 2 \cdot 2 \cdot 3 = 12$

51. F = number of individuals who contributed to the first campaign
S = number of individuals who contributed to the second campaign.

Then $n(F) = 1,475$, $n(S) = 2,350$ and $n(F \cap S) = 920$.

Now, $n(F \cup S) = n(F) + n(S) - n(F \cap S)$
$$= 1,475 + 2,350 - 920$$
$$= 2,905$$

Thus, 2,905 individuals contributed to either the first campaign or the second campaign.

EXERCISE 7-2

Things to remember:

1. FACTORIAL

For n a natural number,
$$n! = n(n - 1)(n - 2) \cdot \ldots \cdot 3 \cdot 2 \cdot 1$$
$$0! = 1$$
$$n! = n(n - 1)!$$

[<u>NOTE</u>: The $\boxed{n!}$ key appears on many calculators.]

2. PERMUTATIONS

A PERMUTATION of a set of distinct objects is an arrangement of the objects in a specific order, without repetitions. The number of permutations of n distinct objects without repetition, denoted by $P_{n,n}$, is:

$$P_{n,n} = n(n - 1) \cdot \ldots \cdot 3 \cdot 2 \cdot 1 = n! \quad (n \text{ factors})$$

3. PERMUTATIONS OF n OBJECTS TAKEN r AT A TIME

A permutation of a set of n distinct objects taken r at a time without repetition is an arrangement of the r objects in a specific order. The number of permutations of n objects taken r at a time, denoted by $P_{n,r}$, is given by:

$$P_{n,r} = n(n - 1)(n - 2) \cdot \ldots \cdot (n - r + 1)$$
$$(r \text{ factors})$$

or $P_{n,r} = \dfrac{n!}{(n - r)!} \qquad 0 \le r \le n$

[Note: $P_{n,n} = \dfrac{n!}{(n - n)!} = \dfrac{n!}{0!} = n!$, the number of permutations of n objects taken n at a time. Remember, by definition, $0! = 1$.]

4. COMBINATIONS OF n OBJECTS TAKEN r AT A TIME

A combination of a set of n distinct objects taken r at a time without repetition is an r-element subset of the set of n objects. (The arrangement of the elements in the subset does not matter.) The number of combinations of n objects taken r at a time, denoted by $C_{n,r}$ or by $\binom{n}{r}$ is given by:

$$C_{n,r} = \binom{n}{r} = \frac{P_{n,r}}{r!} = \frac{n!}{r!(n - r)!} \qquad 0 \le r \le n$$

5. NOTE: In a permutation, the ORDER of the objects counts. In a combination, order does not count.

---

1. $4! = 4 \cdot 3 \cdot 2 \cdot 1 = 24$

3. $\dfrac{9!}{8!} = \dfrac{9 \cdot 8!}{8!} = 9$

5. $\dfrac{11!}{8!} = \dfrac{11 \cdot 10 \cdot 9 \cdot 8!}{8!} = 990$

7. $\dfrac{5!}{2!3!} = \dfrac{5 \cdot 4 \cdot 3!}{2 \cdot 1 \cdot 3!} = 10$

9. $\dfrac{7!}{4!(7 - 4)!} = \dfrac{7!}{4!3!} = \dfrac{7 \cdot 6 \cdot 5 \cdot 4!}{4! \cdot 3 \cdot 2 \cdot 1} = 35$

11. $\dfrac{7!}{7!(7 - 7)!} = \dfrac{7!}{7!0!} = \dfrac{1}{1} = 1$

13. $P_{5,3} = \dfrac{5!}{(5 - 3)!} = \dfrac{5!}{2!} = \dfrac{5 \cdot 4 \cdot 3 \cdot 2!}{2!} = 60$

15. $P_{52,4} = \dfrac{52!}{(52 - 4)!} = \dfrac{52!}{48!} = \dfrac{52 \cdot 51 \cdot 50 \cdot 49 \cdot 48!}{48!} = 6,497,400$

17. $C_{5,3} = \dfrac{5!}{3!(5 - 3)!} = \dfrac{5!}{3!2!} = \dfrac{5 \cdot 4 \cdot 3!}{3! \cdot 2 \cdot 1} = 10$

19. $C_{52,4} = \dfrac{52!}{4!(52 - 4)!} = \dfrac{52!}{4!48!} = \dfrac{52 \cdot 51 \cdot 50 \cdot 49 \cdot 48!}{4 \cdot 3 \cdot 2 \cdot 1 \cdot 48!} = 270,725$

21. The number of different finishes (win, place, show) for the ten horses is the number of permutations of 10 objects 3 at a time. This is:

$$P_{10,3} = \frac{10!}{(10 - 3)!} = \frac{10!}{7!} = \frac{10 \cdot 9 \cdot 8 \cdot 7!}{7!} = 720$$

23. (A) The number of ways that a three-person subcommittee can be selected from a seven-member committee is the number of combinations (since order *is not* important in selecting a subcommittee) of 7 objects 3 at a time. This is:

$$C_{7,3} = \frac{7!}{3!(7-3)!} = \frac{7!}{3!4!} = \frac{7 \cdot 6 \cdot 5 \cdot 4!}{3 \cdot 2 \cdot 1 \cdot 4!} = 35$$

(B) The number of ways a president, vice-president, and secretary can be chosen from a committee of 7 people is the number of permutations (since order *is* important in choosing 3 people for the positions) of 7 objects 3 at a time. This is:

$$P_{7,3} = \frac{7!}{(7-3)!} = \frac{7!}{4!} = \frac{7 \cdot 6 \cdot 5 \cdot 4!}{4!} = 7 \cdot 6 \cdot 5 = 210$$

25. This is a "combinations" problem; we want the number of different ways of selecting two teams from the ten teams.

$$C_{10,2} = \frac{10!}{2!(10-2)!} = \frac{10!}{2!8!} = \frac{10 \cdot 9 \cdot 8!}{2 \cdot 1 \cdot 8!} = 45$$

27. This is a "combinations" problem. We want the number of ways to select 5 objects from 13 objects with order not counting. This is:

$$C_{13,5} = \frac{13!}{5!(13-5)!} = \frac{13!}{5!8!} = \frac{13 \cdot 12 \cdot 11 \cdot 10 \cdot 9 \cdot 8!}{5 \cdot 4 \cdot 3 \cdot 2 \cdot 1 \cdot 8!} = 1287$$

29. The five spades can be selected in $C_{13,5}$ ways and the two hearts can be selected in $C_{13,2}$ ways. Applying the Multiplication Principle, we have:

$$\text{Total number of hands} = C_{13,5} \cdot C_{13,2} = \frac{13!}{5!(13-5)!} \cdot \frac{13!}{2!(13-2)!}$$

$$= \frac{13!}{5!8!} \cdot \frac{13!}{2!11!} = 100,386$$

31. The three appetizers can be selected in $C_{8,3}$ ways. The four main courses can be selected in $C_{10,4}$ ways. The two desserts can be selected in $C_{7,2}$ ways. Now, applying the Multiplication Principle, the total number of ways in which the above can be selected is given by:

$$C_{8,3} \cdot C_{10,4} \cdot C_{7,2} = \frac{8!}{3!(8-3)!} \cdot \frac{10!}{4!(10-4)!} \cdot \frac{7!}{2!(7-2)!} = 246,960$$

33. (A) A chord joins two distinct points. Thus, the total number of chords is given by:

$$C_{8,2} = \frac{8!}{2!(8-2)!} = \frac{8!}{2!6!} = \frac{8 \cdot 7 \cdot 6!}{2 \cdot 1 \cdot 6!} = 28$$

(B) Each triangle requires three distinct points. Thus, there are

$$C_{8,3} = \frac{8!}{3!(8-3)!} = \frac{8!}{3!5!} = \frac{8 \cdot 7 \cdot 6 \cdot 5!}{3 \cdot 2 \cdot 1 \cdot 5!} = 56 \text{ triangles.}$$

(C) Each quadrilateral requires four distinct points. Thus, there are

$$C_{8,4} = \frac{8!}{4!(8-4)!} = \frac{8!}{4!4!} = \frac{8 \cdot 7 \cdot 6 \cdot 5 \cdot 4!}{4 \cdot 3 \cdot 2 \cdot 1 \cdot 4!} = 70 \text{ quadrilaterals.}$$

35. (A) Two people.

O_1: First person selects a chair O_2: Second person selects a chair
N_1: 5 ways N_2: 4 ways

Thus, there are
$$N_1 \cdot N_2 = 5 \cdot 4 = 20$$
ways to seat two people in a row of 5 chairs. Note that this is $P_{5,2}$.

(B) Three people. There will be $P_{5,3}$ ways to seat 3 people in a row of 5 chairs:
$$P_{5,3} = \frac{5!}{(5-3)!} = \frac{5!}{2!} = \frac{5 \cdot 4 \cdot 3 \cdot 2!}{2!} = 60$$

(C) Four people. The number of ways to seat 4 people in a row of 5 chairs is given by:
$$P_{5,4} = \frac{5!}{(5-4)!} = \frac{5!}{1!} = 5 \cdot 4 \cdot 3 \cdot 2 = 120$$

(D) Five people. The number of ways to seat 5 people in a row of 5 chairs is given by:
$$P_{5,5} = \frac{5!}{(5-1)!} = \frac{5!}{0!} = 5! = 120$$

37. (A) The distinct positions are taken into consideration. The number of starting teams is given by:
$$P_{8,5} = \frac{8!}{(8-5)!} = \frac{8!}{3!} = \frac{8 \cdot 7 \cdot 6 \cdot 5 \cdot 4 \cdot 3!}{3!} = 6720$$

(B) The distinct positions are not taken into consideration. The number of starting teams is given by:
$$C_{8,5} = \frac{8!}{5!(8-5)!} = \frac{8!}{5!3!} = \frac{8 \cdot 7 \cdot 6 \cdot 5!}{5! \cdot 3 \cdot 2 \cdot 1} = 56$$

(C) Either Mike or Ken, but not both, must start; distinct positions are not taken into consideration.

O_1: Select either Mike or Ken
N_1: 2 ways

O_2: Select 4 players from the remaining 6
N_2: $C_{6,4}$

Thus, the number of starting teams is given by:
$$N_1 \cdot N_2 = 2 \cdot C_{6,4} = 2 \cdot \frac{6!}{4!(6-4)!} = 2 \cdot \frac{6 \cdot 5 \cdot 4!}{4! \cdot 2 \cdot 1} = 30$$

39. (A) Three printers are to be selected for the display. The *order* of selection does not count. Thus, the number of ways to select the 3 printers from 24 is:
$$C_{24,3} = \frac{24!}{3!(24-3)!} = \frac{24 \cdot 23 \cdot 22(21!)}{3 \cdot 2 \cdot 1(21!)} = 2,024$$

(B) Nineteen of the 24 printers are not defective. Thus, the number of ways to select 3 non-defective printers is:

$$C_{19,3} = \frac{19!}{3!(19-3)!} = \frac{19 \cdot 18 \cdot 17(16!)}{3 \cdot 2 \cdot 1(16!)} = 969$$

41. (A) There are 8 + 12 + 10 = 30 stores in all. The jewelry store chain will select 10 of these stores to close. Since order does not count here, the total number of ways to select the 10 stores to close is:

$$C_{30,10} = \frac{30!}{10!(30-10)!} = \frac{30 \cdot 29 \cdot 28 \cdot 27 \cdot 26 \cdot 25 \cdot 24 \cdot 23 \cdot 22 \cdot 21(20!)}{10 \cdot 9 \cdot 8 \cdot 7 \cdot 6 \cdot 5 \cdot 4 \cdot 3 \cdot 2 \cdot 1(20!)}$$
$$= 30,045,015$$

(B) The number of ways to close 2 stores in Georgia is: $C_{8,2}$
The number of ways to close 5 stores in Florida is: $C_{12,5}$
The number of ways to close 3 stores in Alabama is: $C_{10,3}$

By the multiplication principle, the total number of ways to select the 10 stores for closing is:

$$C_{8,2} \cdot C_{12,5} \cdot C_{10,3} = \frac{8!}{2!(8-2)!} \cdot \frac{12!}{5!(12-5)!} \cdot \frac{10!}{3!(10-3)!}$$
$$= \frac{8 \cdot 7 \cdot 6!}{2 \cdot 1 \cdot 6!} \cdot \frac{12 \cdot 11 \cdot 10 \cdot 9 \cdot 8(7!)}{5 \cdot 4 \cdot 3 \cdot 2 \cdot 1(7!)} \cdot \frac{10 \cdot 9 \cdot 8(7!)}{3 \cdot 2 \cdot 1(7!)}$$
$$= 28 \cdot 792 \cdot 120 = 2,661,120$$

43. (A) Three females can be selected in $C_{6,3}$ ways. Two males can be selected in $C_{5,2}$ ways. Applying the Multiplication Principle, we have:

$$\text{Total number of ways} = C_{6,3} \cdot C_{5,2} = \frac{6!}{3!(6-3)!} \cdot \frac{5!}{2!(5-2)!} = 200$$

(B) Four females and one male can be selected in $C_{6,4} \cdot C_{5,1}$ ways. Thus,

$$C_{6,4} \cdot C_{5,1} = \frac{6!}{4!(6-4)!} \cdot \frac{5!}{1!(5-1)!} = 75$$

(C) Number of ways in which 5 females can be selected is:

$$C_{6,5} = \frac{6!}{5!(6-5)!} = 6$$

(D) Number of ways in which 5 people can be selected is:

$$C_{6+5,5} = C_{11,5} = \frac{11!}{5!(11-5)!} = 462$$

(E) At least four females includes four females and five females. Four females and one male can be selected in 75 ways [see part (B)]. Five females can be selected in 6 ways [see part (C)]. Thus,

$$\text{Total number of ways} = C_{6,4} \cdot C_{5,1} + C_{6,5} = 75 + 6 = 81$$

45. (A) Select 3 samples from 8 blood types, no two samples having the same type. This is a permutation problem. The number of different examinations is:

$$P_{8,3} = \frac{8!}{(8-3)!} = \frac{8!}{5!} = \frac{8 \cdot 7 \cdot 6 \cdot 5!}{5!} = 336$$

(B) Select 3 samples from 8 blood types; repetition is allowed.

O_1: Select the first sample O_3: Select the third sample

N_1: 8 ways N_3: 8 ways

O_2: Select the second sample

N_2: 8 ways

Thus, the number of different examinations in this case is:

$N_1 \cdot N_2 \cdot N_3 = 8 \cdot 8 \cdot 8 = 8^3 = 512$

47. This is a permutations problem. The number of buttons is given by:

$$P_{4,2} = \frac{4!}{(4-2)!} = \frac{4!}{2!} = \frac{4 \cdot 3 \cdot 2!}{2!} = 12$$

Things to remember:

<u>1</u>. SAMPLE SPACE

A set S is a SAMPLE SPACE for an experiment if:

(a) Each element of S is an outcome of the experiment.

(b) Each outcome of the experiment corresponds to one and only one element of S.

Each element in the sample space is called a SIMPLE OUTCOME or SIMPLE EVENT.

<u>2</u>. EVENT

Given a sample space S. An EVENT E is any subset of S (including the empty set $\emptyset$ and the sample space S). An event with only one element is called a SIMPLE EVENT; an event with more than one element is a COMPOUND EVENT. An event E *occurs* if the result of performing the experiment is one of the simple events in E.

<u>3</u>. There is no one correct sample space for a given experiment. When specifying a sample space for an experiment, include as much detail as necessary to answer all questions of interest regarding the outcomes of the experiment. When in doubt, choose a sample space with more elements rather than fewer.

<u>4</u>. PROBABILITIES FOR SIMPLE EVENTS

Given a sample space
$$S = \{e_1, \ e_2, \ \ldots, \ e_n\}.$$
To each simple event e_i assign a real number denoted by $P(e_i)$, called the PROBABILITY OF THE EVENT e_i. These numbers can be assigned in an arbitrary manner provided the following two conditions are satisfied:

(a) $0 \le P(e_i) \le 1$

    (The probability of a simple event is a number between 0 and 1, inclusive.)

(b) $P(e_1) + P(e_2) + \ldots + P(e_n) = 1$

    (The sum of the probabilities of all simple events in the sample space is 1.)

Any probability assignment that meets these two conditions is called an ACCEPTABLE PROBABILITY ASSIGNMENT.

<u>5</u>. PROBABILITY OF AN EVENT E

Given an acceptable probability assignment for the simple events in a sample space S, the probability of an arbitrary event E, denoted $P(E)$, is defined as follows:

(a) $P(E) = 0$ if E is the empty set.

(b) If E is a simple event, then $P(E)$ has already been assigned.

(c) If E is a compound event, then $P(E)$ is the sum of the probabilities of all the simple events in E.

(d) If $E = S$, then $P(E) = P(S) = 1$ [this follows from 4(b)].

<u>6</u>. STEPS FOR FINDING THE PROBABILITY OF AN EVENT E

(a) Set up an appropriate sample space S for the experiment.

(b) Assign acceptable probabilities to the simple events in S.

(c) To obtain the probability of an arbitrary event E, add the probabilities of the simple events in E.

<u>7</u>. PROBABILITIES UNDER AN EQUALLY LIKELY ASSUMPTION

If, in a sample space
$$S = \{e_1, \ e_2, \ \ldots, \ e_n\},$$
each simple event is as likely to occur as any other, then $P(e_i) = \dfrac{1}{n}$, for $i = 1, \ 2, \ \ldots, \ n$, i.e., assign the same probability, $1/n$, to each simple event. The probability of an arbitrary event E in this case is:
$$P(E) = \frac{\text{Number of elements in } E}{\text{Number of elements in } S} = \frac{n(E)}{n(S)}$$

1. $P(E) = 1$ means that the occurrence of E is certain.

3. Let B = boy and G = girl. Then

$$S = \{(B, B), (B, G), (G, B), (G, G)\}$$

where (B, B) means both children are boys, (B, G) means the first child is a boy, the second is a girl, and so on. The event E corresponding to having two children of opposite sex is $E = \{(B, G), (G, B)\}$. Since the simple events are equally likely,

$$P(E) = \frac{n(E)}{n(S)} = \frac{2}{4} = \frac{1}{2}.$$

5. We reject (A) because $P(G) = -0.35$, and probability cannot be negative. We reject (B) because $P(R) + P(G) + P(Y) + P(B) = .32 + .28 + .24 + .30$
$$= 1.14 \neq 1.$$
(C) is acceptable.

7. $E = \{R, Y\}$
$$P(E) = P(R) + P(Y) = .26 + .30 = .56$$

9. $S = \{(B, B, B), (B, B, G), (B, G, B), (B, G, G), (G, B, B), (G, B, G),$
$\qquad\qquad (G, G, B), (G, G, G)\}$

$E = \{(B, B, G)\}$

Since the events are equally likely and $n(S) = 8$, $P(E) = \frac{1}{8}$.

11. The number of three-digit sequences with no digit repeated is $P_{10,3}$.
Since the possible opening combinations are equally likely, the probability of guessing the right combination is:

$$\frac{1}{P_{10,3}} = \frac{1}{10 \cdot 9 \cdot 8} = \frac{1}{720} \approx 0.0014$$

13. Let S = the set of five-card hands. Then $n(S) = C_{52,5}$.
Let A = "five black cards." Then $n(A) = C_{26,5}$.
Since individual hands are equally likely to occur:

$$P(A) = \frac{n(A)}{n(S)} = \frac{C_{26,5}}{C_{52,5}} = \frac{\dfrac{26!}{5!21!}}{\dfrac{52!}{5!47!}} = \frac{26 \cdot 25 \cdot 24 \cdot 23 \cdot 22}{52 \cdot 51 \cdot 50 \cdot 49 \cdot 48} \approx 0.025$$

15. S = set of five-card hands; $n(S) = C_{52,5}$.
F = "five face cards"; $n(F) = C_{12,5}$.
Since individual hands are equally likely to occur:

$$P(F) = \frac{n(F)}{n(S)} = \frac{C_{12,5}}{C_{52,5}} = \frac{\dfrac{12!}{5!7!}}{\dfrac{52!}{5!47!}} = \frac{12 \cdot 11 \cdot 10 \cdot 9 \cdot 8}{52 \cdot 51 \cdot 50 \cdot 49 \cdot 48} \approx 0.000305$$

17. The thousands digit can be selected in 2 ways (1 and 3).
The hundreds digit can be selected in 5 ways.
The tens digit can be selected in 5 ways.
The ones digit can be selected in 5 ways.

Thus, $n(S) = 2 \cdot 5 \cdot 5 \cdot 5 = 250$, where S is the set of five-digit numbers less than 5000 formed from 1, 3, 5, 7, and 9.

To form a number that is divisible by 5 from these five digits, the last digit (the ones digit) must be a 5. Thus, $n(A) = 2 \cdot 5 \cdot 5 \cdot 1 = 50$, where A is the set of elements of S that are divisible by 5.

Since the simple events are equally likely:

$$P(A) = \frac{n(A)}{n(S)} = \frac{50}{250} = .2$$

19. $n(S) = P_{5,5} = 5! = 120$

Let A = all notes inserted into the correct envelopes. Then $n(A) = 1$ and
$$P(A) = \frac{n(A)}{n(S)} = \frac{1}{120} \approx 0.00833$$

21. Using the sample space shown in Figure 3, we have

$n(S) = 36$, $n(A) = 1$,

where Event A = "Sum being 2":
$$P(A) = \frac{n(A)}{n(S)} = \frac{1}{36}$$

23. Let E = "Sum being 6." Then $n(E) = 5$. Thus, $P(E) = \frac{n(E)}{n(S)} = \frac{5}{36}$.

25. Let E = "Sum being less than 5." Then $n(E) = 6$. Thus,
$$P(E) = \frac{n(E)}{n(S)} = \frac{6}{36} = \frac{1}{6}.$$

27. Let E = "Sum not 7 or 11." Then $n(E) = 28$ and $P(E) = \frac{n(E)}{n(S)} = \frac{28}{36} = \frac{7}{9}$.

29. E = "Sum being 1" is not possible. Thus, $P(E) = 0$.

31. Let E = "Sum is divisible by 3" = "Sum is 3, 6, 9, or 12." Then
$n(E) = 12$ and $P(E) = \frac{n(E)}{n(S)} = \frac{12}{36} = \frac{1}{3}$.

33. Let E = "Sum is 7 or 11." Then $n(E) = 8$. Thus, $P(E) = \frac{n(E)}{n(S)} = \frac{8}{36} = \frac{2}{9}$.

35. Let E = "Sum is divisible by 2 or 3" = "Sum is 2, 3, 4, 6, 8, 9, 10, 12." Then $n(E) = 24$, and $P(E) = \frac{n(E)}{n(S)} = \frac{24}{36} = \frac{2}{3}$.

For Problems 37–41, the sample space S is given by:
$$S = \{(H, H, H), (H, H, T), (H, T, H), (H, T, T)\}$$
The outcomes are equally likely and $n(S) = 4$.

37. Let E = "1 head." Then $n(E) = 1$ and $P(E) = \frac{n(E)}{n(S)} = \frac{1}{4}$.

39. Let E = "3 heads." Then $n(E) = 1$ and $P(E) = \frac{n(E)}{n(S)} = \frac{1}{4}$.

41. Let E = "More than 1 head." Then $n(E) = 3$ and $P(E) = \dfrac{n(E)}{n(S)} = \dfrac{3}{4}$.

For Problems 43–49, the sample space S is given by:
$$S = \begin{Bmatrix} (1, 1), & (1, 2), & (1, 3) \\ (2, 1), & (2, 2), & (2, 3) \\ (3, 1), & (3, 2), & (3, 3) \end{Bmatrix}$$
The outcomes are equally likely and $n(S) = 9$.

43. Let E = "Sum is 2." Then $n(E) = 1$ and $P(E) = \dfrac{n(E)}{n(S)} = \dfrac{1}{9}$.

45. Let E = "Sum is 4." Then $n(E) = 3$ and $P(E) = \dfrac{n(E)}{n(S)} = \dfrac{3}{9} = \dfrac{1}{3}$.

47. Let E = "Sum is 6." Then $n(E) = 1$ and $P(E) = \dfrac{n(E)}{n(S)} = \dfrac{1}{9}$.

49. Let E = "Sum is odd" = "Sum is 3 or 5." Then $n(E) = 4$ and
$$P(E) = \dfrac{n(E)}{n(S)} = \dfrac{4}{9}.$$

For Problems 51–57, the sample space S is the set of all 5-card hands. Then $n(S) = C_{52,5}$. The outcomes are equally likely.

51. Let E = "5 cards, jacks through aces." Then $n(E) = C_{16,5}$. Thus,
$$P(E) = \dfrac{C_{16,5}}{C_{52,5}} = \dfrac{\dfrac{16!}{5!11!}}{\dfrac{52!}{5!47!}} = \dfrac{16 \cdot 15 \cdot 14 \cdot 13 \cdot 12}{52 \cdot 51 \cdot 50 \cdot 49 \cdot 48} \approx 0.00168.$$

53. Let E = "4 aces." Then $n(E) = 48$ (the remaining card can be any one of the 48 cards which are not aces). Thus,
$$P(E) = \dfrac{48}{C_{52,5}} = \dfrac{48}{\dfrac{52!}{5!47!}} = \dfrac{48 \cdot 5!}{52 \cdot 51 \cdot 50 \cdot 49 \cdot 48} = \dfrac{5 \cdot 4 \cdot 3 \cdot 2}{52 \cdot 51 \cdot 50 \cdot 49} \approx 0.0000185$$

55. Let E = "Straight flush, ace high." Then $n(E) = 4$ (one such hand in each suit). Thus,
$$P(E) = \dfrac{4}{C_{52,5}} = \dfrac{4 \cdot 5!}{52 \cdot 51 \cdot 50 \cdot 49 \cdot 48} = \dfrac{480}{52 \cdot 51 \cdot 50 \cdot 49 \cdot 48} \approx 0.0000015$$

57. Let E = "2 aces and 3 queens." The number of ways to get 2 aces is $C_{4,2}$ and the number of ways to get 3 queens is $C_{4,3}$.
Thus, $n(E) = C_{4,2} \cdot C_{4,3} = \dfrac{4!}{2!2!} \cdot \dfrac{4!}{3!1!} = \dfrac{4 \cdot 3}{2} \cdot \dfrac{4}{1} = 24$
and
$$P(E) = \dfrac{n(E)}{n(S)} = \dfrac{24}{C_{52,5}} = \dfrac{24 \cdot 5!}{52 \cdot 51 \cdot 50 \cdot 59 \cdot 48} \approx 0.000009.$$

59. (A) The sample space S is the set of all possible permutations of the 12 brands taken 4 at a time, and $n(S) = P_{12,4}$. Thus, the probability of selecting 4 brands and identifying them correctly, with no answer repeated, is:

$$P(E) = \frac{1}{P_{12,4}} = \frac{1}{\dfrac{12!}{(12-4)!}} = \frac{1}{12 \cdot 11 \cdot 10 \cdot 9} \approx 0.000084$$

(B) Allowing repetition, $n(S) = 12^4$ and the probability of identifying them correctly is:

$$P(F) = \frac{1}{12^4} \approx 0.000048$$

61. (A) Total number of applicants = 6 + 5 = 11.

$$n(S) = C_{11,5} = \frac{11!}{5!(11-5)!} = 462$$

The number of ways that three females and two males can be selected is:

$$C_{6,3} \cdot C_{5,2} = \frac{6!}{3!(6-3)!} \cdot \frac{5!}{2!(5-2)!} = 20 \cdot 10 = 200$$

Thus, $P(A) = \dfrac{C_{6,3} \cdot C_{5,2}}{C_{11,5}} = \dfrac{200}{462} = 0.433$

(B) $P(4 \text{ females and } 1 \text{ male}) = \dfrac{C_{6,4} \cdot C_{5,1}}{C_{11,5}} = 0.162$

(C) $P(5 \text{ females}) = \dfrac{C_{6,5}}{C_{11,5}} = 0.013$

(D) $P(\text{at least four females}) = P(4 \text{ females and } 1 \text{ male}) + P(5 \text{ females})$

$$= \frac{C_{6,4} \cdot C_{5,1}}{C_{11,5}} + \frac{C_{6,5}}{C_{11,5}}$$

$$= 0.162 + 0.013 \text{ [refer to parts (B) and (C)]}$$

$$= 0.175$$

63. (A) The sample space S consists of the number of permutations of the 8 blood types chosen 3 at a time. Thus, $n(S) = P_{8,3}$ and the probability of guessing the three types in a sample correctly is:

$$P(E) = \frac{1}{P_{8,3}} = \frac{1}{\dfrac{8!}{(8-3)!}} = \frac{1}{8 \cdot 7 \cdot 6} \approx 0.0030$$

(B) Allowing repetition, $n(S) = 8^3$ and the probabilty of guessing the three types in a sample correctly is:

$$P(E) = \frac{1}{8^3} \approx 0.0020$$

65. (A) The total number of ways of selecting a president and a vice-president from the 11 members of the council is:

$P_{11,2}$, i.e., $n(S) = P_{11,2}$.

The total number of ways of selecting the president and the vice-president from the 6 democrats is $P_{6,2}$. Thus, if E is the event "The president and vice-president are both Democrats," then

$$P(E) = \frac{P_{6,2}}{P_{11,2}} = \frac{\dfrac{6!}{(6-2)!}}{\dfrac{11!}{(11-2)!}} = \frac{6 \cdot 5}{11 \cdot 10} = \frac{30}{110} \approx 0.273.$$

(B) The total number of ways of selecting a committee of 3 from the 11 members of the council is:

$C_{11,3}$, i.e., $n(S) = C_{11,3} = \dfrac{11!}{3!(11-3)!} = \dfrac{11 \cdot 10 \cdot 9 \cdot 8!}{3 \cdot 2 \cdot 1 \cdot 8!} = 165$

If we let F be the event "The majority are Republicans," which is the same as having either 2 Republicans and 1 Democrat or all 3 Republicans, then

$$n(F) = C_{5,2} \cdot C_{6,1} + C_{5,3} = \frac{5!}{2!(5-2)!} \cdot \frac{6!}{1!(6-1)!} + \frac{5!}{3!(5-3)!}$$

$$= 10 \cdot 6 + 10 = 70.$$

Thus,

$$P(F) = \frac{n(F)}{n(S)} = \frac{70}{165} \approx 0.424.$$

EXERCISE 7-4
_____

Things to remember:

1. EMPIRICAL PROBABILITY OF EVENT $E = P(E) \approx \dfrac{f(E)}{n}$, where

   $f(E)$ = frequency of event E, and n = total number of trials. The larger n is, the better the approximation.

---

1. Total number of trials: $n = 250$ 3. $f(E) = 189$
   Frequency of event E: $f(E) = 25$ $n = 420$
   Hence, $P(E) \approx \dfrac{25}{250} = .1$ (using 1)   Hence, $P(E) \approx \dfrac{189}{420} = .45$

5. Event E_1 = "point down," $f(E_1)$ = 389
Event E_2 = "point up," $f(E_2)$ = 611
Total number of trials,

n = 389 + 611 = 1000

Thus, $P(E_1) \approx \dfrac{f(E_1)}{n} = \dfrac{389}{1000}$
and

$P(E_1) \approx .389$ (1)

$P(E_2) \approx \dfrac{f(E_2)}{n} = \dfrac{611}{1000}$
and

$P(E_2) \approx .611$ (2)

From (1) and (2), we conclude that the outcomes *are not* "equally likely."

7. (A) Empirical probabilities are as follows:

$P(2\ girls) \approx \dfrac{2351}{10,000} = .2351$

$P(1\ girl) \approx \dfrac{5435}{10,000} = .5435$

$P(0\ girls) \approx \dfrac{2214}{10,000} = .2214$

(B) Theoretical probabilities are as follows:

$n(S) = 4$, $S = \{GG,\ GB,\ BG,\ BB\}$

$P(2\ girls) = \dfrac{1}{4} = .25$

$P(1\ girl) = \dfrac{2}{4} = .5$

$P(0\ girls) = \dfrac{1}{4} = .25$

9. (A) Event E_1 = "3 heads," $f(E_1)$ = 132
Event E_2 = "2 heads," $f(E_2)$ = 368
Event E_3 = "1 head," $f(E_3)$ = 380
Event E_4 = "0 heads," $f(E_4)$ = 120

Total number of trials,

n = 132 + 368 + 380 + 120 = 1000

Thus, $P(E_1) \approx \dfrac{132}{1000} = .132$

$P(E_2) \approx \dfrac{368}{1000} = .368$

$P(E_3) \approx \dfrac{380}{1000} = .38$

$P(E_4) \approx \dfrac{120}{1000} = .12$

(B) Sample space S = {HHH, HTH, THH, HHT, TTH, THT, HTT, TTT}. Thus, the theoretical probabilities are as follows:

$P(3\ heads) = \dfrac{1}{8} = .125$

$P(2\ heads) = \dfrac{3}{8} = .375$

$P(1\ head) = \dfrac{3}{8} = .375$

$P(0\ heads) = \dfrac{1}{8} = .125$

(C) Using the results from part (B), the expected frequencies for each outcome are as follows:

3 heads = 1000(.125) = 125

2 heads = 1000(.375) = 375

1 head = 1000(.375) = 375

0 heads = 1000(.125) = 125

11. Sample space S = {HHHH, THHH, HTHH, HHTH, HHHT, TTHH, THTH, HTTH, HTHT, HHTT, THHT, TTTH, TTHT, THTT, HTTT, TTTT}.

Thus, the theoretical probabilities are as follows:

$P(4 \text{ heads}) = \dfrac{1}{16}$

$P(3 \text{ heads}) = \dfrac{4}{16} = \dfrac{1}{4}$

$P(2 \text{ heads}) = \dfrac{6}{16} = \dfrac{3}{8}$

$P(1 \text{ head}) = \dfrac{4}{16} = \dfrac{1}{4}$

$P(0 \text{ heads}) = \dfrac{1}{16}$

The expected frequencies for each outcome are as follows:

$4 \text{ heads} = 80 \cdot \dfrac{1}{16} = 5$

$3 \text{ heads} = 80 \cdot \dfrac{1}{4} = 20$

$2 \text{ heads} = 80 \cdot \dfrac{3}{8} = 30$

$1 \text{ head} = 80 \cdot \dfrac{1}{4} = 20$

$0 \text{ heads} = 80 \cdot \dfrac{1}{16} = 5$

13. First, calculate the totals in the table

	NUMBER OF VIDEO CASSETTES RENTED EACH MONTH						
AGE	1	2	3	4	5	Over 5	Totals
Under 18	48	42	47	15	8	11	171
18-25	81	102	97	58	24	17	379
26-35	145	161	135	81	47	25	594
36-55	96	113	94	86	46	19	454
Over 55	91	108	92	62	35	14	402
Totals	461	526	465	302	160	86	2000

(A) Let A = "customer is 36-55."

$P(A) = \dfrac{454}{2000} = .227$

(B) Let B = "customer rents 2 cassettes per month."

$P(B) = \dfrac{526}{2000} = .263$

(C) Let C = "customer is 36-55 *and* rents 2 cassettes per month."

$P(C) = \dfrac{113}{2000} = .0565$

(D) Let D = "customer is 36-55 *or* rents 2 cassettes per month."

$P(D) = \dfrac{454 + 526 - 113}{2000} = \dfrac{867}{2000} = .4335$

(E) Let E = "customer is 18-25 *and* rents more than 3 cassettes per month."

$P(E) = \dfrac{58 + 24 + 17}{2000} = .0495$

15. (A) $n(A) = 15$

$P(A) = \dfrac{15}{1000} = .015$

(B) $n(B) = 130 + 80 + 12 = 222$

$P(B) = \dfrac{222}{1000} = .222$

(C) Event C = "Earning more than \$60,000 per year or owning more than three television sets."

$n(C) = 30 + 32 + 28 + 25 + 20 + 1 + 12 + 21 = 169$

$P(C) = \dfrac{169}{1000} = .169$

(D) $n(D) = 1000 - (2 + 10 + 30) = 958$ (958 families own at least one
                                              television set)

$$P(D) = \frac{958}{1000} = .958$$

17. (A) $P(\text{red}) \approx \frac{300}{1000} = .3$

    $P(\text{pink}) \approx \frac{440}{1000} = .44$

    $P(\text{white}) \approx \frac{260}{1000} = .260$

    (B) $P(\text{red}) = \frac{1}{4};\ P(\text{pink}) = \frac{1}{2};$

         $P(\text{white}) = \frac{1}{4}.$

    The expected frequencies for
    each color are as follows:

    $$P(\text{red}) = 1000 \cdot \frac{1}{4} = 250$$

    $$P(\text{pink}) = 1000 \cdot \frac{1}{2} = 500$$

    $$P(\text{white}) = 1000 \cdot \frac{1}{4} = 250$$

EXERCISE 7-5

Things to remember:

1. RANDOM VARIABLE

    A random variable is a function that assigns a numerical value
    to each simple event in a sample space S.

2. PROBABILITY DISTRIBUTION OF A RANDOM VARIABLE X

    A probability function $P(X = x) = p(x)$ is a PROBABILITY
    DISTRIBUTION OF THE RANDOM VARIABLE X if

    (a) $0 \leq p(x) \leq 1,\ x \in \{x_1,\ x_2,\ \ldots,\ x_n\}$,

    (b) $p(x_1) + p(x_2) + \cdots + p(x_n) = 1$,
    where $\{x_1,\ x_2,\ \ldots,\ x_n\}$ are values of X.

3. EXPECTED VALUE OF A RANDOM VARIABLE X

    Given the probability distribution for the random variable X:

    $$\left.\begin{array}{l} x_i\text{: } x_1,\ x_2,\ \ldots,\ x_m \\ p_i\text{: } p_1,\ p_2,\ \ldots,\ p_m \end{array}\right\} p_i = p(x_i)$$

    The expected value of X, denoted by $E(X)$, is given by the
    formula:

    $$E(X) = x_1 p_1 + x_2 p_2 + \cdots + x_m p_m$$

4. Steps for computing the expected value of a random variable X.

    (a) Form the probability distribution for the random variable X.

    (b) Multiply each image value of X, x_i, by its corresponding probability of occurrence, p_i, then add the results.

---

1. Expected value of X:

$$E(X) = -3(.3) + 0(.5) + 4(.2) = -0.1$$

3. Assign the number 0 to the event of observing zero heads, the number 1 to the event of observing one head, and the number 2 to the event of observing two heads. The probability distribution for X, then, is:

x_i	0	1	2
p_i	$\frac{1}{4}$	$\frac{1}{2}$	$\frac{1}{4}$

[Note: One head can occur two ways out of a total of four different ways (HT, TH).]

Hence, $E(X) = 0 \cdot \frac{1}{4} + 1 \cdot \frac{1}{2} + 2 \cdot \frac{1}{4} = 1$.

5. Assign a payoff of $1 to the event of observing a head and -$1 to the event of observing a tail. Thus, the probability distribution for X is:

x_i	1	-1
p_i	$\frac{1}{2}$	$\frac{1}{2}$

Hence, $E(X) = 1 \cdot \frac{1}{2} + (-1) \cdot \frac{1}{2} = 0$. The game is fair.

7. The table shows a payoff or probability distribution for the game.

Net gain	x_i	-3	-2	-1	0	1	2
	p_i	$\frac{1}{6}$	$\frac{1}{6}$	$\frac{1}{6}$	$\frac{1}{6}$	$\frac{1}{6}$	$\frac{1}{6}$

[Note: A payoff valued at -$3 is assigned to the event of observing a "1" on the die, resulting in a net gain of -$3, and so on.]

Hence, $E(X) = -3 \cdot \frac{1}{6} - 2 \cdot \frac{1}{6} - 1 \cdot \frac{1}{6} + 0 \cdot \frac{1}{6} + 1 \cdot \frac{1}{6} + 2 \cdot \frac{1}{6} = -\frac{1}{2}$ or -$0.50.

The game is not fair.

9. The probability distribution is:

Number of Heads	Gain, x_i	Probability, p_i
0	2	$\frac{1}{4}$
1	-3	$\frac{1}{2}$
2	2	$\frac{1}{4}$

The expected value is:

$$E(X) = 2 \cdot \frac{1}{4} + (-3) \cdot \frac{1}{2} + 2 \cdot \frac{1}{4} = 1 - \frac{3}{2} = -\frac{1}{2} \text{ or } -\$0.50.$$

11. In 4 rolls of a die, the total number of possible outcomes is $6 \cdot 6 \cdot 6 \cdot 6 = 6^4$. Thus, $n(S) = 6^4 = 1296$. The total number of outcomes that contain *no* 6's is $5 \cdot 5 \cdot 5 \cdot 5 = 5^4$. Thus, if E is the event "At least one 6," then $n(E) = 6^4 - 5^4 = 671$ and

$$P(E) = \frac{n(E)}{n(S)} = \frac{671}{1296} \approx 0.5178.$$

First, we compute the expected value to you.

The payoff table is:

x_i	-\$1	\$1
P_i	0.5178	0.4822

The expected value to you is:

$E(X) = (-1)(0.5178) + 1(0.4822) = -0.0356$ or -\$0.036

The expected value to her is:

$E(X) = 1(0.5178) + (-1)(0.4822) = 0.0356$ or \$0.036

13. $P(\text{sum} = 7) = \dfrac{6}{36} = \dfrac{1}{6}$

$P(\text{sum} = 11 \text{ or } 12) = P(\text{sum} = 11) + P(\text{sum} = 12) = \dfrac{2}{36} + \dfrac{1}{36} = \dfrac{3}{36} = \dfrac{1}{12}$

$P(\text{sum other than } 7, 11, \text{ or } 12) = 1 - P(\text{sum} = 7, 11, \text{ or } 12)$
$$= 1 - \frac{9}{36} = \frac{27}{36} = \frac{3}{4}$$

Let x_1 = sum is 7, x_2 = sum is 11 or 12, x_3 = sum is not 7, 11, or 12, and let t denote the amount you "win" if x_3 occurs. Then the payoff table is:

x_i	-\$10	\$11	t
p_i	$\frac{1}{6}$	$\frac{1}{12}$	$\frac{3}{4}$

The expected value is:

$$E(X) = -10\left(\frac{1}{6}\right) + 11\left(\frac{1}{12}\right) + t\left(\frac{3}{4}\right) = \frac{-10}{6} + \frac{11}{12} + \frac{3t}{4}$$

The game is fair if $E(X) = 0$, i.e., if

$$\frac{-10}{6} + \frac{11}{12} + \frac{3}{4}t = 0 \quad \text{or} \quad \frac{3}{4}t = \frac{10}{6} - \frac{11}{12} = \frac{20}{12} - \frac{11}{12} = \frac{9}{12} = \frac{3}{4}$$

Therefore, $t = \$1$.

15. Course A_1: $E(X) = (-200)(.1) + 100(.2) + 400(.4) + 100(.3)$
$\qquad\qquad\quad = -20 + 20 + 160 + 30$
$\qquad\qquad\quad = \$190$

Course A_2: $E(X) = (-100)(.1) + 200(.2) + 300(.4) + 200(.3)$
$\qquad\qquad\quad = -10 + 40 + 120 + 60$
$\qquad\qquad\quad = \$210$

A_2 will produce the largest expected value, and that value is \$210.

17. The probability of winning $35 is $\frac{1}{38}$ and the probability of losing $1 is $\frac{37}{38}$. Thus, the payoff table is:

x_i	$35	-$1
P_i	$\frac{1}{38}$	$\frac{37}{38}$

The expected value of the game is:

$E(X) = 35\left(\frac{1}{38}\right) + (-1)\left(\frac{37}{38}\right) = \frac{35 - 37}{38} = \frac{-1}{19} \approx -0.0526$ or $E(X) = -5.26$¢ or -5¢

19.

p_i	x_i
$\frac{1}{5000}$ chance of winning	$499
$\frac{3}{5000}$ chance of winning	$99
$\frac{5}{5000}$ chance of winning	$19
$\frac{20}{5000}$ chance of winning	$4
$\frac{4971}{5000}$ chance of losing	$1 [Note: $5000 - (1 + 3 + 5 + 20) = 4971$.]

The payoff table is:

x_i	$499	$99	$19	$4	-$1
P_i	0.0002	0.0006	0.001	0.004	0.9942

Thus,

$E(X) = 499(0.0002) + 99(0.0006) + 19(0.001) + 4(0.004) - 1(0.9942)$
$ = -0.80$

or

$E(X) = -\$0.80$ or -80¢

21. (A) Total number of simple events $= n(S) = C_{10,2} = \dfrac{10!}{2!(10-2)!}$

$$= \frac{10!}{2!8!} = \frac{10 \cdot 9}{2} = 45$$

$P(\text{zero defective}) = P(0) = \dfrac{C_{7,2}}{45}$ [Note: None defective means 2 selected fom 7 nondefective.]

$$= \frac{\frac{7!}{2!5!}}{45} = \frac{21}{45} = \frac{7}{15}$$

$P(\text{one defective}) = P(1) = \dfrac{C_{3,1} \cdot C_{7,1}}{45} = \dfrac{21}{45} = \dfrac{7}{15}$

$P(\text{two defective}) = P(2) = \dfrac{C_{3,2}}{45}$ [Note: Two defectives selected from 3 defectives.]

$$= \frac{3}{45} = \frac{1}{15}$$

The probability distribution is as follows:

x_i	0	1	2
P_i	$\frac{7}{15}$	$\frac{7}{15}$	$\frac{1}{15}$

(B) $E(X) = 0\left(\frac{7}{15}\right) + 1\left(\frac{7}{15}\right) + 2\left(\frac{1}{15}\right) = \frac{9}{15} = \frac{3}{5} = 0.6$

23. (A) The total number of simple events = $n(S) = C_{1000,5}$.

$P(0 \text{ winning tickets}) = P(0) = \dfrac{C_{997,5}}{C_{1000,5}} = \dfrac{997 \cdot 996 \cdot 995 \cdot 994 \cdot 993}{1000 \cdot 999 \cdot 998 \cdot 997 \cdot 996} \approx 0.985$

$P(1 \text{ winning ticket}) = P(1) = \dfrac{C_{3,1} \cdot C_{997,4}}{C_{1000,5}} = \dfrac{3 \cdot \dfrac{997!}{4!(993)!}}{\dfrac{1000!}{5!(995)!}} \approx 0.0149$

$P(2 \text{ winning tickets}) = P(2) = \dfrac{C_{3,2} \cdot C_{997,3}}{C_{1000,5}} = \dfrac{3 \cdot \dfrac{997!}{3!(994)!}}{\dfrac{1000!}{5!(995)!}} \approx 0.0000599$

$P(3 \text{ winning tickets}) = P(3) = \dfrac{C_{3,3} \cdot C_{997,2}}{C_{1000,5}} = \dfrac{1 \cdot \dfrac{997!}{2!(995)!}}{\dfrac{1000!}{5!(995)!}} \approx 0.00000006$

The payoff table is shown on the following page.

x_i	-\$5	\$195	\$395	\$595
P_i	0.985	0.0149	0.0000599	0.00000006

(B) The expected value to you is:

$E(X) = (-5)(0.985) + 195(0.0149) + 395(0.0000599) + 595(0.00000006)$
$\approx -\$2.00$

25. The payoff table is as follows:

Gain	x_i	\$4850	-\$150
	p_i	0.01	0.99

[Note: 5000 - 150 = 4850, the gain with a probability of 0.01 if stolen.]

Hence, $E(X) = 4850(0.01) - 150(0.99) = -\$100$

27. The payoff table for site A is as follows:

x_i	30 million	-3 million
p_i	0.2	0.8

The payoff table for site B is as follows:

x_i	70 million	-4 million
p_i	0.1	0.9

Hence $E(X) = 30(0.2) - 3(0.8)$
$= 6 - 2.4$
$= \$3.6$ million

Hence, $E(X) = 70(0.1) + (-4)(0.9)$
$= 7 - 3.6$
$= \$3.4$ million

The company should choose site A with $E(X) = \$3.6$ million.

29. Using 4,

$E(X) = 0(0.12) + 1(0.36) + 2(0.38) + 3(0.14) = 1.54$

31. Action A_1: $E(X) = 10(0.3) + 5(0.2) + 0(0.5) = \$4.00$
Action A_2: $E(X) = 15(0.3) + 3(0.1) + 0(0.6) = \$4.80$
Action A_2 is the better choice.

EXERCISE 7-6 CHAPTER REVIEW

---

1. (A) We construct the following tree diagram for the experiment:

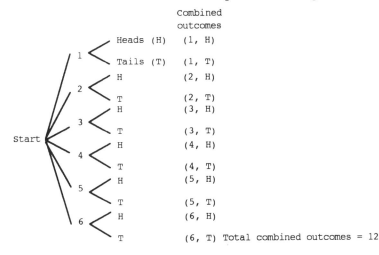

(B) Operation 1: Six possible outcomes, 1, 2, 3, 4, 5, or 6; $N_1 = 6$.
Operation 2: Two possible outcomes, heads (H) or tails (T); $N_2 = 2$.

Using the Multiplication Principle, the total combined outcomes = $N_1 \cdot N_2 = 6 \cdot 2 = 12$.

2. (A) $n(A) = 30 + 35 = 65$ (B) $n(B) = 35 + 40 = 75$
(C) $n(A \cap B) = 35$ (D) $n(A \cup B) = 65 + 75 - 35 = 105$
                    or $n(A \cup B) = 30 + 35 + 40 = 105$
(E) $n(U) = 30 + 35 + 40 + 45 = 150$ (F) $n(A') = n(U) - n(A) = 150 - 65 = 85$
(G) $n([A \cap B]') = n(U) - n(A \cap B)$ (H) $n([A \cup B]') = n(U) - n(A \cup B)$
              $= 150 - 35 = 115$ $= 150 - 105 = 45$

3. $C_{6,2} = \dfrac{6!}{2!(6-2)!} = \dfrac{6!}{2!4!}$ $P_{6,2} = \dfrac{6!}{(6-2)!} = \dfrac{6!}{4!}$

$= \dfrac{6 \cdot 5 \cdot 4!}{2 \cdot 1 \cdot 4!} = 15$ $= \dfrac{6 \cdot 5 \cdot 4!}{4!} = 30$

4. Operation 1: First person can choose the seat in 6 different ways; $N_1 = 6$.
Operation 2: Second person can choose the seat in 5 different ways; $N_2 = 5$.
Operation 3: Third person can choose the seat in 4 different ways; $N_3 = 4$.
Operation 4: Fourth person can choose the seat in 3 different ways; $N_4 = 3$.
Operation 5: Fifth person can choose the seat in 2 different ways; $N_5 = 2$.
Operation 6: Sixth person can choose the seat in 1 way; $N_6 = 1$.
Using the Multiplication Principle, the total number of different
arrangements that can be made is $6 \cdot 5 \cdot 4 \cdot 3 \cdot 2 \cdot 1 = 720$.

5. This is a permutations problem. The permutations of 6 objects taken 6 at a time is:

$$P_{6,6} = \frac{6!}{(6-6)!} = 6! = 720$$

6. First, we calculate the number of 5-card combinations that can be dealt from 52 cards:

$$n(S) = C_{52,5} = \frac{52!}{5! \cdot 47!} = 2,598,960$$

We then calculate the number of 5-club combinations that can be obtained from 13 clubs:

$$n(E) = C_{13,5} = \frac{13!}{5! \cdot 8!} = 1287$$

Thus,

$$P(5 \text{ clubs}) = P(E) = \frac{n(E)}{n(S)} = \frac{1287}{2,598,960} \approx 0.0005$$

7. $n(S)$ is computed by using the permutation formula:

$$n(S) = P_{15,2} = \frac{15!}{(15-2)!} = 15 \cdot 14 = 210$$

Thus, the probability that Betty will be president and Bill will be treasurer is:

$$\frac{n(E)}{n(S)} = \frac{1}{210} \approx 0.0048$$

8. (A) The total number of ways of drawing 3 cards from 10 with order taken into account is given by:

$$P_{10,3} = \frac{10!}{(10-3)!} = \frac{10 \cdot 9 \cdot 8 \cdot 7!}{7!} = 720$$

Thus, the probability of drawing the code word "dig" is:

$$P(\text{"dig"}) = \frac{1}{720} \approx 0.0014$$

(B) The total number of ways of drawing 3 cards from 10 without regard to order is given by:

$$C_{10,3} = \frac{10!}{3!(10-3)!} = \frac{10 \cdot 9 \cdot 8 \cdot 7!}{3!7!} = 120$$

Thus, the probability of drawing the 3 cards "d," "i," and "g" (in some order) is:

$$P(\text{"d," "i," "g"}) = \frac{1}{120} \approx 0.0083$$

9. $P(\text{person having side effects}) = \frac{f(E)}{n} = \frac{50}{1000} = 0.05$

10. The payoff table is as follows:

x_i	-$2	-$1	$0	$1	$2
p_i	$\frac{1}{5}$	$\frac{1}{5}$	$\frac{1}{5}$	$\frac{1}{5}$	$\frac{1}{5}$

Hence,

$$E(X) = (-2) \cdot \frac{1}{5} + (-1) \cdot \frac{1}{5} + 0 \cdot \frac{1}{5} + 1 \cdot \frac{1}{5} + 2 \cdot \frac{1}{5} = 0$$

The game is fair.

11. The function P cannot be a probability function because:

(a) P cannot be negative. [Note: $P(e_2) = -0.2.$]

(b) P cannot have a value greater than 1. [Note: $P(e_4) = 2.$]

(c) The sum of the values of P must equal 1. [Note: $P(e_1) + P(e_2)$ $+ P(e_3) + P(e_4) = 0.1 + (-0.2) + 0.6 + 2 = 2.5 \neq 1.$]

12. Since $n(A \cup B) = n(A) + n(B) - n(A \cap B)$, we have

$80 = 50 + 45 - n(A \cap B)$

and $n(A \cap B) = 15$

Now, $n(B') = n(U) - n(B) = 100 - 45 = 55$

$n(A') = n(U) - n(A) = 100 - 50 = 50$

$n(A \cap B') = 50 - 15 = 35$

$n(B \cap A') = 45 - 15 = 30$

$n(A' \cap B') = 55 - 35 = 20$

Thus,

	A	A'	Totals
B	15	30	45
B'	35	20	55
Totals	50	50	100

13. Each triangle requires 3 distinct points without regard to order. Thus, the total number of triangles that can be formed from the 6 points is:

$$C_{6,3} = \frac{6!}{3!(6-3)!} = \frac{6 \cdot 5 \cdot 4 \cdot 3!}{3 \cdot 2 \cdot 1 \cdot 3!} = 20$$

14.

	Number of ways of completing operation under condition:		
Operation	No letter repeated	Letters can be repeated	Adjacent letters not alike
O_1	8	8	8
O_2	7	8	7
O_3	6	8	7

Total outcomes, without repeating letters = $8 \cdot 7 \cdot 6 = 336$.
Total outcomes, with repeating letters = $8 \cdot 8 \cdot 8 = 512$.
Total outcomes, with adjacent letters not alike = $8 \cdot 7 \cdot 7 = 392$.

15. (A) This is a permutations problem.

$$P_{6,3} = \frac{6!}{(6-3)!} = \frac{6 \cdot 5 \cdot 4 \cdot 3!}{3!} = 120$$

(B) This is a combinations problem.

$$C_{5,2} = \frac{5!}{2!(5-2)!} = \frac{5\cdot4\cdot3!}{2\cdot1\cdot3!} = 10$$

16. Event E_1 = 2 heads; $f(E_1)$ = 210.
    Event E_2 = 1 head; $f(E_2)$ = 480.
    Event E_3 = 0 heads; $f(E_3)$ = 310.
    Total number of trials = 1000.

    (A) The empirical probabilities for the events above are as follows:

    $$P(E_1) = \frac{210}{1000} = 0.21$$

    $$P(E_2) = \frac{480}{1000} = 0.48$$

    $$P(E_3) = \frac{310}{1000} = 0.31$$

    (B) Sample space S = {HH, HT, TH, TT}.

    $$P(2 \text{ heads}) = \frac{1}{4} = 0.25$$

    $$P(1 \text{ head}) = \frac{2}{4} = 0.5$$

    $$P(0 \text{ heads}) = \frac{1}{4} = 0.25$$

    (C) Using part (B), the expected frequencies for each outcome are as follows:

    $$2 \text{ heads} = 1000 \cdot \frac{1}{4} = 250$$

    $$1 \text{ head} = 1000 \cdot \frac{2}{4} = 500$$

    $$0 \text{ heads} = 1000 \cdot \frac{1}{4} = 250$$

17. $n(S) = C_{52,5}$

    (A) Let A be the event "all diamonds." Then $n(A) = C_{13,5}$. Thus,

    $$P(A) = \frac{n(A)}{n(S)} = \frac{C_{13,5}}{C_{52,5}}.$$

    (B) Let B be the event "3 diamonds and 2 spades." Then
    $n(B) = C_{13,3} \cdot C_{13,2}$. Thus,

    $$P(B) = \frac{n(B)}{n(S)} = \frac{C_{13,3} \cdot C_{13,2}}{C_{52,5}}.$$

18. $n(S) = C_{10,4} = \dfrac{10!}{4!(10-4)!} = \dfrac{10\cdot9\cdot8\cdot7\cdot6!}{4\cdot3\cdot2\cdot1\cdot6!} = 210$

    Let A be the event "The married couple is in the group of 4 people."
    Then

    $$n(A) = C_{2,2} \cdot C_{8,2} = 1 \cdot \frac{8!}{2!(8-2)!} = \frac{8\cdot7\cdot6!}{2\cdot1\cdot6!} = 28.$$

    Thus, $P(A) = \dfrac{n(A)}{n(S)} = \dfrac{28}{210} = \dfrac{2}{15} \approx 0.1333$.

19. $S = \{HH, HT, TH, TT\}$.

The probabilities for 2 "heads," 1 "head," and 0 "heads" are, respectively, $\frac{1}{4}$, $\frac{1}{2}$, and $\frac{1}{4}$. Thus, the payoff table is:

x_i	\$5	-\$4	\$2
P_i	0.25	0.5	0.25

$E(X) = 0.25(5) + 0.5(-4) + 0.25(2) = -0.25$ or $-\$0.25$

The game is not fair.

20. $S = \{(1,1), (2,2), (3,3), (1,2), (2,1), (1,3), (3,1), (2,3), (3,2)\}$

$n(S) = 3 \cdot 3 = 9$

(A) $P(A) = \dfrac{n(A)}{n(S)} = \dfrac{3}{9} = \dfrac{1}{3}$ $[A = \{(1,1), (2,2), (3,3)\}]$

(B) $P(B) = \dfrac{n(B)}{n(S)} = \dfrac{2}{9}$ $[B = \{(2,3), (3,2)\}]$

21. (A) The sample space S is given by:

$$S = \{(1,1), (1,2), (1,3), (1,4), (1,5), (1,6),$$
Sum 2
$$(2,1), (2,2), (2,3), (2,4), (2,5), (2,6),$$
Sum 3
$$(3,1), (3,2), (3,3), (3,4), (3,5), (3,6),$$
Sum 4
$$(4,1), (4,2), (4,3), (4,4), (4,5), (4,6),$$
Sum 5
$$(5,1), (5,2), (5,3), (5,4), (5,5), (5,6),$$
$$(6,1), (6,2), (6,3), (6,4), (6,5), (6,6)\}$$

[<u>Note</u>: Event $(2,3)$ means 2 on the the first die and 3 on the second die.]

The probability distribution corresponding to this sample space is:

Sum x_i	2	3	4	5	6	7	8	9	10	11	12
Probability p_i	$\frac{1}{36}$	$\frac{2}{36}$	$\frac{3}{36}$	$\frac{4}{36}$	$\frac{5}{36}$	$\frac{6}{36}$	$\frac{5}{36}$	$\frac{4}{36}$	$\frac{3}{36}$	$\frac{2}{36}$	$\frac{1}{36}$

(B) $E(X) = 2\left(\dfrac{1}{36}\right) + 3\left(\dfrac{2}{36}\right) + 4\left(\dfrac{3}{36}\right) + 5\left(\dfrac{4}{36}\right) + 6\left(\dfrac{5}{36}\right) + 7\left(\dfrac{6}{36}\right) + 8\left(\dfrac{5}{36}\right)$

$\qquad + 9\left(\dfrac{4}{36}\right) + 10\left(\dfrac{3}{36}\right) + 11\left(\dfrac{2}{36}\right) + 12\left(\dfrac{1}{36}\right) = 7$

22. Operation 1: Two possible outcomes, boy or girl, $N_1 = 2$.
Operation 2: Two possible outcomes, boy or girl, $N_2 = 2$.
Operation 3: Two possible outcomes, boy or girl, $N_3 = 2$.
Operation 4: Two possible outcomes, boy or girl, $N_4 = 2$.
Operation 5: Two possible outcomes, boy or girl, $N_5 = 2$.

Using the Multiplication Principle, the total combined outcomes is:
$N_1 \cdot N_2 \cdot N_3 \cdot N_4 \cdot N_5 = 2 \cdot 2 \cdot 2 \cdot 2 \cdot 2 = 32$.

If order pattern is not taken into account, there would be only 6 possible outcomes: families with 0, 1, 2, 3, 4, or 5 boys.

23. Draw a Venn diagram with: A = Chess players, B = Checker players.

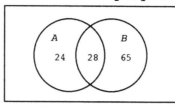

Now, $n(A \cap B) = 28$, $n(A \cap B') = n(A) - n(A \cap B) = 52 - 28 = 24$

$n(B \cap A') = n(B) - n(A \cap B) = 93 - 28 = 65$

Since there are 150 people in all,

$n(A' \cap B') = n(U) - [n(A \cap B') + n(A \cap B) + n(B \cap A')]$
$= 150 - (24 + 28 + 65) = 150 - 117 = 33$

24. The total number of ways that 3 people can be selected from a group of 10 is:

$$C_{10,3} = \frac{10!}{3!(10-3)!} = \frac{10 \cdot 9 \cdot 8 \cdot 7!}{3 \cdot 2 \cdot 1 \cdot 7!} = 120$$

The number of ways of selecting *no* women is:

$$C_{7,3} = \frac{7!}{3!(7-3)!} = \frac{7 \cdot 6 \cdot 5 \cdot 4!}{3 \cdot 2 \cdot 1 \cdot 4!} = 35$$

Thus, the number of samples of 3 people that contain at least one woman is $120 - 35 = 85$.

Therefore, if event A is "At least one woman is selected," then

$$P(A) = \frac{n(A)}{n(S)} = \frac{85}{120} = \frac{17}{24} \approx 0.708.$$

25. (A) This is a permutations problem.

$$P_{10,3} = \frac{10!}{(10-3)!} = \frac{10!}{7!} = 10 \cdot 9 \cdot 8 = 720$$

(B) The number of ways in which women are selected for all three positions is given by:

$$P_{6,3} = \frac{6!}{(6-3)!} = \frac{6!}{3!} = 6 \cdot 5 \cdot 4 = 120$$

Thus, $P(\text{three women are selected}) = \dfrac{P_{6,3}}{P_{10,3}} = \dfrac{120}{720} = \dfrac{1}{6}$

(C) This is a combinations problem.

$$C_{10,3} = \frac{10!}{3!(10-3)!} = \frac{10 \cdot 9 \cdot 8 \cdot 7!}{3 \cdot 2 \cdot 1 \cdot 7!} = 120$$

(D) Let Event D = majority of team members will be women. Then

$n(D)$ = team has 3 women + team has 2 women
$= C_{6,3} + C_{6,2} \cdot C_{4,1}$
$= \dfrac{6!}{3!(6-3)!} + \dfrac{6!}{2!(6-2)!} \cdot \dfrac{4!}{1!(4-1)!} = 20 + 15 \cdot 4 = 80$

Thus,

$$P(D) = \frac{n(D)}{n(S)} = \frac{C_{6,3} + C_{6,2} \cdot C_{4,1}}{C_{10,3}} = \frac{80}{120} = \frac{2}{3}$$

26. The number of ways the 2 people can be seated in a row of 4 chairs is:

$$P_{4,2} = \frac{4!}{(4-2)!} = \frac{4 \cdot 3 \cdot 2!}{2!} = 12$$

27. Let E_2 be the event "2 heads."

(A) From the table, $f(E_2) = 350$. Thus, the approximate empirical probability of obtaining 2 heads is:

$$P(E_2) \approx \frac{f(E_2)}{n} = \frac{350}{1000} = 0.350$$

(B) $S = \{HHH, HHT, HTH, HTT, THH, THT, TTH, TTT\}$
The theoretical probability of obtaining 2 heads is:

$$P(E_2) = \frac{n(E_2)}{n(S)} = \frac{3}{8} = 0.375$$

(C) The expected frequency of obtaining 2 heads in 1000 tosses of 3 fair coins is:
$$f(E_2) = 1000(0.375) = 375$$

28. On one roll of the dice, the probability of getting a double six is $\frac{1}{36}$ and the probability of not getting a double six is $\frac{35}{36}$.

On two rolls of the dice there are $(36)^2$ possible outcomes. There are 71 ways to get at least one double six, namely, a double six on the first roll and any one of the 35 other outcomes on the second roll, or a double six on the second roll and any one of the 35 other outcomes on the first roll, or a double six on both rolls. Thus, the probability of at least one double six on two rolls is $\frac{71}{(36)^2}$ and the probability of no double sixes is:

$$1 - \frac{71}{(36)^2} = \frac{(36)^2 - 2\cdot 36 + 1}{(36)^2} = \frac{(36-1)^2}{(36)^2} = \left(\frac{35}{36}\right)^2$$

Let E be the event "At least one double six." Then E' is the event "No double sixes." Continuing with the reasoning above, we conclude that, in 24 rolls of the die,

$$P(E') = \left(\frac{35}{36}\right)^{24} \approx 0.5086$$

Therefore, $P(E) = 1 - 0.5086 = 0.4914$.

The payoff table is:

x_i	1	−1
P_i	0.4914	0.5086

and $E(X) = 1(0.4914) + (-1)(0.5086)$
$$= 0.4914 - 0.5086$$
$$= -0.0172$$

Thus, your expectation is −$0.0172.
Your friend's expectation is $0.0172.
The game is not fair.

29. Since each die has 6 faces, there are 6·6 = 36 possible pairs for the two up faces.

A sum of 2 corresponds to having (1, 1) as the up faces. This sum can be obtained in 3·3 = 9 ways (3 faces on the first die, 3 faces on the second). Thus,

$$P(2) = \frac{9}{36} = \frac{1}{4}.$$

A sum of 3 corresponds to the two pairs (2, 1) and (1, 2). The number of such pairs is 2·3 + 3·2 = 12. Thus,

$$P(3) = \frac{12}{36} = \frac{1}{3}.$$

A sum of 4 corresponds to the pairs (3, 1), (2, 2), (1, 3). There are 1·3 + 2·2 + 3·1 = 10 such pairs. Thus,

$$P(4) = \frac{10}{36}.$$

A sum of 5 corresponds to the pairs (2, 3) and (3, 2). There are 2·1 + 1·2 = 4 such pairs. Thus,

$$P(5) = \frac{4}{36} = \frac{1}{9}.$$

A sum of 6 corresponds to the pair (3, 3) and there is one such pair. Thus,

$$P(6) = \frac{1}{36}.$$

(A) The probability distribution for X is:

x_i	2	3	4	5	6
P_i	$\frac{9}{36}$	$\frac{12}{36}$	$\frac{10}{36}$	$\frac{4}{36}$	$\frac{1}{36}$

(B) The expected value is:

$$E(X) = 2\left(\frac{9}{36}\right) + 3\left(\frac{12}{36}\right) + 4\left(\frac{10}{36}\right) + 5\left(\frac{4}{36}\right) + 6\left(\frac{1}{36}\right) = \frac{120}{36} = \frac{10}{3}$$

30. The payoff table is:

x_i	-$1.50	-$0.50	$0.50	$1.50	$2.50
P_i	$\frac{9}{36}$	$\frac{12}{36}$	$\frac{10}{36}$	$\frac{4}{36}$	$\frac{1}{36}$

and $E(X) = \frac{9}{36}(-1.50) + \frac{12}{36}(-0.50) + \frac{10}{36}(0.50) + \frac{4}{36}(1.50) + \frac{1}{36}(2.50)$

$= -0.375 - 0.167 + 0.139 + 0.167 + 0.069$

$= -0.167$ or $-\$0.167$

The game is not fair.

31. The number of routes starting from A and visiting each of the 5 stores exactly once is the number of permutations of 5 objects taken 5 at a time, i.e.,

$$P_{5,5} = \frac{5!}{(5 - 5)!} = 120.$$

32. Draw a Venn diagram with:

S = people who have invested in stocks, and

B = people who have invested in bonds.

Then $n(U) = 1000$, $n(S) = 340$, $n(B) = 480$ and $n(S \cap B) = 210$

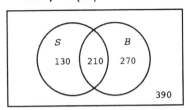

$n(S \cap B') = 340 - 210 = 130$ $n(B \cap S') = 480 - 210 = 270$

$n(S' \cap B') = 1000 - (130 + 210 + 270) = 1000 - 610 = 390$

(A) $n(S \cup B) = n(S) + n(B) - n(S \cap B) = 340 + 480 - 210 = 610$.

(B) $n(S' \cap B') = 390$

(C) $n(B \cap S') = 270$

33. (A) From the table: $P(A) = \dfrac{40}{1000} = 0.04$

(B) From the table, the number of people between 12 and 18 who buy more than one pair of jeans annually is $100 + 60 = 160$. Thus,

$P(B) = \dfrac{160}{1000} = 0.16$.

(C) The number of people who are either between 12 and 18 or buy more than one pair of jeans is $290 + 290 + 120 - 100 - 60 = 540$. Thus,

$P(C) = \dfrac{540}{1000} = 0.54$.

34. The payoff table for plan A is:

x_i	10 million	-2 million
P_i	0.8	0.2

Hence, $E(X) = 10(0.8) - 2(0.2) = 8 - 0.4 = \$7.6$ million.

The payoff table for plan B is:

x_i	12 million	-2 million
P_i	0.7	0.3

Hence, $E(X) = 12(0.7) - 2(0.3) = 8.4 - 0.6 = \$7.8$ million.

Plan B should be chosen.

35. The payoff table is:

Gain	x_i	\$270	-\$30
	P_i	0.08	0.92

[<u>Note</u>: $300 - 30 = 270$ is the "gain" if the bicycle is stolen.]

Hence, $E(X) = 270(0.08) - 30(0.92) = 21.6 - 27.6 = -\$6$.

36. $n(S) = C_{12,4} = \dfrac{12!}{4!(12-4)!} = \dfrac{12 \cdot 11 \cdot 10 \cdot 9 \cdot 8!}{4 \cdot 3 \cdot 2 \cdot 1 \cdot 8!} = 495$

The number of samples that contain *no* substandard parts is:

$C_{10,4} = \dfrac{10!}{4!(10-4)!} = \dfrac{10 \cdot 9 \cdot 8 \cdot 7 \cdot 6!}{4 \cdot 3 \cdot 2 \cdot 1 \cdot 6!} = 210$

Thus, the number of samples that have at least one defective part is 495 − 210 = 285. If E is the event "The shipment is returned," then

$P(E) = \dfrac{n(E)}{n(S)} = \dfrac{285}{495} \approx 0.576.$

37. $n(S) = C_{12,3} = \dfrac{12!}{3!(12-3)!} = \dfrac{12 \cdot 11 \cdot 10 \cdot 9!}{3 \cdot 2 \cdot 1 \cdot 9!} = 220$

A sample will either have 0, 1, or 2 defective circuit boards.

$P(0) = \dfrac{C_{10,3}}{C_{12,3}} = \dfrac{\dfrac{10!}{3!(10-3)!}}{220} = \dfrac{\dfrac{10 \cdot 9 \cdot 8 \cdot 7!}{3 \cdot 2 \cdot 1 \cdot 7!}}{220} = \dfrac{120}{220} = \dfrac{12}{22}$

$P(1) = \dfrac{C_{2,1} \cdot C_{10,2}}{C_{12,3}} = \dfrac{2 \cdot \dfrac{10!}{2!(10-2)!}}{220} = \dfrac{90}{220} = \dfrac{9}{22}$

$P(2) = \dfrac{C_{2,2} \cdot C_{10,1}}{220} = \dfrac{10}{220} = \dfrac{1}{22}$

(A) The probability distribution of X is:

x_i	0	1	2
P_i	$\frac{12}{22}$	$\frac{9}{22}$	$\frac{1}{22}$

(B) $E(X) = 0\left(\dfrac{12}{22}\right) + 1\left(\dfrac{9}{22}\right) + 2\left(\dfrac{1}{22}\right) = \dfrac{11}{22} = \dfrac{1}{2}$

EXERCISE 8-1

Things to remember:

<u>1</u>. PROBABILITY OF A UNION OF TWO EVENTS

For any events A and B,

(a) $P(A \cup B) = P(A) + P(B) - P(A \cap B)$.

If A and B are MUTUALLY EXCLUSIVE ($A \cap B = \varnothing$), then

(b) $P(A \cup B) = P(A) + P(B)$.

<u>2</u>. PROBABILITY OF COMPLEMENTS

For any event E, $E \cup E' = S$ and $E \cap E' = \varnothing$. Thus,

$$P(E) = 1 - P(E')$$
$$P(E') = 1 - P(E)$$

<u>3</u>. PROBABILITY TO ODDS

If $P(E)$ is the probability of the event E, then:

(a) Odds for $E = \dfrac{P(E)}{1 - P(E)} = \dfrac{P(E)}{P(E')}$ $[P(E) \neq 1]$

(b) Odds against $E = \dfrac{P(E')}{P(E)}$ $[P(E) \neq 0]$

[NOTE: When possible, odds are expressed as ratios of whole numbers.]

<u>4</u>. ODDS TO PROBABILTY

If the odds for an event E are $\dfrac{a}{b}$, then the probability of E is:

$$P(E) = \frac{a}{a + b}$$

---

1. Let E be the event "failing within 90 days."
Then E' = "not failing within 90 days."

$\begin{aligned} P(E') &= 1 - P(E) \quad \text{(using } \underline{2}\text{)} \\ &= 1 - .003 \\ &= .997 \end{aligned}$

3. Let Event A = "a number less than 3" = {1, 2}.
Let Event B = "a number greater than 7" = {8, 9, 10}.

Since $A \cap B = \varnothing$, A and B are mutually exclusive. So, using $\underline{1}$(b),

$$P(A \cup B) = P(A) + P(B) = \frac{n(A)}{n(S)} + \frac{n(B)}{n(S)} = \frac{2}{10} + \frac{3}{10} = \frac{1}{2}$$

5. Let Event A = "an even number" = {2, 4, 6, 8, 10}.
Let Event B = "a number divisible by 3" = {3, 6, 9}.

Since $A \cap B$ = {6} $\neq \emptyset$, A and B are not mutually exclusive. So, using $\underline{1}$(a),

$$P(A \cup B) = P(A) + P(B) - P(A \cap B) = \frac{5}{10} + \frac{3}{10} - \frac{1}{10} = \frac{7}{10}$$

7. $P(A) = \dfrac{35 + 5}{35 + 5 + 20 + 40}$

$= \dfrac{40}{100} = .4$

9. $P(B) = \dfrac{5 + 20}{35 + 5 + 20 + 40}$

$= \dfrac{25}{100} = .25$

11. $P(A \cap B) = \dfrac{5}{35 + 5 + 20 + 40}$

$= \dfrac{5}{100} = .05$

13. $P(A' \cap B) = \dfrac{20}{35 + 5 + 20 + 40}$

$= \dfrac{20}{100} = .2$

15. $P(A \cup B) = \dfrac{35 + 5 + 20}{35 + 5 + 20 + 40}$

$= \dfrac{60}{100} = .6$

17. $P(A' \cup B) = \dfrac{20 + 40 + 5}{35 + 5 + 20 + 40}$

$= \dfrac{65}{100} = .65$

19. P(sum of 5 or 6) = P(sum of 5) + P(sum of 6)   [using $\underline{1}$(b)]

$$= \frac{4}{36} + \frac{5}{36} = \frac{9}{36} = \frac{1}{4} \text{ or } .25$$

21. P(1 on first die or 1 on second die)   [using $\underline{1}$(a)]
= P(1 on first die) + P(1 on second die) − P(1 on both dice)
$= \dfrac{6}{36} + \dfrac{6}{36} - \dfrac{1}{36} = \dfrac{11}{36}$

23. Use $\underline{3}$ to find the odds for Event E.

(A) $P(E) = \dfrac{3}{8}$, $P(E') = 1 - P(E) = \dfrac{5}{8}$

Odds for $E = \dfrac{P(E)}{P(E')}$

$= \dfrac{3/8}{5/8} = \dfrac{3}{5}$ (3 to 5)

Odds against $E = \dfrac{P(E')}{P(E)}$

$= \dfrac{5/8}{3/8} = \dfrac{5}{3}$ (5 to 3)

(B) $P(E) = \dfrac{1}{4}$, $P(E') = 1 - P(E) = \dfrac{3}{4}$

Odds for $E = \dfrac{P(E)}{P(E')}$

$= \dfrac{1/4}{3/4} = \dfrac{1}{3}$ (1 to 3)

Odds against $E = \dfrac{P(E')}{P(E)}$

$= \dfrac{3/4}{1/4} = \dfrac{3}{1}$ (3 to 1)

(C) $P(E) = .4$, $P(E') = 1 - P(E) = .6$

Odds for $E = \dfrac{P(E)}{P(E')}$

$= \dfrac{.4}{.6} = \dfrac{2}{3}$ (2 to 3)

Odds against $E = \dfrac{P(E')}{P(E)}$

$= \dfrac{.6}{.4} = \dfrac{3}{2}$ (3 to 2)

(D) $P(E) = .55$, $P(E') = 1 - P(E) - .45$

Odds for $E = \dfrac{P(E)}{P(E')}$

$= \dfrac{.55}{.45} = \dfrac{11}{9}$ (11 to 9)

Odds against $E = \dfrac{P(E')}{P(E)}$

$= \dfrac{.45}{.55} = \dfrac{9}{11}$ (9 to 11)

25. Use $\underline{4}$ to find the probabilty of event E.

(A) Odds for $E = \dfrac{3}{8}$

$P(E) = \dfrac{3}{3 + 8} = \dfrac{3}{11}$

(B) Odds for $E = \dfrac{11}{7}$

$P(E) = \dfrac{11}{11 + 7} = \dfrac{11}{18}$

(C) Odds for $E = \dfrac{4}{1}$

$P(E) = \dfrac{4}{4 + 1} = \dfrac{4}{5} = .8$

(D) Odds for $E = \dfrac{49}{51}$

$P(E) = \dfrac{49}{49 + 51} = \dfrac{49}{100} = .49$

27. Odds for $E = \dfrac{P(E)}{P(E')} = \dfrac{1/2}{1/2} = 1.$

The odds in favor of getting a head in a single toss of a coin are 1 to 1.

29. The sample space for this problem is:

$S = \{HHH, HHT, THH, HTH, TTH, HTT, THT, TTT\}$

Let Event $E = $ "getting at least 1 head."
Let Event $E' = $ "getting no heads."

Thus, $\dfrac{P(E)}{P(E')} = \dfrac{7/8}{1/8} = \dfrac{7}{1}$

The odds in favor of getting at least 1 head are 7 to 1.

31. Let Event $E = $ "getting a number greater than 4."
Let Event $E' = $ "not getting a number greater than 4."

Thus, $\dfrac{P(E')}{P(E)} = \dfrac{4/6}{2/6} = \dfrac{2}{1}$

The odds against getting a number greater than 4 in a single roll of a die are 2 to 1.

33. Let Event $E = $ "getting 3 or an even number" $= \{2, 3, 4, 6\}$.
Let Event $E' = $ "not getting 3 or an even number" $= \{1, 5\}$.

Thus, $\dfrac{P(E')}{P(E)} = \dfrac{2/6}{4/6} = \dfrac{1}{2}$

The odds against getting 3 or an even number are 1 to 2.

35. Let $E = $ "rolling a five." Then $P(E) = \dfrac{n(E)}{n(S)} = \dfrac{4}{36} = \dfrac{1}{9}$ and $P(E') = \dfrac{8}{9}$.

(A) Odds for $E = \dfrac{1/9}{8/9} = \dfrac{1}{8}$ (1 to 8)

(B) Let k be the amount the house should pay for the game to be fair. Then

$E(X) = k\left(\dfrac{1}{9}\right) + (-1)\left(\dfrac{8}{9}\right) = 0$

$\dfrac{k}{9} = \dfrac{8}{9}$ and $k = 8$

The house should pay \$8.

37. (A) Let $E = $ "sum is less than 4 or greater than 9." Then

$P(E) = \dfrac{10 + 30 + 120 + 80 + 70}{1000} = \dfrac{310}{1000} = \dfrac{31}{100} = .31$ and $P(E') = \dfrac{69}{100}$.

Thus,

Odds for $E = \dfrac{31/100}{69/100} = \dfrac{31}{69}$

(B) Let F = "sum is even or divisible by 5." Then

$$P(F) = \frac{10 + 50 + 110 + 170 + 120 + 70 + 70}{1000} = \frac{600}{1000} = \frac{6}{10} = .6$$

and $P(F') = \frac{4}{10}$. Thus,

Odds for $F = \frac{6/10}{4/10} = \frac{6}{4} = \frac{3}{2}$

39. Let A = "drawing a face card" (Jack, Queen, King)
and B = "drawing a club."

Then $P(A \cup B) = P(A) + P(B) - P(A \cap B) = \frac{12}{52} + \frac{13}{52} - \frac{3}{52} = \frac{22}{52} = \frac{11}{26}$

$P[(A \cup B)'] = \frac{15}{26}$

Odds for $A \cup B = \frac{11/26}{15/26} = \frac{11}{15}$

41. Let A = "drawing a black card"
and B = "drawing an ace."

$P(A \cup B) = P(A) + P(B) - P(A \cap B) = \frac{26}{52} + \frac{4}{52} - \frac{2}{52} = \frac{28}{52} = \frac{7}{13}$

$P[(A \cup B)'] = \frac{6}{13}$

Odds for $A \cup B = \frac{7/13}{6/13} = \frac{7}{6}$

43. The sample space S is the set of all 5-card hands and $n(S) = C_{52,5}$

Let E = "getting at least one diamond."
Then E' = "no diamonds" and $n(E) = C_{39,5}$.

Thus, $P(E') = \frac{C_{39,5}}{C_{52,5}}$, and

$$P(E) = 1 - \frac{C_{39,5}}{C_{52,5}} = 1 - \frac{\frac{39!}{5!34!}}{\frac{52!}{5!47!}} = 1 - \frac{39 \cdot 38 \cdot 37 \cdot 36 \cdot 35}{52 \cdot 51 \cdot 50 \cdot 49 \cdot 48} \approx 1 - .22 = .78.$$

45. The number of numbers less than or equal to 1000 which are divisible by 6 is the largest integer in $\frac{1000}{6}$ or 166.

The number of numbers less than or equal to 1000 which are divisible by 8 is the largest integer in $\frac{1000}{8}$ or 125.

The number of numbers less than or equal to 1000 which are divisible by both 6 and 8 is the same as the number of numbers which are divisible by 24. This is the largest integer in $\frac{1000}{24}$ or 41.

Thus, if A is the event "selecting a number which is divisible by either 6 or 8," then

$n(A) = 166 + 125 - 41 = 250$ and $P(A) = \frac{250}{1000} = .25.$

47. Let S be the set of all groups of 5 integers selected from the first 50 integers. Then $n(S) = C_{50,5}$.

Let E be the event "a group contains at least one number divisible by 3." Then E' is the event "a group contains no number divisible by 3." There are 16 integers less than 50 which are divisible by 3. Thus, 34 integers less than or equal to 50 are not divisible by 3 and $n(E') = C_{34,5}$. Therefore,

$$P(E') = \frac{C_{34,5}}{C_{50,5}} = \frac{\dfrac{34!}{5!29!}}{\dfrac{50!}{5!45!}} = \frac{34 \cdot 33 \cdot 32 \cdot 31 \cdot 30}{50 \cdot 49 \cdot 48 \cdot 47 \cdot 46} \approx .13$$

and

$$P(E) = 1 - P(E') = 1 - .13 = .87$$

49. Let R be the event "landing on red." Then

Odds for $R = \dfrac{9}{10}$ and $P(R) = \dfrac{9}{9 + 10} = \dfrac{9}{19} \approx .4737$

The payoff table is:

x_i	1	−1
P_i	.4737	.5263

$E(X) = 1(.4737) + (-1)(.5263) = -.0526$ or $-\$0.0526$

51. S = set of all lists of n birth months, $n \le 12$. Then $n(S) = 12 \cdot 12 \cdot \ldots \cdot 12$ (n times) $= 12^n$.

Let E = "at least two people have the same birth month."
Then E' = "no two people have the same birth month."

$$n(E') = 12 \cdot 11 \cdot 10 \cdot \ldots \cdot [12 - (n - 1)]$$

$$= \frac{12 \cdot 11 \cdot 10 \cdot \ldots \cdot [12 - (n - 1)](12 - n)[12 - (n + 1)] \cdot \ldots \cdot 3 \cdot 2 \cdot 1}{(12 - n)[12 - (n + 1)] \cdot \ldots \cdot 3 \cdot 2 \cdot 1}$$

$$= \frac{12!}{(12 - n)!}$$

Thus, $P(E') = \dfrac{\dfrac{12!}{(12 - n)!}}{12^n} = \dfrac{12!}{12^n(12 - n)!}$ and $P(E) = 1 - \dfrac{12!}{12^n(12 - n)!}$.

53. Odds for $E = \dfrac{P(E)}{P(E')} = \dfrac{P(E)}{1 - P(E)} = \dfrac{a}{b}$. Therefore,

$bP(E) = a[1 - P(E)] = a - aP(E)$.
Thus, $aP(E) + bP(E) = a$
$(a + b)P(E) = a$
$$P(E) = \frac{a}{a + b}$$

55. Venn diagram:

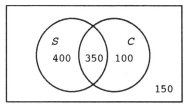

Let S be the event that the student owns a stereo and C be the event that the student owns a car.

The table corresponding to the given data is as follows:

	C	C'	Total
S	350	400	750
S'	100	150	250
Total	450	550	1000

The corresponding probabilities are:

	C	C'	Total
S	.35	.40	.75
S'	.10	.15	.25
Total	.45	.55	1.00

From the above table:

(A) $P(C \text{ or } S) = P(C \cup S) = P(C) + P(S) - P(C \cap S)$
$$= .45 + .75 - .35$$
$$= .85$$

(B) $P(C' \cap S') = .15$

57. (A) Using the table, we have:

$$P(M_1 \text{ or } A) = P(M_1 \cup A) = P(M_1) + P(A) - P(M_1 \cap A)$$
$$= .2 + .3 - .05$$
$$= .45$$

(B) $P[(M_2 \cap A') \cup (M_3 \cap A')] = P(M_2 \cap A') + P(M_3 \cap A')$
$$= .2 + .35 \text{ (from the table)}$$
$$= .55$$

59. Let K = "defective keyboard"
and D = "defective disk drive."

Then $K \cup D$ = "either a defective keyboard or a defective disk drive"
and $(K \cup D)'$ = "neither the keyboard nor the disk drive is defective"
$$= K' \cap D'$$

$P(K \cup D) = P(K) + P(D) - P(K \cap D) = .06 + .05 - .01 = .1$

Thus, $P(K' \cap D') = 1 - .1 = .9$.

61. The sample space S is the set of all possible 10-element samples from the 60 watches, and $n(S) = C_{60,10}$. Let E be the event that a sample contains at least one defective watch. Then E' is the event that a sample contains no defective watches. Now, $n(E') = C_{51,10}$.

Thus, $P(E') = \dfrac{C_{51,10}}{C_{60,10}} = \dfrac{\frac{51!}{10!41!}}{\frac{60!}{10!50!}} \approx .17$ and $P(E) \approx 1 - .17 = .83$.

Therefore, the probabilty that a sample will be returned is .83.

63. The given information is displayed in the Venn diagram:

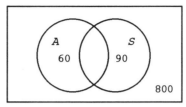

A = suffers from loss of appetite

S = suffers from loss of sleep

Thus, we can conclude that $n(A \cap S) = 1000 - (60 + 90 + 800) = 50$.

$$P(A \cap S) = \frac{50}{1000} = .05$$

65. (A) "Unaffiliated or no preference" $= U \cup N$.

$$P(U \cup N) = P(U) + P(N) - P(U \cap N)$$
$$= \frac{150}{1000} + \frac{85}{1000} - \frac{15}{1000} = \frac{220}{1000} = \frac{11}{50} = .22$$

Therefore, $P[(U \cup N)'] = 1 - \frac{11}{50} = \frac{39}{50}$ and

Odds for $U \cup N = \frac{11/50}{39/50} = \frac{11}{39}$

(B) "Affliated with a party and prefers candidate A" $= (D \cup R) \cap A$.

$$P[(D \cup R) \cap A] = \frac{300}{1000} = \frac{3}{10} = .3$$

The odds against this event are:

$$\frac{1 - 3/10}{3/10} = \frac{7/10}{3/10} = \frac{7}{3}$$

67. Let S = the set of all three-person groups from the total group.
Let E = the set of all three-person groups with at least one black.
Let E' = the set of all three-person groups with no blacks.
First, find $P(E')$, then use $P(E) = 1 - P(E')$ to find $P(E)$.

$$P(E') = \frac{n(E')}{n(E)} = \frac{C_{15,3}}{C_{20,3}} \approx .4$$

$$P(E) = 1 - P(E') \approx .6$$

EXERCISE 8-2

Things to remember:

1. **CONDITIONAL PROBABILITY**

   For events A and B in a sample space S, the CONDITIONAL PROBABILITY of A given B, denoted $P(A \mid B)$, is defined by

   $$P(A \mid B) = \frac{P(A \cap B)}{P(B)}, \quad P(B) \neq 0$$

2. **PRODUCT RULE**

   For events A and B, $P(A) \neq 0$, $P(B) \neq 0$, in a sample space S,

   $$P(A \cap B) = P(A) \cdot P(B \mid A) = P(B) \cdot P(A \mid B).$$

   [<u>Note</u>: We can use either $P(A) \cdot P(B \mid A)$ or $P(B) \cdot P(A \mid B)$ to compute $P(A \cap B)$.]

<u>3</u>. INDEPENDENCE

Let A and B be any events in a sample space S. Then A and B are INDEPENDENT if and only if

$$P(A \cap B) = P(A) \cdot P(B).$$

Otherwise, A and B are DEPENDENT.

<u>4</u>. INDEPENDENT SET OF EVENTS

A FINITE SET OF EVENTS is said to be INDEPENDENT if the probability of each possible intersection of events in the set is the product of the probabilities of the events in the intersection.

Events A, B, and C are independent if and only if

$$P(A \cap B) = P(A)P(B) \qquad P(A \cap C) = P(A)P(C)$$
$$P(B \cap C) = P(B)P(C) \qquad P(A \cap B \cap C) = P(A)P(B)P(C)$$

---

1. $P(A) = .50$
See the given table.

3. $P(D) = .20$
See the given table.

5. $P(A \cap D) = .10$
See the given table for occurrences of both A and D.

7. $P(C \cap D)$ = probability of occurrences of both C and $D = .06$.

9. $P(A \mid D) = \dfrac{P(A \cap D)}{P(D)} = \dfrac{0.10}{0.20} = .50$

11. $P(C \mid D) = \dfrac{P(C \cap D)}{P(D)} = \dfrac{0.06}{0.20} = .30$

13. Events A and D are independent if $P(A \cap D) = P(A) \cdot P(D)$:
$P(A \cap D) = .10$
$P(A) \cdot P(D) = (.50)(.20) = .10$
Thus, A and D are independent.

15. $P(C \cap D) = .06$
$P(C) \cdot P(D) = (.20)(.20) = .04$
Since $P(C \cap D) \neq P(C) \cdot P(D)$, C and D are dependent.

17. (A) Let H_8 = "a head on the eighth toss." Since each toss is independent of the other tosses, $P(H_8) = \dfrac{1}{2}$.

(B) Let H_i = "a head on the ith toss." Since the tosses are independent,
$P(H_1 \cap H_2 \cap \cdots \cap H_8) = P(H_1)P(H_2) \cdots P(H_8) = \left(\dfrac{1}{2}\right)^8 = \dfrac{1}{2^8} = \dfrac{1}{256}$.
Similarly, if T_i = "a tail on the ith toss," then
$P(T_1 \cap T_2 \cap \cdots \cap T_8) = P(T_1)P(T_2) \cdots P(T_8) = \dfrac{1}{2^8} = \dfrac{1}{256}$. Finally, if
H = "all heads" and T = "all tails," then $H \cap T = \varnothing$ and
$P(H \cup T) = P(H) + P(T) = \dfrac{1}{256} + \dfrac{1}{256} = \dfrac{2}{256} = \dfrac{1}{128} \approx .00781$.

19. Given the table:

e_i	1	2	3	4	5
P_i	.3	.1	.2	.3	.1

E = "pointer lands on an even number" = {2, 4}.
F = "pointer lands on a number less than 4" = {1, 2, 3}.

(A) $P(F \mid E) = \dfrac{P(F \cap E)}{P(E)} = \dfrac{P(2)}{P(2) + P(4)} = \dfrac{.1}{.1 + .3} = \dfrac{.1}{.4} = \dfrac{1}{4}$

(B) $P(E \cap F) = P(2) = .1$

$P(E) = .4$,
$P(F) = P(1) + P(2) + P(3) = .3 + .1 + .2 = .6$,
and
$P(E)P(F) = (.4)(.6) = .24 \neq P(E \cap F)$.
Thus, E and F are dependent.

21. From the probability tree,

(A) $P(M \cap S) = (.3)(.6) = .18$

(B) $P(R) = P(N \cap R) + P(M \cap R) = (.7)(.2) + (.3)(.4)$
$\qquad\qquad = .14 + .12$
$\qquad\qquad = .26$

23. $E_1 = \{HH, HT\}$ and $P(E_1) = \dfrac{1}{2}$

$E_2 = \{TH, TT\}$ and $P(E_2) = \dfrac{1}{2}$

$E_3 = \{HT, TT\}$ and $P(E_3) = \dfrac{1}{2}$

(A) $P(E_1 \cap E_3) = P(HT) = \dfrac{1}{4}$

$P(E_1) \cdot P(E_3) = \dfrac{1}{2} \cdot \dfrac{1}{2} = \dfrac{1}{4}$

Thus, E_1 and E_3 are independent.

(B) Two events, A and B, are mutually exclusive if $A \cap B = \varnothing$. Since $E_1 \cap E_3 = \{HT\} \neq \varnothing$, E_1 and E_3 are *not* mutually exclusive.

25. Let E_i = "even number on the ith throw," $i = 1, 2$, and O_i = "odd number on the ith throw," $i = 1, 2$.

Then $P(E_i) = \dfrac{1}{2}$ and $P(O_i) = \dfrac{1}{2}$, $i = 1, 2$.

The probability tree for this experiment is shown at the right.

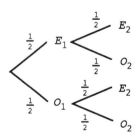

$P(E_1 \cap E_2) = \left(\dfrac{1}{2}\right)\left(\dfrac{1}{2}\right) = \dfrac{1}{4}$

$P(E_1 \cup E_2) = P(E_1) + P(E_2) - P(E_1 \cap E_2) = \dfrac{1}{2} + \dfrac{1}{2} - \dfrac{1}{4} = \dfrac{3}{4}$.

27. Let C = "first card is a club,"
and H = "second card is a heart."

   (A) Without replacement, the probability tree is as shown at the right.

      Thus, $P(C \cap H) = \left(\frac{1}{4}\right)\left(\frac{13}{51}\right) \approx .0637$.

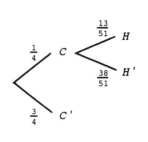

   (B) With replacement, the draws are independent and
$P(C \cap H) = \left(\frac{1}{4}\right)\left(\frac{1}{4}\right) = \frac{1}{16} = .0625$.

29. G = "the card is black" = {spade or club} and $P(G) = \frac{1}{2}$.

H = "the card is divisible by 3" = {3, 6, or 9}. $P(H) = \frac{12}{52} = \frac{3}{13}$

$P(H \cap G)$ = {3, 6, or 9 of clubs or spades} $= \frac{6}{52} = \frac{3}{26}$

   (A) $P(H \mid G) = \dfrac{P(H \cap G)}{P(G)} = \dfrac{3/26}{1/2} = \dfrac{6}{26} = \dfrac{3}{13}$

   (B) $P(H \cap G) = \dfrac{3}{26} = P(H) \cdot P(G)$

      Thus, H and G are independent.

31. (A) S = {BB, BG, GB, GG}

      A = {BB, GG} and $P(A) = \dfrac{2}{4} = \dfrac{1}{2}$

      B = {BG, GB, GG} and $P(B) = \dfrac{3}{4}$

      $A \cap B$ = {GG}.

      $P(A \cap B) = \dfrac{1}{4}$ and $P(A) \cdot P(B) = \dfrac{1}{2} \cdot \dfrac{3}{4} = \dfrac{3}{8}$

      Thus, $P(A \cap B) \neq P(A) \cdot P(B)$ and the events are dependent.

   (B) S = {BBB, BBG, BGB, BGG, GBB, GBG, GGB, GGG}

      A = {BBB, GGG}

      B = {BGG, GBG, GGB, GGG}

      $A \cap B$ = {GGG}

      $P(A) = \dfrac{2}{8} = \dfrac{1}{4}$, $P(B) = \dfrac{4}{8} = \dfrac{1}{2}$, and $P(A \cap B) = \dfrac{1}{8}$

      Since $P(A \cap B) = \dfrac{1}{8} = P(A) \cdot P(B)$, A and B are independent.

33. (A) The probability tree with replacement is as follows:

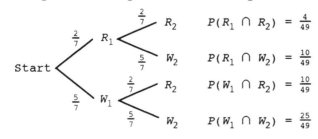

(B) The probability tree without replacement is as follows:

$$P(R_1 \cap R_2) = \tfrac{1}{21}$$

$$P(R_1 \cap W_2) = \tfrac{5}{21}$$

$$P(W_1 \cap R_2) = \tfrac{5}{21}$$

$$P(W_1 \cap W_2) = \tfrac{10}{21}$$

35. Let E = At least one ball was red = $\{R_1 \cap R_2,\ R_1 \cap W_2,\ W_1 \cap R_2\}$.

(A) With replacement [see the probability tree in Problem 33(A)]:
$$P(E) = P(R_1 \cap R_2) + P(R_1 \cap W_2) + P(W_1 \cap R_2)$$
$$= \frac{4}{49} + \frac{10}{49} + \frac{10}{49} = \frac{24}{49}$$

(B) Without replacement [see the probability tree in Problem 33(B)]:
$$P(E) = P(R_1 \cap R_2) + P(R_1 \cap W_2) + P(W_1 \cap R_2) = \frac{1}{21} + \frac{5}{21} + \frac{5}{21} = \frac{11}{21}$$

37. (A) The probability tree with replacement is as follows:

$$P(A_1 \cap A_2) = \tfrac{4}{52} \cdot \tfrac{4}{52} = \tfrac{1}{169}$$

$$P(A_1 \cap A_2') = \tfrac{4}{52} \cdot \tfrac{48}{52} = \tfrac{12}{169}$$

$$P(A_1' \cap A_2) = \tfrac{48}{52} \cdot \tfrac{4}{52} = \tfrac{12}{169}$$

$$P(A_1' \cap A_2') = \tfrac{48}{52} \cdot \tfrac{48}{52} = \tfrac{144}{169}$$

Let E = Exactly one ace = $\{A_1 \cap A_2',\ A_1' \cap A_2\}$.
$$P(E) = P(A_1 \cap A_2') + P(A_1' \cap A_2) = \frac{12}{169} + \frac{12}{169} = \frac{24}{169}$$

(B) The probability tree without replacement is as follows:

$$P(A_1 \cap A_2) = \tfrac{4}{52} \cdot \tfrac{3}{51} = \tfrac{1}{221}$$

$$P(A_1 \cap A_2') = \tfrac{4}{52} \cdot \tfrac{48}{51} = \tfrac{16}{221}$$

$$P(A_1' \cap A_2) = \tfrac{48}{52} \cdot \tfrac{4}{51} = \tfrac{16}{221}$$

$$P(A_1' \cap A_2') = \tfrac{48}{52} \cdot \tfrac{47}{51} = \tfrac{188}{221}$$

$$P(E) = P(A_1 \cap A_2') + P(A_1' \cap A_2) = \frac{16}{221} + \frac{16}{221} = \frac{32}{221}$$

39. $n(S) = C_{9,2} = \dfrac{9!}{2!(9-2)!}$ (total number of balls $= 2 + 3 + 4 = 9$)

$$= \frac{9 \cdot 8 \cdot 7!}{2 \cdot 1 \cdot 7!} = 36$$

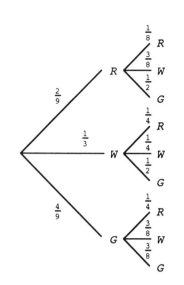

Let A = Both balls are the same color.

$n(A)$ = (No. of ways 2 red balls are selected)
      + (No. of ways 2 white balls are selected)
      + (No. of ways 2 green balls are selected)

$$= C_{2,2} + C_{3,2} + C_{4,2}$$

$$= \frac{2!}{2!(2-2)!} + \frac{3!}{2!(3-2)!} + \frac{4!}{2!(4-2)!}$$

$$= 1 + 3 + 6 = 10$$

$P(A) = \dfrac{n(A)}{n(S)} = \dfrac{10}{36} = \dfrac{5}{18}$

Alternatively, the probability tree for this experiment is shown at the right.

And $P(RR,\ WW,\ \text{or } GG) = P(RR) + P(WW) + P(GG)$

$$= \left(\frac{2}{9}\right)\left(\frac{1}{8}\right) + \left(\frac{1}{3}\right)\left(\frac{1}{4}\right) + \left(\frac{4}{9}\right)\left(\frac{3}{8}\right) = \frac{2}{72} + \frac{1}{12} + \frac{12}{72} = \frac{20}{72} = \frac{5}{18}$$

41. The probability tree for this experiment is:

(A) $P(\$16) = \left(\frac{1}{4}\right)\left(\frac{2}{3}\right)\left(\frac{1}{2}\right) + \left(\frac{1}{2}\right)\left(\frac{1}{3}\right)\left(\frac{1}{2}\right)$

$$= \frac{1}{12} + \frac{1}{12} = \frac{1}{6} \approx .167$$

(B) $P(\$17) = \left(\frac{1}{2}\right)\left(\frac{1}{3}\right)\left(\frac{1}{2}\right) + \left(\frac{1}{2}\right)\left(\frac{1}{3}\right)\left(\frac{1}{2}\right) + \left(\frac{1}{4}\right)\left(\frac{2}{3}\right)\left(\frac{1}{2}\right)$

$$= \frac{1}{12} + \frac{1}{12} + \frac{1}{12} = \frac{1}{4} = .25$$

(C) Let A = "$\$10$ on second draw." Then

$$P(A) = \left(\frac{1}{4}\right)\left(\frac{1}{3}\right) + \left(\frac{1}{2}\right)\left(\frac{1}{3}\right) = \frac{1}{12} + \frac{1}{6} = \frac{1}{4} = .25$$

(D) The payoff table is:

x_i	\$10	\$11	\$12	\$15	\$16	\$17
P_i	.25	.167	.083	.083	.167	.25

Thus,

$E(X) = 10(.25) + 11(.167) + 12(.083) + 15(.083) + 16(.167) + 17(.25)$
      $= \$13.50$

A player should pay \$13.50 for the game to be fair.

43. Assume $P(A) \neq 0$. Then $P(A \mid A) = \dfrac{P(A \cap A)}{P(A)} = \dfrac{P(A)}{P(A)} = 1.$

45. If A and B are mutually exclusive, then $A \cap B = \emptyset$ and $P(A \cap B) = P(\emptyset) = 0$. Also, if $P(A) \neq 0$ and $P(B) \neq 0$, then $P(A) \cdot P(B) \neq 0$. Therefore, $P(A \cap B) = 0 \neq P(A) \cdot P(B)$, and events A and B are dependent.

47. (A)

To strike	Hourly H	Salary S	Salary + bonus B	Total
Yes (Y)	.400	.180	.020	.600
No (N)	.150	.120	.130	.400
Total	.550	.300	.150	1.000

[Note: The probability table above was derived from the table given in the problem by dividing each entry by 1000.]

Referring to the table in part (A):

(B) $P(Y \mid H) = \dfrac{P(Y \cap H)}{P(H)} = \dfrac{.400}{.55} \approx .727$

(C) $P(Y \mid B) = \dfrac{P(Y \cap B)}{P(B)} = \dfrac{.02}{.15} \approx .133$

(D) $P(S) = .300$

$P(S \mid Y) = \dfrac{P(S \cap Y)}{P(Y)} = \dfrac{.180}{.60} = .300$

(E) $P(H) = .550$

$P(H \mid Y) = \dfrac{P(H \cap Y)}{P(Y)} = \dfrac{.400}{.600} \approx .667$

(F) $P(B \cap N) = .130$

(G) S and Y are independent since $P(S \mid Y) = P(S) = .300$

(H) H and Y are dependent since $P(H \mid Y) \approx .667$ is not equal to $P(H) = .550$.

(I) $P(B \mid N) = \dfrac{P(B \cap N)}{P(N)}$

$= \dfrac{.130}{.400}$ (from table)

$= .325$

and $P(B) = .150$. Since $P(B \mid N) \neq P(B)$, B and N are dependent.

49. The probability tree for this experiment is:

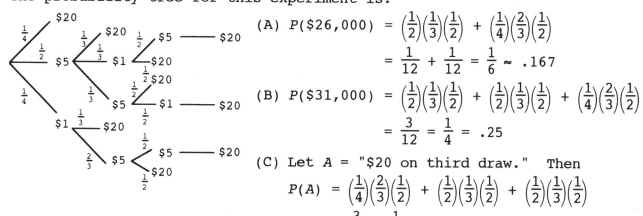

(A) $P(\$26{,}000) = \left(\dfrac{1}{2}\right)\left(\dfrac{1}{3}\right)\left(\dfrac{1}{2}\right) + \left(\dfrac{1}{4}\right)\left(\dfrac{2}{3}\right)\left(\dfrac{1}{2}\right)$

$= \dfrac{1}{12} + \dfrac{1}{12} = \dfrac{1}{6} \approx .167$

(B) $P(\$31{,}000) = \left(\dfrac{1}{2}\right)\left(\dfrac{1}{3}\right)\left(\dfrac{1}{2}\right) + \left(\dfrac{1}{2}\right)\left(\dfrac{1}{3}\right)\left(\dfrac{1}{2}\right) + \left(\dfrac{1}{4}\right)\left(\dfrac{2}{3}\right)\left(\dfrac{1}{2}\right)$

$= \dfrac{3}{12} = \dfrac{1}{4} = .25$

(C) Let A = "$20 on third draw." Then

$P(A) = \left(\dfrac{1}{4}\right)\left(\dfrac{2}{3}\right)\left(\dfrac{1}{2}\right) + \left(\dfrac{1}{2}\right)\left(\dfrac{1}{3}\right)\left(\dfrac{1}{2}\right) + \left(\dfrac{1}{2}\right)\left(\dfrac{1}{3}\right)\left(\dfrac{1}{2}\right)$

$= \dfrac{3}{12} = \dfrac{1}{4} = .25$

(D) The payoff table is:

x_i	$20	$21	$25	$26	$30	$31
P_i	$\frac{1}{4}$	$\frac{1}{12}$	$\frac{1}{6}$	$\frac{1}{6}$	$\frac{1}{12}$	$\frac{1}{4}$

Thus,

$$E(X) = \frac{1}{4}(20) + \frac{1}{12}(21) + \frac{1}{6}(25) + \frac{1}{6}(26) + \frac{1}{12}(30) + \frac{1}{4}(31)$$

$$= \frac{60 + 21 + 50 + 52 + 30 + 93}{12} = \frac{306}{12} = 25.5$$

and the expected value of the game is $25,500.

51. Let Event A = Adverse reaction was a loss of appetite
and Event S = Adverse reaction was a loss of sleep.

$n(A \cap S) = 1000 - (60 + 90 + 800) = 50$ subjects with both
adverse reactions

$n(A) = 60 + 50 = 110$

$n(S) = 90 + 50 = 140$

$P(A) = \frac{110}{1000} = \frac{11}{100}$

$P(S) = \frac{140}{1000} = \frac{14}{100}$

$P(A \cap S) = \frac{50}{1000} = \frac{5}{100}$

(A) $P(S \mid A) = \dfrac{P(S \cap A)}{P(A)} = \dfrac{\frac{5}{100}}{\frac{11}{100}} = \dfrac{5}{11}$ (B) $P(A \mid S) = \dfrac{P(A \cap S)}{P(S)} = \dfrac{\frac{5}{100}}{\frac{14}{100}} = \dfrac{5}{14}$

(C) $P(S \mid A') = \dfrac{P(S \cap A')}{P(A')} = \dfrac{P(\text{only reaction was loss of sleep})}{1 - P(A)}$

$$= \dfrac{\frac{90}{1000}}{1 - \frac{11}{1000}} = \dfrac{\frac{90}{1000}}{\frac{890}{1000}} = \dfrac{9}{89}$$

(D) $P(A \mid S') = \dfrac{P(A \cap S')}{P(S')} = \dfrac{P(\text{only reaction was loss of appetite})}{1 - P(S)}$

$$= \dfrac{\frac{60}{1000}}{1 - \frac{14}{1000}} = \dfrac{\frac{60}{1000}}{\frac{86}{1000}} = \dfrac{6}{86} = \dfrac{3}{43}$$

53. (A)

	Below 90 A	90–120 B	Above 120 C	Total
Female (F)	.130	.286	.104	.520
Male (F')	.120	.264	.096	.480
Total	.250	.550	.200	1.000

[Note: The probability table above was derived from the table given
in the problem by dividing each entry by 1000.]

Referring to the table in part (A):

(B) $P(A \mid F) = \dfrac{P(A \cap F)}{P(F)} = \dfrac{.130}{.520} \approx .250$ (C) $P(C \mid F) = \dfrac{P(C \cap F)}{P(F)} = \dfrac{.104}{.520} \approx .200$

$P(A \mid F') = \dfrac{P(A \cap F')}{P(F')} = \dfrac{.120}{.480} = .250$ $P(C \mid F') = \dfrac{P(C \cap F')}{P(F')} = \dfrac{.096}{.480} = .200$

(D) $P(A) = .25$ (E) $P(B) = .55$

$P(A \mid F) = \dfrac{P(A \cap F)}{P(F)} = \dfrac{.130}{.520} = .250$ $P(B \mid F') = \dfrac{P(B \cap F')}{P(F')} = \dfrac{.264}{.480} = .550$

(F) $P(F \cap C) = .104$ (G) No, the results in parts (B), (C), (D), and (E) imply that A, B, and C are independent of F and F'.

EXERCISE 8-3

Things to remember:

1. BAYES' FORMULA

    Let U_1, U_2, ..., U_n be n mutually exclusive events whose union is the sample space S. Let E be an arbitrary event in S such that $P(E) \neq 0$. Then

    $$P(U_1 \mid E) = \frac{P(U_1 \cap E)}{P(E)}$$

    $$= \frac{P(U_1 \cap E)}{P(U_1 \cap E) + P(U_2 \cap E) + \cdots + P(U_n \cap E)}$$

    $$= \frac{P(E \mid U_1)P(U_1)}{P(E \mid U_1)P(U_1) + \cdots + P(E \mid U_n)P(U_n)}$$

    Similar results hold for U_2, U_3, ..., U_n.

2. BAYES' FORMULA AND PROBABILITY TREES

    $$P(U_1 \mid E) = \frac{\text{Product of branch probabilities leading to } E \text{ through } U_1}{\text{Sum of all branch probabilities leading to } E}$$

    Similar results hold for U_2, U_3, ..., U_n.

---

1. $P(M \cap A) = P(M) \cdot P(A \mid M) = (.6)(.8) = .48$

3. $P(A) = P(M \cap A) + P(N \cap A) = P(M)P(A \mid M) + P(N)P(A \mid N)$
$$= (.6)(.8) + (.4)(.3) = .60$$

5. $P(M \mid A) = \dfrac{P(M \cap A)}{P(M \cap A) + P(N \cap A)} = \dfrac{.48}{.60}$ (see Problems 1 and 3)

$$= .80$$

7. Referring to the Venn diagram:

$$P(U_1 \mid R) = \frac{P(U_1 \cap R)}{P(R)} = \frac{\dfrac{25}{100}}{\dfrac{60}{100}} = \frac{25}{60} = \frac{5}{12} \approx .417$$

Using Bayes' formula:

$$P(U_1 \mid R) = \frac{P(U_1 \cap R)}{P(U_1 \cap R) + P(U_2 \cap R)} = \frac{P(U_1)P(R \mid U_1)}{P(U_1)P(R \mid U_1) + P(U_2)P(R \mid U_2)}$$

$$= \frac{\left(\dfrac{40}{100}\right)\left(\dfrac{25}{40}\right)}{\left(\dfrac{40}{100}\right)\left(\dfrac{25}{40}\right) + \left(\dfrac{60}{100}\right)\left(\dfrac{35}{60}\right)} = \frac{.25}{.25 + .35} = \frac{.25}{.60} = \frac{5}{12} \approx .417$$

9. $P(U_1 \mid R') = \dfrac{P(U_1 \cap R')}{P(R')} = \dfrac{\dfrac{15}{100}}{1 - P(R)}$ (from the Venn diagram)

$$= \frac{\dfrac{15}{100}}{1 - \dfrac{60}{100}} = \frac{\dfrac{15}{100}}{\dfrac{40}{100}} = \frac{3}{8} = .375$$

Using Bayes' formula:

$$P(U_1 \mid R') = \frac{P(U_1 \cap R')}{P(R')} = \frac{P(U_1)P(R' \mid U_1)}{P(U_1 \cap R') + P(U_2 \cap R')}$$

$$= \frac{P(U_1)P(R' \mid U_1)}{P(U_1)P(R' \mid U_1) + P(U_2)P(R' \mid U_2)} = \frac{\left(\dfrac{40}{100}\right)\left(\dfrac{15}{40}\right)}{\left(\dfrac{40}{100}\right)\left(\dfrac{15}{40}\right) + \left(\dfrac{60}{100}\right)\left(\dfrac{25}{60}\right)}$$

$$= \frac{.15}{.15 + .25} = \frac{15}{40} = \frac{3}{8} = .375$$

11. $P(U \mid C) = \dfrac{P(U \cap C)}{P(C)} = \dfrac{P(U \cap C)}{P(U \cap C) + P(V \cap C) + P(W \cap C)}$

$$= \frac{(.2)(.4)}{(.2)(.4) + (.5)(.2) + (.3)(.6)} \quad \text{[\underline{Note}: Recall } P(A \cap B) = P(A) \cdot P(B \mid A).]$$

$$= \frac{.08}{.36} \approx .222$$

13. $P(W \mid C) = \dfrac{P(W \cap C)}{P(C)} = \dfrac{P(W \cap C)}{P(W \cap C) + P(V \cap C) + P(U \cap C)}$

$$= \frac{(.3)(.6)}{(.3)(.6) + (.5)(.2) + (.2)(.4)} \quad \text{(see Problem 11)}$$

$$= \frac{.18}{.36} = .5$$

15. $P(V \mid C) = \dfrac{P(V \cap C)}{P(C)} = \dfrac{P(V \cap C)}{P(V \cap C) + P(W \cap C) + P(U \cap C)}$

$$= \dfrac{(.5)(.2)}{(.5)(.2) + (.3)(.6) + (.2)(.4)} = \dfrac{.1}{.36} = .278$$

17. From the Venn diagram,

$P(U_1 \mid R) = \dfrac{5}{5 + 15 + 20} = \dfrac{5}{40} = \dfrac{1}{8} = .125$

or $\quad = \dfrac{P(U_1 \cap R)}{P(R)} = \dfrac{\frac{5}{100}}{\frac{40}{100}} = .125$

Using Bayes' formula:

$P(U_1 \mid R) = \dfrac{P(U_1 \cap R)}{P(U_1 \cap R) + P(U_2 \cap R) + P(U_3 \cap R)} = \dfrac{\frac{5}{100}}{\frac{5}{100} + \frac{15}{100} + \frac{20}{100}}$

$$= \dfrac{.05}{.05 + .15 + .2} = \dfrac{.05}{.40} = .125$$

19. From the Venn diagram,

$P(U_3 \mid R) = \dfrac{20}{5 + 15 + 20} = \dfrac{20}{40} = .5$

Using Bayes' formula:

$P(U_3 \mid R) = \dfrac{P(U_3 \cap R)}{P(U_1 \cap R) + P(U_2 \cap R) + P(U_3 \cap R)} = \dfrac{\frac{20}{100}}{\frac{5}{100} + \frac{15}{100} + \frac{20}{100}}$

$$= \dfrac{.2}{.05 + .15 + .2} = \dfrac{.2}{.4} = .5$$

21. From the Venn diagram,

$P(U_2 \mid R) = \dfrac{15}{5 + 15 + 20} = \dfrac{15}{40} = .375$

Using Bayes' formula:

$P(U_2 \mid R) = \dfrac{P(U_2 \cap R)}{P(U_1 \cap R) + P(U_2 \cap R) + P(U_3 \cap R)} = \dfrac{\frac{15}{100}}{\frac{5}{100} + \frac{15}{100} + \frac{20}{100}}$

$$= \dfrac{.15}{.05 + .15 + .2} = \dfrac{.15}{.40}$$

$$= \dfrac{3}{8} = .375$$

23. From the given tree diagram, we have:

$$P(A) = \dfrac{1}{4} \qquad\qquad P(A') = \dfrac{3}{4}$$

$$P(B \mid A) = \dfrac{1}{5} \qquad\qquad P(B \mid A') = \dfrac{3}{5}$$

$$P(B' \mid A) = \dfrac{4}{5} \qquad\qquad P(B' \mid A') = \dfrac{2}{5}$$

We want to find the following:

$$P(B) = P(B \cap A) + P(B \cap A') = P(A)P(B \mid A) + P(A')P(B \mid A')$$

$$= \left(\tfrac{1}{4}\right)\left(\tfrac{1}{5}\right) + \left(\tfrac{3}{4}\right)\left(\tfrac{3}{5}\right) = \tfrac{1}{20} + \tfrac{9}{20} = \tfrac{10}{20} = \tfrac{1}{2}$$

$$P(B') = 1 - P(B) = 1 - \tfrac{1}{2} = \tfrac{1}{2}$$

$$P(A \mid B) = \frac{P(A \cap B)}{P(B)} = \frac{P(A)P(B \mid A)}{P(B)} = \frac{\left(\tfrac{1}{4}\right)\left(\tfrac{1}{5}\right)}{\tfrac{1}{2}} = \frac{\tfrac{1}{20}}{\tfrac{1}{2}} = \tfrac{1}{10}$$

Thus, $P(A' \mid B) = 1 - P(A \mid B) = 1 - \tfrac{1}{10} = \tfrac{9}{10}.$

$$P(A \mid B') = \frac{P(A \cap B')}{P(B')} = \frac{P(A)P(B' \mid A)}{P(B')} = \frac{\left(\tfrac{1}{4}\right)\left(\tfrac{4}{5}\right)}{\tfrac{1}{2}} = \frac{\tfrac{4}{20}}{\tfrac{1}{2}} = \tfrac{2}{5}$$

Thus, $P(A' \mid B') = 1 - P(A \mid B') = 1 - \tfrac{2}{5} = \tfrac{3}{5}.$

Therefore, the tree diagram for this problem is as shown at the right.

The following tree diagram is to be used for Problems 25 and 27.

25. $P(U_1 \mid W) = \dfrac{P(U_1 \cap W)}{P(W)} = \dfrac{P(U_1 \cap W)}{P(U_1 \cap W) + P(U_2 \cap W)}$

$$= \frac{P(U_1)P(W \mid U_1)}{P(U_1)P(W \mid U_1) + P(U_2)P(W \mid U_2)} = \frac{(.5)(.2)}{(.5)(.2) + (.5)(.6)}$$

$$= \frac{.1}{.4} = .25$$

27. $P(U_2 \mid R) = \dfrac{P(U_2 \cap R)}{P(R)} = \dfrac{P(U_2 \cap R)}{P(U_2 \cap R) + P(U_1 \cap R)}$

$$= \frac{P(U_2)P(R \mid U_2)}{P(U_2)P(R \mid U_2) + P(U_1)P(R \mid U_1)} = \frac{(.5)(.4)}{(.5)(.4) + (.5)(.8)}$$

$$= \frac{.4}{1.2} = \frac{1}{3} \approx .333$$

29. $P(W_1 \mid W_2) = \dfrac{P(W_1 \cap W_2)}{P(W_2)} = \dfrac{P(W_1)P(W_2 \mid W_1)}{P(R_1 \cap W_2) + P(W_1 \cap W_2)}$

$$= \dfrac{P(W_1)P(W_2 \mid W_1)}{P(R_1)P(W_2 \mid R_1) + P(W_1)P(W_2 \mid W_1)} = \dfrac{\left(\frac{5}{9}\right)\left(\frac{4}{8}\right)}{\left(\frac{4}{9}\right)\left(\frac{5}{8}\right) + \left(\frac{5}{9}\right)\left(\frac{4}{8}\right)} = \dfrac{\frac{20}{72}}{\frac{20}{72} + \frac{20}{72}}$$

$$= \dfrac{20}{40} = \dfrac{1}{2} \text{ or } .5$$

31. $P(U_{R_1} \mid U_{R_2}) = \dfrac{P(U_{R_1} \cap U_{R_2})}{P(U_{R_2})} = \dfrac{P(U_{R_1})P(U_{R_2} \mid U_{R_1})}{P(U_{W_1} \cap U_{R_2}) + P(U_{R_1} \cap U_{R_2})}$

$$= \dfrac{P(U_{R_1})P(U_{R_2} \mid U_{R_1})}{P(U_{W_1})P(U_{R_2} \mid U_{W_1}) + P(U_{R_1})P(U_{R_2} \mid U_{R_1})}$$

$$= \dfrac{\left(\frac{7}{10}\right)\left(\frac{5}{10}\right)}{\left(\frac{3}{10}\right)\left(\frac{4}{10}\right) + \left(\frac{7}{10}\right)\left(\frac{5}{10}\right)} = \dfrac{.35}{.12 + .35}$$

$$= \dfrac{.35}{.47} = \dfrac{35}{47} \approx .745$$

The tree diagram follows:

where U_{R_1} is red from urn one,

U_{R_2} is red from urn two,

U_{W_1} is white from urn one,

and U_{W_2} is white from urn two.

33.

$P(H_1 \mid H_2) = \dfrac{P(H_1 \cap H_2)}{P(H_2)} = \dfrac{P(H_1 \cap H_2)}{P(H_1 \cap H_2) + P(\overline{H}_1 \cap H_2)}$

$$= \dfrac{P(H_1)P(H_2 \mid H_1)}{P(H_1)P(H_2 \mid H_1) + P(\overline{H}_1)P(H_2 \mid \overline{H}_1)} = \dfrac{\frac{13}{52} \cdot \frac{12}{51}}{\frac{13}{52} \cdot \frac{12}{51} + \frac{39}{52} \cdot \frac{13}{51}}$$

$$= \dfrac{13(12)}{13(12) + 39(13)} = \dfrac{12}{51} \approx .235$$

35. Consider the following Venn diagram:

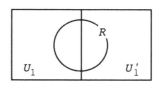

$$P(U_1 \mid R) = \frac{P(U_1 \cap R)}{P(U_1 \cap R) + P(U_1' \cap R)} \quad \text{and} \quad P(U_1' \mid R) = \frac{P(U_1' \cap R)}{P(U_1 \cap R) + P(U_1' \cap R)}$$

Adding these two equations, we obtain:

$$P(U_1 \mid R) + P(U_1' \mid R) = \frac{P(U_1 \cap R)}{P(U_1 \cap R) + P(U_1' \cap R)} + \frac{P(U_1' \cap R)}{P(U_1 \cap R) + P(U_1' \cap R)}$$

$$= \frac{P(U_1 \cap R) + P(U_1' \cap R)}{P(U_1 \cap R) + P(U_1' \cap R)} = 1$$

37. Consider the following tree diagram:

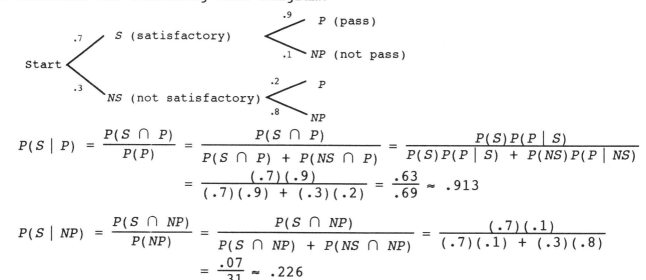

$$P(S \mid P) = \frac{P(S \cap P)}{P(P)} = \frac{P(S \cap P)}{P(S \cap P) + P(NS \cap P)} = \frac{P(S)P(P \mid S)}{P(S)P(P \mid S) + P(NS)P(P \mid NS)}$$

$$= \frac{(.7)(.9)}{(.7)(.9) + (.3)(.2)} = \frac{.63}{.69} \approx .913$$

$$P(S \mid NP) = \frac{P(S \cap NP)}{P(NP)} = \frac{P(S \cap NP)}{P(S \cap NP) + P(NS \cap NP)} = \frac{(.7)(.1)}{(.7)(.1) + (.3)(.8)}$$

$$= \frac{.07}{.31} \approx .226$$

39. Consider the following tree diagram:

$$P(A \mid D) = \frac{P(A \cap D)}{P(D)}, \text{ where}$$

$$P(D) = P(A \cap D) + P(B \cap D) + P(C \cap D)$$
$$= P(A)P(D \mid A) + P(B)P(D \mid B) + P(C)P(D \mid C)$$
$$= (.2)(.01) + (.40)(.03) + (.40)(.02)$$
$$= .002 + .012 + .008$$
$$= .022$$

Thus, $P(A \mid D) = \dfrac{P(A \cap D)}{P(D)} = \dfrac{P(A)P(D \mid A)}{P(D)} = \dfrac{(.20)(.01)}{.022} = \dfrac{.002}{.022} = \dfrac{2}{22}$ or .091

Similarly,

$$P(B \mid D) = \frac{P(B \cap D)}{P(D)} = \frac{P(B)P(D \mid B)}{P(D)} = \frac{(.40)(.03)}{.022} = \frac{.012}{.022} = \frac{6}{11} \text{ or } .545,$$

$$\text{and } P(C \mid D) = \frac{P(C \cap D)}{P(D)} = \frac{P(C)P(D \mid C)}{P(D)} = \frac{(.40)(.02)}{.022} = \frac{.008}{.022} = \frac{4}{11} \text{ or } .364.$$

41. Consider the following tree diagram:

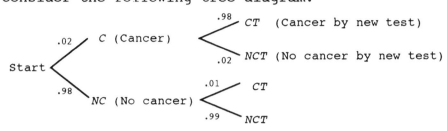

$$P(C \mid CT) = \frac{P(C \cap CT)}{P(CT)} = \frac{P(C)P(CT \mid C)}{P(C \cap CT) + P(NC \cap CT)} = \frac{P(C)P(CT \mid C)}{P(C)P(CT \mid C) + P(NC)P(CT \mid NC)}$$

$$= \frac{(.02)(.98)}{(.02)(.98) + (.98)(.01)} = \frac{.0196}{.0196 + .0098} = \frac{.0196}{.0294} = .6667$$

$$P(C \mid NCT) = \frac{P(C \cap NCT)}{P(NCT)} = \frac{P(C)P(NCT \mid C)}{P(C)P(NCT \mid C) + P(NC)P(NCT \mid NC)}$$

$$= \frac{(.02)(.02)}{(.02)(.02) + (.98)(.99)} \approx .000412$$

43. Consider the following tree diagram.

```
                              .4    HD (Heavy drinker)
            .07   L (Liver ailment)  .5  MD (Moderate drinker)
                                     .1  ND (Nondrinker)
   Start
            .93                      .1    HD
                  NL (No liver ailment)  .7  MD
                                      .2  ND
```

$$P(L \mid HD) = \frac{P(L \cap HD)}{P(HD)} = \frac{P(L)P(HD \mid L)}{P(L \cap HD) + P(NL \cap HD)} = \frac{P(L)P(HD \mid L)}{P(L)P(HD \mid L) + P(NL)P(HD \mid NL)}$$

$$= \frac{(.07)(.4)}{(.07)(.4) + (.93)(.1)} \quad \text{(from the tree diagram)}$$

$$= \frac{.028}{.028 + .093} = \frac{.028}{.121} = \frac{28}{121} \approx .231$$

$$P(L \mid ND) = \frac{P(L \cap ND)}{P(ND)} = \frac{P(L)P(ND \mid L)}{P(L \cap ND) + P(NL \cap ND)} = \frac{P(L)P(ND \mid L)}{P(L)P(ND \mid L) + P(NL)P(ND \mid NL)}$$

$$= \frac{(.07)(.1)}{(.07)(.1) + (.93)(.2)} \quad \text{(from the tree diagram)}$$

$$= \frac{.007}{.007 + .186} = \frac{.007}{.194} = \frac{7}{194} \approx .036$$

45. Consider the following tree diagram.

$$P(L \mid LT) = \frac{P(L \cap LT)}{P(LT)} = \frac{P(L \cap LT)}{P(L \cap LT) + P(\overline{L} \cap LT)} = \frac{(.5)(.8)}{(.5)(.8) + (.5)(.05)}$$

$$= \frac{.4}{.425} \approx .941$$ If the test indicates that the subject was lying, then he was lying with a probability of 0.941.

$$P(\overline{L} \mid LT) = \frac{P(\overline{L} \cap LT)}{P(LT)} = \frac{(.5)(.05)}{(.5)(.8) + (.5)(.05)}$$

$$= \frac{.05}{.85} \approx .0588$$ If the test indicates that the subject was lying, there is still a probability of 0.0588 that he was not lying.

EXERCISE 8-4   CHAPTER REVIEW

1. $P(A) = .3$, $P(B) = .4$, $P(A \cap B) = .1$

(A) $P(A') = 1 - P(A) = 1 - .3 = .7$

(B) $P(A \cup B) = P(A) + P(B) - P(A \cap B) = .3 + .4 - .1 = .6$

2. Since the spinner cannot land on R and G simultaneously, $R \cap G = \varnothing$. Thus,

$P(R \cup G) = P(R) + P(G) = .3 + .5 = .8$

The odds for an event E are: $\frac{P(E)}{P(E')}$

Thus, the odds for landing on either R or G are: $\frac{P(R \cup G)}{P[(R \cup G)']} = \frac{.8}{.2} = \frac{8}{2}$ or the odds are 8 to 2.

3. If the odds for an event E are a to b, then $P(E) = \frac{a}{a + b}$. Thus, the probability of rolling an 8 before rolling a 7 is: $\frac{5}{11} \approx .455$.

4. $P(T) = .27$ **5.** $P(Z) = .20$ **6.** $P(T \cap Z) = .02$ **7.** $P(R \cap Z) = .03$

8. $P(R \mid Z) = \frac{P(R \cap Z)}{P(Z)} = \frac{.03}{.20} = .15$ **9.** $P(Z \mid R) = \frac{P(Z \cap R)}{P(R)} = \frac{.03}{.23} \approx .1304$

10. $P(T \mid Z) = \frac{P(T \cap Z)}{P(Z)} = \frac{.02}{.20} = .10$

11. No, because $P(T \cap Z) = .02 \neq P(T) \cdot P(Z) = (.27)(.20) = .054$.

12. Yes, because $P(S \cap X) = .10 = P(S) \cdot P(X) = (.5)(.2)$.

13. $P(A) = .4$ from the tree diagram.   **14.** $P(B \mid A) = .2$ from the tree diagram.

15. $P(B \mid A') = .3$ from the tree diagram

16. $P(A \cap B) = P(A)P(B \mid A) = (.4)(.2) = .08$

17. $P(A' \cap B) = P(A')P(B \mid A') = (.6)(.3) = .18$

18. $\begin{aligned} P(B) &= P(A \cap B) + P(A' \cap B) \\ &= P(A)P(B \mid A) + P(A')P(B \mid A') \\ &= (.4)(.2) + (.6)(.3) \\ &= .08 + .18 \\ &= .26 \end{aligned}$

19. $P(A \mid B) = \dfrac{P(A \cap B)}{P(B)} = \dfrac{P(A)P(B \mid A)}{P(A \cap B) + P(A' \cap B)} = \dfrac{P(A)P(B \mid A)}{P(A)P(B \mid A) + P(A')P(B \mid A')}$

$= \dfrac{(.4)(.2)}{(.4)(.2) + (.6)(.3)}$ (from the tree diagram)

$= \dfrac{.08}{.26} = \dfrac{8}{26}$ or $.307 \approx .31$

20. $P(A \mid B') = \dfrac{P(A \cap B')}{P(B')} = \dfrac{P(A)P(B' \mid A)}{1 - P(B)} = \dfrac{(.4)(.8)}{1 - .26}$ $[P(B) = .26$, see Problem 18.$]$

$= \dfrac{.32}{.74} = \dfrac{16}{37}$ or $.432$

21. (A) $P(\text{jack or queen}) = P(\text{jack}) + P(\text{queen}) = \dfrac{4}{52} + \dfrac{4}{52} = \dfrac{8}{52} = \dfrac{2}{13}$

[<u>Note</u>: jack $\cap$ queen $= \varnothing$.]

The odds for drawing a jack or queen are 2 to 11.

(B) $P(\text{jack or spade}) = P(\text{jack}) + P(\text{spade}) - P(\text{jack and spade})$
$= \dfrac{4}{52} + \dfrac{13}{52} - \dfrac{1}{52} = \dfrac{16}{52} = \dfrac{4}{13}$

The odds for drawing a jack or a spade are 4 to 9.

(C) $P(\text{ace}) = \dfrac{4}{52} = \dfrac{1}{13}$.   Thus,

$P(\text{card other than an ace}) = 1 - P(\text{ace}) = 1 - \dfrac{1}{13} = \dfrac{12}{13}$

The odds for drawing a card other than an ace are 12 to 1.

22. (A) The probability of rolling a 5 is $\dfrac{4}{36} = \dfrac{1}{9}$.

Thus, the odds for rolling a five are 1 to 8.

(B) Let x = amount house should pay (and return the \$1 bet). Then, for the game to be fair,

$$E(X) = x\left(\frac{1}{9}\right) + (-1)\left(\frac{8}{9}\right) = 0$$

$$\frac{x}{9} - \frac{8}{9} = 0$$

$$x = 8$$

Thus, the house should pay \$8.

23. The event A that corresponds to the sum being divisible by 4 includes sums 4, 8, and 12. This set is:

$A = \{(1, 3), (2, 2), (3, 1), (2, 6), (3, 5), (4, 4), (5, 3), (6, 2), (6, 6)\}$

The event B that corresponds to the sum being divisible by 6 includes sums 6 and 12. This set is:

$B = \{(1, 5), (2, 4), (3, 3), (4, 2), (5, 1), (6, 6)\}$

$$P(A) = \frac{n(A)}{n(S)} = \frac{9}{36} = \frac{1}{4}$$

$$P(B) = \frac{n(B)}{n(S)} = \frac{6}{36} = \frac{1}{6}$$

$$P(A \cap B) = \frac{1}{36} \quad [\underline{\text{Note}}:\ A \cap B = \{(6, 6)\}]$$

$$P(A \cup B) = \frac{14}{36} \ \text{ or } \ \frac{7}{18} \qquad [\underline{\text{Note}}:\ A \cup B = \{(1, 3), (2, 2), (3, 1), (2, 6),$$
$$(3, 5), (4, 4), (5, 3), (6, 2), (6, 6),$$
$$(1, 5), (2, 4), (3, 3), (4, 2), (5, 1)\}]$$

24. (A) $P(\text{odd number}) = P(1) + P(3) + P(5) = .2 + .3 + .1 = .6$

(B) Let E = "number less than 4,"
and F = "odd number."
Now, $E \cap F = \{1, 3\}$, $F = \{1, 3, 5\}$.

$$P(E \mid F) = \frac{P(E \cap F)}{P(F)} = \frac{.2 + .3}{.6} = \frac{5}{6}$$

25. Let E = "card is red" and F = "card is an ace." Then $F \cap E$ = "card is a red ace."

(A) $P(F \mid E) = \dfrac{P(F \cap E)}{P(E)} = \dfrac{2/52}{26/52} = \dfrac{1}{13}$

(B) $P(F \cap E) = \dfrac{1}{26}$, and $P(E) = \dfrac{1}{2}$, $P(F) = \dfrac{1}{13}$. Thus,
$P(F \cap E) = P(E) \cdot P(F)$, and E and F are independent.

26. (A) The tree diagram with replacement is:

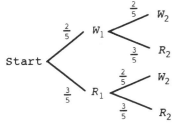

$$P(W_1 \cap R_2) = P(W_1)P(R_2 \mid W_1)$$
$$= \frac{2}{5} \cdot \frac{3}{5} = \frac{6}{25} \approx .24$$

(B) The tree diagram without replacement is:

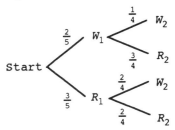

$$P(W_1 \cap R_2) = P(W_1)P(R_2 \mid W_1)$$
$$= \frac{2}{5} \cdot \frac{3}{4} = \frac{6}{20} = .3$$

27. Part (B) involves dependent events because

$$P(R_2 \mid W_1) = \frac{3}{4}$$

$$P(R_2) = P(W_1 \cap R_2) + P(R_1 \cap R_2) = \frac{6}{20} + \frac{6}{20} = \frac{12}{20} = \frac{3}{5}$$

and $P(R_2 \mid W_1) \neq P(R_2)$

The events in part (A) are independent.

28. (A) Using the tree diagram in Problem 26(A), we have:

$$P(\text{zero red balls}) = P(W_1 \cap W_2) = P(W_1)P(W_2) = \frac{2}{5} \cdot \frac{2}{5} = \frac{4}{25} = .16$$

$$P(\text{one red ball}) = P(W_1 \cap R_2) + P(R_1 \cap W_2)$$
$$= P(W_1)P(R_2) + P(R_1)P(W_2)$$
$$= \frac{2}{5} \cdot \frac{3}{5} + \frac{3}{5} \cdot \frac{2}{5} = \frac{12}{25} = .48$$

$$P(\text{two red balls}) = P(R_1 \cap R_2) = P(R_1)P(R_2) = \frac{3}{5} \cdot \frac{3}{5} = \frac{9}{25} = .36$$

Thus, the probability distribution is:

Number of red balls x_i	Probability p_i
0	.16
1	.48
2	.36

The expected number of red balls is:

$$E(X) = 0(.16) + 1(.48) + 2(.36) = .48 + .72 = 1.2$$

(B) Using the tree diagram in Problem 26(B), we have:

$$P(\text{zero red balls}) = P(W_1 \cap W_2) = P(W_1)P(W_2 \mid W_1) = \frac{2}{5} \cdot \frac{1}{4} = \frac{1}{10} = .1$$

$$P(\text{one red ball}) = P(W_1 \cap R_2) + P(R_1 \cap W_2)$$
$$= P(W_1)P(R_2 \mid W_1) + P(R_1)P(W_2 \mid R_1)$$
$$= \frac{2}{5} \cdot \frac{3}{4} + \frac{3}{5} \cdot \frac{2}{4} = \frac{12}{20} = \frac{3}{5} = .6$$

$$P(\text{two red balls}) = P(R_1 \cap R_2) = P(R_1)P(R_2 \mid R_1) = \frac{3}{5} \cdot \frac{2}{4} = \frac{6}{20} = .3$$

Thus, the probability distribution is:

Number of red balls x_i	Probability p_i
0	.1
1	.6
2	.3

The expected number of red balls is:

$$E(X) = 0(.1) + 1(.6) + 2(.3) = 1.2$$

29. The tree diagram for this problem is as follows:

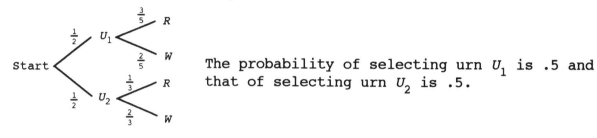

The probability of selecting urn U_1 is .5 and that of selecting urn U_2 is .5.

(A) $P(R \mid U_1) = \dfrac{3}{5}$ (B) $P(R \mid U_2) = \dfrac{1}{3}$

(C) $P(R) = P(R \cap U_1) + P(R \cap U_2)$
$= P(U_1)P(R \mid U_1) + P(U_2)P(R \mid U_2)$
$= \dfrac{1}{2} \cdot \dfrac{3}{5} + \dfrac{1}{2} \cdot \dfrac{1}{3} = \dfrac{28}{60} = \dfrac{7}{15} \approx .4667$

(D) $P(U_1 \mid R) = \dfrac{P(U_1 \cap R)}{P(R)} = \dfrac{P(U_1)P(R \mid U_1)}{P(U_1)P(R \mid U_1) + P(U_2)P(R \mid U_2)}$

$= \dfrac{\dfrac{1}{2} \cdot \dfrac{3}{5}}{\dfrac{1}{2} \cdot \dfrac{3}{5} + \dfrac{1}{2} \cdot \dfrac{1}{3}} = \dfrac{\dfrac{3}{10}}{\dfrac{7}{15}} = \dfrac{9}{14} \approx .6429$

(E) $P(U_2 \mid W) = \dfrac{P(U_2 \cap W)}{P(W)} = \dfrac{P(U_2)P(W \mid U_2)}{P(U_2)P(W \mid U_2) + P(U_1)P(W \mid U_1)}$

$= \dfrac{\dfrac{1}{2} \cdot \dfrac{2}{3}}{\dfrac{1}{2} \cdot \dfrac{2}{3} + \dfrac{1}{2} \cdot \dfrac{2}{5}} = \dfrac{\dfrac{2}{3}}{\dfrac{16}{15}} = \dfrac{5}{8} = .625$

(F) $P(U_1 \cap R) = P(U_1)P(R \mid U_1) = \dfrac{1}{2} \cdot \dfrac{3}{5} = .3$

[<u>Note</u>: In parts (A)—(F), we derived the values of the probabilities from the tree diagram.]

30. No, because $P(R \mid U_1) \neq P(R)$. (See Problem 28.)

31. Let A = "number selected is divisible by 3." Then

$n(A) = \dfrac{200}{3} = 66.67$ or 66.

Let B = "number selected is divisible by 5." Then

$n(B) = \dfrac{200}{5} = 40$.

Now, $A \cap B$ = "number selected is divisible by 3 and 5"

= "number selected is divisible by 15."

Thus, $n(A \cap B) = \dfrac{200}{15} = 13.33$ or 13.

$$P(A \cup B) = P(A) + P(B) - P(A \cap B)$$
$$= \frac{66}{200} + \frac{40}{200} - \frac{13}{200} = \frac{93}{200} = .465$$

32. The number of numbers between 1 and 30 (inclusive) which are divisible by 4 is:

$\dfrac{30}{4} = 7.5$ or 7

Let A_1 = "first number is divisible by 4"
A_2 = "second number is divisible by 4"
A_3 = "third number is divisible by 4."

The tree diagram for this experiment is shown at the right.

Let B = "at least one number is divisible by 4." Then

$$P(B) = 1 - P(A_1') \cdot P(A_2') \cdot P(A_3') = 1 - \frac{23}{30} \cdot \frac{22}{29} \cdot \frac{21}{28}$$
$$\approx 1 - .436 = .564.$$

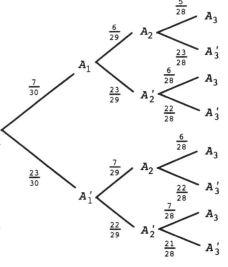

33. $P(\text{second heart} \mid \text{first heart}) = P(H_2 \mid H_1) = \dfrac{12}{51} \approx .235$

[<u>Note</u>: One can see that $P(H_2 \mid H_1) = \dfrac{12}{51}$ directly.]

34. $P(\text{first heart} \mid \text{second heart}) = P(H_1 \mid H_2)$

$$= \frac{P(H_1 \cap H_2)}{P(H_2)} = \frac{P(H_1)P(H_2 \mid H_1)}{P(H_2)}$$

$$= \frac{P(H_1)P(H_2 \mid H_1)}{P(H_1 \cap H_2) + P(H_1' \cap H_2)}$$

$$= \frac{P(H_1)P(H_2 \mid H_1)}{P(H_1)P(H_2 \mid H_1) + P(H_1')P(H_2 \mid H_1')}$$

$$= \frac{\dfrac{13}{52} \cdot \dfrac{12}{51}}{\dfrac{13}{52} \cdot \dfrac{12}{51} + \dfrac{39}{52} \cdot \dfrac{13}{51}} = \frac{12}{51} \approx .235$$

35. The tree diagram for this experiment is:

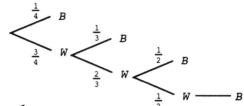

(A) P(black on the fourth draw) $= \dfrac{3}{4} \cdot \dfrac{2}{3} \cdot \dfrac{1}{2} = \dfrac{1}{4}$

The odds for black on the fourth draw are 1 to 3.

(B) Let x = amount house should pay (and return the $1 bet). Then, for the game to be fair:

$$E(X) = x\left(\dfrac{1}{4}\right) + (-1)\left(\dfrac{3}{4}\right) = 0$$

$$\dfrac{x}{4} - \dfrac{3}{4} = 0$$

$$x = 3$$

Thus, the house should pay $3.

36. $n(S) = 10 \cdot 10 \cdot 10 \cdot 10 \cdot 10 = 10^5$

Let event A = "at least two people identify the same book." Then A' = "each person identifies a different book," and

$$n(A') = 10 \cdot 9 \cdot 8 \cdot 7 \cdot 6 = \dfrac{10!}{5!}$$

Thus, $P(A') = \dfrac{\dfrac{10!}{5!}}{10^5} = \dfrac{10!}{5! \, 10^5}$ and $P(A) = 1 - \dfrac{10!}{5! \, 10^5} \approx 1 - .3 = .7.$

37. Venn diagram:

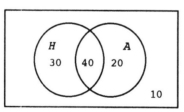

Event H = video games played at home.
Event A = video games played at arcades.

(A) $P(H$ or $A) = P(H \cup A) = \dfrac{90}{100}$ (from the Venn diagram)

$= .9$

or $= P(H) + P(A) - P(H \cap A)$

$= \dfrac{70}{100} + \dfrac{60}{100} - \dfrac{40}{100} = \dfrac{90}{100} = .9$

(B) P(played only at home) $= P(H \cap A')$

$= \dfrac{30}{100}$ (from the Venn diagram)

$= .3$

38.

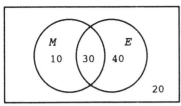

Event M = Reads the morning paper.
Event E = Reads the evening paper.

(A) $P(\text{reads a daily paper}) = P(M \text{ or } E) = P(M \cup E)$
$$= P(M) + P(E) - P(M \cap E)$$
$$= \frac{40}{100} + \frac{70}{100} - \frac{30}{100} = .8$$

(B) $P(\text{does not read a daily paper}) = \frac{20}{100}$ (from the Venn diagram)
$$= .2$$
$$\text{or} = 1 - P(M \cup E) \quad [\text{i.e., } P((M \cup E)')]$$
$$= 1 - .8 = .2$$

(C) $P(\text{reads exactly one daily paper}) = \frac{10 + 40}{100}$ (from the Venn diagram)
$$= .5$$
$$\text{or} \quad P((M \cap E') \text{ or } (M' \cap E))$$
$$= P(M \cap E') + P(M' \cap E)$$
$$= \frac{10}{100} + \frac{40}{100} = .5$$

39. $n(S) = C_{30,10} = \dfrac{30!}{10!20!}$

Let A = "at least one defective part." Then A' = "no defective parts,"
and $n(A') = \dfrac{25!}{10!15!}$.

Thus,

$$P(A) = 1 - P(A') = 1 - \frac{\dfrac{25!}{10!15!}}{\dfrac{30!}{10!20!}} = 1 - \frac{25!20!}{15!30!} \approx 1 - .109 = .891$$

40. Let A be the event that a person has seen the advertising and P be the event that the person purchased the product. Given:

$P(A) = .4$ and $P(P \mid A) = .85$

We want to find:

$P(A \cap P) = P(A)P(P \mid A) = (.4)(.85) = .34$

41. Let Event NH = individual with normal heart,
     Event MH = individual with minor heart problem,
     Event SH = individual with severe heart problem,
and Event P = individual passes the cardiogram test.

Then, using the notation given above, we have:

$$P(NH) = .82$$
$$P(MH) = .11$$
$$P(SH) = .07$$
$$P(P \mid NH) = .95$$
$$P(P \mid MH) = .30$$
$$P(P \mid SH) = .05$$

We want to find $P(NH \mid P) = \dfrac{P(NH \cap P)}{P(P)} = \dfrac{P(NH)\,P(P \mid NH)}{P(NH \cap P) + P(MH \cap P) + P(SH \cap P)}$

$$= \dfrac{P(NH)\,P(P \mid NH)}{P(NH)\,P(P \mid NH) + P(MH)\,P(P \mid MH) + P(SH)\,P(P \mid SH)}$$

$$= \dfrac{(.82)(.95)}{(.82)(.95) + (.11)(.30) + (.07)(.05)} = .955$$

42. The tree diagram for this problem is as follows:

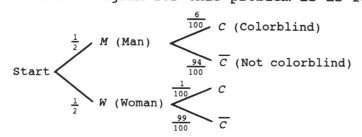

We now compute

$$P(M \mid C) = \dfrac{P(M \cap C)}{P(C)} = \dfrac{P(M \cap C)}{P(M \cap C) + P(W \cap C)} = \dfrac{P(M)\,P(C \mid M)}{P(M)\,P(C \mid M) + P(W)\,P(C \mid W)}$$

$$= \dfrac{\dfrac{1}{2} \cdot \dfrac{6}{100}}{\dfrac{1}{2} \cdot \dfrac{6}{100} + \dfrac{1}{2} \cdot \dfrac{1}{100}} = \dfrac{6}{7} \approx .857$$

EXERCISE 9-1

Things to remember:

1. LIMIT

   We write

   $$\lim_{x \to c} f(x) = L \text{ or } f(x) \to L \text{ as } x \to c$$

   if the functional value $f(x)$ is close to the single real number L whenever x is close to but not equal to c (on either side of c).

   [Note: The existence of a limit at c has nothing to do with the value of the function at c. In fact, c may not even be in the domain of f (see Example 3). However, the function must be defined on both sides of c.]

2. ONE-SIDED LIMITS

   We write $\lim\limits_{x \to c^-} f(x) = K$ [$x \to c^-$ is read "x approaches c from the left" and means $x \to c$ and $x < c$] and call K the LIMIT FROM THE LEFT or LEFT-HAND LIMIT if $f(x)$ is close to K whenever x is close to c, but to the left of c on the real number line.

   We write $\lim\limits_{x \to c^+} f(x) = L$ [$x \to c^+$ is read "x approaches c from the right" and means $x \to c$ and $x > c$] and call L the LIMIT FROM THE RIGHT or RIGHT-HAND LIMIT if $f(x)$ is close to L whenever x is close to c, but to the right of c on the real number line.

3. EXISTENCE OF A LIMIT

   In order for a limit to exist, the limit from the left and the limit from the right must both exist, and must be equal.

4. CONTINUITY

   A function f is CONTINUOUS AT THE POINT $x = c$ if:

   (a) $\lim\limits_{x \to c} f(x)$ exists;

   (b) $f(c)$ exists;

   (c) $\lim\limits_{x \to c} f(x) = f(c)$

   A function is CONTINUOUS ON THE OPEN INTERVAL (a, b) if it is continuous at each point on the interval.

5. ONE-SIDED CONTINUITY

A function f is CONTINUOUS ON THE LEFT AT $x = c$ if $\lim\limits_{x \to c^-} f(x) = f(c)$; f is CONTINUOUS ON THE RIGHT AT $x = c$ if $\lim\limits_{x \to c^+} f(x) = f(c)$.

The function f is continuous on the closed interval $[a, b]$ if it is continuous on the open interval (a, b), and is continuous on the right at a and continuous on the left at b.

6. CONTINUITY PROPERTIES OF SOME SPECIFIC FUNCTIONS

(a) A constant function, $f(x) = k$, is continuous for all x.

(b) For n a positive integer, $f(x) = x^n$ is continuous for all x.

(c) A polynomial function
$$P(x) = a_n x^n + a_{n-1} x^{n-1} + \ldots + a_1 x + a_0$$
is continuous for all x.

(d) A rational function
$$R(x) = \frac{P(x)}{Q(x)},$$
P and Q polynomial functions, is continuous for all x except those numbers $x = c$ such that $Q(c) = 0$.

(e) For n an odd positive integer, $n > 1$, $\sqrt[n]{f(x)}$ is continuous wherever f is continuous.

(f) For n an even positive integer, $\sqrt[n]{f(x)}$ is continuous wherever f is continuous and non-negative.

---

1. (A) $\lim\limits_{x \to 0} f(x) = 2$ (B) $\lim\limits_{x \to 1} f(x) = 2$

(C) $\lim\limits_{x \to 2} f(x)$ does not exist   (D) $\lim\limits_{x \to 4} f(x) = 4$
[Note: $\lim\limits_{x \to 2^-} f(x) = 1$, $\lim\limits_{x \to 2^+} f(x) = 2$.]

3. (A) $f(0) = 2$ (C) $f(2) = 2$
(B) $f(1) = 2$ (D) $f(4)$ is not defined

5. $\lim\limits_{x \to c} f(x) = f(c)$ for $c = 0$ and $c = 1$.

7. f is discontinuous at $x = 2$ because $\lim\limits_{x \to 2^-} f(x) \neq \lim\limits_{x \to 2^+} f(x)$, i.e., $\lim\limits_{x \to 2} f(x)$ does not exits.
f is discontinuous at $x = 4$ since f is not defined at 4.

9. $f(x) = 2x - 3$ is a polynomial function. Therefore, f is continuous for all x [6(c)].

11. $h(x) = \dfrac{2}{x - 5}$ is a rational function and the denominator $x - 5$ is 0 when $x = 5$. Thus, h is continuous for all x except $x = 5$ [$\underline{6}$(d)].

13. $g(x) = \dfrac{x - 5}{(x - 3)(x + 2)}$ is a rational function and the denominator $(x - 3)(x + 2)$ is 0 when $x = 3$ or $x = -2$. Thus, g is continuous for all x except $x = 3$, $x = -2$.

15. (A) $\lim\limits_{x \to 0^-} f(x) = 1$, $\lim\limits_{x \to 0^+} f(x) = 1$, $\lim\limits_{x \to 0} f(x) = 1$, $f(0) = 1$.

 (B) f **is** continuous at $x = 0$ since $\lim\limits_{x \to 0} f(x)$ exists, $f(0)$ exists and $\lim\limits_{x \to 0} f(x) = f(0)$.

17. (A) $\lim\limits_{x \to 1^-} f(x) = 2$, $\lim\limits_{x \to 1^+} f(x) = 1$, $\lim\limits_{x \to 1} f(x)$ does not exist, $f(1) = 1$.

 (B) f **is not** continuous at $x = 1$ since $\lim\limits_{x \to 1} f(x)$ does not exist.

19. (A) $\lim\limits_{x \to -2^-} f(x) = 1$, $\lim\limits_{x \to -2^+} f(x) = 1$, $\lim\limits_{x \to -2} f(x) = 1$, $f(-2) = 3$.

 (B) f **is not** continuous at $x = -2$ since $\lim\limits_{x \to -2} f(x) \neq f(-2)$.

21. $f(x) = \begin{cases} 2 & \text{if } x \text{ is an integer} \\ 1 & \text{if } x \text{ is not an integer} \end{cases}$

 (A) The graph of f is:

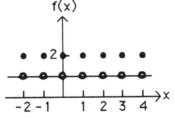

 (B) $\lim\limits_{x \to 2} f(x) = 1$ (C) $f(2) = 2$

 (D) f is not continuous at $x = 2$ since $\lim\limits_{x \to 2} f(x) \neq f(2)$.

 (E) f is discontinuous at $x = n$ for all integers n.

23. $\lim\limits_{x \to 2}(2x + 1) = 2 \cdot 2 + 1 = 5$ **25.** $\lim\limits_{x \to 2} 7 = 7$ [see $\underline{6}$(a)]

27. Since $\dfrac{x^2 - 9}{x + 3} = \dfrac{(x + 3)(x - 3)}{x + 3} = x - 3$, $x \neq -3$,

 $\lim\limits_{x \to -3} \dfrac{x^2 - 9}{x + 3} = \lim\limits_{x \to -3} (x - 3) = -6$.

29. For $x > 1$, $|x - 1| = x - 1$. Thus, $\dfrac{|x - 1|}{x - 1} = \dfrac{x - 1}{x - 1} = 1$ for $x > 1$;

 and $\lim\limits_{x \to 1^+} \dfrac{|x - 1|}{x - 1} = 1$.

31. $F(x) = 2x^8 - 3x^4 + 5$ is a polynomial function. Thus F is continuous fo all x, i.e., F is continuous on $(-\infty, \infty)$ [see $\underline{6}$(c)].

33. Since $f(x) = x - 5$ is a polynomial function, it is continuous for all x Thus, $g(x) = \sqrt{f(x)} = \sqrt{x - 5}$ is continuous for all x such that $x - 5 \geq 0$, i.e., g is continuous on $[5, \infty)$ [see $\underline{6}$(f)].

35. Since $f(x) = x - 5$ is continuous for all x, $K(x) = \sqrt[3]{x - 5}$ is continuous for all x, i.e., K is continuous on $(-\infty, \infty)$ [see $\underline{6}$(e)].

37. $f(x) = \dfrac{x^2 - 1}{x^2 - 3x + 2} = \dfrac{x^2 - 1}{(x - 1)(x - 2)}$

Since f is a rational function, it is continuous for all x except the numbers $x = c$ at which the denominator is 0. Thus, f is continuous for all x except $x = 1$ and $x = 2$, i.e., f is continuous on $(-\infty, 1)$, $(1, 2)$, and $(2, \infty)$.

39.

x	0.9	0.99	0.999	→ 1 ←	1.001	1.01	1.1
$f(x)$	-1	-1	-1	→ ? ←	1	1	1

(A) $\lim\limits_{x \to 1^-} f(x) = -1$ (B) $\lim\limits_{x \to 1^+} f(x) = 1$

(C) $\lim\limits_{x \to 1} f(x)$ does not exist.

41.

x	0.9	0.99	0.999	→ 1 ←	1.001	1.01	1.1
$f(x)$	2.71	2.97	2.997	→ 3 ←	3.003	3.03	3.31

(A) $\lim\limits_{x \to 1^-} f(x) = 3$ (B) $\lim\limits_{x \to 1^+} f(x) = 3$

(C) $\lim\limits_{x \to 1} f(x) = 3$

43. The graph of f is shown at the right. This function is discontinuous at $x = 1$. [$\lim\limits_{x \to 1} f(x)$ does not exist.]

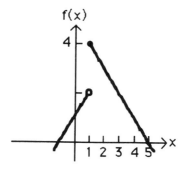

45. The graph of f is:

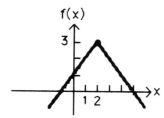

This function is continuous for all x. [$\underline{\text{Note}}$: $\lim\limits_{x \to 2} f(x) = f(2) = 3$.]

47. The graph of f is:

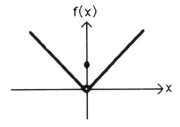

This function is discontinuous at $x = 0$.
[<u>Note</u>: $\lim\limits_{x \to 0} f(x) = 0 \neq f(0) = 1$.]

49. (A) g is continuous on $(-1, 2)$.

(B) Since $\lim\limits_{x \to -1^+} g(x) = -1 = g(-1)$, g is continuous from the right at $x = -1$.

(C) Since $\lim\limits_{x \to 2^-} g(x) = 2 = g(2)$, g is continuous from the left at $x = 2$.

(D) g is continuous on the closed interval $[-1, 2]$.

51. (A) Since $\lim\limits_{x \to 0^+} f(x) = f(0) = 0$, f is continuous from the right at $x = 0$.

(B) Since $\lim\limits_{x \to 0^-} f(x) = -1 \neq f(0) = 0$, f is not continuous from the left at $x = 0$.

(C) f is continuous on the open interval $(0, 1)$.

(D) f is *not* continuous on the closed interval $[0, 1]$ since $\lim\limits_{x \to 1^-} f(x) = 0 \neq f(1) = 1$, i.e., f is not continuous from the left at $x = 1$.

(E) f is continuous on the half-closed interval $[0, 1)$.

53. Since f is continuous and nonzero on $(1, 5)$, and since $f(2) = 3$, we can conclude that $f(x) > 0$ for all x on $(1, 5)$. Therefore, $f(4)$ *cannot* be negative; $f(x)$ cannot be negative for any x on the interval $(1, 5)$.

55. $\lim\limits_{x \to 0^-} f(x) = -3$, $\lim\limits_{x \to 0^+} f(x) = 3$

57. $\lim\limits_{x \to 0^-} f(x) = -4$, $\lim\limits_{x \to 0^+} f(x) = 4$

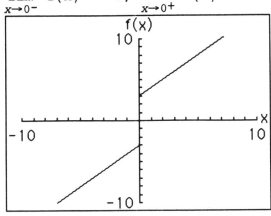

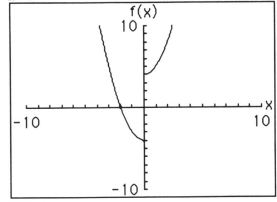

59. $\lim\limits_{x \to 2^-} f(x) = -4$, $\lim\limits_{x \to 2^+} f(x) = 4$ **61.** $\lim\limits_{x \to -3^-} f(x) = -3$, $\lim\limits_{x \to -3^+} f(x) = 3$,

$\lim\limits_{x \to 3^-} f(x) = -3$, $\lim\limits_{x \to 3^+} f(x) = 3$

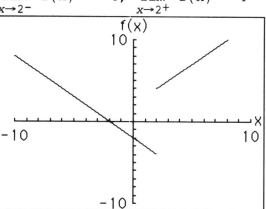

63. (A) $P(x) = \begin{cases} \$0.25 & 0 < x \le 1 \\ \$0.45 & 1 < x \le 2 \\ \$0.65 & 2 < x \le 3 \\ \$0.85 & 3 < x \le 4 \\ \$1.05 & 4 < x \le 5 \end{cases}$

The graph is:

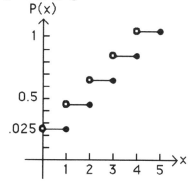

(B) $\lim\limits_{x \to 4.5} P(x) = \$1.05$ and $P(4.5) = \$1.05$

(C) $\lim\limits_{x \to 4} P(x)$ does not exist since $\lim\limits_{x \to 4^-} P(x) = \$0.85$ and $\lim\limits_{x \to 4^+} P(x) = \$1.05$.
$P(4) = \$0.85$.

(D) P is continous at $x = 4.5$; P is *not* continuous at $x = 4$.

65. (A) $E(s) = \begin{cases} 1000, & 0 \le s \le 10,000 \\ 1000 + 0.05(s - 10,000), & 10,000 < s < 20,000 \\ 1500 + 0.05(s - 10,000), & s \ge 20,000 \end{cases}$

The graph of E is:

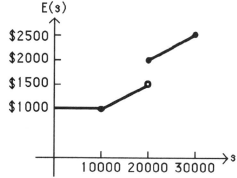

(B) From the graph, $\lim\limits_{s \to 10,000} E(s) = \$1000$ and $E(10,000) = \$1000$.

(C) From the graph, $\lim\limits_{s \to 20,000} E(s)$ does not exist. $E(20,000) = \$2000$.

(D) E is continuous at 10,000; E is not continuous at 20,000.

67. (A) From the graph, N is discontinuous at $t = t_2$, $t = t_3$, $t = t_4$, $t = t_6$, and $t = t_7$.

(B) From the graph, $\lim\limits_{t \to t_5} N(t) = 7$ and $N(t_5) = 7$.

(C) From the graph, $\lim\limits_{t \to t_3} N(t)$ does not exist; $N(t_3) = 4$.

EXERCISE 9-2

Things to remember:

1. LIMIT AT A POINT OF CONTINUITY

   A function f is continuous at $x = c$ if and only if
   $$\lim_{x \to c} f(x) = f(c)$$

2. VERTICAL ASYMPTOTES

   The vertical line $x = a$ is a vertical asymptote for the graph of $y = f(x)$ if $f(x)$ either increases or decreases without bound as x approaches a from either the left or the right. That is, $x = a$ is a VERTICAL ASYMPTOTE for the graph of $y = f(x)$ if any of the following holds:
   $$\lim_{x \to a^-} f(x) = \infty \quad (\text{or } -\infty);$$
   $$\lim_{x \to a^+} f(x) = \infty \quad (\text{or } -\infty);$$
   $$\lim_{x \to a} f(x) = \infty \quad (\text{or } -\infty).$$

3. PROPERTIES OF LIMITS

   Let f and g be two functions and assume that
   $$\lim_{x \to c} f(x) = L \qquad \lim_{x \to c} g(x) = M$$

   where L and M are real numbers (both limits exist). Then,

   (a) $\lim\limits_{x \to c}[f(x) + g(x)] = \lim\limits_{x \to c} f(x) + \lim\limits_{x \to c} g(x) = L + M,$

   (b) $\lim\limits_{x \to c}[f(x) - g(x)] = \lim\limits_{x \to c} f(x) - \lim\limits_{x \to c} g(x) = L - M,$

(c) $\lim\limits_{x \to c} kf(x) = k \lim\limits_{x \to c} f(x) = kL$ for any constant k,

(d) $\lim\limits_{x \to c} [f(x)g(x)] = \left(\lim\limits_{x \to c} f(x)\right)\left(\lim\limits_{x \to c} g(x)\right) = LM$,

(e) $\lim\limits_{x \to c} \dfrac{f(x)}{g(x)} = \dfrac{\lim\limits_{x \to c} f(x)}{\lim\limits_{x \to c} g(x)} = \dfrac{L}{M}$ if $M \neq 0$,

(f) $\lim\limits_{x \to c} \sqrt[n]{f(x)} = \sqrt[n]{\lim\limits_{x \to c} f(x)} = \sqrt[n]{L}$ ($L \geq 0$ for n even).

4. HORIZONTAL ASYMPTOTES

The horizontal line $y = b$ is a HORIZONTAL ASYMPTOTE for the graph of $y = f(x)$ if $f(x)$ approaches b as x either increases or decreases without bound. That is, $y = b$ is a horizontal asymptote for the graph of $y = f(x)$ if either

$$\lim\limits_{x \to -\infty} f(x) = b \quad \text{or} \quad \lim\limits_{x \to \infty} f(x) = b.$$

5. THREE SPECIAL LIMITS AT INFINITY

(a) If p is a positive real number, and k is any real constant,

then $\quad \lim\limits_{x \to -\infty} \dfrac{k}{x^p} = 0, \quad$ provided that x^p is defined for negative values of x.

and $\quad \lim\limits_{x \to \infty} \dfrac{k}{x^p} = 0$

(b) If $\lim\limits_{x \to \infty} f(x) = \infty$ and $\lim\limits_{x \to \infty} g(x) = L > 0$, then $\lim\limits_{x \to \infty} (f(x)g(x)) = \infty$. This statement also holds if $x \to \infty$ is replaced by $x \to -\infty$ or $x \to c$, for c any real number.

---

1. $\lim\limits_{x \to 3} [f(x) - g(x)] = \lim\limits_{x \to 3} f(x) - \lim\limits_{x \to 3} g(x) \quad$ [Property 3(b)]

$\qquad = 5 - 9 = -4$

3. $\lim\limits_{x \to 3} 4g(x) = 4 \lim\limits_{x \to 3} g(x) \quad$ [Property 3(c)]

$\qquad = 4 \cdot 9 = 36$

5. $\lim\limits_{x \to 3} \dfrac{f(x)}{g(x)} = \dfrac{\lim\limits_{x \to 3} f(x)}{\lim\limits_{x \to 3} g(x)} \quad$ [since $\lim\limits_{x \to 3} g(x) \neq 0$, Property 3(e)]

$\qquad = \dfrac{5}{9}$

7. $\lim\limits_{x \to 3} \sqrt{f(x)} = \sqrt{\lim\limits_{x \to 3} f(x)} = \sqrt{5} \quad$ [Property 3(f)]

9. $\displaystyle\lim_{x\to 3}\frac{f(x)+g(x)}{2f(x)}=\frac{\displaystyle\lim_{x\to 3}[f(x)+g(x)]}{\displaystyle\lim_{x\to 3}2f(x)}$ [since $\displaystyle\lim_{x\to 3}2f(x)\neq 0$]

$$=\frac{\displaystyle\lim_{x\to 3}f(x)+\lim_{x\to 3}g(x)}{2\cdot\displaystyle\lim_{x\to 3}f(x)}=\frac{5+9}{2\cdot 5}=\frac{14}{10}=\frac{7}{5}$$

11. $\displaystyle\lim_{x\to 5}(2x^2-3)=2(5)^2-3$ [since $f(x)=2x^2-3$ is continuous at $x=5$]

$$=47$$

13. $\displaystyle\lim_{x\to 4}(x^2-5x)=4^2-5(4)$ [$f(x)=x^2-5x$ is continuous at $x=4$]

$$=-4$$

15. $\displaystyle\lim_{x\to 2}\frac{5x}{2+x^2}=\frac{5(2)}{2+(2)^2}$ $\left[f(x)=\dfrac{5x}{2+x^2}\text{ is continuous at }x=2\right]$

$$=\frac{10}{6}=\frac{5}{3}$$

17. $\displaystyle\lim_{x\to 2}(x+1)^3(2x-1)^2=\lim_{x\to 2}(x+1)^3\cdot\lim_{x\to 2}(2x-1)^2$

$$=(2+1)^3\cdot(2\cdot 2-1)^2=3^3\cdot 3^2=3^5=243$$

19. $\dfrac{x^2-3x}{x}=\dfrac{x(x-3)}{x}=x-3,\ x\neq 0$

Thus, $\displaystyle\lim_{x\to 0}\frac{x^2-3x}{x}=\lim_{x\to 0}(x-3)=-3$

21. $\displaystyle\lim_{x\to\infty}\frac{3}{x^2}=0$ [see $\underline{5}$(a)]

23. $\displaystyle\lim_{x\to\infty}\left(5-\frac{3}{x}+\frac{2}{x^2}\right)=\lim_{x\to\infty}5-\lim_{x\to\infty}\frac{3}{x}+\lim_{x\to\infty}\frac{2}{x^2}=5-0+0=5$

25. $\dfrac{2(3+h)-2(3)}{h}=\dfrac{6+2h-6}{h}=\dfrac{2h}{h}=2,\ h\neq 0$

Thus, $\displaystyle\lim_{h\to 0}\frac{2(3+h)-2(3)}{h}=\lim_{h\to 0}2=2$

27. $\displaystyle\lim_{x\to 1}(3x^4-2x^2+x-2)=3\cdot 1^4-2\cdot 1^2+1-2=0$

29. $\displaystyle\lim_{x\to 1}\frac{x-2}{x^2-2x}=\frac{1-2}{1-2}=1$ $\left[\underline{\text{Note}}:\ f(x)=\dfrac{x-2}{x^2-2x}\text{ is continuous at }x=1.\right]$

31. $\displaystyle\lim_{x\to 2}\frac{x-2}{x^2-2x}$ is a 0/0 indeterminate form.

Thus, we try to manipulate the expression algebraically.

$$\frac{x-2}{x^2-2x}=\frac{x-2}{x(x-2)}=\frac{1}{x},\ x\neq 2$$

Now, $\displaystyle\lim_{x\to 2}\frac{x-2}{x^2-2x}=\lim_{x\to 2}\frac{1}{x}=\frac{1}{2}.$

33. $\displaystyle\lim_{x\to\infty}\frac{x-2}{x^2-2x} = \lim_{x\to\infty}\frac{x\left[1-\dfrac{2}{x}\right]}{x^2\left[1-\dfrac{2}{x}\right]} = \lim_{x\to\infty}\frac{1}{x} = 0$

35. $\displaystyle\lim_{x\to 2}\frac{x^2-x-6}{x+2} = \frac{2^2-2-6}{2+2} = -1$

37. $\displaystyle\lim_{x\to-2}\frac{x^2-x-6}{x+2}$ is a 0/0 indeterminate form.

$\dfrac{x^2-x-6}{x+2} = \dfrac{(x-3)(x+2)}{x+2} = x-3, \ x\neq -2$

Thus, $\displaystyle\lim_{x\to-2}\frac{x^2-x-6}{x+2} = \lim_{x\to-2}(x-3) = -5.$

39. $\displaystyle\lim_{x\to\infty}\frac{x^2-x-6}{x+2} = \lim_{x\to\infty}\frac{x^2\left(1-\dfrac{1}{x}-\dfrac{6}{x^2}\right)}{x\left(1+\dfrac{2}{x}\right)} = \lim_{x\to\infty}x\cdot\lim_{x\to\infty}\frac{1-\dfrac{1}{x}-\dfrac{6}{x^2}}{1+\dfrac{2}{x}} = \infty,$

since $\displaystyle\lim_{x\to\infty}x = \infty$ and $\displaystyle\lim_{x\to\infty}\frac{1-\dfrac{1}{x}-\dfrac{6}{x^2}}{1-\dfrac{2}{x}} = 1;$ see $\underline{5}$(b).

41. $\displaystyle\lim_{x\to\infty}\frac{2x+4}{x} = \lim_{x\to\infty}\frac{x\left(2+\dfrac{4}{x}\right)}{x} = \lim_{x\to\infty}\left(2+\dfrac{4}{x}\right) = 2$

43. $\displaystyle\lim_{x\to\infty}\frac{3x^3-x+1}{5x^3-7} = \lim_{x\to\infty}\frac{x^3\left(3-\dfrac{1}{x^2}+\dfrac{1}{x^3}\right)}{x^3\left(5-\dfrac{7}{x^3}\right)} = \lim_{x\to\infty}\frac{3-\dfrac{1}{x^2}+\dfrac{1}{x^3}}{5-\dfrac{7}{x^3}} = \frac{3}{5}$

45. $\dfrac{(2+h)^2-2^2}{h} = \dfrac{4+4h+h^2-4}{h} = \dfrac{4h+h^2}{h} = 4+h, \ h\neq 0$

Thus, $\displaystyle\lim_{h\to 0}\frac{(2+h)^2-2^2}{h} = \lim_{h\to 0}(4+h) = 4.$

47. $\displaystyle\lim_{x\to 3}\left(\frac{x}{x+3}+\frac{x-3}{x^2-9}\right) = \lim_{x\to 3}\frac{x}{x+3}+\lim_{x\to 3}\frac{x-3}{x^2-9} = \frac{3}{6}+\lim_{x\to 3}\frac{x-3}{x^2-9}$

Now, $\dfrac{x-3}{x^2-9} = \dfrac{x-3}{(x-3)(x+3)} = \dfrac{1}{x+3}, \ x\neq 3.$

Thus, $\displaystyle\lim_{x\to 3}\left(\frac{x}{x+3}+\frac{x-3}{x^2-9}\right) = \frac{1}{2}+\lim_{x\to 3}\frac{1}{x+3} = \frac{1}{2}+\frac{1}{6} = \frac{2}{3}.$

49. $\displaystyle\lim_{x\to 0^-}\frac{x-2}{x^2-2x} = \lim_{x\to 0^-}\frac{x-2}{x(x-2)} = \lim_{x\to 0^-}\frac{1}{x} = -\infty$

51. $\displaystyle\lim_{x\to 0^+}\frac{x-2}{x^2-2x} = \lim_{x\to 0^+}\frac{x-2}{x(x-2)} = \lim_{x\to 0^+}\frac{1}{x} = \infty$

53. $\lim\limits_{x \to 0} \dfrac{x - 2}{x^2 - 2x}$ does not exist.

55. (A) From the graph, $\lim\limits_{x \to -\infty} f(x) = 0$.

    (B) From the graph, $\lim\limits_{x \to \infty} f(x) = 0$.

    (C) $y = 0$ is a horizontal asymptote.

57. (A) From the graph, $\lim\limits_{x \to -1^-} g(x) = \infty$.

    (B) From the graph, $\lim\limits_{x \to -1^+} g(x) = \infty$.

    (C) $\lim\limits_{x \to -1} g(x) = \infty$.

    (D) $x = -1$ is a vertical asymptote.

59. $f(x) = 3x + 1$

$$\lim\limits_{h \to 0} \frac{f(2 + h) - f(2)}{h} = \lim\limits_{h \to 0} \frac{3(2 + h) + 1 - (3 \cdot 2 + 1)}{h}$$

$$= \lim\limits_{h \to 0} \frac{6 + 3h + 1 - 7}{h} = \lim\limits_{h \to 0} \frac{3h}{h} = \lim\limits_{h \to 0} 3 = 3$$

61. $f(x) = x^2 + 1$

$$\lim\limits_{h \to 0} \frac{f(2 + h) - f(2)}{h} = \lim\limits_{h \to 0} \frac{(2 + h)^2 + 1 - (2^2 + 1)}{h}$$

$$= \lim\limits_{h \to 0} \frac{4 + 4h + h^2 + 1 - 5}{h} = \lim\limits_{h \to 0} \frac{4h + h^2}{h}$$

$$= \lim\limits_{h \to 0} (4 + h) = 4$$

63. $f(x) = 5$

$$\lim\limits_{h \to 0} \frac{f(2 + h) - f(2)}{h} = \lim\limits_{h \to 0} \frac{5 - 5}{h} = \lim\limits_{h \to 0} 0 = 0$$

65. $f(x) = \sqrt{x} - 2$

$$\lim\limits_{h \to 0} \frac{f(2 + h) - f(2)}{h} = \lim\limits_{h \to 0} \frac{\sqrt{2 + h} - 2 - (\sqrt{2} - 2)}{h} = \lim\limits_{h \to 0} \frac{\sqrt{2 + h} - \sqrt{2}}{h}$$

$$= \lim\limits_{h \to 0} \frac{\sqrt{2 + h} - \sqrt{2}}{h} \cdot \frac{\sqrt{2 + h} + \sqrt{2}}{\sqrt{2 + h} + \sqrt{2}} = \lim\limits_{h \to 0} \frac{2 + h - 2}{h(\sqrt{2 + h} + \sqrt{2})}$$

$$= \lim\limits_{h \to 0} \frac{h}{h(\sqrt{2 + h} + \sqrt{2})} = \lim\limits_{h \to 0} \frac{1}{\sqrt{2 + h} + \sqrt{2}} = \frac{1}{2\sqrt{2}}$$

67. $f(x) = |x - 2| - 3$

$$\lim\limits_{h \to 0} \frac{f(2 + h) - f(2)}{h} = \lim\limits_{h \to 0} \frac{|(2 + h) - 2| - 3 - (|2 - 2| - 3)}{h}$$

$$= \lim\limits_{h \to 0} \frac{|h| - 3 + 3}{h} = \lim\limits_{h \to 0} \frac{|h|}{h} \text{ does not exist.}$$

69. $\lim\limits_{x \to 1} \sqrt{x^2 + 2x} = \sqrt{\lim\limits_{x \to 1}(x^2 + 2x)} = \sqrt{3}$

71. $\lim\limits_{x \to 4} \sqrt[3]{x^2 - 3x} = \sqrt[3]{\lim\limits_{x \to 4}(x^2 - 3x)} = \sqrt[3]{4}$

73. $\lim\limits_{x \to 2} \dfrac{4}{(x - 2)^2} = \lim\limits_{x \to 2} \dfrac{1}{(x - 2)^2} \cdot \lim\limits_{x \to 2} 4 = \infty$

75. $\lim\limits_{x \to \infty}\left(\dfrac{1}{x^2} + \dfrac{1}{\sqrt{x}}\right) = \lim\limits_{x \to \infty} \dfrac{1}{x^2} + \lim\limits_{x \to \infty} \dfrac{1}{\sqrt{x}} = 0 + 0 = 0$

77. $\lim\limits_{x \to 0}\left(\sqrt{x^2 + 9} - \dfrac{x^2 + 3x}{x}\right) = \lim\limits_{x \to 0} \sqrt{x^2 + 9} - \lim\limits_{x \to 0} \dfrac{x(x + 3)}{x}$

$\qquad\qquad = \sqrt{\lim\limits_{x \to 0}(x^2 + 9)} - \lim\limits_{x \to 0}(x + 3) = \sqrt{9} - 3 = 0$

79. $\lim\limits_{x \to 4} \dfrac{\sqrt{x} - 2}{x - 4} = \lim\limits_{x \to 4} \dfrac{\sqrt{x} - 2}{(\sqrt{x} - 2)(\sqrt{x} + 2)} = \lim\limits_{x \to 4} \dfrac{1}{\sqrt{x} + 2} = \dfrac{1}{4}$

81. $\lim\limits_{x \to 2} \dfrac{x^3 - 8}{x - 2} = \lim\limits_{x \to 2} \dfrac{(x - 2)(x^2 + 2x + 4)}{x - 2} = \lim\limits_{x \to 2}(x^2 + 2x + 4) = 12$

83. (A) $\lim\limits_{x \to -2^-} \dfrac{2}{x + 2} = -\infty$ $\qquad\qquad$ (B) $\lim\limits_{x \to -2^+} \dfrac{2}{x + 2} = \infty$

$\quad\,$ (C) $\lim\limits_{x \to -2} \dfrac{2}{x + 2}$ does not exist $\qquad$ (D) $x = -2$ is a vertical asymptote

85. $\lim\limits_{h \to 0} \dfrac{(a + h)^2 - a^2}{h} = \lim\limits_{h \to 0} \dfrac{a^2 + 2ah + h^2 - a^2}{h}$

$\qquad\qquad = \lim\limits_{h \to 0} \dfrac{2ah + h^2}{h} = \lim\limits_{h \to 0}(2a + h) = 2a$

87. $\lim\limits_{h \to 0} \dfrac{\sqrt{a + h} - \sqrt{a}}{h} = \lim\limits_{h \to 0} \dfrac{\sqrt{a + h} - \sqrt{a}}{h} \cdot \dfrac{\sqrt{a + h} + \sqrt{a}}{\sqrt{a + h} + \sqrt{a}} = \lim\limits_{h \to 0} \dfrac{(a + h) - a}{h(\sqrt{a + h} + \sqrt{a})}$

$\qquad\qquad = \lim\limits_{h \to 0} \dfrac{1}{\sqrt{a + h} + \sqrt{a}} = \dfrac{1}{2\sqrt{a}}$

89. $C(x) = 20{,}000 + 3x$ and $\overline{C}(x) = \dfrac{20{,}000 + 3x}{x}$

$\quad\,$ (A) $\overline{C}(1000) = \dfrac{20{,}000 + 3(1000)}{1000} = \dfrac{23{,}000}{1000} = 23$ or \$23.00

$\quad\,$ (B) $\overline{C}(100{,}000) = \dfrac{20{,}000 + 3(100{,}000)}{100{,}000} = \dfrac{320{,}000}{100{,}000} = 3.2$ or \$3.20

$\quad\,$ (C) $\lim\limits_{x \to 10{,}000} \overline{C}(x) = \lim\limits_{x \to 10{,}000} \dfrac{20{,}000 + 3x}{x} = \dfrac{20{,}000 + 3(10{,}000)}{10{,}000}$

$\qquad\qquad = \dfrac{50{,}000}{10{,}000} = 5$ or \$5.00

(D) $\displaystyle\lim_{x\to\infty} \overline{C}(x) = \lim_{x\to\infty} \frac{20{,}000 + 3x}{x}$ (divide numerator and denominator by x)

$$= \lim_{x\to\infty} \frac{\dfrac{20{,}000}{x} + 3}{1} = \frac{3}{1} = 3 \text{ or } \$3.00$$

91. (A)

COMPOUNDED	n	$A(n)$
Annually	1	$108.00
Semiannually	2	$108.16
Quarterly	4	$108.24
Monthly	12	$108.30
Weekly	52	$108.32
Daily	365	$108.33
Hourly	8760	$108.33

(B) From the table, we conclude that:

$$\lim_{n\to\infty} A(n) = \$108.33$$

93. $C(t) = \dfrac{0.14t}{t^2 + 1}$

(A) $C(0.5) = \dfrac{0.14(0.5)}{(0.5)^2 + 1} = \dfrac{0.07}{1.25} = 0.056$

(B) $C(1) = \dfrac{0.14(1)}{1^2 + 1} = \dfrac{0.14}{2} = 0.07$

(C) $\displaystyle\lim_{t\to1} C(t) = \lim_{t\to1} \frac{0.14t}{t^2 + 1} = \frac{0.14(1)}{1^2 + 1} = \frac{0.14}{2} = 0.07$

(D) $\displaystyle\lim_{t\to\infty} C(t) = \lim_{t\to\infty} \frac{0.14t}{t^2 + 1}$ (divide numerator and denominator by t^2)

$$= \lim_{t\to\infty} \frac{\dfrac{0.14}{t}}{1 + \dfrac{1}{t^2}} = \frac{0}{1} = 0$$

95. $f(x) = \dfrac{60(x + 1)}{x + 5}$

(A) $f(3) = \dfrac{60(3 + 1)}{3 + 5} = \dfrac{60(4)}{8} = 30$

(B) $f(10) = \dfrac{60(10 + 1)}{10 + 5} = \dfrac{60(11)}{15} = 44$

(C) $\displaystyle\lim_{x\to10} f(x) = \lim_{x\to10} \frac{60(x + 1)}{x + 5} = \frac{60(11)}{15} = 44$

(D) $\displaystyle\lim_{x\to\infty} f(x) = \lim_{x\to\infty} \frac{60(x + 1)}{x + 5}$

$$= \lim_{x\to\infty} \frac{60x + 60}{x + 5} \text{ (divide numerator and denominator by } x)$$

$$= \lim_{x\to\infty} \frac{60 + \dfrac{60}{x}}{1 + \dfrac{5}{x}} = \frac{60}{1} = 60$$

Things to remember:

1. TANGENT LINE

    Given the graph of $y = f(x)$, the TANGENT LINE at $x = x_1$ is the line that passes through the point $(x_1, f(x_1))$ with slope

    $$\text{SLOPE OF TANGENT LINE } = \lim_{h \to 0} \frac{f(x_1 + h) - f(x_1)}{h}$$

    if the limit exists. The slope of the tangent line is also referred to as the SLOPE OF THE GRAPH at $(x_1, f(x_1))$. The expression $\dfrac{f(x_1 + h) - f(x_1)}{h}$ is called the DIFFERENCE QUOTIENT.

2. AVERAGE AND INSTANTANEOUS RATES OF CHANGE

    For $y = f(x)$,

    AVERAGE RATE OF CHANGE FROM $x = x_1$ TO $x = x_1 + h$ is

    $$\frac{f(x_1 + h) - f(x_1)}{h} \qquad h \neq 0$$

    INSTANTANEOUS RATE OF CHANGE AT $x = x_1$ is

    $$\lim_{h \to 0} \frac{f(x_1 + h) - f(x_1)}{h}$$

    if the limit exists.

3. THE DERIVATIVE

    For $y = f(x)$, we define THE DERIVATIVE OF f AT x, denoted by $f'(x)$, to be

    $$f'(x) = \lim_{h \to 0} \frac{f(x + h) - f(x)}{h} \quad \text{if the limit exists.}$$

    If $f'(x)$ exists for each x in the open interval (a, b), then f is said to be DIFFERENTIABLE OVER (a, b).

    [Note: The derivative of a function f is a new function that gives:

    a. The slope of the tangent line to the graph of $y = f(x)$ for each x;

    b. The instantaneous rate of change of $y = f(x)$ with respect to x.

    The domain of f' is a subset of the domain of f.]

4. CONTINUITY AND DIFFERENTIABILITY

    Given a function f:

    (a) If f is not continuous at $x = a$, then $f'(a)$ does not exist;

    (b) If f is differentiable at $x = b$, then f must be continuous at $x = b$.

1. $f(x) = 3x - 2$

(A) $\dfrac{f(1 + h) - f(1)}{h} = \dfrac{3(1 + h) - 2 - (3 \cdot 1 - 2)}{h}$

$\qquad\qquad\qquad = \dfrac{3 + 3h - 2 - 1}{h} = \dfrac{3h}{h} = 3$

(B) $\lim\limits_{h \to 0} \dfrac{f(1 + h) - f(1)}{h} = \lim\limits_{h \to 0} 3 = 3$

3. $f(x) = 2x^2$

(A) $\dfrac{f(2 + h) - f(2)}{h} = \dfrac{2(2 + h)^2 - 2 \cdot 2^2}{h} = \dfrac{2(4 + 4h + h^2) - 8}{h}$

$\qquad\qquad\qquad = \dfrac{8 + 8h + 2h^2 - 8}{h} = 8 + 2h$

(B) $\lim\limits_{h \to 0} \dfrac{f(2 + h) - f(2)}{h} = \lim\limits_{h \to 0} (8 + 2h) = 8$

5. $y = f(x) = x^2 - 1$

(A) $f(0) = -1,\ f(1) = 0$

Slope of secant line: $\dfrac{f(1) - f(0)}{1 - 0} = \dfrac{0 - (-1)}{1} = 1$

(B) $f(1) = 0,\ f(1 + h) = (1 + h)^2 - 1 = 1 + 2h + h^2 - 1 = 2h + h^2$

Slope of secant line: $\dfrac{f(1 + h) - f(1)}{h} = \dfrac{2h + h^2}{h} = 2 + h$

(C) Slope of tangent line at $x = 1$:

$\lim\limits_{h \to 0} \dfrac{f(1 + h) - f(1)}{h} = \lim\limits_{h \to 0}(2 + h) = 2$

7. $f(x) = 2x - 3$

Step 1. Simplify $\dfrac{f(x + h) - f(x)}{h}$.

$\dfrac{f(x + h) - f(x)}{h} = \dfrac{2(x + h) - 3 - (2x - 3)}{h}$

$\qquad\qquad\qquad = \dfrac{2x + 2h - 3 - 2x + 3}{h} = \dfrac{2h}{h} = 2$

Step 2. Evaluate $\lim\limits_{h \to 0} \dfrac{f(x + h) - f(x)}{h}$.

$\lim\limits_{h \to 0} \dfrac{f(x + h) - f(x)}{h} = \lim\limits_{h \to 0} 2 = 2$

Thus, $f'(x) = 2$. Now $f'(1) = 2,\ f'(2) = 2,\ f'(3) = 2$.

9. $f(x) = 2 - x^2$

Step 1. Simplify $\dfrac{f(x + h) - f(x)}{h}$.

$\dfrac{f(x + h) - f(x)}{h} = \dfrac{2 - (x + h)^2 - (2 - x^2)}{h}$

$\qquad\qquad\qquad = \dfrac{2 - (x^2 + 2xh + h^2) - 2 + x^2}{h}$

$\qquad\qquad\qquad = \dfrac{-2xh - h^2}{h} = -2x - h$

Step 2. Evaluate $\lim\limits_{h \to 0} \dfrac{f(x + h) - f(x)}{h}$.

$$\lim\limits_{h \to 0} \dfrac{f(x + h) - f(x)}{h} = \lim\limits_{h \to 0}(-2x - h) = -2x$$

Thus, $f'(x) = -2x$. Now $f'(1) = -2$, $f'(2) = -4$, $f'(3) = -6$.

11. $y = f(x) = x^2 + x$

  (A) $f(1) = 1^2 + 1 = 2$, $f(3) = 3^2 + 3 = 12$

     Slope of secant line: $\dfrac{f(3) - f(1)}{3 - 1} = \dfrac{12 - 2}{2} = 5$

  (B) $f(1) = 2$, $f(1 + h) = (1 + h)^2 + (1 + h) = 1 + 2h + h^2 + 1 + h$
                       $= 2 + 3h + h^2$

     Slope of secant line: $\dfrac{f(1 + h) - f(1)}{h} = \dfrac{2 + 3h + h^2 - 2}{h} = 3 + h$

  (C) Slope of tangent line at $(1, f(1))$:

     $\lim\limits_{h \to 0} \dfrac{f(1 + h) - f(1)}{h} = \lim\limits_{h \to 0}(3 + h) = 3$

  (D) Equation of tangent line at $(1, f(1))$:
     $y - f(1) = f'(1)(x - 1)$ or $y - 2 = 3(x - 1)$ and $y = 3x - 1$.

13. $f(x) = x^2 + x$

  (A) Average velocity: $\dfrac{f(3) - f(1)}{3 - 1} = \dfrac{3^2 + 3 - (1^2 + 1)}{2} = \dfrac{12 - 2}{2}$
                                $= 5$ meters/sec.

  (B) Average velocity: $\dfrac{f(1 + h) - f(1)}{h} = \dfrac{(1 + h)^2 + (1 + h) - (1^2 + 1)}{h}$

                        $= \dfrac{1 + 2h + h^2 + 1 + h - 2}{h}$

                        $= \dfrac{3h + h^2}{h} = 3 + h$ meters/sec.

  (C) Instantaneous velocity: $\lim\limits_{h \to 0} \dfrac{f(1 + h) - f(1)}{h}$

                         $= \lim\limits_{h \to 0}(3 + h) = 3$ meters/sec.

15. $f(x) = 6x - x^2$

  Step 1. Simplify $\dfrac{f(x + h) - f(x)}{h}$.

     $\dfrac{f(x + h) - f(x)}{h} = \dfrac{6(x + h) - (x + h)^2 - (6x - x^2)}{h}$

                   $= \dfrac{6x + 6h - (x^2 + 2xh + h^2) - 6x + x^2}{h}$

                   $= \dfrac{6h - 2xh - h^2}{h} = 6 - 2x - h$

  Step 2. Evaluate $\lim\limits_{h \to 0} \dfrac{f(x + h) - f(x)}{h}$.

     $\lim\limits_{h \to 0} \dfrac{f(x + h) - f(x)}{h} = \lim\limits_{h \to 0}(6 - 2x - h) = 6 - 2x$

  Therefore, $f'(x) = 6 - 2x$. $f'(1) = 6 - 2(1) = 4$, $f'(2) = 6 - 2(2) = 2$,
  $f'(3) = 6 - 2(3) = 0$.

17. $f(x) = \sqrt{x} - 3$

Step 1. Simplify $\dfrac{f(x + h) - f(x)}{h}$

$$\dfrac{f(x + h) - f(x)}{h} = \dfrac{\sqrt{x + h} - 3 - (\sqrt{x} - 3)}{h}$$

$$= \dfrac{\sqrt{x + h} - \sqrt{x}}{h} \cdot \dfrac{\sqrt{x + h} + \sqrt{x}}{\sqrt{x + h} + \sqrt{x}}$$

$$= \dfrac{x + h - x}{h(\sqrt{x + h} + \sqrt{x})} = \dfrac{1}{\sqrt{x + h} + \sqrt{x}}$$

Step 2. Evaluate $\lim\limits_{h \to 0} \dfrac{f(x + h) - f(x)}{h}$.

$$\lim_{h \to 0} \dfrac{f(x + h) - f(x)}{h} = \lim_{h \to 0} \dfrac{1}{\sqrt{x + h} + \sqrt{x}} = \dfrac{1}{2\sqrt{x}}$$

Therefore, $f'(x) = \dfrac{1}{2\sqrt{x}}$. $f'(1) = \dfrac{1}{2\sqrt{1}} = \dfrac{1}{2}$, $f'(2) = \dfrac{1}{2\sqrt{2}}$, $f'(3) = \dfrac{1}{2\sqrt{3}}$.

19. $f(x) = -\dfrac{1}{x}$

Step 1. Simplify $\dfrac{f(x + h) - f(x)}{h}$.

$$\dfrac{f(x + h) - f(x)}{h} = \dfrac{-\dfrac{1}{x + h} - \left(-\dfrac{1}{x}\right)}{h} = \dfrac{-\dfrac{1}{x + h} + \dfrac{1}{x}}{h}$$

$$= \dfrac{\dfrac{-x + x + h}{x(x + h)}}{h} = \dfrac{1}{x(x + h)}$$

Step 2. Evaluate $\lim\limits_{h \to 0} \dfrac{f(x + h) - f(x)}{h}$.

$$\lim_{h \to 0} \dfrac{f(x + h) - f(x)}{h} = \lim_{h \to 0} \dfrac{1}{x(x + h)} = \dfrac{1}{x^2}$$

Therefore, $f'(x) = \dfrac{1}{x^2}$. $f'(1) = \dfrac{1}{1^2} = 1$, $f'(2) = \dfrac{1}{2^2} = \dfrac{1}{4}$, $f'(3) = \dfrac{1}{3^2} = \dfrac{1}{9}$.

21. $F'(x)$ does exist at $x = a$.

23. $F'(x)$ does not exist at $x = c$; the graph has a vertical tangent line at $(c, F(c))$.

25. $F'(x)$ does not exist at $x = e$; F is not defined at $x = e$.

27. $F'(x)$ does exist at $x = g$.

29. $f(x) = x^2 - 4x$

(A) Step 1. Simplify $\dfrac{f(x + h) - f(x)}{h}$.

$$\dfrac{f(x + h) - f(x)}{h} = \dfrac{(x + h)^2 - 4(x + h) - (x^2 - 4x)}{h}$$

$$= \dfrac{x^2 + 2xh + h^2 - 4x - 4h - x^2 + 4x}{h}$$

$$= \dfrac{2xh + h^2 - 4h}{h} = 2x - 4 + h$$

Step 2. Evaluate $\lim\limits_{h \to 0} \dfrac{f(x + h) - f(x)}{h}$.

$$\lim\limits_{h \to 0} \dfrac{f(x + h) - f(x)}{h} = \lim\limits_{h \to 0}(2x - 4 + h) = 2x - 4$$

Therefore, $f'(x) = 2x - 4$.

(B) $f'(0) = -4$, $f'(2) = 0$,
$f'(4) = 4$

(C) Since f is a quadratic
function, the graph of f is
a parabola.

y intercept: $y = 0$
x intercepts: $x = 0$, $x = 4$
Vertex: $(2, -4)$

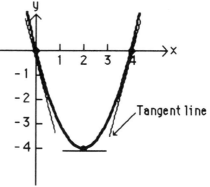
Tangent line

31. To find $v = f'(x)$, use the two-step process for the given distance
function, $f(x) = 4x^2 - 2x$.

Step 1. $f(x + h) = 4(x + h)^2 - 2(x + h)$
$\qquad\qquad\quad = 4(x^2 + 2xh + h^2) - 2(x + h)$
$\qquad\qquad\quad = 4x^2 + 8xh + 4h^2 - 2x - 2h$

$f(x + h) - f(x) = (4x^2 + 8xh + 4h^2 - 2x - 2h) - (4x^2 - 2x)$
$\qquad\qquad\qquad\quad = 4x^2 + 8xh + 4h^2 - 2x - 2h - 4x^2 + 2x$
$\qquad\qquad\qquad\quad = 8xh + 4h^2 - 2h$
$\qquad\qquad\qquad\quad = h(8x + 4h - 2)$

$\dfrac{f(x + h) - f(x)}{h} = \dfrac{h(8x + 4h - 2)}{h}$
$\qquad\qquad\qquad\quad = 8x + 4h - 2,\ h \ne 0$

Step 2. $\lim\limits_{h \to 0} \dfrac{f(x + h) - f(x)}{h} = \lim\limits_{h \to 0}(8x + 4h - 2) = 8x - 2$

Thus, the velocity, $v = f'(x) = 8x - 2$
$\qquad\qquad\qquad\qquad\quad f'(1) = 8 \cdot 1 - 2 = 6$ feet per second
$\qquad\qquad\qquad\qquad\quad f'(3) = 8 \cdot 3 - 2 = 22$ feet per second
$\qquad\qquad\qquad\qquad\quad f'(5) = 8 \cdot 5 - 2 = 38$ feet per second

33. The graph of $f(x) = \begin{cases} 2x, & x < 1 \\ 2, & x \ge 1 \end{cases}$ is:

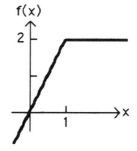

f is not differentiable at $x = 1$
because the graph of f has a sharp
corner at this point.

35. $f(x) = |x|$

$$\lim\limits_{h \to 0} \dfrac{f(0 + h) - f(0)}{h} = \lim\limits_{h \to 0} \dfrac{|0 + h| - |0|}{h} = \lim\limits_{h \to 0} \dfrac{|h|}{h}$$

The limit does not exist. Thus, f is not differentiable at $x = 0$.

37. $f(x) = \sqrt[3]{x} = x^{1/3}$

$$\lim_{h \to 0} \frac{f(0 + h) - f(0)}{h} = \lim_{h \to 0} \frac{(0 + h)^{1/3} - 0^{1/3}}{h} = \lim_{h \to 0} \frac{h^{1/3}}{h} = \lim_{h \to 0} \frac{1}{h^{2/3}}$$

The limit does not exist. Thus, f is not differentiable at $x = 0$.

39. $f(x) = 2x - x^2$, $0 \le x \le 2$

(A) For $0 < x < 2$:

$$\lim_{h \to 0} \frac{f(x + h) - f(x)}{h} = \lim_{h \to 0} \frac{2(x + h) - (x + h)^2 - (2x - x^2)}{h}$$

$$= \lim_{h \to 0} \frac{2x + 2h - x^2 - 2xh - h^2 - 2x + x^2}{h}$$

$$= \lim_{h \to 0} \frac{2h - 2xh - h^2}{h} = \lim_{h \to 0}(2 - 2x - h) = 2 - 2x$$

Thus, $f'(x) = 2 - 2x$, $0 < x < 2$.

(B) For $x = 0$:

$$\lim_{h \to 0^+} \frac{f(0 + h) - f(0)}{h} = \lim_{h \to 0^+} \frac{2(0 + h) - (0 + h)^2 - 0}{h}$$

$$= \lim_{h \to 0^+} \frac{2h - h^2}{h} = 2$$

(C) For $x = 2$:

$$\lim_{h \to 0^-} \frac{f(2 + h) - f(2)}{h} = \lim_{h \to 0^-} \frac{2(2 + h) - (2 + h)^2 - (2 \cdot 2 - 2^2)}{h}$$

$$= \lim_{h \to 0^-} \frac{4 + 2h - 4 - 4h - h^2 - 0}{h}$$

$$= \lim_{h \to 0^-} -\frac{2h + h^2}{h} = -2$$

41. $C(x) = 3 + 10x - x^2$, $0 \le x \le 5$

(A) $\dfrac{C(4) - C(3)}{4 - 3} = \dfrac{3 + 10(4) - 4^2 - (3 + 10 \cdot 3 - 3^2)}{1}$

$$= \frac{27 - 24}{1} = 3 \quad \text{(or \$300 per board)}$$

(B) $\dfrac{C(3 + h) - C(3)}{h} = \dfrac{3 + 10(3 + h) - (3 + h)^2 - (3 + 10 \cdot 3 - 3^2)}{h}$

$$= \frac{3 + 30 + 10h - 9 - 6h - h^2 - 24}{h}$$

$$= \frac{4h - h^2}{h} = 4 - h$$

(C) $C'(3) = \lim_{h \to 0} \dfrac{C(3 + h) - C(3)}{h} = \lim_{h \to 0}(4 - h) = 4$ (\$400 per board)

(D) <u>Step 1</u>. Simplify $\dfrac{C(x + h) - C(x)}{h}$.

$$\frac{C(x + h) - C(x)}{h} = \frac{3 + 10(x + h) - (x + h)^2 - (3 + 10x - x^2}{h}$$

$$= \frac{3 + 10x + 10h - x^2 - 2xh - h^2 - 3 - 10x + x^2}{h}$$

$$= \frac{10h - 2xh - h^2}{h} = 10 - 2x - h$$

<u>Step 2</u>. Evaluate $\lim\limits_{h \to 0} \dfrac{C(x + h) - C(x)}{h}$.

$$\lim_{h \to 0} \frac{C(x + h) - C(x)}{h} = \lim_{h \to 0}(10 - 2x - h) = 10 - 2x$$

Thus, $C'(x) = 10 - 2x$.

(E) $C'(1) = 10 - 2(1) = 8$. The rate of total cost per day increase at the one-board-per-day level is $800 per board.

$C'(2) = 10 - 2(2) = 6$. The rate of increase at the two-board-per-day level is $600 per board.

$C'(3) = 10 - 2(3) = 4$. The rate of increase at the three-board-per-day level is $400 per board.

$C'(4) = 10 - 2(4) = 2$. The rate of increase at the four-board-per-day level is $200 per board.

EXERCISE 9-4

Things to remember:

1. **DERIVATIVE NOTATION**

   Given $y = f(x)$, then

   $$f'(x), \quad y', \quad \frac{dy}{dx}, \quad D_x f(x)$$

   all represent the derivative of f at x.

2. **DERIVATIVE OF A CONSTANT**

   If $f(x) = C$, C a constant, then $f'(x) = 0$. Also

   $$y' = 0, \quad \frac{dy}{dx} = 0, \quad \text{and} \quad D_x C = 0.$$

3. **POWER RULE**

   If $f(x) = x^n$, n any real number, then

   $$f'(x) = nx^{n-1}.$$

   [<u>Note</u>: The domain of f and/or f' might have to be restricted to avoid division by 0 and/or even roots of negative numbers.]

4. CONSTANT TIMES A FUNCTION RULE

If $y = f(x) = ku(x)$, where k is a constant, then

$\quad f'(x) = ku'(x)$.

Also,

$\quad y' = ku'$, $\dfrac{dy}{dx} = k\dfrac{du}{dx}$, $D_x ku(x) = kD_x u(x)$.

5. SUM AND DIFFERENCE RULE

If $y = f(x) = u(x) \pm v(x)$, then

$\quad f'(x) = u'(x) \pm v'(x)$.

[Note: This rule generalizes to the sum and difference of any given number of functions.]

---

1. $f(x) = 12$

$\quad f'(x) = (12)' = 0 \quad$ (using 2)

3. $D_x 23 = 0 \quad$ (23 is a constant)

5. $y = x^{12}$

$\quad \dfrac{dy}{dx} = 12x^{12-1} \quad$ (using 3)

$\quad\quad = 12x^{11}$

7. $f(x) = x$

$\quad f'(x) = 1x^{1-1} \quad$ (using 3)

$\quad\quad = x^0 \quad (x^0 = 1)$

$\quad\quad = 1$

9. $y = x^{-7}$

$\quad y' = -7x^{-7-1}$

$\quad\quad = -7x^{-8}$

11. $y = x^{5/2}$

$\quad \dfrac{dy}{dx} = \dfrac{5}{2}x^{5/2-1} = \dfrac{5}{2}x^{5/2-2/2} = \dfrac{5}{2}x^{3/2}$

13. $f(x) = \dfrac{1}{x^5} = x^{-5}$

$\quad D_x(x^{-5}) = -5x^{-5-1} = -5x^{-6}$

15. $f(x) = 2x^4$

$\quad f'(x) = 2 \cdot 4x^3 \quad$ (using 4)

$\quad\quad = 8x^3$

17. $D_x\left(\dfrac{1}{3}x^6\right) = \dfrac{1}{3}D_x(x^6) \quad$ (using 4)

$\quad\quad = \dfrac{1}{3} \cdot 6x^5 = 2x^5$

19. $y = \dfrac{x^5}{15} = \dfrac{1}{15}x^5$

$\quad \dfrac{dy}{dx} = \dfrac{1}{15} \cdot 5x^4 = \dfrac{x^4}{3}$

21. $D_x(2x^{-5}) = 2D_x(x^{-5}) = 2(-5)x^{-6} = -10x^{-6}$

23. $f(x) = \dfrac{4}{x^4} = 4x^{-4}$

$\quad f'(x) = 4(-4)x^{-5} = -16x^{-5}$

25. $D_x\left(\dfrac{-1}{2x^2}\right) = D_x\left(-\dfrac{1}{2}x^{-2}\right) = -\dfrac{1}{2}D_x(x^{-2})$

$\quad\quad = -\dfrac{1}{2}(-2)x^{-3} = x^{-3}$

27. $f(x) = -3x^{1/3}$

$\quad f'(x) = -3\left(\dfrac{1}{3}\right)x^{1/3-1} = -x^{-2/3}$

29. $D_x(2x^2 - 3x + 4) = D_x(2x^2) - D_x(3x) + D_x(4)$ (using $\underline{5}$)

$$= 2D_x(x^2) - 3D_x(x) + D_x(4)$$
$$= 2 \cdot 2x - 3 \cdot 1 = 4x - 3$$

31. $y = 3x^5 - 2x^3 + 5$

$$\frac{dy}{dx} = (3x^5)' - (2x^3)' + (5)' = 15x^4 - 6x^2$$

33. $D_x(3x^{-4} + 2x^{-2}) = D_x(3x^{-4}) + D_x(2x^{-2}) = -12x^{-5} - 4x^{-3}$

35. $y = \dfrac{1}{2x} - \dfrac{2}{3x^3} = \dfrac{1}{2}x^{-1} - \dfrac{2}{3}x^{-3}$

$$\frac{dy}{dx} = -\frac{1}{2}x^{-2} - \frac{2}{3}(-3)x^{-4} = -\frac{1}{2}x^{-2} + 2x^{-4}$$

37. $D_x(3x^{2/3} - 5x^{1/3}) = D_x(3x^{2/3}) - D_x(5x^{1/3})$

$$= 3\left(\frac{2}{3}\right)x^{-1/3} - 5\left(\frac{1}{3}\right)x^{-2/3} = 2x^{-1/3} - \frac{5}{3}x^{-2/3}$$

39. $D_x\left(\dfrac{3}{x^{3/5}} - \dfrac{6}{x^{1/2}}\right) = D_x(3x^{-3/5} - 6x^{-1/2}) = D_x(3x^{-3/5}) - D_x(6x^{-1/2})$

$$= 3\left(\frac{-3}{5}\right)x^{-8/5} - 6\left(-\frac{1}{2}\right)x^{-3/2} = \frac{-9}{5}x^{-8/5} + 3x^{-3/2}$$

41. $D_x \dfrac{1}{\sqrt[3]{x}} = D_x(x^{-1/3}) = -\dfrac{1}{3}x^{-4/3}$

43. $y = \dfrac{12}{\sqrt{x}} - 3x^{-2} + x = 12x^{-1/2} - 3x^{-2} + x$

$$\frac{dy}{dx} = 12\left(-\frac{1}{2}\right)x^{-3/2} - 3(-2)x^{-3} + 1 = -6x^{-3/2} + 6x^{-3} + 1$$

45. $f(x) = 6x - x^2$

(A) $f'(x) = 6 - 2x$

(B) Slope of the graph of f at $x = 2$: $f'(2) = 6 - 2(2) = 2$
    Slope of the graph of f at $x = 4$: $f'(4) = 6 - 2(4) = -2$

(C) Tangent line at $x = 2$: $y - y_1 = m(x - x_1)$
    $x_1 = 2$
    $y_1 = f(2) = 6(2) - 2^2 = 8$
    $m = f'(2) = 2$
    Thus, $y - 8 = 2(x - 2)$ or $y = 2x + 4$.
    Tangent line at $x = 4$: $y - y_1 = m(x - x_1)$
    $x_1 = 4$
    $y_1 = f(4) = 6(4) - 4^2 = 8$
    $m = f'(4) = -2$
    Thus, $y - 8 = -2(x - 4)$ or $y = -2x + 16$

(D) The tangent line is horizontal at the values $x = c$ such that $f'(c) = 0$. Thus, we must solve the following:
$$f'(x) = 6 - 2x = 0$$
$$2x = 6$$
$$x = 3$$

47. $f(x) = 3x^4 - 6x^2 - 7$

(A) $f'(x) = 12x^3 - 12x$

(B) Slope of the graph of $x = 2$: $f'(2) = 12(2)^3 - 12(2) = 72$
Slope of the graph of $x = 4$: $f'(4) = 12(4)^3 - 12(4) = 720$

(C) Tangent line at $x = 2$: $y - y_1 = m(x - x_1)$, where $x_1 = 2$,
$y_1 = f(2) = 3(2)^4 - 6(2)^2 - 7 = 17$, $m = 72$.
$y - 17 = 72(x - 2)$ or $y = 72x - 127$

Tangent line at $x = 4$: $y - y_1 = m(x - x_1)$, where $x_1 = 4$,
$y_1 = f(4) = 3(4)^4 - 6(4)^2 - 7 = 665$, $m = 720$.
$y - 665 = 720(x - 4)$ or $y = 720x - 2215$

(D) Solve $f'(x) = 0$ for x:
$$12x^3 - 12x = 0$$
$$12x(x^2 - 1) = 0$$
$$12x(x - 1)(x + 1) = 0$$
$$x = -1, \; x = 0, \; x = 1$$

49. $f(x) = 176x - 16x^2$

(A) $v = f'(x) = 176 - 32x$

(B) $v\Big|_{x=0} = f'(0) = 176$ feet/sec.

$v\Big|_{x=3} = f'(3) = 176 - 32(3) = 80$ feet/sec.

(C) Solve $v = f'(x) = 0$ for x:
$$176 - 32x = 0$$
$$32x = 176$$
$$x = 5.5 \text{ seconds}$$

51. $f(x) = x^3 - 9x^2 + 15x$

(A) $v = f'(x) = 3x^2 - 18x + 15$

(B) $v\Big|_{x=0} = f'(0) = 15$ feet/sec.

$v\Big|_{x=3} = f'(3) = 3(3)^2 - 18(3) + 15 = -12$ feet/sec.

(C) Solve $v = f'(x) = 0$ for x:
$$3x^2 - 18x + 15 = 0$$
$$3(x^2 - 6x + 5) = 0$$
$$3(x - 5)(x - 1) = 0$$
$$x = 1, \quad x = 5$$

53. $f(x) = \dfrac{10x + 20}{x} = 10 + \dfrac{20}{x} = 10 + 20x^{-1}$

$f'(x) = -20x^{-2}$

55. $D_x\left(\dfrac{x^4 - 3x^3 + 5}{x^2}\right) = D_x\left(x^2 - 3x + \dfrac{5}{x^2}\right)$

$$= D_x(x^2) - D_x(3x) + D_x(5x^{-2}) = 2x - 3 - 10x^{-3}$$

57. Let $f(x) = x^3$

Step 1. Simplify $\dfrac{f(x + h) - f(x)}{h}$.

$$\dfrac{f(x + h) - f(x)}{h} = \dfrac{(x + h)^3 - x^3}{h} = \dfrac{x^3 + 3x^2h + 3xh^2 + h^3 - x^3}{h}$$

$$= \dfrac{3x^2h + 3xh^2 + h^3}{h} = \dfrac{h(3x^2 + 3xh + h^2)}{h} = 3x^2 + 3xh + h^2 \quad (h \neq 0)$$

Step 2. Evaluate $\lim\limits_{h \to 0} \dfrac{f(x + h) - f(x)}{h}$.

$$\lim\limits_{h \to 0} \dfrac{f(x + h) - f(x)}{h} = \lim\limits_{h \to 0}(3x^2 + 3xh + h^2) = 3x^2$$

Therefore, $D_x x^3 = 3x^2$.

59. $f(x) = x^2 - 3x - 4\sqrt{x} = x^2 - 3x - 4x^{1/2}$

$f'(x) = 2x - 3 - 2x^{-1/2}$

The graph of f has a horizontal tangent line at the value(s) of x where $f'(x) = 0$. Thus, we need to solve the equation
$$2x - 3 - 2x^{-1/2} = 0$$
By graphing the function $y = 2x - 3 - 2x^{-1/2}$, we see that there is one zero. To two decimal places, it is $x = 2.18$.

61. $f(x) = 3\sqrt[3]{x^4} - 1.5x^2 - 3x = 3x^{4/3} - 1.5x^2 - 3x$

$f'(x) = 4x^{1/3} - 3x - 3$

The graph of f has a horizontal tangent line at the value(s) of x where $f'(x) = 0$. Thus, we need to solve the equation
$$4x^{1/3} - 3x - 3 = 0$$
Graphing the function $y = 4x^{1/3} - 3x - 3$, we see that there is one zero. To two decimal places, it is $x = -2.90$.

63. $f(x) = 0.05x^4 - 0.1x^3 - 1.5x^2 - 1.6x + 3$

$f'(x) = 0.2x^3 + 0.3x^2 - 3x - 1.6$

The graph of f has a horizontal tangent line at the value(s) of x where $f'(x) = 0$. Thus, we need to solve the equation

$$0.2x^3 + 0.3x^2 - 3x - 1.6 = 0$$

By graphing the function $y = 0.2x^3 + 0.3x^2 - 3x - 1.6$, we see that there are three zeros. To two decimal places, they are
$$x_1 = -4.46, \quad x_2 = -0.52, \quad x_3 = 3.48$$

65. $f(x) = 0.2x^4 - 3.12x^3 + 16.25x^2 - 28.25x + 7.5$
$f'(x) = 0.8x^3 - 9.36x^2 + 32.5x - 28.25$
The graph of f has a horizontal tangent line at the value(s) of x where $f'(x) = 0$. Thus, we need to solve the equation
$$0.8x^3 - 9.36x^2 + 32.5x - 28.25 = 0$$
Graphing the function $y = 0.8x^3 - 9.36x^2 + 32.5x - 28.25$, we see that there is one zero. To two decimal places, it is $x = 1.30$.

67. $C(x) = 800 + 60x - \dfrac{x^2}{4}, \quad 0 \leq x \leq 120$

(A) Marginal cost $= C'(x) = 60 - \dfrac{1}{2}x$

(B) $C'(60) = 60 - \dfrac{1}{2}(60) = 30 \quad$ or $\quad \$30$ per racket

Interpretation: At a production level of 60 rackets, the rate of change of total costs with respect to production is $30 per racket. Thus, the cost of producing one more racket at this level of production is approximately $30.

(C) $C(61) - C(60) = 800 + 60(61) - \dfrac{(61)^2}{4} - \left[800 + 60(60) - \dfrac{(60)^2}{4} \right]$
$$= 3529.75 - 3500 = 29.75$$
The actual cost of producing the 61st racket is $29.75.

(D) $C'(80) = 60 - \dfrac{1}{2}(80) = 20 \quad$ or $\quad \$20$ per racket

Interpretation: At a production level of 80 rackets, the rate of change of total cost with respect to production is $20 per racket. Thus, the cost of producing the 81st racket is approximately $20.

69. $N(x) = 60x - x^2, \quad 5 \leq x \leq 30$

(A) $N'(x) = 60 - 2x$

(B) $N'(10) = 60 - 2 \cdot 10 = 40$ units of increase in sales per $1000 increase in advertising at the $10,000 level.

$N'(20) = 60 - 2 \cdot 20 = 20$ units of increase in sales per $1000 increase in advertising at the $20,000 level.

The effect of advertising decreases as the amount spent increases.

71. $y = 590x^{-1/2}$, $30 \le x \le 75$

First, find $\frac{dy}{dx} = \frac{d}{dx}590x^{-1/2} = -295x^{-3/2} = \frac{-295}{x^{3/2}}$, the instantaneous rate of change of pulse when a person is x inches tall.

(A) The instantaneous rate of change of pulse rate at $x = 36$ is:
$$\frac{-295}{(36)^{3/2}} = \frac{-295}{216} = -1.37 \text{ (1.37 decrease in pulse rate)}$$

(B) The instantaneous rate of change of pulse rate at $x = 64$ is:
$$\frac{-295}{(64)^{3/2}} = \frac{-295}{512} = -0.58 \text{ (0.58 decrease in pulse rate)}$$

73. $y = 50\sqrt{x}$, $0 \le x \le 9$

First, find $y' = (50\sqrt{x})' = (50x^{1/2})' = 25x^{-1/2}$
$$= \frac{25}{\sqrt{x}}, \text{ the rate of learning at the end of } x \text{ hours.}$$

(A) Rate of learning at the end of 1 hour:
$$\frac{25}{\sqrt{1}} = 25 \text{ items per hour}$$

(B) Rate of learning at the end of 9 hours:
$$\frac{25}{\sqrt{9}} = \frac{25}{3} = 8.33 \text{ items per hour}$$

EXERCISE 9-5

Things to remember:

1. PRODUCT RULE

If
$$y = f(x) = F(x)S(x)$$
and if $F'(x)$ and $S'(x)$ exist, then
$$f'(x) = F(x)S'(x) + S(x)F'(x).$$

Also,
$$y' = FS' + SF';$$
$$\frac{dy}{dx} = F\frac{dS}{dx} + S\frac{dF}{dx};$$
$$D_x[F(x)S(x)] = F(x)D_xS(x) + S(x)D_xF(x).$$

2. QUOTIENT RULE

If
$$y = f(x) = \frac{T(x)}{B(x)}$$
and if $T'(x)$ and $B'(x)$ exist, then
$$f'(x) = \frac{B(x)T'(x) - T(x)B'(x)}{[B(x)]^2}.$$

Also,

$$y' = \frac{BT' - TB'}{B^2};$$

$$\frac{dy}{dx} = \frac{B\left(\frac{dT}{dx}\right) - T\left(\frac{dB}{dx}\right)}{B^2};$$

$$D_x \frac{T(X)}{B(x)} = \frac{B(x)D_xT(x) - T(x)D_xB(x)}{[B(x)]^2}$$

---

1. $f(x) = 2x^3(x^2 - 2)$

$\quad f'(x) = 2x^3(x^2 - 2)' + (x^2 - 2)(2x^3)'$ [using 1 with $F(x) = 2x^3$,

$\qquad\quad = 2x^3(2x) + (x^2 - 2)6x^2$ $S(x) = x^2 - 2$]

$\qquad\quad = 4x^4 + 6x^4 - 12x^2$

$\qquad\quad = 10x^4 - 12x^2$

3. $f(x) = (x - 3)(2x - 1)$

$\quad f'(x) = (x - 3)(2x - 1)' + (2x - 1)(x - 3)'$ (using 1)

$\qquad\quad = (x - 3)(2) + (2x - 1)(1)$

$\qquad\quad = 2x - 6 + 2x - 1$

$\qquad\quad = 4x - 7$

5. $f(x) = \dfrac{x}{x - 3}$

$\quad f'(x) = \dfrac{(x - 3)(x)' - x(x - 3)'}{(x - 3)^2}$ [using 2 with $T(x) = x$, $B(x) = x - 3$]

$\qquad\quad = \dfrac{(x - 3)(1) - x(1)}{(x - 3)^2} = \dfrac{-3}{(x - 3)^2}$

7. $f(x) = \dfrac{2x + 3}{x - 2}$

$\quad f'(x) = \dfrac{(x - 2)(2x + 3)' - (2x + 3)(x - 2)'}{(x - 2)^2}$ (using 2)

$\qquad\quad = \dfrac{(x - 2)(2) - (2x + 3)(1)}{(x - 2)^2} = \dfrac{2x - 4 - 2x - 3}{(x - 2)^2} = \dfrac{-7}{(x - 2)^2}$

9. $f(x) = (x^2 + 1)(2x - 3)$

$\quad f'(x) = (x^2 + 1)(2x - 3)' + (2x - 3)(x^2 + 1)'$ (using 1)

$\qquad\quad = (x^2 + 1)(2) + (2x - 3)(2x)$

$\qquad\quad = 2x^2 + 2 + 4x^2 - 6x$

$\qquad\quad = 6x^2 - 6x + 2$

11. $f(x) = \dfrac{x^2 + 1}{2x - 3}$

$\quad f'(x) = \dfrac{(2x - 3)(x^2 + 1)' - (x^2 + 1)(2x - 3)'}{(2x - 3)^2}$ (using 2)

$\qquad\quad = \dfrac{(2x - 3)(2x) - (x^2 + 1)(2)}{(2x - 3)^2}$

$\qquad\quad = \dfrac{4x^2 - 6x - 2x^2 - 2}{(2x - 3)^2} = \dfrac{2x^2 - 6x - 2}{(2x - 3)^2}$

13. $f(x) = (x^2 + 2)(x^2 - 3)$

$\begin{aligned} f'(x) &= (x^2 + 2)(x^2 - 3)' + (x^2 - 3)(x^2 + 2)' \\ &= (x^2 + 2)(2x) + (x^2 - 3)(2x) \\ &= 2x^3 + 4x + 2x^3 - 6x \\ &= 4x^3 - 2x \end{aligned}$

15. $f(x) = \dfrac{x^2 + 2}{x^2 - 3}$

$\begin{aligned} f'(x) &= \frac{(x^2 - 3)(x^2 + 2)' - (x^2 + 2)(x^2 - 3)'}{(x^2 - 3)^2} \\ &= \frac{(x^2 - 3)(2x) - (x^2 + 2)(2x)}{(x^2 - 3)^2} = \frac{2x^3 - 6x - 2x^3 - 4x}{(x^2 - 3)^2} = \frac{-10x}{(x^2 - 3)^2} \end{aligned}$

17. $f(x) = (2x + 1)(x^2 - 3x)$

$\begin{aligned} f'(x) &= (2x + 1)(x^2 - 3x)' + (x^2 - 3x)(2x + 1)' \\ &= (2x + 1)(2x - 3) + (x^2 - 3x)(2) \\ &= 6x^2 - 10x - 3 \end{aligned}$

19. $y = (2x - x^2)(5x + 2)$

$\begin{aligned} \frac{dy}{dx} &= (2x - x^2)\frac{d}{dx}(5x + 2) + (5x + 2)\frac{d}{dx}(2x - x^2) \\ &= (2x - x^2)(5) + (5x + 2)(2 - 2x) \\ &= -15x^2 + 16x + 4 \end{aligned}$

21. $y = \dfrac{5x - 3}{x^2 + 2x}$

$\begin{aligned} y' &= \frac{(x^2 + 2x)(5x - 3)' - (5x - 3)(x^2 + 2x)'}{(x^2 + 2x)^2} \\ &= \frac{(x^2 + 2x)(5) - (5x - 3)(2x + 2)}{(x^2 + 2x)^2} = \frac{-5x^2 + 6x + 6}{(x^2 + 2x)^2} \end{aligned}$

23. $\begin{aligned} D_x\left[\dfrac{x^2 - 3x + 1}{x^2 - 1}\right] &= \frac{(x^2 - 1)D_x(x^2 - 3x + 1) - (x^2 - 3x + 1)D_x(x^2 - 1)}{(x^2 - 1)^2} \\ &= \frac{(x^2 - 1)(2x - 3) - (x^2 - 3x + 1)(2x)}{(x^2 - 1)^2} \\ &= \frac{3x^2 - 4x + 3}{(x^2 - 1)^2} \end{aligned}$

25. $f(x) = (1 + 3x)(5 - 2x)$

First find $f'(x)$:

$\begin{aligned} f'(x) &= (1 + 3x)(5 - 2x)' + (5 - 2x)(1 + 3x)' \\ &= (1 + 3x)(-2) + (5 - 2x)(3) \\ &= -2 - 6x + 15 - 6x \\ &= 13 - 12x \end{aligned}$

An equation for the tangent line at $x = 2$ is:

$y - y_1 = m(x - x_1)$

where $x_1 = 2$, $y_1 = f(x_1) = f(2) = 7$, and $m = f'(x_1) = f'(2) = -11$.

Thus, we have:

$y - 7 = -11(x - 2)$ or $y = -11x + 29$

27. $f(x) = \dfrac{x - 8}{3x - 4}$

First find $f'(x)$:

$f'(x) = \dfrac{(3x - 4)(x - 8)' - (x - 8)(3x - 4)'}{(3x - 4)^2}$

$= \dfrac{(3x - 4)(1) - (x - 8)(3)}{(3x - 4)^2} = \dfrac{20}{(3x - 4)^2}$

An equation for the tangent line at $x = 2$ is:

$y - y_1 = m(x - x_1)$

where $x_1 = 2$, $y_1 = f(x_1) = f(2) = -3$, and $m = f'(x_1) = f'(2) = 5$.

Thus, we have: $y - (-3) = 5(x - 2)$ or $y = 5x - 13$

29. $f(x) = (2x - 15)(x^2 + 18)$

$f'(x) = (2x - 15)(x^2 + 18)' + (x^2 + 18)(2x - 15)'$

$= (2x - 15)(2x) + (x^2 + 18)(2)$

$= 6x^2 - 30x + 36$

To find the values of x where $f'(x) = 0$, set: $f'(x) = 6x^2 - 30x + 36 = 0$

or $\quad x^2 - 5x + 6 = 0$

$(x - 2)(x - 3) = 0$

Thus, $x = 2$, $x = 3$.

31. $f(x) = \dfrac{x}{x^2 + 1}$

$f'(x) = \dfrac{(x^2 + 1)(x)' - x(x^2 + 1)'}{(x^2 + 1)^2} = \dfrac{(x^2 + 1)(1) - x(2x)}{(x^2 + 1)^2} = \dfrac{1 - x^2}{(x^2 + 1)^2}$

Now, set $f'(x) = \dfrac{1 - x^2}{(x^2 + 1)^2} = 0$

or $\quad 1 - x^2 = 0$

$(1 - x)(1 + x) = 0$

Thus, $x = 1$, $x = -1$.

33. $f(x) = x^3(x^4 - 1)$

First, we use the product rule:

$f'(x) = x^3(x^4 - 1)' + (x^4 - 1)(x^3)'$

$= x^3(4x^3) + (x^4 - 1)(3x^2)$

$= 7x^6 - 3x^2$

Next, simplifying $f(x)$, we have $f(x) = x^7 - x^3$. Thus, $f'(x) = 7x^6 - 3x^2$.

35. $f(x) = \dfrac{x^3 + 9}{x^3}$

First, we use the quotient rule:

$f'(x) = \dfrac{x^3(x^3 + 9)' - (x^3 + 9)(x^3)'}{(x^3)^2} = \dfrac{x^3(3x^2) - (x^3 + 9)(3x^2)}{x^6}$

$= \dfrac{-27x^2}{x^6} = \dfrac{-27}{x^4}$

Next, simplifying $f(x)$, we have $f(x) = \dfrac{x^3 + 9}{x^3} = 1 + \dfrac{9}{x^3} = 1 + 9x^{-3}$

Thus, $f'(x) = -27x^{-4} = -\dfrac{27}{x^4}$.

37. $f(x) = (2x^4 - 3x^3 + x)(x^2 - x + 5)$

$f'(x) = (2x^4 - 3x^3 + x)(x^2 - x + 5)' + (x^2 - x + 5)(2x^4 - 3x^3 + x)'$

$\qquad = (2x^4 - 3x^3 + x)(2x - 1) + (x^2 - x + 5)(8x^3 - 9x^2 + 1)$

39. $D_x \dfrac{3x^2 - 2x + 3}{4x^2 + 5x - 1}$

$\quad = \dfrac{(4x^2 + 5x - 1)D_x(3x^2 - 2x + 3) - (3x^2 - 2x + 3)D_x(4x^2 + 5x - 1)}{(4x^2 + 5x - 1)^2}$

$\quad = \dfrac{(4x^2 + 5x - 1)(6x - 2) - (3x^2 - 2x + 3)(8x + 5)}{(4x^2 + 5x - 1)^2}$

41. $y = 9x^{1/3}(x^3 + 5)$

$\dfrac{dy}{dx} = 9x^{1/3}\dfrac{d}{dx}(x^3 + 5) + (x^3 + 5)\dfrac{d}{dx}(9x^{1/3})$

$\qquad = 9x^{1/3}(3x^2) + (x^3 + 5)\left(9 \cdot \dfrac{1}{3}x^{-2/3}\right) = 27x^{7/3} + (x^3 + 5)(3x^{-2/3})$

43. $f(x) = \dfrac{6\sqrt[3]{x}}{x^2 - 3} = \dfrac{6x^{1/3}}{x^2 - 3}$

$f'(x) = \dfrac{(x^2 - 3)(6x^{1/3})' - 6x^{1/3}(x^2 - 3)'}{(x^2 - 3)^2}$

$\quad = \dfrac{(x^2 - 3)\left(6 \cdot \dfrac{1}{3}x^{-2/3}\right) - 6x^{1/3}(2x)}{(x^2 - 3)^2} = \dfrac{(x^2 - 3)(2x^{-2/3}) - 12x^{4/3}}{(x^2 - 3)^2}$

45. $D_x \dfrac{x^3 - 2x^2}{\sqrt[3]{x^2}} = D_x \dfrac{x^3 - 2x^2}{x^{2/3}}$

$\qquad = \dfrac{x^{2/3}D_x(x^3 - 2x^2) - (x^3 - 2x^2)D_x x^{2/3}}{(x^{2/3})^2}$

$\qquad = \dfrac{x^{2/3}(3x^2 - 4x) - (x^3 - 2x^2)\left(\dfrac{2}{3}x^{-1/3}\right)}{x^{4/3}}$

$\qquad = x^{-2/3}(3x^2 - 4x) - \dfrac{2}{3}x^{-5/3}(x^3 - 2x^2)$

47. $f(x) = \dfrac{(2x^2 - 1)(x^2 + 3)}{x^2 + 1}$

$f'(x) = \dfrac{(x^2 + 1)[(2x^2 - 1)(x^2 + 3)]' - (2x^2 - 1)(x^2 + 3)(x^2 + 1)'}{(x^2 + 1)^2}$

$\quad = \dfrac{(x^2 + 1)[(2x^2 - 1)(x^2 + 3)' + (x^2 + 3)(2x^2 - 1)'] - (2x^2 - 1)(x^2 + 3)(2x)}{(x^2 + 1)^2}$

$\quad = \dfrac{(x^2 + 1)[(2x^2 - 1)(2x) + (x^2 + 3)(4x)] - (2x^2 - 1)(x^2 + 3)(2x)}{(x^2 + 1)^2}$

49. $S(t) = \dfrac{200t}{t^2 + 36}$

(A) $S'(t) = \dfrac{(t^2 + 36)(200t)' - 200t(t^2 + 36)'}{(t^2 + 36)^2}$

$= \dfrac{(t^2 + 36)(200) - 200t(2t)}{(t^2 + 36)^2} = \dfrac{7200 - 200t^2}{(t^2 + 36)^2}$

[handwritten: $200t^2 + 7200 - 200t^2$ / $7200 - 200t^2$]

(B) $S(2) = \dfrac{200(2)}{2^2 + 36} = \dfrac{400}{40} = 10$

$S'(2) = \dfrac{7200 - 200(2)^2}{(2^2 + 36)^2} = \dfrac{7200 - 800}{1600} = \dfrac{6400}{1600} = 4$

Interpretation: At $t = 2$ months, the monthly sales are 10,000 and are increasing at the rate of 4000 albums per month.

(C) $S(8) = \dfrac{200(8)}{8^2 + 36} = \dfrac{1600}{100} = 16$

$S'(8) = \dfrac{7200 - 200(8)^2}{(8^2 + 36)^2} = \dfrac{7200 - 12,800}{10,000} = \dfrac{-5,600}{10,000} = -0.56$

Interpretation: At $t = 8$ months, the monthly sales are 16,000 and are decreasing at the rate of 560 albums per month.

51. $d(x) = \dfrac{50,000}{x^2 + 10x + 25}$, $2 \le x \le 25$

(A) $d'(x) = \dfrac{(x^2 + 10x + 25)(50,000)' - 50,000(x^2 + 10x + 25)'}{(x^2 + 10x + 25)^2}$

$= \dfrac{(x^2 + 10x + 25)(0) - 50,000(2x + 10)}{(x^2 + 10x + 25)^2}$

$= \dfrac{-100,000(x + 5)}{[(x + 5)^2]^2}$ [Note: $x^2 + 10x + 25 = (x + 5)^2$.]

$= \dfrac{-100,000(x + 5)}{(x + 5)^4} = \dfrac{-100,000}{(x + 5)^3}$

(B) $d'(5) = \dfrac{-100,000}{(10)^3} = \dfrac{-100,000}{1000} = -100$ radios per \$1 increase in price.

$d'(15) = \dfrac{-100,000}{(20)^3} = \dfrac{-100,000}{8000} \approx -12.5$ radios per \$1 increase in price.

53. $S(t) = \dfrac{200t}{t^2 + 36}$

$S'(t) = \dfrac{(t^2 + 36)(200) - 200t(2t)}{(t^2 + 36)^2} = \dfrac{7200 - 200t^2}{(t^2 + 36)^2}$

The equation of the tangent line at $t = 2$ is: $y - S(2) = S'(2)(t - 2)$

Now $S(2) = \dfrac{200(2)}{2^2 + 36} = \dfrac{400}{40} = 10$ and

$S'(2) = \dfrac{7200 - 200(2)^2}{(2^2 + 36)^2} = \dfrac{7200 - 800}{(40)^2} = \dfrac{6400}{1600} = 4$

Thus, the equation is: $y - 10 = 4(t - 2)$ or $y = 4t + 2$

The tangent line at $t = 8$ is:

$y - S(8) = S'(8)(t - 8)$

Now $S(8) = \dfrac{200(8)}{8^2 + 36} = \dfrac{1600}{100} = 16$ and

$S'(8) = \dfrac{7200 - 200(8)^2}{(8^2 + 36)^2}$

$= \dfrac{7200 - 12,800}{10,000}$

$= \dfrac{-5600}{10,000} = -0.56$

Thus, $y - 16 = -0.56(t - 8)$
    or $y = -0.56t + 20.48$

The graphs of S and the tangent lines are shown at the right.

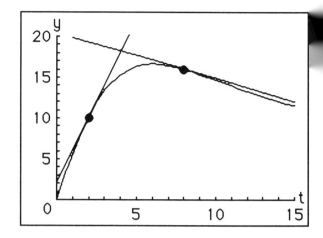

55. $d(x) = \dfrac{50,000}{x^2 + 10x + 25} = \dfrac{50,000}{(x + 5)^2}$ and

$d'(x) = \dfrac{-100,000}{(x + 5)^3}$

The equation of the tangent line at $x = 5$ is
    $y - d(5) = d'(5)(x - 5)$

Now, $d(5) = \dfrac{50,000}{(10)^2} = 500$, $d'(5) = \dfrac{-100,000}{(10)^3} = -100$

Thus, $y - 500 = -100(x - 5)$
    or $y = -100x + 1000$

The equation of the tangent line at $x = 15$ is
    $y - d(15) = d'(15)(x - 15)$

Now $d(15) = \dfrac{50,000}{(20)^2} = \dfrac{50,000}{400} = 125$

and $d'(15) = \dfrac{-100,000}{(20)^3} = \dfrac{-100,000}{8000}$

$= -12.5$

Thus, $y - 125 = -12.5(x - 15)$
    or $y = -12.5x + 312.5$

The graphs of d and the tangent lines are shown at the right.

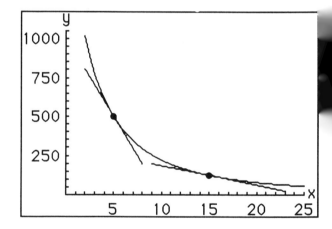

57. $C(t) = \dfrac{0.14t}{t^2 + 1}$

(A) $C'(t) = \dfrac{(t^2 + 1)(0.14t)' - (0.14t)(t^2 + 1)'}{(t^2 + 1)^2}$

$= \dfrac{(t^2 + 1)(0.14) - (0.14t)(2t)}{(t^2 + 1)^2} = \dfrac{0.14 - 0.14t^2}{(t^2 + 1)^2} = \dfrac{0.14(1 - t^2)}{(t^2 + 1)^2}$

(B) $C'(0.5) = \dfrac{0.14(1 - [0.5]^2)}{([0.5]^2 + 1)^2} = \dfrac{0.14(1 - 0.25)}{(1.25)^2} = 0.0672$

Interpretation: At $t = 0.5$ hours, the concentration is increasing at the rate of 0.0672 units per hour.

$C'(3) = \dfrac{0.14(1 - 3^2)}{(3^2 + 1)^2} = \dfrac{0.14(-8)}{100} = -0.0112$

Interpretation: At $t = 3$ hours, the concentration is decreasing at the rate of 0.0112 units per hour.

59. $N(x) = \dfrac{100x + 200}{x + 32}$

(A) $N'(x) = \dfrac{(x + 32)(100x + 200)' - (100x + 200)(x + 32)'}{(x + 32)^2}$

$= \dfrac{(x + 32)(100) - (100x + 200)(1)}{(x + 32)^2}$

$= \dfrac{100x + 3200 - 100x - 200}{(x + 32)^2} = \dfrac{3000}{(x + 32)^2}$

(B) $N'(4) = \dfrac{3000}{(36)^2} = \dfrac{3000}{1296} \approx 2.31;\quad N'(68) = \dfrac{3000}{(100)^2} = \dfrac{3000}{10,000} = \dfrac{3}{10} = 0.30$

EXERCISE 9-6

Things to remember:

1. GENERAL POWER RULE

    If n is any real number, then

    $$D_x[u(x)]^n = n[u(x)]^{n-1}u'(x),$$

    provided $u'(x)$ exists. [Note: The domains of $u(x)$ and/or $u'(x)$ might have to be restricted in order to avoid even roots of negative numbers or division by 0.]

    This rule can also be written as:

    $$\frac{dy}{dx} = n[u(x)]^{n-1}\frac{du}{dx}, \text{ or } D_x u^n = nu^{n-1}\frac{du}{dx}$$

---

1. $f(x) = (2x + 5)^3$

$f'(x) = 3(2x + 5)^2(2x + 5)'$

$= 3(2x + 5)^2(2)$

$= 6(2x + 5)^2$

3. $f(x) = (5 - 2x)^4$

$f'(x) = 4(5 - 2x)^3(5 - 2x)'$

$= 4(5 - 2x)^3(-2)$

$= -8(5 - 2x)^3$

5. $f(x) = (3x^2 + 5)^5$

$f'(x) = 5(3x^2 + 5)^4(3x^2 + 5)'$

$= 5(3x^2 + 5)^4(6x)$

$= 30x(3x^2 + 5)^4$

7. $f(x) = (x^3 - 2x^2 + 2)^8$

$f'(x) = 8(x^3 - 2x^2 + 2)^7(x^3 - 2x^2 + 2)'$

$= 8(x^3 - 2x^2 + 2)^7(3x^2 - 4x)$

9. $f(x) = (2x - 5)^{1/2}$

$f'(x) = \dfrac{1}{2}(2x - 5)^{-1/2}(2x - 5)'$

$= \dfrac{1}{2}(2x - 5)^{-1/2}(2) = \dfrac{1}{(2x - 5)^{1/2}}$

11. $f(x) = (x^4 + 1)^{-2}$

$f'(x) = -2(x^4 + 1)^{-3}(x^4 + 1)'$

$= -2(x^4 + 1)^{-3}(4x^3)$

$= -8x^3(x^4 + 1)^{-3} = \dfrac{-8x^3}{(x^4 + 1)^3}$

13. $f(x) = (2x - 1)^3$

$f'(x) = 3(2x - 1)^2(2) = 6(2x - 1)^2$

Tangent line at $x = 1$: $y - y_1 = m(x - x_1)$ where $x_1 = 1$, $y_1 = f(1) = (2(1) - 1)^3 = 1$, $m = f'(1) = 6[2(1) - 1]^2 = 6$. Thus, $y - 1 = 6(x - 1)$ or $y = 6x - 5$.

The tangent line is horizontal at the value(s) of x such that $f'(x) = 0$:

$$6(2x - 1)^2 = 0$$
$$2x - 1 = 0$$
$$x = \frac{1}{2}$$

15. $f(x) = (4x - 3)^{1/2}$

$f'(x) = \frac{1}{2}(4x - 3)^{-1/2}(4) = \frac{2}{(4x - 3)^{1/2}}$

Tangent line at $x = 3$: $y - y_1 = m(x - x_1)$ where $x_1 = 3$, $y_1 = f(3) = (4 \cdot 3 - 3)^{1/2} = 3$, $f'(3) = \frac{2}{(4 \cdot 3 - 3)^{1/2}} = \frac{2}{3}$. Thus, $y - 3 = \frac{2}{3}(x - 3)$ or $y = \frac{2}{3}x + 1$.

The tangent line is horizontal at the value(s) of x such that $f'(x) = 0$. Since $\frac{2}{(4x - 3)^{1/2}} \neq 0$ for all x $\left(x \neq \frac{3}{4}\right)$, there are no values of x where the tangent line is horizontal.

17. $y = 3(x^2 - 2)^4$

$\frac{dy}{dx} = 3 \cdot 4(x^2 - 2)^3(2x) = 24x(x^2 - 2)^3$

19. $y = 2(x^2 + 3x)^{-3}$

$\frac{dy}{dx} = 2 \cdot (-3)(x^2 + 3x)^{-4}(2x + 3) = -6(x^2 + 3x)^{-4}(2x + 3)$

21. $y = \sqrt{x^2 + 8} = (x^2 + 8)^{1/2}$

$\frac{dy}{dx} = \frac{1}{2}(x^2 + 8)^{-1/2}(2x) = \frac{x}{\sqrt{x^2 + 8}}$

23. $y = \sqrt[3]{3x + 4} = (3x + 4)^{1/3}$

$\frac{dy}{dx} = \frac{1}{3}(3x + 4)^{-2/3}(3) = (3x + 4)^{-2/3} = \frac{1}{(3x + 4)^{2/3}} = \frac{1}{\sqrt[3]{(3x + 4)^2}}$

25. $y = (x^2 - 4x + 2)^{1/2}$

$\frac{dy}{dx} = \frac{1}{2}(x^2 - 4x + 2)^{-1/2}(2x - 4)$

$= \frac{2(x - 2)}{2(x^2 - 4x + 2)^{1/2}} = \frac{x - 2}{(x^2 - 4x + 2)^{1/2}}$

27. $y = \frac{1}{2x + 4} = (2x + 4)^{-1}$

$\frac{dy}{dx} = -1(2x + 4)^{-2}(2) = \frac{-2}{(2x + 4)^2}$

29. $y = \dfrac{1}{(x^3 + 4)^5} = (x^3 + 4)^{-5}$

$\dfrac{dy}{dx} = -5(x^3 + 4)^{-6}(3x^2) = -15x^2(x^3 + 4)^{-6} = \dfrac{-15x^2}{(x^3 + 4)^6}$

31. $y = \dfrac{1}{4x^2 - 4x + 1} = (4x^2 - 4x + 1)^{-1}$

$\dfrac{dy}{dx} = -1(4x^2 - 4x + 1)^{-2}(8x - 4) = \dfrac{-4(2x - 1)}{(4x^2 - 4x + 1)^2}$

$= \dfrac{-4(2x - 1)}{(2x - 1)^4} = \dfrac{-4}{(2x - 1)^3}$

33. $y = \dfrac{4}{\sqrt{x^2 - 3x}} = \dfrac{4}{(x^2 - 3x)^{1/2}} = 4(x^2 - 3x)^{-1/2}$

$\dfrac{dy}{dx} = 4\left[-\dfrac{1}{2}(x^2 - 3x)^{-3/2}\right](2x - 3) = \dfrac{-2(2x - 3)}{(x^2 - 3x)^{3/2}}$

35. $f(x) = x(4 - x)^3$

$f'(x) = x[(4 - x)^3]' + (4 - x)^3(x)'$

$\quad\quad = x(3)(4 - x)^2(-1) + (4 - x)^3(1)$

$\quad\quad = (4 - x)^3 - 3x(4 - x)^2$

An equation for the tangent line to the graph of f at $x = 2$ is:

$y - y_1 = m(x - x_1)$ where $x_1 = 2$, $y_1 = f(x_1) = f(2) = 16$, and

$m = f'(x_1) = f'(2) = -16$.   Thus, $y - 16 = -16(x - 2)$ or $y = -16x + 48$.

37. $f(x) = \dfrac{x}{(2x - 5)^3}$

$f'(x) = \dfrac{(2x - 5)^3(1) - x(3)(2x - 5)^2(2)}{[(2x - 5)^3]^2}$

$\quad\quad = \dfrac{(2x - 5)^3 - 6x(2x - 5)^2}{(2x - 5)^6} = \dfrac{(2x - 5) - 6x}{(2x - 5)^4} = \dfrac{-5 - 4x}{(2x - 5)^4}$

An equation for the tangent line to the graph of f at $x = 3$ is:

$y - y_1 = m(x - x_1)$ where $x_1 = 3$, $y_1 = f(x_1) = f(3) = 3$, and

$m = f'(x_1) = f'(3) = -17$.   Thus, $y - 3 = -17(x - 3)$ or $y = -17x + 54$.

39. $f(x) = x\sqrt{2x + 2} = x(2x + 2)^{1/2}$

$f'(x) = x[(2x + 2)^{1/2}]' + (2x + 2)^{1/2}(x)'$

$\quad\quad = x\left(\dfrac{1}{2}\right)(2x + 2)^{-1/2}(2) + (2x + 2)^{1/2}(1) = \dfrac{x}{(2x + 2)^{1/2}} + (2x + 2)^{1/2}$

An equation for the tangent line to the graph of f at $x = 1$ is:

$y - y_1 = m(x - x_1)$ where $x_1 = 1$, $y_1 = f(x_1) = f(1) = 2$, and

$m = f'(x_1) = f'(1) = \dfrac{5}{2}$.   Thus, $y - 2 = \dfrac{5}{2}(x - 1)$ or $y = \dfrac{5}{2}x - \dfrac{1}{2}$.

41. $f(x) = x^2(x - 5)^3$

$f'(x) = x^2[(x - 5)^3]' + (x - 5)^3(x^2)'$

$\quad\quad = x^2(3)(x - 5)^2(1) + (x - 5)^3(2x)$

$\quad\quad = 3x^2(x - 5)^2 + 2x(x - 5)^3$

The tangent line to the graph of f is horizontal at the values of x such that $f'(x) = 0$. Thus, we set $3x^2(x - 5)^2 + 2x(x - 5)^3 = 0$

$$x(x - 5)^2[3x + 2(x - 5)] = 0$$
$$x(x - 5)^2(5x - 10) = 0$$
$$5x(x - 2)(x - 5)^2 = 0$$

and $x = 0$, $x = 2$, $x = 5$.

43. $f(x) = \dfrac{x}{(2x + 5)^2}$

$$f'(x) = \frac{(2x + 5)^2(x)' - x[(2x + 5)^2]'}{[(2x + 5)^2]^2}$$

$$= \frac{(2x + 5)^2(1) - x(2)(2x + 5)(2)}{(2x + 5)^4} = \frac{2x + 5 - 4x}{(2x + 5)^3} = \frac{5 - 2x}{(2x + 5)^3}$$

The tangent line to the graph of f is horizontal at the values of x such that $f'(x) = 0$. Thus, we set

$$\frac{5 - 2x}{(2x + 5)^3} = 0$$
$$5 - 2x = 0$$

and $x = \dfrac{5}{2}$.

45. $f(x) = \sqrt{x^2 - 8x + 20} = (x^2 - 8x + 20)^{1/2}$

$$f'(x) = \frac{1}{2}(x^2 - 8x + 20)^{-1/2}(2x - 8)$$

$$= \frac{x - 4}{(x^2 - 8x + 20)^{1/2}} = \frac{x - 4}{\sqrt{x^2 - 8x + 20}}$$

The tangent line to the graph of f is horizontal at the values of x such that $f'(x) = 0$. Thus, we set

$$\frac{x - 4}{(x^2 - 8x + 20)^{1/2}} = 0$$
$$x - 4 = 0$$

and $x = 4$.

47. $D_x[3x(x^2 + 1)^3] = 3xD_x(x^2 + 1)^3 + (x^2 + 1)^3 D_x 3x$

$$= 3x \cdot 3(x^2 + 1)^2(2x) + (x^2 + 1)^3(3)$$
$$= 18x^2(x^2 + 1)^2 + 3(x^2 + 1)^3$$
$$= (x^2 + 1)^2[18x^2 + 3(x^2 + 1)]$$
$$= (x^2 + 1)^2(21x^2 + 3)$$
$$= 3(x^2 + 1)^2(7x^2 + 1)$$

49. $D_x \dfrac{(x^3 - 7)^4}{2x^3} = \dfrac{2x^3 D_x(x^3 - 7)^4 - (x^3 - 7)^4 D_x 2x^3}{(2x^3)^2}$

$$= \frac{2x^3 \cdot 4(x^3 - 7)^3(3x^2) - (x^3 - 7)^4 6x^2}{4x^6}$$

$$= \frac{3(x^3 - 7)^3 x^2[8x^3 - 2(x^3 - 7)]}{4x^6}$$

$$= \frac{3(x^3 - 7)^3(6x^3 + 14)}{4x^4} = \frac{3(x^3 - 7)^3(3x^3 + 7)}{2x^4}$$

51. $D_x[(2x - 3)^2(2x^2 + 1)^3] = (2x - 3)^2 D_x(2x^2 + 1)^3 + (2x^2 + 1)^3 D_x(2x - 3)^2$

$\qquad = (2x - 3)^2 3(2x^2 + 1)^2(4x) + (2x^2 + 1)^3 2(2x - 3)(2)$

$\qquad = 12x(2x - 3)^2(2x^2 + 1)^2 + 4(2x^2 + 1)^3(2x - 3)$

$\qquad = 4(2x - 3)(2x^2 + 1)^2[3x(2x - 3) + (2x^2 + 1)]$

$\qquad = 4(2x - 3)(2x^2 + 1)^2(6x^2 - 9x + 2x^2 + 1)$

$\qquad = 4(2x - 3)(2x^2 + 1)^2(8x^2 - 9x + 1)$

53. $D_x[4x^2\sqrt{x^2 - 1}] = D_x[\sqrt{16x^4(x^2 - 1)}]$

$\qquad = D_x[(16x^6 - 16x^4)^{1/2}]$

$\qquad = \frac{1}{2}(16x^6 - 16x^4)^{-1/2}(96x^5 - 64x^3)$

$\qquad = \frac{96x^5 - 64x^3}{2\sqrt{16x^6 - 16x^4}} = \frac{8x^2(12x^3 - 8x)}{2 \cdot 4x^2\sqrt{x^2 - 1}} = \frac{(12x^3 - 8x)}{\sqrt{x^2 - 1}}$

or

$D_x[4x^2\sqrt{x^2 - 1}] = D_x[4x^2(x^2 - 1)^{1/2}]$

$\qquad = 4x^2 \cdot \frac{1}{2}(x^2 - 1)^{-1/2}(2x) + (x^2 - 1)^{1/2}(8x)$

$\qquad = \frac{4x^3}{(x^2 - 1)^{1/2}} + 8x(x^2 - 1)^{1/2}$

$\qquad = \frac{4x^3 + 8x(x^2 - 1)}{(x^2 - 1)^{1/2}} = \frac{4x^3 + 8x^3 - 8x}{(x^2 - 1)^{1/2}} = \frac{12x^3 - 8x}{(x^2 - 1)^{1/2}}$

55. $D_x\dfrac{2x}{\sqrt{x - 3}} = \dfrac{(x - 3)^{1/2}(2) - 2x \cdot \frac{1}{2}(x - 3)^{-1/2}}{(x - 3)}$

$\qquad = \dfrac{2(x - 3)^{1/2} - \dfrac{x}{(x - 3)^{1/2}}}{(x - 3)} = \dfrac{2(x - 3) - x}{(x - 3)(x - 3)^{1/2}}$

$\qquad = \dfrac{2x - 6 - x}{(x - 3)^{3/2}} = \dfrac{x - 6}{(x - 3)^{3/2}}$

57. $D_x\sqrt{(2x - 1)^3(x^2 + 3)^4} = D_x[(2x - 1)^3(x^2 + 3)^4]^{1/2}$

$\qquad = D_x(2x - 1)^{3/2}(x^2 + 3)^2$

$\qquad = (2x - 1)^{3/2}D_x(x^2 + 3)^2 + (x^2 + 3)^2 D_x(2x - 1)^{3/2}$

$\qquad = (2x - 1)^{3/2}2(x^2 + 3)(2x) + (x^2 + 3)^2 \cdot \frac{3}{2}(2x - 1)^{1/2}(2)$

$\qquad = (2x - 1)^{1/2}(x^2 + 3)[4x(2x - 1) + 3(x^2 + 3)]$

$\qquad = (2x - 1)^{1/2}(x^2 + 3)(8x^2 - 4x + 3x^2 + 9)$

$\qquad = (2x - 1)^{1/2}(x^2 + 3)(11x^2 - 4x + 9)$

59. $C(x) = 10 + \sqrt{2x + 16} = 10 + (2x + 16)^{1/2}, \quad 0 \le x \le 50$

(A) $C'(x) = \frac{1}{2}(2x + 16)^{-1/2}(2) = \frac{1}{\sqrt{2x + 16}}$

(B) $C'(24) = \frac{1}{\sqrt{2(24) + 16}} = \frac{1}{\sqrt{64}} = \frac{1}{8}$ or \$12.50 per calculator.

Interpretation: The rate of change of cost with respect to production at a production level of 24 calculators is \$12.50 per calculator. The cost of producing the 25th calculator is approximately \$12.50.

$C'(42) = \frac{1}{\sqrt{2(42) + 16}} = \frac{1}{\sqrt{100}} = \frac{1}{10}$ or \$10.00 per calculator.

Interpretation: The rate of change of cost with respect to production at a production level of 42 calculators is \$10.00 per calculator. The cost of producing the 43rd calculator is approximately \$10.00.

61. $A = 1000\left(1 + \frac{1}{12}r\right)^{48}$

$\frac{dA}{dr} = 1000(48)\left(1 + \frac{1}{12}r\right)^{47}\left(\frac{1}{12}\right) = 4000\left(1 + \frac{1}{12}r\right)^{47}$

63. $y = (3 \times 10^6)\left[1 - \frac{1}{\sqrt[3]{(x^2 - 1)^2}}\right] = (3 \times 10^6)[1 - (x^2 - 1)^{-2/3}]$

$\frac{dy}{dx} = -(3 \times 10^6)\left(-\frac{2}{3}\right)(x^2 - 1)^{-5/3}(2x) = \frac{(4 \times 10^6)x}{(x^2 - 1)^{5/3}}$

65. $T = f(n) = 2n\sqrt{n - 2} = 2n(n - 2)^{1/2}$

(A) $\frac{dT}{dn} = 2n[(n - 2)^{1/2}]' + (n - 2)^{1/2}(2n)'$

$= 2n\left(\frac{1}{2}\right)(n - 2)^{-1/2}(1) + (n - 2)^{1/2}(2)$

$= \frac{n}{(n - 2)^{1/2}} + 2(n - 2)^{1/2}$

$= \frac{n + 2(n - 2)}{(n - 2)^{1/2}} = \frac{3n - 4}{(n - 2)^{1/2}}$

(B) $f'(11) = \frac{3 \cdot 11 - 4}{(11 - 2)^{1/2}} = \frac{29}{3}$; learning is increasing at the rate of $\frac{29}{3}$ units per minute at the $n = 11$ level.

$f'(27) = \frac{3 \cdot 27 - 4}{(27 - 2)^{1/2}} = \frac{77}{5}$; learning is increasing at the rate of $\frac{77}{5}$ units per minute at the $n = 27$ level.

Things to remember:

1. MARGINAL COST, REVENUE, AND PROFIT

    If x is the number of units of a product produced in some time interval, then:

    Total Cost = $C(x)$
    Marginal Cost = $C'(x)$
    Total Revenue = $R(x)$
    Marginal Revenue = $R'(x)$
    Total Profit = $P(x) = R(x) - C(x)$
    Marginal Profit = $P'(x) = R'(x) - C'(x)$
    $\qquad\qquad$ = (Marginal Revenue) - (Marginal Cost)

    Marginal cost (or revenue or profit) is the instantaneous rate of change of cost (or revenue or profit) relative to production at a given production level.

2. MARGINAL AVERAGE COST, REVENUE, AND PROFIT
    If x is the number of units of a product produced in some time interval, then:

    Average Cost = $\overline{C}(x) = \dfrac{C(x)}{x}$ $\qquad$ Cost per unit

    Marginal Average Cost = $\overline{C}'(x)$

    Average Revenue = $\overline{R}(x) = \dfrac{R(x)}{x}$ $\qquad$ Revenue per unit

    Marginal Average Revenue = $\overline{R}'(x)$

    Average Profit = $\overline{P}(x) = \dfrac{P(x)}{x}$ $\qquad$ Profit per unit

    Marginal Average Profit = $\overline{P}'(x)$

---

1. $C(x) = 2000 + 50x - \dfrac{x^2}{2}$

    (A) The exact cost of producing the 21st food processor is:

    $C(21) - C(20) = 2000 + 50(21) - \dfrac{(21)^2}{2} - \left[2000 + 50(20) - \dfrac{(20)^2}{2}\right]$
    $\qquad\qquad\qquad = 2829.50 - 2800$
    $\qquad\qquad\qquad = 29.50$ or \$29.50

    (B) $C'(x) = 50 - x$
    $C'(20) = 50 - 20 = 30$ or \$30

3. $C(x) = 60,000 + 300x$

    (A) $\overline{C}(x) = \dfrac{60,000 + 300x}{x} = \dfrac{60,000}{x} + 300 = 60,000x^{-1} + 300$

    $\overline{C}(500) = \dfrac{60,000 + 300(500)}{500} = \dfrac{210,000}{500} = 420$ or \$420

(B) $\overline{C}'(x) = -60,000x^{-2} = \dfrac{-60,000}{x^2}$

$\overline{C}'(500) = \dfrac{-60,000}{(500)^2} = -0.24$ or $0.24

Interpretation: At a production level of 500 units, a unit increase in production will decrease average cost by approximately 24¢.

5. $R(x) = 100x - \dfrac{x^2}{40}$

$R'(x) = 100 - \dfrac{x}{20}$

(A) $R'(1600) = 100 - \dfrac{1600}{20} = 100 - 80 = 20$ or $20

Interpretation: At a production level of 1600, a unit increase in production will increase revenue by approximately $20.

(B) $R'(2500) = 100 - \dfrac{2500}{20} = -25$ or -$25

Interpretation: At a production level of 2500, a unit increase in production will decrease revenue by approximately $25.

7. $P(x) = 30x - \dfrac{x^2}{2} - 250$

(A) The exact profit from the sale of the 26th skateboard is:

$P(26) - P(25) = 30(26) - \dfrac{(26)^2}{2} - 250 - \left[30(25) - \dfrac{(25)^2}{2} - 250 \right]$

$= 192 - 187.50$
$= 4.50$ or $4.50

(B) $P'(x) = 30 - x$
$P'(25) = 30 - 25 = 5$ or $5.00

9. $P(x) = 5x - \dfrac{x^2}{200} - 450$

$P'(x) = 5 - \dfrac{x}{100}$

(A) $P'(450) = 5 - \dfrac{450}{100} = 0.5$ or $0.50

Interpretation: At a production level of 450, a unit increase in production will increase profits by approximately 50¢.

(B) $P'(750) = 5 - \dfrac{750}{100} = -2.5$ or -$2.50

Interpretation: At a production level of 750, a unit increase in production will decrease profits by approximately $2.50.

11. (A) $\overline{P}(x) = \dfrac{P(x)}{x} = \dfrac{5x - \dfrac{x^2}{200} - 450}{x} = 5 - \dfrac{x}{200} - \dfrac{450}{x}$

$\overline{P}(150) = 5 - \dfrac{150}{200} - \dfrac{450}{150} = 1.25$ or $1.25

(B) $\overline{P}(x) = 5 - \dfrac{x}{200} - 450x^{-1}$

$\overline{P}'(x) = -\dfrac{1}{200} + 450x^{-2} = -\dfrac{1}{200} + \dfrac{450}{x^2}$

$$\bar{P}'(150) = -\frac{1}{200} + \frac{450}{(150)^2} = 0.015 \text{ or } \$0.015$$

Interpretation: At a production level of 150 units, a unit increase in production will increase the average profit by approximately 1.5¢.

13. Cost: $C(x) = 72,000 + 60x$

(A) $C'(x) = 60$

(B) Revenue: $R(x) = x \cdot p = 200x - \frac{x^2}{30}$

(C) $R'(x) = 200 - \frac{x}{15}$

(D) $R'(1500) = 200 - \frac{1500}{15} = 100 \text{ or } \$100$

Interpretation: At a production level of 1500 units, a unit increase in production will increase revenue by approximately $100.

$$R'(4500) = 200 - \frac{4500}{15} = -100 \text{ or } -\$100$$

Interpretation: At a production level of 4500 units, a unit increase in production will decrease revenue by approximately $100.

(E) The graphs of $C(x)$ and $R(x)$ are shown below:

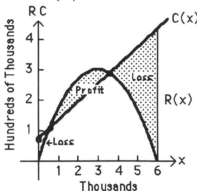

To find the break-even points, set $C(x) = R(x)$.

$$72,000 + 60x = 200x - \frac{x^2}{30}$$

$$\frac{x^2}{30} - 140x + 72,000 = 0$$

$$x^2 - 4200x + 2,160,000 = 0$$

$$(x - 600)(x - 3600) = 0$$

$$x = 600 \text{ or } x = 3600$$

Now, $C(600) = 72,000 + 60(600) = 108,000$;
$C(3600) = 72,000 + 60(3600) = 288,000$.
Thus, the break-even points are (600, 108,000) and (3600, 288,000).

(F) $P(x) = R(x) - C(x)$

$$= 200x - \frac{x^2}{30} - (72,000 + 60x)$$

$$= -\frac{x^2}{30} + 140x - 72,000$$

(G) $P(x) = \frac{-x^2}{30} + 140x - 72,000$

$$P'(x) = \frac{-x}{15} + 140$$

(H) $P'(1500) = \frac{-1500}{15} + 140 = 40 \text{ or } \$40$

Interpretation: At a production level of 1500 units, a unit increase in production will increase profits by approximately $40.

$$P'(3000) = \frac{-3000}{15} + 140 = -60 \text{ or } -\$60$$

Interpretation: At a production level of 3000 units, a unit increase in production will decrease profits by approximately $60.

15. (A) We are given $p = 16$ when $x = 200$ and $p = 14$ when $x = 300$. Thus, we have the pair of equations:

$16 = 200m + b$

$14 = 300m + b$

Subtracting the second equation from the first, we get $-100m = 2$. Thus,

$$m = -\frac{1}{50}.$$

Substituting this into either equation yields $b = 20$. Therefore,

$$p = -\frac{x}{50} + 20 \text{ or } p = 20 - \frac{x}{50}.$$

(B) $R(x) = x \cdot p(x) = 20x - \dfrac{x^2}{50}$

(C) From the financial department's estimates, $m = 4$ and $b = 1400$. Thus, $C(x) = 4x + 1400$.

(D) The graphs of $R(x)$ and $C(x)$ are shown below:

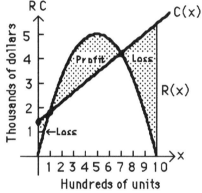

To find the break-even points, set $C(x) = R(x)$.

$$4x + 1400 = 20x - \frac{x^2}{50}$$

$$\frac{x^2}{50} - 16x + 1400 = 0$$

$$x^2 - 800x + 70{,}000 = 0$$

$$(x - 100)(x - 700) = 0$$

$$x = 100 \text{ or } x = 700$$

Now, $C(100) = 1800$ and $C(700) = 4200$. Thus, the break-even points are $(100, 1800)$ and $(700, 4200)$.

(E) $P(x) = R(x) - C(x) = 20x - \dfrac{x^2}{50} - (4x + 1400) = 16x - \dfrac{x^2}{50} - 1400$

(F) $P(x) = 16x - \dfrac{x^2}{50} - 1400$

$P'(x) = 16 - \dfrac{x}{25}$

$$P'(250) = 16 - \frac{250}{25} = 6 \text{ or } \$6$$

Interpretation: At a production level of 250 units, a unit increase in production will increase profits by approximately $6.

$$P'(475) = 16 - \frac{475}{25} = -3 \text{ or } -\$3$$

Interpretation: At a production level of 475 units, a unit increase in production will decrease profits by approximately $3.

17. Total cost: $C(x) = 24x + 21,900$
    Total revenue: $R(x) = 200x - 0.2x^2$, $0 \le x \le 1,000$

(A) $R'(x) = 200 - 0.4x$
    The graph of R has a horizontal tangent line at the value(s) of x where $R'(x) = 0$, i.e.,
    $$200 - 0.4x = 0$$
    $$\text{or } x = 500$$

(B) $P(x) = R(x) - C(x) = 200x - 0.2x^2 - (24x + 21,900)$
    $$= 176x - 0.2x^2 - 21,900$$

(C) $P'(x) = 176 - 0.4x$. Setting $P'(x) = 0$, we have
    $$176 - 0.4x = 0$$
    $$\text{or } x = 440$$

(D) The graphs of C, R and P are shown below.

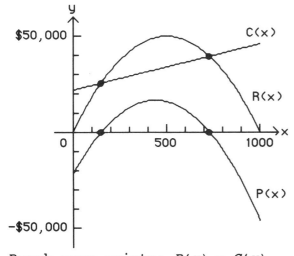

Break-even points: $R(x) = C(x)$
$$200x - 0.2x^2 = 24x + 21,900$$
$$0.2x^2 - 176x + 21,900 = 0$$
$$x = \frac{176 \pm \sqrt{(176)^2 - (4)(0.2)(21,900)}}{2(0.2)} \quad \text{(quadratic formula)}$$
$$= \frac{176 \pm \sqrt{30,976 - 17,520}}{0.4}$$
$$= \frac{176 \pm \sqrt{13,456}}{0.4} = \frac{176 \pm 116}{0.4} = 730, \ 150$$

Thus, the break-even points are: (730, 39,420) and (150, 25,500).

x-intercepts for P: $-0.2x^2 + 176x - 21{,}900 = 0$
or $0.2x^2 - 176x + 21{,}900 = 0$
which is the same as the equation above.

Thus, $x = 150$ and $x = 730$.

19. Demand equation: $p = 20 - \sqrt{x} = 20 - x^{1/2}$
Cost equation: $C(x) = 500 + 2x$

(A) Revenue $R(x) = xp = x(20 - x^{1/2})$
or $R(x) = 20x - x^{3/2}$

(B) The graphs for R and C for $0 \le x \le 400$ are

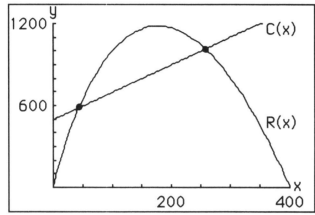

Break-even points $(44, 588)$ and $(258, 1{,}016)$.

EXERCISE 9-8   CHAPTER REVIEW

1. $f(x) = 3x^4 - 2x^2 + 1$
   $f'(x) = 12x^3 - 4x$

2. $f(x) = 2x^{1/2} - 3x$
   $f'(x) = 2 \cdot \dfrac{1}{2} x^{-1/2} - 3 = \dfrac{1}{x^{1/2}} - 3$

3. $f(x) = 5$
   $f'(x) = 0$

4. $f(x) = \dfrac{1}{2x^2} + \dfrac{x^2}{2}$
   $= \dfrac{1}{2} x^{-2} + \dfrac{x^2}{2}$
   $f'(x) = -x^{-3} + x$

5. $f(x) = (2x - 1)(3x + 2)$
   $f'(x) = (2x - 1)3 + (3x + 2)2$
   $= 6x - 3 + 6x + 4$
   $= 12x + 1$

6. $f(x) = (x^2 - 1)(x^3 - 3)$
   $f'(x) = (x^2 - 1)(3x^2) + (x^3 - 3)(2x) = 3x^4 - 3x^2 + 2x^4 - 6x = 5x^4 - 3x^2 - 6x$

7. $f(x) = \dfrac{2x}{x^2 + 2}$
   $f'(x) = \dfrac{(x^2 + 2)(2) - 2x(2x)}{(x^2 + 2)^2} = \dfrac{2x^2 + 4 - 4x^2}{(x^2 + 2)^2} = \dfrac{4 - 2x^2}{(x^2 + 2)^2}$

8. $f(x) = \dfrac{1}{3x + 2} = (3x + 2)^{-1}$
   $f'(x) = -1(3x + 2)^{-2}(3) = \dfrac{-3}{(3x + 2)^2}$

9. $f(x) = (2x - 3)^3$

$f'(x) = 3(2x - 3)^2(2) = 6(2x - 3)^2$

10. $f(x) = (x^2 + 2)^{-2}$

$f'(x) = -2(x^2 + 2)^{-3}(2x) = \dfrac{-4x}{(x^2 + 2)^3}$

11. (A) From the graph, $\lim\limits_{x \to 1} f(x)$ does not exist since

$\lim\limits_{x \to 1^-} f(x) = 2 \neq \lim\limits_{x \to 1^+} f(x) = 3.$

(B) $f(1) = 3$

(C) f is *not* continuous at $x = 1$ since $\lim\limits_{x \to 1} f(x)$ does not exist.

12. (A) $\lim\limits_{x \to 2} f(x) = 2$

(B) $f(2)$ is not defined

(C) f is not continuous at $x = 2$ since $f(2)$ is not defined.

13. (A) $\lim\limits_{x \to 3} f(x) = 1$

(B) $f(3) = 1$

(C) f is continuous at $x = 3$ since $\lim\limits_{x \to 3} f(x) = f(3).$

14. $y = 3x^4 - 2x^{-3} + 5$

$\dfrac{dy}{dx} = 12x^3 + 6x^{-4}$

15. $y = (2x^2 - 3x + 2)(x^2 + 2x - 1)$

$y' = (2x^2 - 3x + 2)(2x + 2) + (x^2 + 2x - 1)(4x - 3)$

$\quad = 4x^3 - 6x^2 + 4x + 4x^2 - 6x + 4 + 4x^3 + 8x^2 - 4x - 3x^2 - 6x + 3$

$\quad = 8x^3 + 3x^2 - 12x + 7$

16. $f(x) = \dfrac{2x - 3}{(x - 1)^2}$

$f'(x) = \dfrac{(x - 1)^2 2 - (2x - 3)2(x - 1)}{(x - 1)^4} = \dfrac{(x - 1)[2(x - 1) - 4x + 6]}{(x - 1)^4}$

$\quad = \dfrac{(2x - 2 - 4x + 6)}{(x - 1)^3} = \dfrac{4 - 2x}{(x - 1)^3}$

17. $y = 2\sqrt{x} + \dfrac{4}{\sqrt{x}} = 2x^{1/2} + 4x^{-1/2}$

$y' = 2 \cdot \dfrac{1}{2}x^{-1/2} + 4\left(-\dfrac{1}{2}\right)x^{-3/2} = \dfrac{1}{\sqrt{x}} - \dfrac{2}{x^{3/2}}$ or $\dfrac{1}{\sqrt{x}} - \dfrac{2}{\sqrt{x^3}}$

18. $D_x[(x^2 - 1)(2x + 1)^2] = (x^2 - 1)D_x(2x + 1)^2 + (2x + 1)^2 D_x(x^2 - 1)$

$\quad = (x^2 - 1)2(2x + 1)2 + (2x + 1)^2 2x$

$\quad = 2(2x + 1)[2(x^2 - 1) + x(2x + 1)]$

$\quad = 2(2x + 1)(2x^2 - 2 + 2x^2 + x)$

$\quad = 2(2x + 1)(4x^2 + x - 2)$

19. $D_x(x^3 - 5)^{1/3} = \frac{1}{3}(x^3 - 5)^{-2/3}(3x^2) = \frac{x^2}{(x^3 - 5)^{2/3}}$

20. $y = \frac{3x^2 + 4}{x^2}$

$\frac{dy}{dx} = \frac{x^2(6x) - (3x^2 + 4)2x}{[x^2]^2} = \frac{-8x}{x^4} = \frac{-8}{x^3}$

21. $D_x \frac{(x^2 + 2)^4}{2x - 3} = \frac{(2x - 3)4(x^2 + 2)^3 2x - (x^2 + 2)^4 2}{(2x - 3)^2}$

$= \frac{2(x^2 + 2)^3[4x(2x - 3) - (x^2 + 2)]}{(2x - 3)^2}$

$= \frac{2(x^2 + 2)^3(8x^2 - 12x - x^2 - 2)}{(2x - 3)^2} = \frac{2(x^2 + 2)^3(7x^2 - 12x - 2)}{(2x - 3)^2}$

22. $f(x) = x^2 + 4$
$f'(x) = 2x$

(A) The slope of the graph at $x = 1$ is $m = f'(1) = 2$.

(B) $f(1) = 1^2 + 4 = 5$
The tangent line at $(1, 5)$, where the slope $m = 2$, is:
$(y - 5) = 2(x - 1)$ [Note: $(y - y_1) = m(x - x_1)$.]
$\qquad y = 5 + 2x - 2$
$\qquad y = 2x + 3$

23. $f(x) = x^3(x + 1)^2$
$f'(x) = x^3(2)(x + 1)(1) + (x + 1)^2(3x^2)$
$\qquad = 2x^3(x + 1) + 3x^2(x + 1)^2$

(A) The slope of the graph of f at $x = 1$ is:
$f'(1) = 2 \cdot 1^3(1 + 1) + 3 \cdot 1^2(1 + 1)^2 = 16$

(B) $f(1) = 1^3(1 + 1)^2 = 4$

An equation for the tangent line to the graph of f at $x = 1$ is
$y - 4 = 16(x - 1)$ or $y = 16x - 12$.

24. $f(x) = 10x - x^2$
$f'(x) = 10 - 2x$
The tangent line is horizontal at the values of x such that $f'(x) = 0$:
$10 - 2x = 0$
$\qquad x = 5$

25. $f(x) = (x + 3)(x^2 - 45)$
$f'(x) = (x + 3)(2x) + (x^2 - 45)(1) = 3x^2 + 6x - 45$
Set $f'(x) = 0$:
$3x^2 + 6x - 45 = 0$
$x^2 + 2x - 15 = 0$
$(x - 3)(x + 5) = 0$
$\qquad x = 3, \ x = -5$

26. $f(x) = \frac{x}{x^2 + 4}$

$f'(x) = \frac{(x^2 + 4)(1) - x(2x)}{(x^2 + 4)^2} = \frac{4 - x^2}{(x^2 + 4)^2}$

Set $f'(x) = 0$:

$$\frac{4 - x^2}{(x^2 + 4)^2} = 0$$
$$4 - x^2 = 0$$
$$(2 - x)(2 + x) = 0$$
$$x = 2, \quad x = -2$$

27. $f(x) = x^2(2x - 15)^3$

   $f'(x) = x^2(3)(2x - 15)^2(2) + (2x - 15)^3(2x)$
   $\qquad = (2x - 15)^2[6x^2 + 4x^2 - 30x]$
   $\qquad = (2x - 15)^2 10x(x - 3)$

   Set $f'(x) = 0$:

   $10x(x - 3)(2x - 15)^2 = 0$
   $$x = 0, \quad x = 3, \quad x = \frac{15}{2}$$

28. $y = f(x) = 16x^2 - 4x$

   (A) Velocity function: $v(x) = f'(x) = 32x - 4$

   (B) $v(3) = 32(3) - 4 = 92$ ft/sec

29. $y = f(x) = 96x - 16x^2$

   (A) Velocity function: $v(x) = f'(x) = 96 - 32x$

   (B) $v(x) = 0$ when $96 - 32x = 0$
   $$32x = 96$$
   $$x = 3 \text{ sec}$$

30. From the graph:

   (A) $\lim\limits_{x \to 2^-} f(x) = 4$ (B) $\lim\limits_{x \to 2^+} f(x) = 6$

   (C) $\lim\limits_{x \to 2} f(x)$ does not exist, since $\lim\limits_{x \to 2^-} f(x) \neq \lim\limits_{x \to 2^+} f(x)$

   (D) $f(2) = 6$ (E) No, since $\lim\limits_{x \to 2} f(x)$ does not exist.

31. From the graph:

   (A) $\lim\limits_{x \to 5^-} f(x) = 3$ (B) $\lim\limits_{x \to 5^+} f(x) = 3$

   (C) $\lim\limits_{x \to 5} f(x) = 3$, since $\lim\limits_{x \to 5^-} f(x) = \lim\limits_{x \to 5^+} f(x)$

   (D) $f(5) = 3$ (E) Yes, since $\lim\limits_{x \to 5} f(x) = f(5) = 3$.

32. (A) $\lim\limits_{x \to 0^-} \dfrac{1}{|x|} = \infty$ (B) $\lim\limits_{x \to 0^+} \dfrac{1}{|x|} = \infty$ (C) $\lim\limits_{x \to 0} \dfrac{1}{|x|} = \infty$

33. (A) $\lim\limits_{x \to 1^-} \dfrac{1}{x - 1} = -\infty$ (B) $\lim\limits_{x \to 1^+} \dfrac{1}{x - 1} = \infty$

   (C) $\lim\limits_{x \to 1} \dfrac{1}{x - 1}$ does not exist.

34. $f(x) = 2x^2 - 3x + 1$. Since f is a polynomial, f is continuous for all x; i.e., f is continuous on $(-\infty, \infty)$.

35. $f(x) = \dfrac{1}{x + 5}$. Since f is a rational function, f is continuous for all such that the denominator $x + 5 \neq 0$, i.e., for all x such that $x \neq -5$. Thus, f is continuous on $(-\infty, -5)$ and on $(-5, \infty)$.

36. $f(x) = \dfrac{x - 3}{x^2 - x - 6}$
Since f is a rational function, f is continuous for all x such that the denominator $x^2 - x - 6 \neq 0$. Now $x^2 - x - 6 = (x - 3)(x + 2) = 0$ for $x = -2$ and $x = 3$. Thus, f is continuous on $(-\infty, -2)$, $(-2, 3)$, and $(3, \infty)$.

37. $f(x) = \sqrt{x - 3}$. f is continuous for all x such that $x - 3 \geq 0$, or $x \geq 3$. Thus, f is continuous on $[3, \infty)$.

38. $f(x) = \sqrt[3]{1 - x^2}$. f is continuous for all x such that $g(x) = 1 - x^2$ is continuous. Since g is a polynomial, g is continuous for all x. Thus, f is continuous on $(-\infty, \infty)$.

39. $\displaystyle\lim_{x \to 3} \dfrac{2x - 3}{x + 5} = \dfrac{2(3) - 3}{3 + 5} = \dfrac{6 - 3}{8} = \dfrac{3}{8}$

40. $\displaystyle\lim_{x \to 3}(2x^2 - x + 1) = 2 \cdot 3^2 - 3 + 1 = 16$

41. $\displaystyle\lim_{x \to 0} \dfrac{2x}{3x^2 - 2x} = \lim_{x \to 0} \dfrac{2x}{x(3x - 2)} = \lim_{x \to 0} \dfrac{2}{3x - 2} = \dfrac{2}{-2} = -1$

42. $\displaystyle\lim_{h \to 0} \dfrac{[(2 + h)^2 - 1] - [2^2 - 1]}{h} = \lim_{h \to 0} \dfrac{4 + 4h + h^2 - 1 - 3}{h}$
$$= \lim_{h \to 0} \dfrac{4h + h^2}{h} = \lim_{h \to 0}(4 + h) = 4$$

43. $f(x) = x^2 + 4$
$$\lim_{h \to 0} \dfrac{f(2 + h) - f(2)}{h} = \lim_{h \to 0} \dfrac{[(2 + h)^2 + 4] - [2^2 + 4]}{h}$$
$$= \lim_{h \to 0} \dfrac{4 + 4h + h^2 + 4 - 8}{h} = \lim_{h \to 0} \dfrac{4h + h^2}{h}$$
$$= \lim_{h \to 0}(4 + h) = 4$$

44. $\displaystyle\lim_{x \to 3} \dfrac{x - 3}{x^2 - 9} = \lim_{x \to 3} \dfrac{\cancel{x - 3}}{(x + 3)\cancel{(x - 3)}}$
$$= \dfrac{1}{3 + 3} = \dfrac{1}{6}$$

45. $\displaystyle\lim_{x \to -3} \dfrac{x - 3}{x^2 - 9} = \lim_{x \to -3} \dfrac{\cancel{x - 3}}{(x + 3)\cancel{(x - 3)}} = \lim_{x \to -3} \dfrac{1}{x + 3}$. The limit does not exist.

46. $\displaystyle\lim_{x \to 7} \dfrac{\sqrt{x} - \sqrt{7}}{x - 7} = \lim_{x \to 7} \dfrac{\sqrt{x} - \sqrt{7}}{(\sqrt{x})^2 - (\sqrt{7})^2} = \lim_{x \to 7} \dfrac{\cancel{\sqrt{x} - \sqrt{7}}}{(\sqrt{x} + \sqrt{7})\cancel{(\sqrt{x} - \sqrt{7})}}$
$$= \dfrac{1}{\sqrt{7} + \sqrt{7}} = \dfrac{1}{2\sqrt{7}}$$

47. $\lim\limits_{x \to -2} \sqrt{\dfrac{x^2 + 4}{2 - x}} = \sqrt{\dfrac{(-2)^2 + 4}{2 - (-2)}} = \sqrt{\dfrac{8}{4}} = \sqrt{2}$

48. $\lim\limits_{x \to \infty}\left(3 + \dfrac{1}{x^{1/3}} + \dfrac{2}{x^3}\right) = \lim\limits_{x \to \infty} 3 + \lim\limits_{x \to \infty}\dfrac{1}{x^{1/3}} + \lim\limits_{x \to \infty}\dfrac{2}{x^3} = 3 + 0 + 0 = 3$

49. $\lim\limits_{x \to \infty}(3x^3 - 2x^2 - 10x - 100) = \lim\limits_{x \to \infty}\left[x^3\left(3 - \dfrac{2}{x} - \dfrac{10}{x^2} - \dfrac{100}{x^3}\right)\right]$

$$= \left(\lim\limits_{x \to \infty} x^3\right)\lim\limits_{x \to \infty}\left(3 - \dfrac{2}{x} - \dfrac{10}{x^2} - \dfrac{100}{x^3}\right) = \infty,$$

since the limit of the first factor is ∞ and the limit of the second factor is 3.

50. $\lim\limits_{x \to \infty}\dfrac{2x^2 + 3}{3x^2 + 2} = \lim\limits_{x \to \infty}\dfrac{x^2\left(2 + \dfrac{3}{x^2}\right)}{x^2\left(3 + \dfrac{2}{x^2}\right)} = \lim\limits_{x \to \infty}\dfrac{2 + \dfrac{3}{x^2}}{3 + \dfrac{2}{x^2}} = \dfrac{2}{3}$

51. $\lim\limits_{x \to \infty}\dfrac{2x + 3}{3x^2 + 2} = \lim\limits_{x \to \infty}\dfrac{x\left(2 + \dfrac{3}{x}\right)}{x^2\left(3 + \dfrac{2}{x^2}\right)} = \lim\limits_{x \to \infty}\left[\dfrac{1}{x} \cdot \dfrac{\left(2 + \dfrac{3}{x}\right)}{\left(3 + \dfrac{2}{x^2}\right)}\right]$

$$= \left(\lim\limits_{x \to \infty}\dfrac{1}{x}\right) \cdot \left(\lim\limits_{x \to \infty}\dfrac{2 + \dfrac{3}{x}}{3 + \dfrac{2}{x^2}}\right) = 0 \cdot \dfrac{2}{3} = 0$$

52. $\lim\limits_{x \to \infty}\dfrac{2x^2 + 3}{3x + 2} = \lim\limits_{x \to \infty}\dfrac{x^2\left(2 + \dfrac{3}{x^2}\right)}{x\left(3 + \dfrac{2}{x}\right)} = \lim\limits_{x \to \infty}\dfrac{x\left(2 + \dfrac{3}{x^2}\right)}{3 + \dfrac{2}{x}} = \left(\lim\limits_{x \to \infty} x\right)\left(\lim\limits_{x \to \infty}\dfrac{2 + \dfrac{3}{x^2}}{3 + \dfrac{2}{x}}\right) = \infty$

since $\lim\limits_{x \to \infty} x = \infty$ and $\lim\limits_{x \to \infty}\dfrac{2 + \dfrac{3}{x^2}}{3 + \dfrac{2}{x}} = \dfrac{2}{3}$

53. $f(x) = x^2 - x$

    <u>Step 1</u>. Simplify $\dfrac{f(x + h) - f(x)}{h}$.

$$\dfrac{f(x + h) - f(x)}{h} = \dfrac{[(x + h)^2 - (x + h)] - (x^2 - x)}{h}$$

$$= \dfrac{x^2 + 2xh + h^2 - x - h - x^2 + x}{h}$$

$$= \dfrac{2xh + h^2 - h}{h} = 2x + h - 1$$

Step 2. Evaluate $\lim\limits_{h \to 0} \dfrac{f(x + h) - f(x)}{h}$.

$$\lim_{h \to 0} \frac{f(x + h) - f(x)}{h} = \lim_{h \to 0}(2x + h - 1) = 2x - 1$$

Thus, $f'(x) = 2x - 1$.

54. $f(x) = \sqrt{x} - 3$

Step 1. Simplify $\dfrac{f(x + h) - f(x)}{h}$.

$$\frac{f(x + h) - f(x)}{h} = \frac{[\sqrt{x + h} - 3] - (\sqrt{x} - 3)}{h}$$

$$= \frac{\sqrt{x + h} - \sqrt{x}}{h} \cdot \frac{\sqrt{x + h} + \sqrt{x}}{\sqrt{x + h} + \sqrt{x}} = \frac{x + h - x}{h[\sqrt{x + h} + \sqrt{x}]}$$

$$= \frac{1}{\sqrt{x + h} + \sqrt{x}}$$

Step 2. Evaluate $\lim\limits_{h \to 0} \dfrac{f(x + h) - f(x)}{h}$.

$$\lim_{h \to 0} \frac{f(x + h) - f(x)}{h} = \lim_{h \to 0} \frac{1}{\sqrt{x + h} + \sqrt{x}} = \frac{1}{2\sqrt{x}}$$

55. f is not differentiable at $x = 0$, since f is not continuous at 0.

56. f is not differentiable at $x = 1$; the curve has a vertical tangent line at this point.

57. f is not differentiable at $x = 2$; the curve has a "corner" at this point.

58. f is differentiable at $x = 3$. In fact, $f'(3) = 0$.

59. The graph of f is shown below. f is discontinuous at $x = 0$.

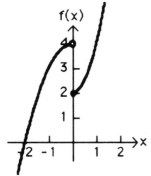

60. The graph of f is shown below. f is continuous for all x.

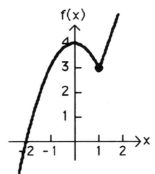

61. $f(x) = (x - 4)^4(x + 3)^3$

$f'(x) = (x - 4)^4(3)(x + 3)^2(1) + (x + 3)^3(4)(x - 4)^3(1)$

$\quad\quad = (x - 4)^3(x + 3)^2[3(x - 4) + 4(x + 3)]$

$\quad\quad = 7x(x - 4)^3(x + 3)^2$

62. $f(x) = \dfrac{x^5}{(2x + 1)^4}$

$f'(x) = \dfrac{(2x + 1)^4(5x^4) - x^5(4)(2x + 1)^3(2)}{[(2x + 1)^4]^2}$

$= \dfrac{(2x + 1)(5x^4) - 8x^5}{(2x + 1)^5} = \dfrac{2x^5 + 5x^4}{(2x + 1)^5} = \dfrac{x^4(2x + 5)}{(2x + 1)^5}$

63. $f(x) = \dfrac{\sqrt{x^2 - 1}}{x} = \dfrac{(x^2 - 1)^{1/2}}{x}$

$f'(x) = \dfrac{x\left(\frac{1}{2}\right)(x^2 - 1)^{-1/2}(2x) - (x^2 - 1)^{1/2}(1)}{x^2} = \dfrac{\dfrac{x^2}{(x^2 - 1)^{1/2}} - (x^2 - 1)^{1/2}}{x^2}$

$= \dfrac{1}{x^2(x^2 - 1)^{1/2}} = \dfrac{1}{x^2\sqrt{x^2 - 1}}$

64. $f(x) = \dfrac{x}{\sqrt{x^2 + 4}} = \dfrac{x}{(x^2 + 4)^{1/2}}$

$f'(x) = \dfrac{(x^2 + 4)^{1/2}(1) - x\left(\frac{1}{2}\right)(x^2 + 4)^{-1/2}(2x)}{[(x^2 + 4)^{1/2}]^2}$

$= \dfrac{(x^2 + 4)^{1/2} - \dfrac{x^2}{(x^2 + 4)^{1/2}}}{(x^2 + 4)} = \dfrac{4}{(x^2 + 4)^{3/2}}$

65. (A) $\lim\limits_{x \to 1^-} f(x) = 1$

(B) $\lim\limits_{x \to 1^+} f(x) = 1$

(C) $\lim\limits_{x \to 1} f(x) = 1$

(D) f is continuous at $x = 1$, since $\lim\limits_{x \to 1} f(x) = f(1)$.

66. (A) Yes

(B) Yes, $\lim\limits_{x \to 0^+} f(x) = 0 = f(0)$.

(C) Yes, $\lim\limits_{x \to 2^-} f(x) = 0 = f(2)$.

(D) Yes

67. $f(x) = 1 - |x - 1|, \ 0 \le x \le 2$

(A) $\lim\limits_{h \to 0^-} \dfrac{f(1 + h) - f(1)}{h} = \lim\limits_{h \to 0^-} \dfrac{1 - |1 + h - 1| - 1}{h} = \lim\limits_{h \to 0^-} \dfrac{-|h|}{h}$

$= \lim\limits_{h \to 0^-} \dfrac{h}{h} = 1 \quad (|h| = -h \text{ if } h < 0)$

(B) $\lim\limits_{h \to 0^+} \dfrac{f(1 + h) - f(1)}{h} = \lim\limits_{h \to 0^+} \dfrac{1 - |1 + h - 1| - 1}{h} = \lim\limits_{h \to 0^+} \dfrac{-|h|}{h}$

$= \lim\limits_{h \to 0^+} \dfrac{-h}{h} = -1 \quad (|h| = h \text{ if } h > 0)$

(C) $\lim\limits_{h \to 0} \dfrac{f(1 + h) - f(1)}{h}$ does not exist, since the left limit and the right limit are not equal.

(D) $f'(1)$ does not exist.

68. $C(x) = 10,000 + 200x - 0.1x^2$

    (A) $C(101) - C(100) = 10,000 + 200(101) - 0.1(101)^2$
$$- [10,000 + 200(100) - 0.1(100)^2]$$
$$= 29,179.90 - 29,000$$
$$= \$179.90$$

    (B) $C'(x) = 200 - 0.2x$
$$C'(100) = 200 - 0.2(100)$$
$$= 200 - 20$$
$$= \$180$$

69. $C(x) = 5,000 + 40x + 0.05x^2$

    (A) Cost of producing 100 bicycles:
$$C(100) = 5,000 + 40(100) + 0.05(100)^2$$
$$= 9000 + 500 = 9500$$
Marginal cost:
$$C'(x) = 40 + 0.1x$$
$$C'(100) = 40 + 0.1(100) = 40 + 10 = 50$$

Interpretation: At a production level of 100 bicycles, the total cost is \$9,500 and is increasing at the rate of \$50 per unit increase in production.

    (B) Average cost: $\bar{C}(x) = \dfrac{C(x)}{x} = \dfrac{5000}{x} + 40 + 0.05x$

$$\bar{C}(100) = \frac{5000}{100} + 40 + 0.05(100) = 50 + 40 + 5 = 95$$

Marginal average cost: $\bar{C}'(x) = -\dfrac{5000}{x^2} + 0.05$

$$\text{and } \bar{C}'(100) = -\frac{5000}{(100)^2} + 0.05 = -0.5 + 0.05 = -0.45$$

Interpretation: At a production level of 100 bicycles, the average cost is \$95 and the marginal average cost is decreasing at a rate of -\$0.45 per unit increase in production.

70. $p = 20 - x$ and $C(x) = 2x + 56$, $0 \le x \le 20$

    (A) Marginal cost: $C'(x) = 2$

Average cost: $\bar{C}(x) = \dfrac{C(x)}{x} = 2 + \dfrac{56}{x} = 2 + 56x^{-1}$

Marginal average cost: $\bar{C}'(x) = -56x^{-2} = -\dfrac{56}{x^2}$

    (B) Revenue: $R(x) = x \cdot p(x) = x(20 - x) = 20x - x^2$

Marginal revenue: $R'(x) = 20 - 2x$

Average revenue: $\bar{R}(x) = \dfrac{20x - x^2}{x} = 20 - x$

Marginal average revenue: $\bar{R}'(x) = -1$

(C) Profit: $P(x) = R(x) - C(x) = 20x - x^2 - (2x + 56) = 18x - x^2 - 56$

Marginal profit: $P'(x) = 18 - 2x$

Average profit: $\overline{P}(x) = \dfrac{18x - x^2 - 56}{x} = 18 - x - 56x^{-1}$

Marginal average profit: $\overline{P}'(x) = -1 + 56x^{-2} = -1 + \dfrac{56}{x^2}$

(D) Break-even points: $R(x) = C(x)$
$$20x - x^2 = 2x + 56$$
$$-x^2 + 18x - 56 = 0$$
$$x^2 - 18x + 56 = 0$$
$$(x - 4)(x - 14) = 0$$
Thus, the break-even points are at $x = 4$ and $x = 14$.

(E) $P'(7) = 18 - 2 \cdot 7 = 4$
Interpretation: At a production level of 7 units, a unit increase in production will increase profits by approximately \$4.

$P'(9) = 18 - 2 \cdot 9 = 0$
Interpretation: At a production level of 9, a unit increase in production will neither increase nor decrease profits.

$P'(11) = 18 - 2 \cdot 11 = -4$
Interpretation: At a production level of 11, a unit increase in production will decrease profits by approximately \$4.

(F) The graphs of $R(x)$ and $C(x)$ are shown in the diagram below:

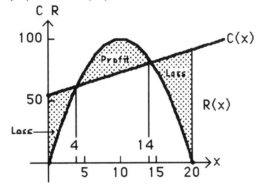

71. $N(t) = \dfrac{40t}{t + 2}$

(A) Average rate of change from $t = 3$ to $t = 6$:
$$\dfrac{N(6) - N(3)}{6 - 3} = \dfrac{\dfrac{40 \cdot 6}{6 + 2} - \dfrac{40 \cdot 3}{3 + 2}}{3} = \dfrac{30 - 24}{3} = 2 \text{ components per day}$$

(B) $N'(t) = \dfrac{(t + 2)(40) - 40t(1)}{(t + 2)^2} = \dfrac{80}{(t + 2)^2}$

$N'(3) = \dfrac{80}{25} = 3.2$ components per day

(C) $\lim\limits_{t \to \infty} N(t) = \lim\limits_{t \to \infty} \dfrac{40t}{t + 2} = \lim\limits_{t \to \infty} \dfrac{40}{1 + \dfrac{2}{t}}$ (divide numerator and denominator by t)
$$= 40 \text{ components per day}$$

72. $N(t) = 5 + t\sqrt{12 - t} = 5 + t(12 - t)^{1/2}$

$N'(t) = t\left(\frac{1}{2}\right)(12 - t)^{-1/2}(-1) + (12 - t)^{1/2}(1)$

$\qquad = \dfrac{-t}{2(12 - t)^{1/2}} + (12 - t)^{1/2}$

$\qquad = \dfrac{-t + 2(12 - t)}{2(12 - t)^{1/2}} = \dfrac{24 - 3t}{2\sqrt{12 - t}}$

Now, $N(3) = 5 + 3\sqrt{12 - 3} = 5 + 3\sqrt{9} = 14$ and

$\qquad N'(3) = \dfrac{24 - 3(3)}{2\sqrt{12 - 3}} = \dfrac{15}{2\sqrt{9}} = \dfrac{15}{6} = \dfrac{5}{2} = 2.5$

Interpretation: Monthly sales during the third month are 14,000 pools and are increasing at the rate of 2,500 pools per month.

73. $C(x) = 500(x + 1)^{-2}$

The instantaneous rate of change of concentration at x meters is:

$C'(x) = 500(-2)(x + 1)^{-3}$

$\qquad = -1000(x + 1)^{-3} = \dfrac{-1000}{(x + 1)^3}$

The rate of change of concentration at 9 meters is:

$C'(9) = \dfrac{-1000}{(9 + 1)^3} = \dfrac{-1000}{10^3} = -1$ part per million per meter

The rate of change of concentration at 99 meters is:

$C'(99) = \dfrac{-1000}{(99 + 1)^3} = \dfrac{-1000}{100^3} = \dfrac{-10^3}{10^6} = -10^{-3}$ or $= -\dfrac{1}{1000}$

$\qquad\qquad\qquad\qquad\qquad\qquad\qquad = -0.001$ parts per million per meter

74. $N(t) = 20\sqrt{t} = 20t^{1/2}$

The rate of learning is $N'(t) = 20\left(\frac{1}{2}\right)t^{-1/2} = 10t^{-1/2} = \dfrac{10}{\sqrt{t}}$.

(A) The rate of learning after one hour is $N'(1) = \dfrac{10}{\sqrt{1}}$

$\qquad\qquad\qquad\qquad\qquad\qquad\qquad = 10$ items per hour.

(B) The rate of learning after four hours is $N'(4) = \dfrac{10}{\sqrt{4}} = \dfrac{10}{2}$

$\qquad\qquad\qquad\qquad\qquad\qquad\qquad = 5$ items per hour.

Things to remember:

1. SIGN PROPERTIES ON AN INTERVAL (a, b)

    If f is continuous on (a, b) and $f(x) \neq 0$ for each $x \in (a, b)$, then either $f(x) > 0$ for all x in (a, b) or $f(x) < 0$ for all x in (a, b).

2. CONSTRUCTING SIGN CHARTS

    Given a function f. The values x such that f is discontinuous at x or $f(x) = 0$ are called PARTITION NUMBERS.

    Step 1. Find all partition numbers.

    Step 2. Plot the numbers found in Step 1 on a real number line, dividing the number line into intervals.

    Step 3. Select a test number in each open interval determined in Step 2 and evaluate $f(x)$ at each test number to determine whether $f(x)$ is positive(+) or negative(-) in each interval.

    Step 4. Construct a sign chart using the real number line in Step 2. This will show the sign of $f(x)$ on each open interval.

    [Note: From the sign chart, it is easy to find the solution for the inequality $f(x) < 0$ or $f(x) > 0$.]

3. INCREASING AND DECREASING FUNCTIONS

    For the interval (a, b):

$f'(x)$	$f(x)$	Graph of f	Examples
+	Increases ↗	Rises ↗	
−	Decreases ↘	Falls ↘	

4. LOCAL EXTREMA

    Given a function f. The value $f(c)$ is a LOCAL MAXIMUM of f if there is an interval (m, n) containing c such that $f(x) \leq f(c)$ for all x in (m, n). The value $f(e)$ is a LOCAL MINIMUM of f if there is an interval (p, q) containing e such that $f(x) \geq f(e)$ for all x in (p, q). Local maxima and local minima are called LOCAL EXTREMA.

<u>5</u>. EXISTENCE OF LOCAL EXTREMA: CRITICAL VALUES

If f is continuous on the interval (a, b) and $f(c)$ is a local extremum, then either $f'(c) = 0$ or $f'(c)$ does not exist (is not defined). The values of x in the domain of f where $f'(x) = 0$ or $f'(x)$ does not exist are called the CRITICAL VALUES OF f.

<u>6</u>. FIRST DERIVATIVE TEST FOR LOCAL EXTREMA

Let c be a critical value of f [$f(c)$ is defined and either $f'(c) = 0$ or $f'(c)$ is not defined.]

Construct a sign chart for $f'(x)$ close to and on either side of c.

Sign Chart	$f(c)$
$f'(x)$ ——$(\!\!\!\!\underset{m}{}\,---\,\underset{c}{|}\,+++\,\underset{n}{}\!\!\!\!)$→ x $f(x)$ Decreasing | Increasing	$f(c)$ is a local minimum. If $f'(x)$ changes from negative to positive at c, then $f(c)$ is a local minimum.
$f'(x)$ ——$(\!\!\!\!\underset{m}{}\,+++\,\underset{c}{|}\,---\,\underset{n}{}\!\!\!\!)$→ x $f(x)$ Increasing | Decreasing	$f(c)$ is a local maximum. If $f'(x)$ changes from positive to negative at c, then $f(c)$ is a local maximum.
$f'(x)$ ——$(\!\!\!\!\underset{m}{}\,---\,\underset{c}{|}\,---\,\underset{n}{}\!\!\!\!)$→ x $f(x)$ Decreasing | Decreasing	$f(c)$ is not a local extremum. If $f'(x)$ does not change sign at c, then $f(c)$ is neither a local maximum nor a local minimum.
$f'(x)$ ——$(\!\!\!\!\underset{m}{}\,+++\,\underset{c}{|}\,+++\,\underset{n}{}\!\!\!\!)$→ x $f(x)$ Increasing | Increasing	$f(c)$ is not a local extremum. If $f'(x)$ does not change sign at c, then $f(c)$ is neither a local maximum nor a local minimum.

---

1. From the graph, $f(x)$ is increasing on the intervals (a, b), (d, f), and (g, h).

3. From the graph, $f'(x) = 0$ when $x = c$, d, f.

5. From the graph, $f(x)$ has a local maximum at $x = b$, f.

7. At $x = a$: $f'(x) > 0$ on the left of a and $f'(x) < 0$ on the right of a. Thus, $f(a)$ is a local maximum.

At $x = b$: $f'(x)$ does not change sign at $x = b$ ($f'(x) < 0$ on the left and right of b). Thus, $f(b)$ is neither a local maximum nor a local minimum.

At $x = c$: $f'(x) < 0$ on the left of c and $f'(x) > 0$ on the right of c. Thus, $f(c)$ is a local minimum.

At $x = d$: $f'(x)$ does not change sign at $x = d$ ($f'(x) > 0$ on the left and right of d). Thus, $f(d)$ is neither a local maximum nor a local minimum.

9.

x	$f'(x)$	$f(x)$	GRAPH OF f
$(-\infty, -1)$	+	Increasing	Rising
$x = -1$	0	Neither local maximum nor local minimum	Horizontal tangent
$(-1, 1)$	+	Increasing	Rising
$x = 1$	0	Local maximum	Horizontal tangent
$(1, \infty)$	–	Decreasing	Falling

Using this information together with the points $(-2, -1)$, $(-1, 1)$, $(0, 2)$, $(1, 3)$, $(2, 1)$ on the graph, we have

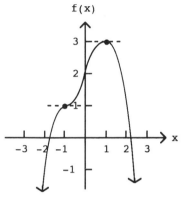

11.

x	$f'(x)$	$f(x)$	GRAPH OF $f(x)$
$(-\infty, -1)$	–	Decreasing	Falling
$x = -1$	0	Local minimum	Horizontal tangent
$(-1, 0)$	+	Increasing	Rising
$x = 0$	Not defined	Local maximum	Vertical tangent line
$(0, 2)$	–	Decreasing	Falling
$x = 2$	0	Neither local maximum nor local minimum	Horizontal tangent
$(2, \infty)$	–	Decreasing	Falling

Using this information together with the points $(-2, 2)$, $(-1, 1)$, $(0, 2)$, $(2, 1)$, $(4, 0)$ on the graph, we have

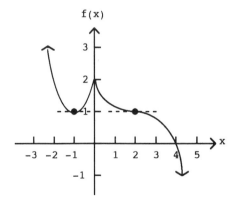

13. $x^2 - x - 12 < 0$

Let $f(x) = x^2 - x - 12 = (x - 4)(x + 3)$. Then f is continuous for all x and $f(-3) = f(4) = 0$. Thus, $x = -3$ and $x = 4$ are partition numbers.

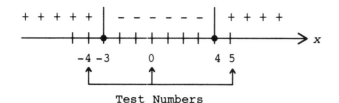

Test Numbers	
x	$f(x)$
-4	8 $(+)$
0	-12 $(-)$
5	8 $(+)$

Thus, $x^2 - x - 12 < 0$ for:
$-3 < x < 4$ (inequality notation)
$(-3, 4)$ (interval notation)

15. $x^2 + 21 > 10x$ or $x^2 - 10x + 21 > 0$
Let $f(x) = x^2 - 10x + 21 = (x - 7)(x - 3)$. Then f is continuous for all x and $f(3) = f(7) = 0$. Thus, $x = 3$ and $x = 7$ are partition numbers.

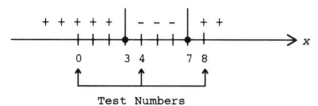

Test Numbers	
x	$f(x)$
0	21 $(+)$
4	-3 $(-)$
8	5 $(+)$

Thus, $x^2 - 10x + 21 > 0$ for:
$x < 3$ or $x > 7$ (inequality notation)
$(-\infty, 3) \cup (7, \infty)$ (interval notation)

17. $\dfrac{x^2 + 5x}{x - 3} > 0$

Let $f(x) = \dfrac{x^2 + 5x}{x - 3} = \dfrac{x(x + 5)}{x - 3}$. Then f is discontinuous at $x = 3$ and $f(0) = f(-5) = 0$. Thus, $x = -5$, $x = 0$, and $x = 3$ are partition numbers.

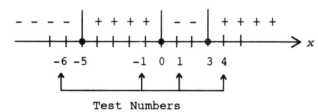

Test Numbers	
x	$f(x)$
-6	$-\frac{2}{3}$ $(-)$
-1	1 $(+)$
1	-3 $(-)$
4	36 $(+)$

Thus, $\dfrac{x^2 + 5x}{x - 3} > 0$ for: $-5 < x < 0$ or $x > 3$ (inequality notation)
$(-5, 0) \cup (3, \infty)$ (interval notation)

19. $f(x) = x^2 - 16x + 12$
$f'(x) = 2x - 16$
f' is continuous for all x and
$f'(x) = 2x - 16 = 0$
         $x = 8$
Thus, $x = 8$ is a partition number for f'.
Next, we construct a sign chart for f' (partition number is 8).

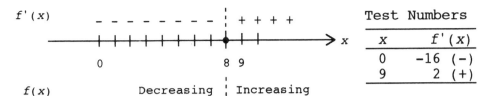

$f(x)$ Decreasing ┊ Increasing

Therefore, f is decreasing on $(-\infty, 8)$ and increasing on $(8, \infty)$; f has a local minimum at $x = 8$.

21. $f(x) = 4 + 10x - x^2$
$f'(x) = 10 - 2x$
f' is continuous for all x and
$f'(x) = 10 - 2x = 0$
$\qquad x = 5$
Thus, $x = 5$ is a partition number for f'.
Next, we construct a sign chart for f' (partition number is 5).

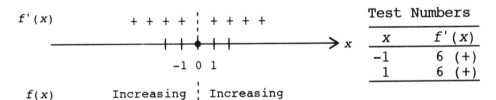

$f(x)$ Increasing ┊ Decreasing

Therefore, f is increasing on $(-\infty, 5)$ and decreasing on $(5, \infty)$; f has a local maximum at $x = 5$.

23. $f(x) = 2x^3 + 4$
$f'(x) = 6x^2$
f' is continuous for all x and
$f'(x) = 6x^2 = 0$
$\qquad x = 0$
Thus, $x = 0$ is a partition number for f'.
Next, we construct a sign chart for f' (partition number is 0).

```
f'(x)          + + + + ┊ + + + +
          ──────┼┼┼●┼┼┼──────────→ x
                -1 0 1

f(x)          Increasing ┊ Increasing
```

Test Numbers	
x	$f'(x)$
-1	6 (+)
1	6 (+)

Therefore, f is increasing for all x; i.e., on $(-\infty, \infty)$; f has no local extrema.

25. $f(x) = 2 - 6x - 2x^3$
$f'(x) = -6 - 6x^2$
f' is continuous for all x and
$f'(x) = -6 - 6x^2 = 0$
$\qquad -6(1 + x^2) = 0$
There are no real numbers that satisfy this equation.
The sign chart for f' is:

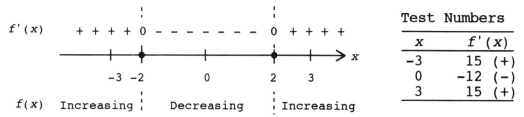

Test Numbers

x	$f'(x)$
0	-6 (-)

$f(x)$ Decreasing

Thus, f is decreasing for all x; i.e., on $(-\infty, \infty)$; f has no local extrema.

27. $f(x) = x^3 - 12x + 8$
$f'(x) = 3x^2 - 12$
f' is continuous for all x and
$f'(x) = 3x^2 - 12 = 0$
        $3(x^2 - 4) = 0$
$3(x - 2)(x + 2) = 0$
Thus, $x = -2$ and $x = 2$ are partition numbers for f'.
Next, we construct the sign chart for f' (partition numbers are -2 and 2).

$f'(x)$ + + + + 0 - - - - - - - 0 + + + +

                -3 -2   0   2   3

$f(x)$ Increasing | Decreasing | Increasing

Test Numbers

x	$f'(x)$
-3	15 (+)
0	-12 (-)
3	15 (+)

Therefore, f is increasing on $(-\infty, -2)$ and on $(2, \infty)$, f is decreasing on $(-2, 2)$; f has a local maximum at $x = -2$ and a local minimum at $x = 2$.

29. $f(x) = x^3 - 3x^2 - 24x + 7$
$f'(x) = 3x^2 - 6x - 24$
f' is continuous for all x and
$f'(x) = 3x^2 - 6x - 24 = 0$
       $3(x^2 - 2x - 8) = 0$
     $3(x + 2)(x - 4) = 0$
Thus, $x = -2$ and $x = 4$ are partition numbers for f'.
The sign chart for f' is:

$f'(x)$ + + + + 0 - - - - - 0 + + + +

            -3 -2   0   4 5

$f(x)$ Increasing| Decreasing |Increasing

Test Numbers

x	$f'(x)$
-3	21 (+)
0	-24 (-)
5	21 (+)

Therefore, f is increasing on $(-\infty, -2)$ and on $(4, \infty)$, f is decreasing on $(-2, 4)$; f has a local maximum at $x = -2$ and a local minimum at $x = 4$.

31. $f(x) = 2x^2 - x^4$

$f'(x) = 4x - 4x^3$

f' is continuous for all x and

$f'(x) = 4x - 4x^3 = 0$

$\qquad 4x(1 - x^2) = 0$

$4x(1 - x)(1 + x) = 0$

Thus, $x = -1$, $x = 0$, and $x = 1$ are partition numbers for f'.

The sign chart for f' is:

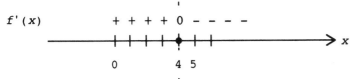

x	f'(x)
-2	24 $(+)$
$-\frac{1}{2}$	$-\frac{3}{2}$ $(-)$
$\frac{1}{2}$	$\frac{3}{2}$ $(+)$
2	-24 $(-)$

Test Numbers

Therefore, f is increasing on $(-\infty, -1)$ and on $(0, 1)$, f is decreasing on $(-1, 0)$ and on $(1, \infty)$; f has local maxima at $x = -1$ and $x = 1$ and a local minimum at $x = 0$.

33. $f(x) = 4 + 8x - x^2$

$f'(x) = 8 - 2x$

f' is continuous for all x and

$f'(x) = 8 - 2x = 0$

$\qquad\qquad x = 4$

Thus, $x = 4$ is a partition number for f'.

The sign chart for f' is:

x	f'(x)
0	8 $(+)$
5	-2 $(-)$

Test Numbers

Therefore, f is increasing on $(-\infty, 4)$ and decreasing on $(4, \infty)$; f has a local maximum at $x = 4$.

x	f'(x)	f	GRAPH OF f
$(-\infty, 4)$	$+$	Increasing	Rising
$x = 4$	0	Local maximum	Horizontal tangent
$(4, \infty)$	$-$	Decreasing	Falling

x	f(x)
0	4
4	20

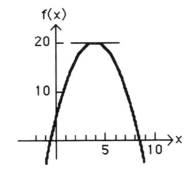

35. $f(x) = x^3 - 3x + 1$

$f'(x) = 3x^2 - 3$

f' is continuous for all x and

$f'(x) = 3x^2 - 3 = 0$

$3(x^2 - 1) = 0$

$3(x + 1)(x - 1) = 0$

Thus, $x = -1$ and $x = 1$ are partition numbers for f'.

The sign chart for f' is:

Test Numbers

x	$f'(x)$
-2	9 (+)
0	-3 (−)
2	9 (+)

$f'(x)$ + + + + + 0 − − − − − − 0 + + + + +

$-2 \quad -1 \quad 0 \quad 1 \quad 2$

$f(x)$ Increasing ┊ Decreasing ┊ Increasing

Therefore, f is increasing on $(-\infty, -1)$ and on $(1, \infty)$, f is decreasing o $(-1, 1)$; f has a local maximum at $x = -1$ and a local minimum at $x = 1$.

x	$f'(x)$	f	GRAPH OF f
$(-\infty, -1)$	+	Increasing	Rising
$x = -1$	0	Local maximum	Horizontal tangent
$(-1, 1)$	−	Decreasing	Falling
$x = 1$	0	Local minimum	Horizontal tangent
$(1, \infty)$	+	Increasing	Rising

x	$f(x)$
-1	3
0	1
1	-1

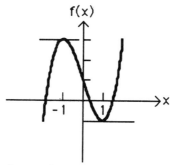

37. $f(x) = 10 - 12x + 6x^2 - x^3$

$f'(x) = -12 + 12x - 3x^2$

f' is continuous for all x and

$f'(x) = -12 + 12x - 3x^2 = 0$

$-3(x^2 - 4x + 4) = 0$

$-3(x - 2)^2 = 0$

Thus, $x = 2$ is a partition number for f'.

The sign chart for f' is:

Test Numbers

x	$f'(x)$
0	-12 (−)
3	-3 (−)

$f'(x)$ − − − − − 0 − − − − −

$0 \quad 1 \quad 2 \quad 3$

$f(x)$ Decreasing ┊ Decreasing

Therefore, f is decreasing for all x, i.e., on $(-\infty, \infty)$, and there is a horizontal tangent line at $x = 2$.

x	$f'(x)$	f	GRAPH OF f
$(-\infty,\ 2)$	$-$	Decreasing	Falling
$x = 2$	0		Horizontal tangent
$x > 2$	$-$	Decreasing	Falling

x	$f(x)$
0	10
2	2

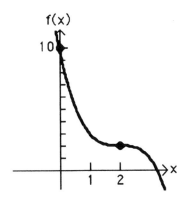

39. $f(x) = \dfrac{x - 1}{x + 2}$ [<u>Note</u>: f is not defined at $x = -2$.]

$f'(x) = \dfrac{(x + 2)(1) - (x - 1)(1)}{(x + 2)^2} = \dfrac{3}{(x + 2)^2}$

Critical values: $f'(x) \neq 0$ for all x, and f' is defined at all points *in the domain of f* (i.e., -2 is not a critical value since -2 is not in the domain of f). Thus, f does not have any critical values; $x = -2$ is a partition number for f'.

The sign chart for f' is:

$f'(x)$ $+ + + +$ ND $+ + + + +$ → x

  -3 -2 -1 0

$f(x)$ Increasing ┊ Increasing

Test Numbers	
x	$f'(x)$
-3	$3\ (+)$
0	$\frac{3}{4}\ (+)$

Therefore, f is increasing on $(-\infty,\ -2)$ and on $(-2,\ \infty)$; f has no local extrema.

41. $f(x) = x + \dfrac{4}{x}$ [<u>Note</u>: f is not defined at $x = 0$.]

$f'(x) = 1 - \dfrac{4}{x^2}$

Critical values: $x = 0$ is *not* a critical value of f since 0 is not in the domain of f, but $x = 0$ is a partition number for f'.

$f'(x) = 1 - \dfrac{4}{x^2} = 0$

$x^2 - 4 = 0$

$(x + 2)(x - 2) = 0$

Thus, the critical values are $x = -2$ and $x = 2$; $x = -2$ and $x = 2$ are also partition numbers for f'.

The sign chart for f' is:

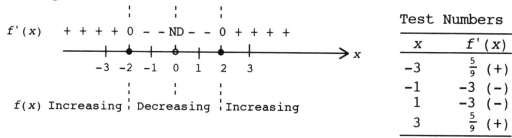

Test Numbers	
x	$f'(x)$
-3	$\frac{5}{9}$ (+)
-1	-3 (−)
1	-3 (−)
3	$\frac{5}{9}$ (+)

Therefore, f is increasing on $(-\infty, -2)$ and on $(2, \infty)$, f is decreasing on $(-2, 0)$ and on $(0, 2)$; f has a local maximum at $x = -2$ and a local minimum at $x = 2$.

43. $f(x) = 1 + \dfrac{1}{x} + \dfrac{1}{x^2}$ [<u>Note</u>: f is not defined at $x = 0$.]

$f'(x) = -\dfrac{1}{x^2} - \dfrac{2}{x^3}$

Critical values: $x = 0$ is not a critical value of f since 0 is not in the domain of f; $x = 0$ is a partition number for f'.

$f'(x) = -\dfrac{1}{x^2} - \dfrac{2}{x^3} = 0$

$\qquad\quad -x - 2 = 0$

$\qquad\qquad\quad x = -2$

Thus, the critical value is $x = -2$; -2 is also a partition number for f'.

The sign chart for f' is:

```
f'(x) - - - - - 0  + + + + + ND - - - - -
      ────────●──────────●────────→ x
       -3    -2    -1     0     1
f(x) Decreasing  Increasing  Decreasing
```

Test Numbers	
x	$f'(x)$
-3	$-\frac{1}{27}$ (−)
-1	1 (+)
1	-3 (−)

Therefore, f is increasing on $(-2, 0)$ and f is decreasing on $(-\infty, -2)$ and on $(0, \infty)$; f has a local minimum at $x = -2$.

45. $f(x) = \dfrac{x^2}{x - 2}$ [<u>Note</u>: f is not defined at $x = 2$.]

$f'(x) = \dfrac{(x - 2)(2x) - x^2(1)}{(x - 2)^2} = \dfrac{x^2 - 4x}{(x - 2)^2}$

Critical values: $x = 2$ is *not* a critical value of f since 2 is not in the domain of f; $x = 2$ is a partition number for f'.

$f'(x) = \dfrac{x^2 - 4x}{(x - 2)^2} = 0$

$\qquad\quad x^2 - 4x = 0$

$\qquad\quad x(x - 4) = 0$

Thus, the critical values are $x = 0$ and $x = 4$; 0 and 4 are also partition numbers for f'.

The sign chart for f' is:

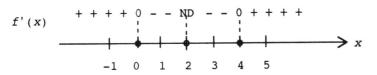

Test Numbers	
x	$f'(x)$
-1	$\frac{5}{9}$ (+)
1	-3 (−)
3	-3 (−)
5	$\frac{5}{9}$ (+)

Therefore, f is increasing on $(-\infty, 0)$ and on $(4, \infty)$, f is decreasing on $(0, 2)$ and on $(2, 4)$; f has a local maximum at $x = 0$ and a local minimum at $x = 4$.

47. $f(x) = x^4(x - 6)^2$

$$f'(x) = x^4(2)(x - 6)(1) + (x - 6)^2(4x^3)$$
$$= 2x^3(x - 6)[x + 2(x - 6)]$$
$$= 2x^3(x - 6)(3x - 12)$$
$$= 6x^3(x - 4)(x - 6)$$

Thus, the critical values of f are $x = 0$, $x = 4$, and $x = 6$.

Now we construct the sign chart for f' ($x = 0$, $x = 4$, $x = 6$ are partition numbers).

Test Numbers	
x	$f'(x)$
-1	-210 (−)
1	90 (+)
5	-750 (−)
7	+

Therefore, f is increasing on $(0, 4)$ and on $(6, \infty)$, f is decreasing on $(-\infty, 0)$ and on $(4, 6)$; f has a local maximum at $x = 4$ and local minima at $x = 0$ and $x = 6$.

49. $f(x) = 3(x - 2)^{2/3} + 4$

$$f'(x) = 3\left(\frac{2}{3}\right)(x - 2)^{-1/3} = \frac{2}{(x - 2)^{1/3}}$$

Critical values: f' is not defined at $x = 2$. [<u>Note</u>: $f(2)$ is defined, $f(2) = 4$.] $f'(x) \neq 0$ for all x. Thus, the critical value for f is $x = 2$; $x = 2$ is also a partition number for f'.

Test Numbers	
x	$f'(x)$
1	-2 (−)
3	2 (+)

Therefore, f is increasing on $(2, \infty)$ and decreasing on $(-\infty, 2)$; f has a local minimum at $x = 2$.

51. $f(x) = 2\sqrt{x} - x = 2x^{1/2} - x, \; x > 0$

$f'(x) = x^{-1/2} - 1 = \dfrac{1}{x^{1/2}} - 1 = \dfrac{1 - \sqrt{x}}{\sqrt{x}}, \; x > 0$

Critical values: $f'(x) = \dfrac{1 - \sqrt{x}}{\sqrt{x}} = 0, \; x > 0$

$$1 - \sqrt{x} = 0$$
$$\sqrt{x} = 1$$
$$x = 1$$

Thus, the critical value for f is $x = 1$; $x = 1$ is also a partition number for f'.

The sign chart for f' is:

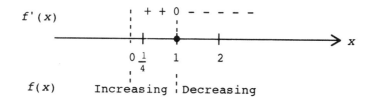

Test Numbers	
x	$f'(x)$
$\frac{1}{4}$	$1 \; (+)$
4	$-\frac{1}{2} \; (-)$

Therefore, f is increasing on $(0, 1)$ and decreasing on $(1, \infty)$; f has a local maximum at $x = 1$.

53. Critical values: $x = -1.26$; increasing on $(-1.26, \infty)$; decreasing on $(-\infty, -1.26)$; local minimum at $x = -1.26$

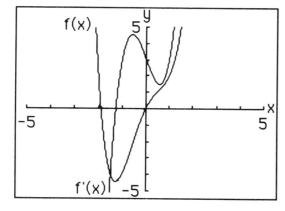

55. Critical values: $x = -0.43$, $x = 0.54$, $x = 2.14$; increasing on $(-0.43, 0.54)$ and $(2.14, \infty)$; decreasing on $(-\infty, -0.43)$ and $(0.54, 2.14)$; local maximum at $x = 0.54$, local minima at $x = -0.43$ and $x = 2.14$

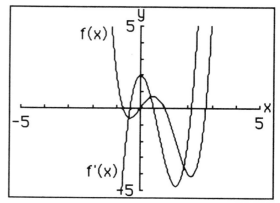

57. $\overline{C}(x) = \dfrac{x}{20} + 20 + \dfrac{320}{x}, \; 0 < x < 150$

$\overline{C}'(x) = \dfrac{1}{20} - \dfrac{320}{x^2}$

Critical values: $\overline{C}'$ is continuous for all x on the interval $(0, 150)$:

$$\bar{C}'(x) = \frac{1}{20} - \frac{320}{x^2} = 0$$
$$x^2 - 320(20) = 0$$
$$x^2 - 6400 = 0$$
$$(x - 80)(x + 80) = 0$$

Thus, the critical value of $\bar{C}$ on the interval $(0, 150)$ is $x = 80$.

Next, construct the sign chart for $\bar{C}'$ ($x = 80$ is a partition number for $\bar{C}'$).

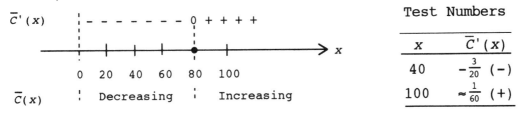

Test Numbers

x	$\bar{C}'(x)$
40	$-\frac{3}{20}$ $(-)$
100	$\approx\frac{1}{60}$ $(+)$

Therefore, $\bar{C}$ is increasing for $80 < x < 150$ and decreasing for $0 < x < 80$; $\bar{C}$ has a local minimum at $x = 80$.

59. $P(x) = R(x) - C(x)$
$P'(x) = R'(x) - C'(x)$
Thus, if $R'(x) > C'(x)$ on the interval (a, b), then $P'(x) = R'(x) - C'(x) > 0$ on this interval and P is increasing.

61. $C(t) = \dfrac{0.14t}{t^2 + 1}$, $0 < t < 24$

$C'(t) = \dfrac{(t^2 + 1)(0.14) - 0.14t(2t)}{(t^2 + 1)^2} = \dfrac{0.14(1 - t^2)}{(t^2 + 1)^2}$

Critical values: C' is continuous for all t on the interval $(0, 24)$:

$C'(t) = \dfrac{0.14(1 - t^2)}{(t^2 + 1)^2} = 0$
$$1 - t^2 = 0$$
$$(1 - t)(1 + t) = 0$$

Thus, the critical value of C on the interval $(0, 24)$ is $t = 1$. The sign chart for C' ($t = 1$ is a partition number for C') is:

Test Numbers

x	$C'(t)$
$\frac{1}{2}$	$+$
2	$-$

Therefore, C is increasing for $0 < t < 1$ and decreasing for $1 < t < 24$; C has a local maximum at $t = 1$.

63. $P(t) = \dfrac{8.4t}{t^2 + 49} + 0.1$, $0 < t < 24$

$P'(t) = \dfrac{(t^2 + 49)(8.4) - 8.4t(2t)}{(t^2 + 49)^2} = \dfrac{8.4(49 - t^2)}{(t^2 + 49)^2}$

Critical values: P is continuous for all t on the interval $(0, 24)$:

$P'(t) = \dfrac{8.4(49 - t^2)}{(t^2 + 49)^2} = 0$

$49 - t^2 = 0$

$(7 - t)(7 + t) = 0$

Thus, the critical value of P on $(0, 24)$ is $t = 7$.

The sign chart for P' ($t = 7$ is a partition number for P') is:

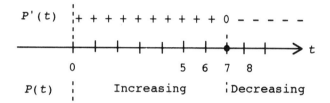

Test Numbers	
x	$P'(t)$
6	+
8	−

Therefore, P is increasing for $0 < t < 7$ and decreasing for $7 < t < 24$; P has a local maximum at $t = 7$.

EXERCISE 10-2

Things to remember:

1. SECOND DERIVATIVE

    For $y = f(x)$, the second derivative of f is:
       $f''(x) = D_x f'(x)$
    Other notations for $f''(x)$ are:
       $\dfrac{d^2y}{dx^2}$, y'', $D_x^2 f(x)$

2. CONCAVITY

    For the interval (a, b):

$f''(x)$	$f'(x)$	Graph of $y = f(x)$	Example
+	Increasing	Concave upward	$\smile$
−	Decreasing	Concave downward	$\frown$

<u>3.</u> SECOND-DERIVATIVE TEST FOR LOCAL MAXIMA AND MINIMA

$f'(c)$	$f''(c)$	Graph of f is	$f(c)$	Example
0	+	Concave upward	Local minimum	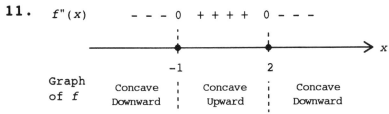
0	−	Concave downward	Local maximum	
0	0	?	Test fails	

The first-derivative test must be used whenever $f''(c) = 0$ [or $f''(c)$ does not exist].

1. From the graph, f is concave upward on (a, c), (c, d), and (e, g).

3. From the graph, f has points of inflection at $x = d$, e, and g.

5. f has a local minimum at $x = 2$.

7. Unable to determine from the given information $(f'(-3) = f''(-3) = 0)$.

9. Neither a local maximum nor a local minimum at $x = 6$; $x = 6$ is not a critical value of f.

11.
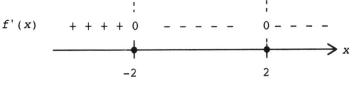

Using this information together with the points $(-4, 0)$, $(-2, 3)$, $(-1, 1.5)$, $(0, 0)$, $(2, -1)$, $(4, -3)$ on the graph, we have

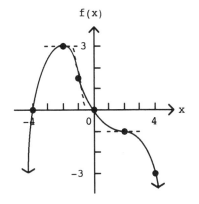

13.

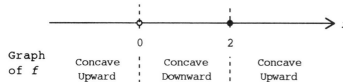

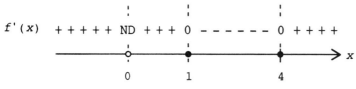

$f(x)$ Increasing¦ Incr. ¦ Decreasing ¦ Increasing

                   Local   Local

                 maximum   minimum

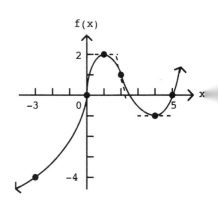

Using this information together with the points $(-3, -4)$, $(0, 0)$, $(1, 2)$, $(2, 1)$, $(4, -1)$, $(5, 0)$ on the graph, we have

15. $f(x) = x^3 - 2x^2 - 1$

    $f'(x) = 3x^2 - 4x$

    $f''(x) = 6x - 4$

17. $y = 2x^5 - 3$

    $\dfrac{dy}{dx} = 10x^4$

    $\dfrac{d^2y}{dx^2} = 40x^3$

19. $f(x) = 1 - 2x + x^3$

    $D_x f(x) = D_x(1 - 2x + x^3) = -2 + 3x^2$

    $D_x^2 f(x) = D_x f'(x) = D_x(-2 + 3x^2) = 6x$

21. $y = (x^2 - 1)^3$

    $y' = 3(x^2 - 1)^2(2x) = 6x(x^2 - 1)^2$

    $y'' = 6x(2)(x^2 - 1)(2x) + (x^2 - 1)^2(6)$

       $= 24x^2(x^2 - 1) + 6(x^2 - 1)^2$

       $= 6(x^2 - 1)[4x^2 + x^2 - 1]$

       $= 6(x^2 - 1)(5x^2 - 1)$

23. $f(x) = 3x^{-1} + 2x^{-2} + 5$

    $f'(x) = -3x^{-2} - 4x^{-3}$

    $f''(x) = 6x^{-3} + 12x^{-4}$

25. $f(x) = 2x^2 - 8x + 6$
$f'(x) = 4x - 8 = 4(x - 2)$
$f''(x) = 4$
Critical value: $x = 2$
Now, $f''(2) = 4 > 0$. Therefore, $f(2) = 2 \cdot 2^2 - 8 \cdot 2 + 6 = -2$ is a local minimum.

27. $f(x) = 2x^3 - 3x^2 - 12x - 5$
$f'(x) = 6x^2 - 6x - 12 = 6(x^2 - x - 2) = 6(x - 2)(x + 1)$
$f''(x) = 12x - 6 = 6(2x - 1)$
Critical values: $x = 2, -1$.
Now, $f''(2) = 6(2 \cdot 2 - 1) = 18 > 0$.
Therefore, $f(2) = 2 \cdot 2^3 - 3 \cdot 2^2 - 12 \cdot 2 - 5 = -25$ is a local minimum.
$f''(-1) = 6[2(-1) - 1] = -18 < 0$.
Therefore, $f(-1) = 2(-1)^3 - 3(-1)^2 - 12(-1) - 5 = 2$ is a local maximum.

29. $f(x) = 3 - x^3 + 3x^2 - 3x$
$f'(x) = -3x^2 + 6x - 3 = -3(x^2 - 2x + 1) = -3(x - 1)^2$
$f''(x) = -6(x - 1)$
Critical value: $x = 1$
Now, $f''(1) = -6(1 - 1) = 0$.
Thus, the second-derivative test fails. Since $f'(x) = -3(x - 1)^2 < 0$ for all $x \neq 1$, $f(x)$ is decreasing on $(-\infty, \infty)$. Therefore, $f(x)$ has no local extrema.

31. $f(x) = x^4 - 8x^2 + 10$
$f'(x) = 4x^3 - 16x = 4x(x^2 - 4) = 4x(x + 2)(x - 2)$
$f''(x) = 12x^2 - 16$
Critical values: $x = 0, -2, 2$
Now $f''(0) = 0 - 16 = -16 < 0$. Therefore, $f(0) = 10$ is a local maximum.
$f''(-2) = 12(-2)^2 - 16 = 32 > 0$. Therefore, $f(-2) = -6$ is a local minimum. $f''(2) = 12 \cdot 2^2 - 16 = 32 > 0$. Therefore, $f(2) = -6$ is a local minimum.

33. $f(x) = x^6 + 3x^4 + 2$
$f'(x) = 6x^5 + 12x^3 = 6x^3(x^2 + 2)$
$f''(x) = 30x^4 + 36x^2 = 6x^2(5x^2 + 6)$
Critical value: $x = 0$ [Note: $x^2 + 2 \neq 0$ for all x.]
Now, $f''(0) = 0$. Thus, the second-derivative test fails, so the first-derivative test must be used.
The sign chart for f' (partition number is 0) is:

	Test Numbers	
	x	$f'(x)$
	-1	-18 $(-)$
	1	18 $(+)$

$f'(x)$ $\quad - - - - 0 + + + +$

$f(x)$ $\quad$ Decreasing $\vdots$ Increasing

Therefore, $f(0) = 2$ is a local minimum of f.

35. $f(x) = x + \dfrac{16}{x}$

$f'(x) = 1 - \dfrac{16}{x^2} = \dfrac{x^2 - 16}{x^2} = \dfrac{(x + 4)(x - 4)}{x^2}$

$f''(x) = \dfrac{32}{x^3}$

Critical values: $x = -4, 4$ [<u>Note</u>: $x = 0$ is not a critical value, since $f(0)$ is not defined.]

$f''(-4) = \dfrac{32}{(-4)^3} = -\dfrac{1}{2} < 0$

Therefore, $f(-4) = -8$ is a local maximum.

$f''(4) = \dfrac{32}{(4)^3} = \dfrac{1}{2} > 0$

Therefore, $f(4) = 8$ is a local minimum.

37. $f(x) = x^2 - 4x + 5$
$f'(x) = 2x - 4$
$f''(x) = 2 > 0$ Thus, the graph of f is concave upward for all x; there are no inflection points.

39. $f(x) = x^3 - 18x^2 + 10x - 11$
$f'(x) = 3x^2 - 36x + 10$
$f''(x) = 6x - 36$
Now, $f''(x) = 6x - 36 = 0$
$$x - 6 = 0$$
$$x = 6$$

The sign chart for f'' (partition number is 6) is:

x	$f''(x)$
5	-6 $(-)$
7	6 $(+)$

Test Numbers

$f''(x)$ $- - - - - - \; 0 + + + +$

Graph of f :   Concave Downward | Concave Upward   (0   5 6 7)

Therefore, the graph of f is concave upward on $(6, \infty)$ and concave downward on $(-\infty, 6)$; there is an inflection point at $x = 6$.

41. $f(x) = x^4 - 24x^2 + 10x - 5$
$f'(x) = 4x^3 - 48x + 10$
$f''(x) = 12x^2 - 48$
Now, $f''(x) = 12x^2 - 48 = 0$
$$12(x^2 - 4) = 0$$
$$12(x + 2)(x - 2) = 0$$
$$x = -2, 2$$

The sign chart for f'' (partition numbers are -2 and 2) is:

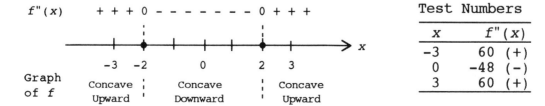

x	$f''(x)$
-3	60 $(+)$
0	-48 $(-)$
3	60 $(+)$

Test Numbers

Thus, the graph of f is concave upward on $(-\infty, -2)$ and on $(2, \infty)$, the graph is concave downward on $(-2, 2)$; the graph has inflection points at $x = -2$ and at $x = 2$.

43. $f(x) = -x^4 + 4x^3 + 3x + 7$
$f'(x) = -4x^3 + 12x^2 + 3$
$f''(x) = -12x^2 + 24x$
Now, $f''(x) = -12x^2 + 24x = 0$
$12x(-x + 2) = 0$
or $-12x(x - 2) = 0$
$x = 0, 2$
The sign chart for f'' (partition numbers are 0 and 2) is:

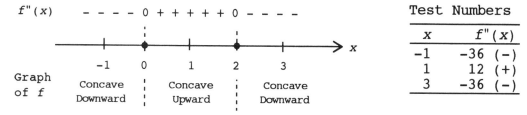

x	f''(x)
-1	-36 (−)
1	12 (+)
3	-36 (−)

Test Numbers

Thus, the graph of f is concave downward on $(-\infty, 0)$ and on $(2, \infty)$. The graph is concave upward on $(0, 2)$; the graph has inflection points at $x = 0$ and at $x = 2$.

45. $f(x) = x^3 - 6x^2 + 16$
$f'(x) = 3x^2 - 12x = 3x(x - 4)$
$f''(x) = 6x - 12 = 6(x - 2)$
Critical values: $x = 0, 4$
$f''(0) = -12 < 0$. Therefore, f has a local maximum at $x = 0$. $f''(4) = 6(4 - 2) = 12 > 0$. Therefore, f has a local minimum at $x = 4$.

The sign chart for $f''(x)$ (partition number is 2) is:

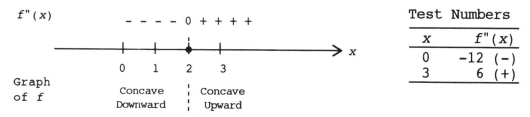

x	f''(x)
0	-12 (−)
3	6 (+)

Test Numbers

The graph of f has an inflection point at $x = 2$. The graph of f is:

x	f(x)
0	16
2	0
4	-16

47. $f(x) = x^3 + x + 2$

$f'(x) = 3x^2 + 1$

$f''(x) = 6x$

Since $f'(x) = 3x^2 + 1 > 0$ for all x, f does not have any critical values. Now, $f''(x) = 6x = 0$

$$x = 0$$

The sign chart for f'' (partition number is 0) is:

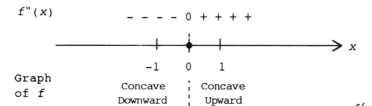

Test Numbers	
x	$f''(x)$
-1	-6 $(-)$
1	6 $(+)$

The graph of f has an inflection point at $x = 0$.
The graph of f is:

x	$f(x)$
-1	0
0	2
1	4

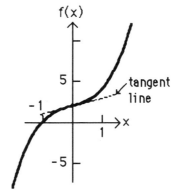

49. $f(x) = (2 - x)^3 + 1$

$f'(x) = 3(2 - x)^2(-1) = -3(2 - x)^2$

$f''(x) = -6(2 - x)(-1) = 6(2 - x)$

Critical value: $x = 2$

$f''(2) = 0$. Thus, the second derivative test fails. Note that $f'(x) = -3(2 - x)^2 < 0$ for all $x \neq 2$. Therefore, f is decreasing on $(-\infty, 2)$ and on $(2, \infty)$, and f does not have any local extrema. The sign chart for f'' (partition number is 2) is:

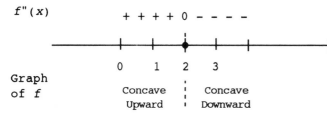

Test Numbers	
x	$f''(x)$
0	12 $(+)$
3	-6 $(-)$

The graph of f has an inflection point at $x = 2$.
The graph of f is:

x	$f(x)$
0	9
2	1
3	0

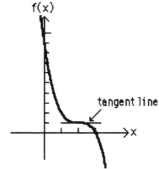

51. $f(x) = x^3 - 12x$
$f'(x) = 3x^2 - 12 = 3(x^2 - 4) = 3(x + 2)(x - 2)$
$f''(x) = 6x$
Critical values: $x = -2$, $x = 2$
$f''(-2) = 6(-2) = -12 < 0$. Therefore, f has a local maximum at $x = -2$.
$f''(2) = 6(2) = 12 > 0$. Therefore, f has a local minimum at $x = 2$. The
sign chart for $f''(x) = 6x$ (partition number is 0) is:

Test Numbers	
x	$f''(x)$
-1	-6 $(-)$
1	6 $(+)$

$f''(x)$ — — — — 0 + + + +

Graph
of f

Concave
Downward

Concave
Upward

The graph of f has an inflection point at $x = 0$. The graph of f is:

x	$f(x)$
-2	$+16$
0	0
2	-16

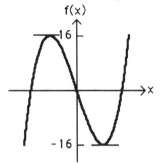

53. $f(x) = \dfrac{1}{x^2 + 12}$

$f'(x) = \dfrac{(x^2 + 12)(0) - 1(2x)}{(x^2 + 12)^2} = \dfrac{-2x}{(x^2 + 12)^2}$

$f''(x) = \dfrac{(x^2 + 12)^2(-2) - (-2x)(2)(x^2 + 12)(2x)}{(x^2 + 12)^4}$

$= \dfrac{(x^2 + 12)(-2) + 8x^2}{(x^2 + 12)^3} = \dfrac{6x^2 - 24}{(x^2 + 12)^3}$

Now $f''(x) = \dfrac{6x^2 - 24}{(x^2 + 12)^3} = 0$

$6x^2 - 24 = 0$
$6(x^2 - 4) = 0$
$6(x + 2)(x - 2) = 0$
$x = -2, 2$

The sign chart for f'' (partition numbers are -2 and 2) is:

Test Numbers	
x	$f''(x)$
-3	$+$
0	$-$
3	$+$

$f''(x)$ + + + 0 — — — — — — — — — — 0 + + + +

Graph Concave
of f Upward

Concave
Downward

Concave
Upward

Thus, the graph of f has inflection points at $x = -2$ and $x = 2$.

55. $f(x) = \dfrac{x}{x^2 + 12}$

$f'(x) = \dfrac{(x^2 + 12)(1) - x(2x)}{(x^2 + 12)^2} = \dfrac{12 - x^2}{(x^2 + 12)^2}$

$f''(x) = \dfrac{(x^2 + 12)^2(-2x) - (12 - x^2)(2)(x^2 + 12)(2x)}{(x^2 + 12)^4}$

$\qquad = \dfrac{(x^2 + 12)(-2x) - 4x(12 - x^2)}{(x^2 + 12)^3} = \dfrac{2x^3 - 72x}{(x^2 + 12)^3} = \dfrac{2x(x^2 - 36)}{(x^2 + 12)^3}$

Now $f''(x) = \dfrac{2x(x^2 - 36)}{(x^2 + 12)^3} = 0$

$\qquad\qquad 2x(x + 6)(x - 6) = 0$

$\qquad\qquad\qquad\qquad x = 0, -6, 6$

The sign chart for f'' (partition numbers are 0, -6, 6) is:

	x	$f''(x)$
Test Numbers	-7	$-$
	-1	$+$
	1	$-$
	7	$+$

$f''(x) \quad - - - \; 0 + + + + + \; 0 - - - - - - \; 0 + + +$

Graph of f: Concave Downward | Concave Upward | Concave Downward | Concave Upward

(-7 -6 $\quad$ -1 0 1 $\quad$ 6 7)

Thus, the graph of f has inflection points at $x = -6$, $x = 0$, and $x = 6$.

57. Inflection point at $x = -1.40$; concave upward on $(-1.40, \infty)$; concave downward on $(-\infty, -1.40)$

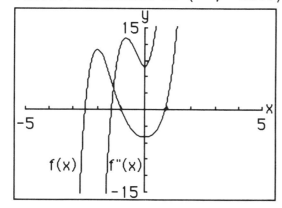

59. Inflection points at $x = -0.61$, $x = 0.66$, and $x = 1.74$; concave upward on $(-0.61, 0.66)$ and $(1.74, \infty)$; concave downward on $(-\infty, -0.61)$ and $(0.66, 1.74)$

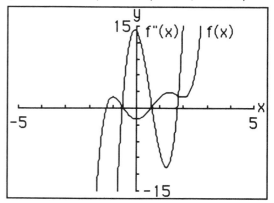

61. $R(x) = xp = 1296x - 0.12x^3$, $0 < x < 80$

$R'(x) = 1296 - 0.36x^2$

Critical values: $R'(x) = 1296 - 0.36x^2 = 0$

$\qquad\qquad\qquad\qquad\qquad x^2 = \dfrac{1296}{0.36} = 3600$

$\qquad\qquad\qquad\qquad\qquad x = \pm 60$

Thus, $x = 60$ is the only critical value on the interval $(0, 80)$.

$R''(x) = -0.72x$

$R''(60) = -43.2 < 0$

(A) R has a local maximum at $x = 60$.

(B) Since $R''(x) = -0.72x < 0$ for $0 < x < 80$, R is concave downward on this interval.

63. $N(x) = -3x^3 + 225x^2 - 3600x + 17{,}000$, $10 \le x \le 40$

$N'(x) = -9x^2 + 450x - 3600$

$N''(x) = -18x + 450$

(A) To determine when N' is increasing or decreasing, we must solve the inequalities $N''(x) > 0$ and $N''(x) < 0$, respectively. Now

$$N''(x) = -18x + 450 = 0$$
$$x = 25$$

The sign chart for N'' (partition number is 25) is:

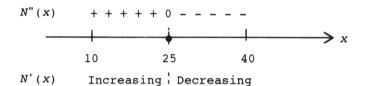

x	$N''(x)$
10	270 (+)
30	-90 (-)

Test Numbers

Thus N' is increasing on $(10, 25)$ and decreasing on $(25, 40)$.

(B) Using the results in (A), the graph of N has an inflection point at $x = 25$.

(C)

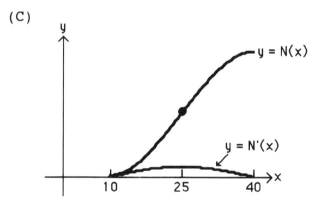

(D) Using the results in (A), N' has a local maximum at $x = 25$:

$N'(25) = 2025$

65. $N(t) = 1000 + 30t^2 - t^3$, $0 \le t \le 20$

$N'(t) = 60t - 3t^2$

$N''(t) = 60 - 6t$

(A) To determine when N' is increasing or decreasing, we must solve the inequalities $N''(t) > 0$ and $N''(t) < 0$, respectively. Now

$$N''(t) = 60 - 6t = 0$$
$$t = 10$$

The sign chart for N'' (partition number is 10) is:

t	$N''(t)$
0	60 (+)
20	-60 (-)

Test Numbers

Thus, N' is increasing on $(0, 10)$ and decreasing on $(10, 20)$.

(B) From the results in (A), the graph of N has an inflection point at $t = 10$.

(C)

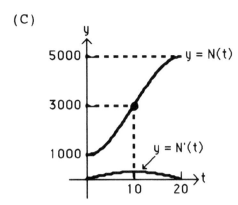

(D) Using the results in (A), N' has a local maximum at $t = 10$:
$N'(10) = 300$

67. $T(n) = \frac{2}{25}n^3 - \frac{6}{5}n^2 + 6n, \ n \geq 0$

$T'(n) = \frac{6}{25}n^2 - \frac{12}{5}n + 6$

$T''(n) = \frac{12}{25}n - \frac{12}{5}$

(A) To determine when the rate of change of T, i.e., T', is increasing or decreasing, we must solve the inequalities $T''(n) > 0$ and $T''(n) < 0$, respectively. Now

$T''(n) = \frac{12}{25}n - \frac{12}{5} = 0$

$n = 5$

The sign chart for T'' (partition number is 5) is:

$T''(n) \qquad \quad - - - - \ 0 + + + +$

```
         |        |        |            → n
         1        5        10
```

Test Numbers	
t	$N''(t)$
1	$-\frac{48}{25}$ (−)
10	$\frac{12}{5}$ (+)

$T'(n) \qquad$ Decreasing ┆ Increasing

Thus, T' is increasing on $(5, \infty)$ and decreasing on $(0, 5)$.

(B) Using the results in (A), the graph of T has an inflection point at $n = 5$. The graphs of T and T' are shown at the right.

(C) Using the results in (A), T' has a local minimum at $n = 5$:
$T'(5) = \frac{6}{25}(5)^2 - \frac{12}{5}(5) + 6 = 0$

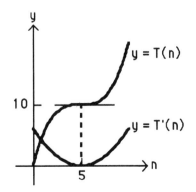

Things to remember:

1. HORIZONTAL AND VERTICAL ASYMPTOTES

   A line $y = b$ is a HORIZONTAL ASYMPTOTE for the graph of $y = f(x)$ if:

   $$\lim_{x \to -\infty} f(x) = b \quad \text{or} \quad \lim_{x \to \infty} f(x) = b$$

   A line $x = a$ is a VERTICAL ASYMPTOTE for the graph of $y = f(x)$ if:

   $$\lim_{x \to a^-} f(x) = \infty \ (\text{or} \ -\infty), \ \lim_{x \to a^+} f(x) = \infty \ (\text{or} \ -\infty),$$
   $$\text{or} \ \lim_{x \to a} f(x) = \infty \ (\text{or} \ -\infty)$$

2. ON VERTICAL ASYMPTOTES

   Let $f(x) = N(x)/D(x)$, where both N and D are continuous at $x = c$. If, at $x = c$, the denominator $D(x)$ is 0 and the numerator $N(x)$ is not 0, then the line $x = c$ is a vertical asymptote for the graph of f.

   [Note: Since a rational function is the ratio of two polynomial functions, and polynomial functions are continuous for all real numbers, this theorem includes rational functions as a special case.]

3. A GRAPHING STRATEGY FOR $y = f(x)$

   Omit any of the following steps if procedures involved are too difficult or impossible.

   Step 1. Use $f(x)$:

   (A) Find the domain of f. [The domain of f is the set of all real numbers x that produce real values for $f(x)$.]

   (B) Find intercepts. [The y intercept is $f(0)$ if it exists; the x intercepts are the solutions to $f(x) = 0$, if they exist.]

   (C) Find asymptotes. Find any horizontal asymptotes by calculating
   $$\lim_{x \to \pm\infty} f(x).$$
   Find any vertical asymptotes by using 2, above.

   Step 2. Use $f'(x)$:

   Find any critical values for $f(x)$ and any partition numbers for $f'(x)$. [Remember, every critical value for $f(x)$ is also a partition number for $f'(x)$, but some partition numbers for $f'(x)$ may not be critical values for $f(x)$.] Construct a sign chart for $f'(x)$, determine the intervals where $f(x)$ is increasing and decreasing, and find local maxima and minima.

   Step 3. Use $f''(x)$:

   Construct a sign chart for $f''(x)$, determine where the graph of f is concave upward and concave downward, and find any inflection points.

Step 4. Sketch the graph of *f*:

Draw asymptotes and locate intercepts, local maxima and minima, and inflection points. Sketch in what you know from Steps 1–3. In regions of uncertainty, use point-by-point plotting to complete the graph.

---

1. From the graph, *f*(*x*) is increasing on (*b*, *d*), (*d*, 0), and (*g*, ∞).

3. From the graph, *f*(*x*) has a local maximum at *x* = 0.

5. The graph of *f* is concave upward on the intervals (*a*, *d*), (*e*, *h*).

7. From the graph, *f* has a point of inflection at *x* = *a* and at *x* = *h*.

9. From the graph, *x* = *d* and *x* = *e* are vertical asymptotes.

11. Step 1. Use *f*(*x*):

(A) Domain: All real numbers

(B) Intercepts: *y*-intercept: 0
    x-intercepts: −4, 0, 4

Step 2. Use *f*'(*x*):

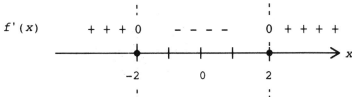

Step 3. Use *f*"(*x*):

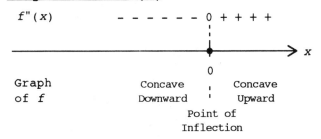

Step 4. Sketch the graph of *f*:

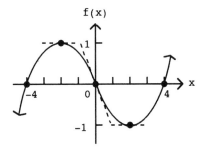

13. Step 1. Use *f*(*x*):

(A) Domain: All real numbers

(B) Intercepts: *y*-intercept: 0
    x-intercepts: −4, 0, 4

(C) Asymptotes: Horizontal asymptote: $y = 2$

Step 2. Use $f'(x)$:

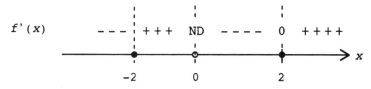

$f(x)$ Decreasing | Incr. | Decreasing | Increasing

Local   Local   Local
minimum   maximum   minimum

Step 3. Use $f''(x)$:

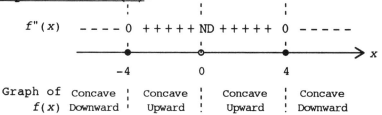

Graph of Concave | Concave | Concave | Concave
$f(x)$ Downward | Upward | Upward | Downward

Step 4. Sketch the graph of f:

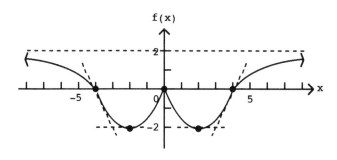

15. Step 1. Use $f(x)$:

(A) Domain: All real numbers except $x = -2$

(B) Intercepts: y-intercept: 0
x-intercepts: -4, 0

(C) Asymptotes: Horizontal asymptote: $y = 1$
Vertical asymptote: $x = -2$

Step 2. Use $f'(x)$:

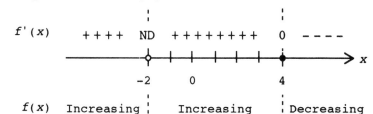

$f(x)$ Increasing | Increasing | Decreasing

Local
maximum

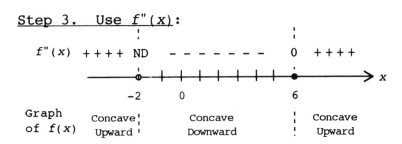

$f''(x)$ $+ + + +$ ND $- - - - - - - -$ 0 $+ + + +$

-2 0   6

| Graph of $f(x)$ | Concave Upward | Concave Downward | Concave Upward |

Step 4. Sketch the graph of f:

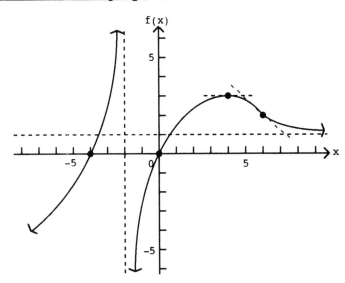

17. $f(x) = \dfrac{2x}{x + 2}$

Horizontal asymptotes: Since f is a rational function, the graph of f has at most one horizontal asymptote.

$$\lim_{x \to \infty} \frac{2x}{x + 2} = \lim_{x \to \infty} \frac{2x}{x\left(1 + \dfrac{2}{x}\right)} = 2$$

Thus, $y = 2$ is a horizontal asymptote.

Vertical asymptotes: Using 2, $D(-2) = 0$ and $N(-2) = 2(-2) = -4 \neq 0$. Thus, the line $x = -2$ is a vertical asymptote.

19. $f(x) = \dfrac{x^2 + 1}{x^2 - 1} = \dfrac{x^2 + 1}{(x - 1)(x + 1)}$

Horizontal asymptotes: Since f is a rational function, its graph has at most one horizontal asymptote.

$$\lim_{x \to \infty} \frac{x^2 + 1}{x^2 - 1} = \lim_{x \to \infty} \frac{x^2\left(1 + \dfrac{1}{x^2}\right)}{x^2\left(1 - \dfrac{1}{x^2}\right)} = 1$$

Thus, $y = 1$ is a horizontal asymptote.

Vertical asymptote: Using 2, $D(-1) = 0$ and $N(-1) = (1)^2 + 1 = 2$. Thus, $x = -1$ is a vertical asymptote. Similarly, $D(1) = 0$ and $N(1) = 1^2 + 1 = 2$, so $x = 1$ is a vertical asymptote.

21. $f(x) = \dfrac{x^3}{x^2 + 6}$

Horizontal asymptotes: The graph of f has at most one horizontal asymptote.

$$\lim_{x \to \infty} \frac{x^3}{x^2 + 6} = \lim_{x \to \infty} \frac{x^3}{x^2\left(1 + \dfrac{6}{x^2}\right)} = \lim_{x \to \infty} \frac{x}{1 + \dfrac{6}{x^2}} = \infty.$$

There are no horizontal asymptotes.

Vertical asymptotes: Since the denominator $D(x) = x^2 + 6$ is never 0, there are no vertical asymptotes.

23. $f(x) = \dfrac{x}{x^2 + 4}$

Horizontal asymptotes: The graph of f has at most one horizontal asymptote.

$$\lim_{x \to \infty} \frac{x}{x^2 + 4} = \lim_{x \to \infty} \frac{x}{x^2\left(1 + \dfrac{4}{x^2}\right)} = \lim_{x \to \infty} \left(\frac{1}{x} \cdot \frac{1}{1 + \dfrac{4}{x^2}}\right) = 0$$

Thus, $y = 0$ is a horizontal asymptote.

Vertical asymptotes: Since the denominator $x^2 + 4$ is never 0, there are no vertical asymptotes.

25. $f(x) = \dfrac{x^2}{x - 3}$

Horizontal asymptotes: The graph of f has at most one horizontal asymptote.

$$\lim_{x \to \infty} \frac{x^2}{x - 3} = \lim_{x \to \infty} \frac{x^2}{x\left(1 - \dfrac{3}{x}\right)} = \lim_{x \to \infty} \left(x \cdot \frac{1}{1 - \dfrac{3}{x}}\right) = \infty$$

Thus, there are no horizontal asymptotes.

Vertical asymptotes: The line $x = 3$ is a vertical asymptote since, for $x = 3$, the denominator is 0 and the numerator is not 0.

27. $f(x) = x^2 - 6x + 5$

Step 1. Use $f(x)$:
(A) Domain: All real numbers, $(-\infty, \infty)$.
(B) Intercepts: y intercept: $f(0) = 5$

x intercepts: $x^2 - 6x + 5 = 0$
$(x - 5)(x - 1) = 0$
$x = 1, 5$

(C) Asymptotes: Since f is a polynomial, there are no horizontal or vertical asymptotes.

Step 2. Use $f'(x)$:
$f'(x) = 2x - 6 = 2(x - 3)$
Critical value: $x = 3$
Partition number: $x = 3$

Sign chart for f':

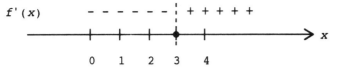

Test Numbers	
x	$f'(x)$
0	-6 $(-)$
4	2 $(+)$

Thus, f is decreasing on $(-\infty, 3)$ and increasing on $(3, \infty)$; f has a local minimum at $x = 3$.

<u>Step 3. Use $f''(x)$</u>:
$f''(x) = 2 > 0$ for all x.
Thus, the graph of f is concave upward on $(-\infty, \infty)$.
<u>Step 4. Sketch the graph of f</u>:

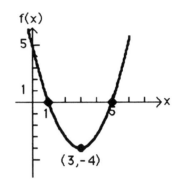

29. $f(x) = x^3 - 6x^2$

<u>Step 1. Use $f(x)$</u>:
(A) Domain: All real numbers, $(-\infty, \infty)$.
(B) Intercepts: y intercept: $f(0) = 0$
  x intercepts: $x^3 - 6x^2 = 0$
  $x^2(x - 6) = 0$
  $x = 0, 6$
(C) Asymptotes: Since f is a polynomial, there are no horizontal or vertical asymptotes.

<u>Step 2. Use $f'(x)$</u>:
$f'(x) = 3x^2 - 12x = 3x(x - 4)$
Critical values: $x = 0$, $x = 4$
Partition numbers: $x = 0$, $x = 4$
The sign chart for f' is:

Test Numbers	
x	$f'(x)$
-1	15 $(+)$
3	-9 $(-)$
5	15 $(+)$

Thus, f is increasing on $(-\infty, 0)$ and on $(4, \infty)$; f is decreasing on $(0, 4)$; f has a local maximum at $x = 0$ and a local minimum at $x = 4$.

<u>Step 3. Use $f''(x)$</u>:
$f''(x) = 6x - 12 = 6(x - 2)$
Partition numbers for f'': $x = 2$

Sign chart for f'':

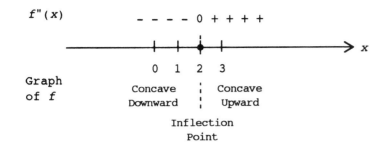

x	$f''(x)$
1	-6 (-)
3	6 (+)

Graph
of f

Concave
Downward

Concave
Upward

Inflection
Point

Thus, the graph of f is concave downward on $(-\infty, 2)$ and concave upward on $(2, \infty)$; there is an inflection point at $x = 2$.

Step 4. Sketch the graph of f:

x	$f(x)$
0	0
2	-16
4	-32
6	0

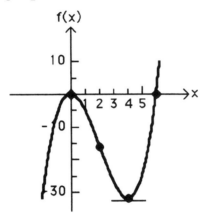

31. $f(x) = (x + 4)(x - 2)^2$

Step 1. Use $f(x)$:
(A) Domain: All real numbers, $(-\infty, \infty)$.
(B) Intercepts: y intercept: $f(0) = 4(-2)^2 = 16$
x intercepts: $(x + 4)(x - 2)^2 = 0$
$x = -4, 2$
(C) Asymptotes: Since f is a polynomial, there are no horizontal or vertical asymptotes.

Step 2. Use $f'(x)$:
$f'(x) = (x + 4)2(x - 2)(1) + (x - 2)^2(1)$
$= (x - 2)[2(x + 4) + (x - 2)]$
$= (x - 2)(3x + 6)$
$= 3(x - 2)(x + 2)$
Critical values: $x = -2$, $x = 2$
Partition numbers: $x = -2$, $x = 2$

Sign chart for f':

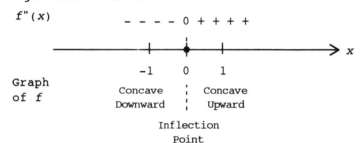

Test Numbers	
x	$f'(x)$
-3	15 (+)
0	-12 (-)
3	15 (+)

Thus, f is increasing on $(-\infty, -2)$ and on $(2, \infty)$; f is decreasing on $(-2, 2)$; f has a local maximum at $x = -2$ and a local minimum at $x = 2$.

Step 3. Use $f''(x)$:

$f''(x) = 3(x + 2)(1) + 3(x - 2)(1) = 6x$

Partition number for f'': $x = 0$

Sign chart for f'':

$f''(x)$ $- - - - 0 + + + +$

                      $\longrightarrow x$

        -1   0   1

Graph   Concave   Concave
of f Downward   Upward

             Inflection
               Point

Test Numbers	
x	$f''(x)$
1	-6 (-)
1	6 (+)

Thus, the graph of f is concave downward on $(-\infty, 0)$ and concave upward on $(0, \infty)$; there is an inflection point at $x = 0$.

Step 4. Sketch the graph of f:

x	$f(x)$
-2	32
0	16
2	0

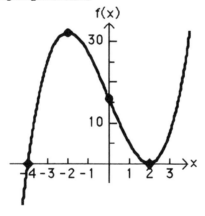

33. $f(x) = 8x^3 - 2x^4$

Step 1. Use $f(x)$:

(A) Domain: All real numbers, $(-\infty, \infty)$.

(B) Intercepts: y intercept: $f(0) = 0$

                x intercepts: $8x^3 - 2x^4 = 0$

                           $2x^3(4 - x) = 0$

                                $x = 0, 4$

(C) Asymptotes: No horizontal or vertical asymptotes.

Step 2. Use $f'(x)$:

$f'(x) = 24x^2 - 8x^3 = 8x^2(3 - x)$

Critical values: $x = 0$, $x = 3$

Partition numbers: $x = 0$, $x = 3$

Sign chart for f':

```
                + + + + 0 + + + + + 0 - - - -
f'(x)
        ─────────┼──────●──┼──┼──┼──●──┼─────────────→ x
                -1   0     1        3   4

  f(x)    Increasing ¦ Increasing ¦ Decreasing
                             ¦
                          Local
                         Maximum
```

Test Numbers	
x	$f'(x)$
-1	32 (+)
1	16 (+)
4	-128 (-)

Thus, f is increasing on $(-\infty, 3)$ and decreasing on $(3, \infty)$; f has a local maximum at $x = 3$.

Step 3. Use $f''(x)$:

$f''(x) = 48x - 24x^2 = 24x(2 - x)$

Partition numbers for f'': $x = 0$, $x = 2$

Sign chart for f'':

```
                 - - - - 0 + + + + 0 - - - -
f''(x)
        ─────────┼──────●──┼──┼──●──┼─────────────→ x
                -1   0     1     2    3

  Graph      Concave ¦ Concave ¦ Concave
  of f       Downward ¦ Upward ¦ Downward
                 Inflection  Inflection
                   Point       Point
```

Test Numbers	
x	$f''(x)$
-1	-72 (-)
1	24 (+)
3	-72 (-)

Thus, the graph of f is concave downward on $(-\infty, 0)$ and on $(2, \infty)$; the graph is concave upward on $(0, 2)$; there are inflection points at $x = 0$ and $x = 2$.

Step 4. Sketch the graph of f:

x	$f(x)$
0	0
2	32
3	54

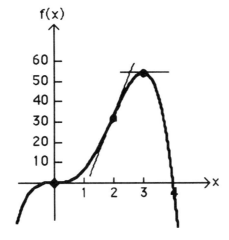

35. $f(x) = \dfrac{x + 3}{x - 3}$

Step 1. Use $f(x)$:

(A) Domain: All real numbers except $x = 3$.

(B) Intercepts: y intercept: $f(0) = \dfrac{3}{-3} = -1$

$$x \text{ intercepts: } \dfrac{x + 3}{x - 3} = 0$$
$$x + 3 = 0$$
$$x = -3$$

(C) Asymptotes:

Horizontal asymptote: $\displaystyle\lim_{x \to \infty} \dfrac{x + 3}{x - 3} = \lim_{x \to \infty} \dfrac{x\left(1 + \dfrac{3}{x}\right)}{x\left(1 - \dfrac{3}{x}\right)} = 1.$

Thus, $y = 1$ is a horizontal asymptote.

Vertical asymptote: The denominator is 0 at $x = 3$ and the numerator is not 0 at $x = 3$. Thus, $x = 3$ is a vertical asymptote.

Step 2. Use $f'(x)$:

$f'(x) = \dfrac{(x - 3)(1) - (x + 3)(1)}{(x - 3)^2} = \dfrac{-6}{(x - 3)^2} = -6(x - 3)^{-2}$

Critical values: None

Partition number: $x = 3$

Sign chart for f':

	Test Numbers	
x	$f'(x)$	
2	−6 (−)	
4	−6 (−)	

$f'(x)$ — − − − − − ND − − − − (0 1 2 3 4) → x

$f(x)$ Decreasing ¦ Decreasing

Thus, f is decreasing on $(-\infty, 3)$ and on $(3, \infty)$; there are no local extrema.

Step 3. Use $f''(x)$:

$f''(x) = 12(x - 3)^{-3} = \dfrac{12}{(x - 3)^3}$

Partition number for f'': $x = 3$

Sign chart for f'':

	Test Numbers	
x	$f''(x)$	
2	−12 (−)	
4	12 (+)	

$f''(x)$ — − − − − ND + + + + (0 1 2 3 4) → x

Graph of f — Concave Downward ¦ Concave Upward

Thus, the graph of f is concave downward on $(-\infty, 3)$ and concave upward on $(3, \infty)$.

Step 4. Sketch the graph of f:

x	$f(x)$
-3	0
0	-1
5	4

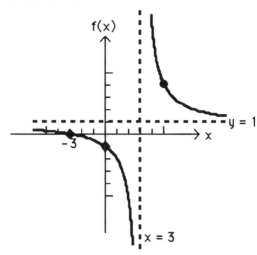

37. $f(x) = \dfrac{x}{x - 2}$

Step 1. Use $f(x)$:

(A) Domain: All real numbers except $x = 2$.

(B) Intercepts: y intercept: $f(0) = \dfrac{0}{-2} = 0$

$\qquad\qquad x$ intercepts: $\dfrac{x}{x - 2} = 0$

$\qquad\qquad\qquad\qquad\qquad x = 0$

(C) Asymptotes:

<u>Horizontal asymptote</u>: $\displaystyle\lim_{x \to \infty} \dfrac{x}{x - 2} = \lim_{x \to \infty} \dfrac{x}{x\left(1 - \dfrac{2}{x}\right)} = 1.$

Thus, $y = 1$ is a horizontal asymptote.

<u>Vertical asymptote</u>: The denominator is 0 at $x = 2$ and the numerator is not 0 at $x = 2$. Thus, $x = 2$ is a vertical asymptote.

Step 2. Use $f'(x)$:

$f'(x) = \dfrac{(x - 2)(1) - x(1)}{(x - 2)^2} = \dfrac{-2}{(x - 2)^2} = -2(x - 2)^{-2}$

Critical values: None

Partition number: $x = 2$

Sign chart for f':

$f'(x)$ $\qquad\qquad$ - - - - ND - - - -

$\qquad$ 0 $\quad$ 1 $\quad$ 2 $\quad$ 3 $\qquad\qquad$ x

$f(x)$ $\qquad$ Decreasing ¦ Decreasing

Test Numbers	
x	$f'(x)$
0	$-\frac{1}{2}$ (−)
3	-2 (−)

Thus, f is decreasing on $(-\infty, 2)$ and on $(2, \infty)$; there are no local extrema.

Step 3. Use $f''(x)$:

$$f''(x) = 4(x - 2)^{-3} = \frac{4}{(x - 2)^3}$$

Partition number for f'': $x = 2$
Sign chart for f'':

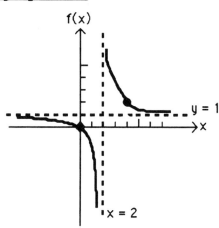

Test Numbers	
x	$f''(x)$
0	$-\frac{1}{2}$ (−)
3	4 (+)

Thus, the graph of f is concave downward on $(-\infty, 2)$ and concave upward on $(2, \infty)$.

Step 4. Sketch the graph of f:

x	$f(x)$
0	0
4	2

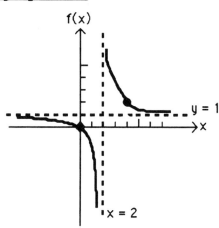

39. For $|x|$ very large, $f(x) = x + \frac{1}{x} \approx x$. Thus, the line $y = x$ is an oblique asymptote.

Step 1. Use $f(x)$:
(A) Domain: All real numbers except $x = 0$.
(B) Intercepts: y intercept: There is no y intercept since f is not defined at $x = 0$.

$$x \text{ intercepts: } x + \frac{1}{x} = 0$$
$$x^2 + 1 = 0$$

Thus, there are no x intercepts.
(C) Asymptotes:
Oblique asymptote: $y = x$

Vertical asymptote: $f(x) = x + \frac{1}{x} = \frac{x^2 + 1}{x}$

The denominator is 0 at $x = 0$ and the numerator is not 0 at $x = 0$. Thus, $x = 0$ is a vertical asymptote.

Step 2. Use $f'(x)$:

$$f'(x) = 1 - \frac{1}{x^2} = \frac{x^2 - 1}{x^2} = \frac{(x + 1)(x - 1)}{x^2}$$

Critical values: $\dfrac{(x - 1)(x + 1)}{x^2} = 0$

$$(x - 1)(x + 1) = 0$$
$$x = 1, -1$$

Partition numbers: $x = 0$, $x = 1$, $x = -1$

Sign chart for f':

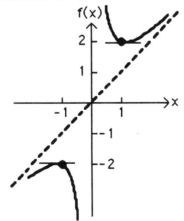

x	f'(x)
-2	$\frac{3}{4}$ (+)
$-\frac{1}{2}$	-3 (−)
$\frac{1}{2}$	-3 (−)
2	$\frac{3}{4}$ (+)

Test Numbers

Thus, f is increasing on $(-\infty, -1)$ and on $(1, \infty)$; f is decreasing on $(-1, 0)$ and on $(0, 1)$; f has a local maximum at $x = -1$ and a local minimum at $x = 1$.

Step 3. Use $f''(x)$:

$$f''(x) = 2x^{-3} = \frac{2}{x^3}$$

Partition number for f'': $x = 0$

Sign chart for f'':

Test Numbers

x	f''(x)
-1	-2 (−)
1	2 (+)

Thus, the graph of f is concave downward on $(-\infty, 0)$ and concave upward on $(0, \infty)$.

Step 4. Sketch the graph of f:

x	f(x)
-1	-2
1	2

41. $f(x) = x^3 - x$

Step 1. Use $f(x)$:

(A) Domain: All real numbers, $(-\infty, \infty)$.

(B) Intercepts: y intercept: $f(0) = 0^3 - 0 = 0$

x intercepts: $x^3 - x = 0$

$$x(x^2 - 1) = 0$$
$$x(x - 1)(x + 1) = 0$$
$$x = 0, 1, -1$$

(C) Asymptotes: There are no asymptotes.

Step 2. Use $f'(x)$:

$f'(x) = 3x^2 - 1 = (\sqrt{3}x + 1)(\sqrt{3}x - 1)$

Critical values: $x = -\dfrac{\sqrt{3}}{3}$, $x = \dfrac{\sqrt{3}}{3}$

Partition numbers: $x = -\dfrac{\sqrt{3}}{3}$, $x = \dfrac{\sqrt{3}}{3}$

Sign chart for f':

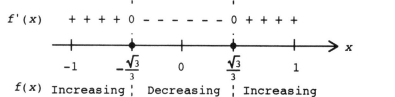

Test Numbers

x	$f'(x)$
-1	2 (+)
0	-1 (-)
1	2 (+)

Thus, f is increasing on $\left(-\infty, -\dfrac{\sqrt{3}}{3}\right)$ and on $\left(\dfrac{\sqrt{3}}{3}, \infty\right)$; f is decreasing on $\left(-\dfrac{\sqrt{3}}{3}, \dfrac{\sqrt{3}}{3}\right)$; f has a local maximum at $x = -\dfrac{\sqrt{3}}{3}$ and a local minimum at $x = \dfrac{\sqrt{3}}{3}$.

Step 3. Use $f''(x)$:

$f''(x) = 6x$

Partition number for f'': $x = 0$

Sign chart for f'':

$f''(x)$

$- - - - 0 + + + +$

Graph
of f

Concave
Downward

Concave
Upward

Thus, the graph of f is concave downward on $(-\infty, 0)$ and concave upward on $(0, \infty)$. There is an inflection point at $x = 0$.

Step 4. Sketch the graph of f:

x	$f(x)$
-1	0
$-\frac{\sqrt{3}}{3}$	≈ 0.4
0	0
$\frac{\sqrt{3}}{3}$	≈ -0.4

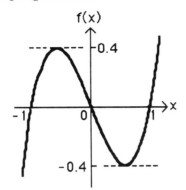

43. $f(x) = (x^2 + 3)(9 - x^2)$

Step 1. Use $f(x)$:

(A) Domain: All real numbers, $(-\infty, \infty)$.
(B) Intercepts: y intercept: $f(0) = 3(9) = 27$
  x intercepts: $(x^2 + 3)(9 - x^2) = 0$
    $(3 - x)(3 + x) = 0$
      $x = 3, -3$

(C) Asymptotes: There are no asymptotes.

Step 2. Use $f'(x)$:
$$f'(x) = (x^2 + 3)(-2x) + (9 - x^2)(2x)$$
$$= 2x[9 - x^2 - (x^2 + 3)]$$
$$= 2x(6 - 2x^2)$$
$$= 4x(\sqrt{3} + x)(\sqrt{3} - x)$$
Critical values: $x = 0$, $x = -\sqrt{3}$, $x = \sqrt{3}$
Partition numbers: $x = 0$, $x = -\sqrt{3}$, $x = \sqrt{3}$
Sign chart for f':

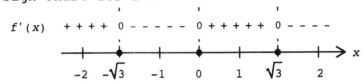

Test Numbers	
x	$f'(x)$
-2	$8\ (+)$
-1	$-8\ (-)$
1	$8\ (+)$
2	$-8\ (-)$

Thus, f is increasing on $(-\infty, -\sqrt{3})$ and on $(0, \sqrt{3})$; f is decreasing on $(-\sqrt{3}, 0)$ and on $(\sqrt{3}, \infty)$; f has local maxima at $x = -\sqrt{3}$ and $x = \sqrt{3}$ and a local minimum at $x = 0$.

Step 3. Use $f''(x)$:
$$f''(x) = 2x(-4x) + (6 - 2x^2)(2) = 12 - 12x^2 = -12(x - 1)(x + 1)$$
Partition numbers for f'': $x = 1$, $x = -1$

Sign chart for f'':

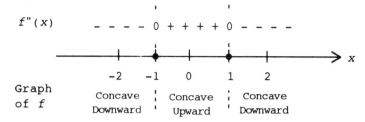

x	$f''(x)$
Test Numbers	
-2	-36 (−)
0	12 (+)
2	-36 (−)

Thus, the graph of f is concave downward on $(-\infty, -1)$ and on $(1, \infty)$; the graph of f is concave upward on $(-1, 1)$; the graph has inflection point at $x = -1$ and $x = 1$.

Step 4.   Sketch the graph of f:

x	$f(x)$
$-\sqrt{3}$	36
−1	32
0	27
1	32
$\sqrt{3}$	36

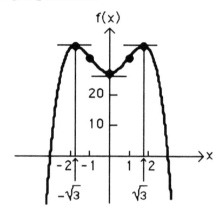

45. $f(x) = (x^2 - 4)^2$

Step 1.   Use $f(x)$:
(A) Domain: All real numbers, $(-\infty, \infty)$.
(B) Intercepts: y intercept: $f(0) = (-4)^2 = 16$
  x intercepts: $(x^2 - 4)^2 = 0$
  $[(x - 2)(x + 2)]^2 = 0$
  $(x - 2)^2(x + 2)^2 = 0$
  $x = 2, -2$
(C) Asymptotes: There are no asymptotes.

Step 2.   Use $f'(x)$:
$f'(x) = 2(x^2 - 4)(2x) = 4x(x - 2)(x + 2)$
Critical values: $x = 0$, $x = 2$, $x = -2$
Partition numbers: Same as critical values.
Sign chart for f':

f'(x)   − − − − 0 + + + + + 0 − − − − − 0 + + + +

-3   -2   -1   0   1   2   3

f(x)   Decreasing ┆ Increasing ┆ Decreasing ┆ Increasing

x	$f'(x)$
Test Numbers	
-3	-60 (−)
-1	12 (+)
1	-12 (−)
3	60 (+)

Thus, f is decreasing on $(-\infty, -2)$ and on $(0, 2)$; f is increasing on $(-2, 0)$ and on $(2, \infty)$; f has local minima at $x = -2$ and $x = 2$ and a local maximum at $x = 0$.

Step 3. Use $f''(x)$:

$$f''(x) = 4x(2x) + (x^2 - 4)(4) = 12x^2 - 16 = 12\left(x^2 - \frac{4}{3}\right)$$

$$= 12\left(x - \frac{2\sqrt{3}}{3}\right)\left(x + \frac{2\sqrt{3}}{3}\right)$$

Partition numbers for f'': $x = \frac{2\sqrt{3}}{3}$, $x = \frac{-2\sqrt{3}}{3}$

Sign chart for f'':

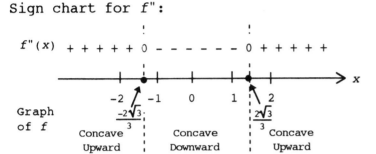

Test Numbers	
x	$f''(x)$
-2	32 (+)
0	-16 (-)
2	32 (+)

Thus, the graph of f is concave upward on $\left(-\infty, \frac{-2\sqrt{3}}{3}\right)$ and on $\left(\frac{2\sqrt{3}}{3}, \infty\right)$; the graph of f is concave downward on $\left(\frac{-2\sqrt{3}}{3}, \frac{2\sqrt{3}}{3}\right)$; the graph has inflection points at $x = \frac{-2\sqrt{3}}{3}$ and $x = \frac{2\sqrt{3}}{3}$.

Step 4. Sketch the graph of f:

x	$f(x)$
-2	0
$-\frac{2\sqrt{3}}{3}$	$\frac{64}{9}$
0	16
$\frac{2\sqrt{3}}{3}$	$\frac{64}{9}$
2	0

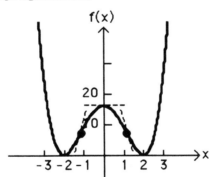

47. $f(x) = 2x^6 - 3x^5$

Step 1. Use $f(x)$:
(A) Domain: All real numbers, $(-\infty, \infty)$.
(B) Intercepts: y intercept: $f(0) = 2 \cdot 0^6 - 3 \cdot 0^5 = 0$
$\qquad\qquad\qquad x$ intercepts: $2x^6 - 3x^5 = 0$
$\qquad\qquad\qquad\qquad\qquad x^5(2x - 3) = 0$
$\qquad\qquad\qquad\qquad\qquad\qquad x = 0, \frac{3}{2}$

(C) Asymptotes: There are no asymptotes.

<u>Step 2.</u> Use $f'(x)$:

$$f'(x) = 12x^5 - 15x^4 = 12x^4\left(x - \frac{5}{4}\right)$$

Critical values: $x = 0$, $x = \frac{5}{4}$

Partition numbers: Same as critical values.

Sign chart for f':

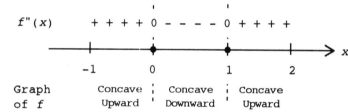

x	f'(x)
−1	−27 (−)
1	−3 (−)
2	144 (+)

Test Numbers

Thus, f is decreasing on $(-\infty,\ 0)$ and $\left(0,\ \frac{5}{4}\right)$; f is increasing on $\left(\frac{5}{4},\ \infty\right)$; f has a local minimum at $x = \frac{5}{4}$.

<u>Step 3.</u> Use $f''(x)$:

$$f''(x) = 60x^4 - 60x^3 = 60x^3(x - 1)$$

Partition numbers for f'': $x = 0$, $x = 1$

Sign chart for f'':

x	f''(x)
−1	120 (+)
$\frac{1}{2}$	$-\frac{15}{4}$ (−)
2	480 (+)

Test Numbers

Thus, the graph of f is concave upward on $(-\infty,\ 0)$ and on $(1,\ \infty)$; the graph of f is concave downward on $(0,\ 1)$; the graph has inflection points at $x = 0$ and $x = 1$.

<u>Step 4.</u> Sketch the graph of f:

x	f(x)
0	0
1	−1
$\frac{5}{4}$	≈ −1.5

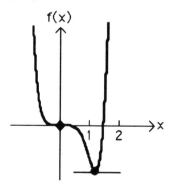

49. $f(x) = \dfrac{x}{x^2 - 4} = \dfrac{x}{(x - 2)(x + 2)}$

Step 1. Use $f(x)$:

(A) Domain: All real numbers except $x = 2$, $x = -2$.

(B) Intercepts: y intercept: $f(0) = \dfrac{0}{-4} = 0$

                    x intercept: $\dfrac{x}{x^2 - 4} = 0$

                              $x = 0$

(C) Asymptotes:
   Horizontal asymptote:

$$\lim_{x \to \infty} \frac{x}{x^2 - 4} = \lim_{x \to \infty} \frac{x}{x^2\left(1 - \dfrac{4}{x^2}\right)} = \lim_{x \to \infty} \frac{1}{x}\left(\frac{1}{1 - \dfrac{4}{x^2}}\right) = 0$$

Thus, $y = 0$ (the x axis) is a horizontal asymptote.

Vertical asymptotes: The denominator is 0 at $x = 2$ and $x = -2$. The numerator is nonzero at each of these points. Thus, $x = 2$ and $x = -2$ are vertical asymptotes.

Step 2. Use $f'(x)$:

$f'(x) = \dfrac{(x^2 - 4)(1) - x(2x)}{(x^2 - 4)^2} = \dfrac{-(x^2 + 4)}{(x^2 - 4)^2}$

Critical values: None ($x^2 + 4 \neq 0$ for all x)
Partition numbers: $x = 2$, $x = -2$
Sign chart for f':

```
                       ¦                ¦
 f'(x)     - - - - - ND - - - - - - - ND - - - - -
         ──────────┼──◇──┼──┼──┼──┼──◇──┼─────────→ x
                  -2   -1  0  1   2
```

$f(x)$ Decreasing ¦ Decreasing ¦ Decreasing

Thus, f is decreasing on $(-\infty, -2)$, on $(-2, 2)$, and on $(2, \infty)$; f has no local extrema.

Step 3. Use $f''(x)$:

$f''(x) = \dfrac{(x^2 - 4)^2(-2x) - [-(x^2 + 4)](2)(x^2 - 4)(2x)}{(x^2 - 4)^4}$

      $= \dfrac{(x^2 - 4)(-2x) + 4x(x^2 + 4)}{(x^2 - 4)^3} = \dfrac{2x^3 + 24x}{(x^2 - 4)^3} = \dfrac{2x(x^2 + 12)}{(x^2 - 4)^3}$

Partition numbers for f'': $x = 0$, $x = 2$, $x = -2$

Sign chart for f'':

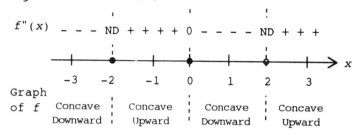

Test Numbers	
x	$f''(x)$
-3	$-\frac{126}{125}$ $(-)$
-1	$\frac{26}{27}$ $(+)$
1	$-\frac{26}{27}$ $(-)$
3	$\frac{126}{127}$ $(+)$

Thus, the graph of f is concave downward on $(-\infty, -2)$ and on $(0, 2)$; the graph of f is concave upward on $(-2, 0)$ and on $(2, \infty)$; the graph has an inflection point at $x = 0$.

Step 4.   Sketch the graph of f:

x	$f(x)$
0	0
1	$-\frac{1}{3}$
-1	$\frac{1}{3}$
3	$\frac{3}{5}$
-3	$-\frac{3}{5}$

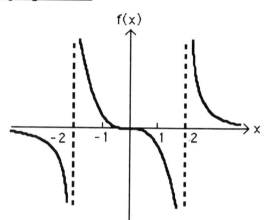

51. $f(x) = \dfrac{1}{1 + x^2}$

Step 1.   Use $f(x)$:

(A) Domain: All real numbers ($1 + x^2 \neq 0$ for all x).

(B) Intercepts: y intercept: $f(0) = 1$

   x intercept: $\dfrac{1}{1 + x^2} \neq 0$ for all x; no x intercepts

(C) Asymptotes:

   <u>Horizontal asymptote</u>: $\lim\limits_{x \to \infty} \dfrac{1}{1 + x^2} = 0$

   Thus, $y = 0$ (the x axis) is a horizontal asymptote.

   <u>Vertical asymptotes</u>: Since $1 + x^2 \neq 0$ for all x, there are no vertical asymptotes.

Step 2.   Use $f'(x)$:

$f'(x) = \dfrac{(1 + x^2)(0) - 1(2x)}{(1 + x^2)^2} = \dfrac{-2x}{(1 + x^2)^2}$

Critical values: $x = 0$

Partition numbers: $x = 0$

Sign chart for f':

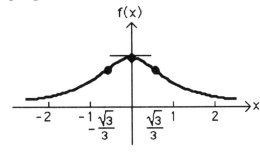

Test Numbers

x	$f'(x)$
-1	$\frac{1}{2}$ (+)
1	$-\frac{1}{2}$ (−)

$f'(x)$ + + + + 0 − − − −

-1 0 1

$f(x)$ Increasing ⋮ Decreasing

Thus, f is increasing on $(-\infty, 0)$; f is decreasing on $(0, \infty)$; f has a local maximum at $x = 0$.

Step 3. Use $f''(x)$:

$$f''(x) = \frac{(1 + x^2)^2(-2) - (-2x)(2)(1 + x^2)2x}{(1 + x^2)^4} = \frac{(-2)(1 + x^2) + 8x^2}{(1 + x^2)^3}$$

$$= \frac{6x^2 - 2}{(1 + x^2)^3} = \frac{6\left(x + \frac{\sqrt{3}}{3}\right)\left(x - \frac{\sqrt{3}}{3}\right)}{(1 + x^2)^3}$$

Partition numbers for f'': $x = -\frac{\sqrt{3}}{3}$, $x = \frac{\sqrt{3}}{3}$

Sign chart for f'':

$f''(x)$ + + + + + 0 − − − 0 + + + +

-2 -1 $-\frac{\sqrt{3}}{3}$ 0 $\frac{\sqrt{3}}{3}$ 1 2

Graph
of f Concave ⋮ Concave ⋮ Concave
          Upward ⋮ Downward ⋮ Upward

Test Numbers

x	$f''(x)$
-1	$\frac{1}{2}$ (+)
0	-2 (−)
1	$\frac{1}{2}$ (+)

Thus, the graph of f is concave upward on $\left(-\infty, \frac{-\sqrt{3}}{3}\right)$ and on $\left(\frac{\sqrt{3}}{3}, \infty\right)$; the graph of f is concave downward on $\left(\frac{-\sqrt{3}}{3}, \frac{\sqrt{3}}{3}\right)$; the graph has inflection points at $x = \frac{-\sqrt{3}}{3}$ and $x = \frac{\sqrt{3}}{3}$.

Step 4. Sketch the graph of f:

x	$f(x)$
$-\frac{\sqrt{3}}{3}$	$\frac{3}{4}$
0	1
$\frac{\sqrt{3}}{3}$	$\frac{3}{4}$

53. Horizontal asymptotes: $y = -1$
and $y = 1$;
vertical asymptote: $x = 0$

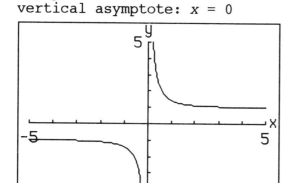

55. Horizontal asymptote: $y = -1$;
vertical asymptote: $x = 1$

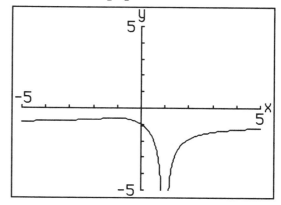

57. Horizontal asymptotes: $y = -1$ and $y = 1$; vertical asymptotes: $x = -1$ and $x = 1$

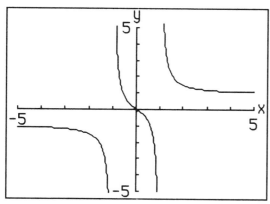

59. $R(x) = xp = 1296x - 0.12x^3$, $0 < x < 80$

Step 1. Use $R(x)$:

(A) Domain: $0 < x < 80$ or $(0, 80)$.

(B) Intercepts: There are no intercepts on $(0, 80)$.

(C) Asymptotes: Since R is a polynomial, there are no asymptotes.

Step 2. Use $R'(x)$:

$$R'(x) = 1296 - 0.36x^2$$
$$= -0.36(x^2 - 3600)$$
$$= -0.36(x - 60)(x + 60), \ 0 < x < 80$$

Critical values: [on $(0, 80)$]: $x = 60$

Partition numbers: $x = 60$

Sign chart for R':

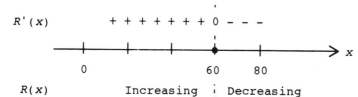

Test Numbers	
x	$R'(x)$
1	1295.64 (+)
61	−43.56 (−)

Thus, R is increasing on $(0, 60)$ and decreasing on $(60, 80)$, R has a local maximum at $x = 60$.

Step 3.   Use $R''(x)$:
$R''(x) = -0.72x < 0$ for $0 < x < 80$
Thus, the graph of R is concave downward on $(0, 80)$.

Step 4.   Sketch the graph of R:

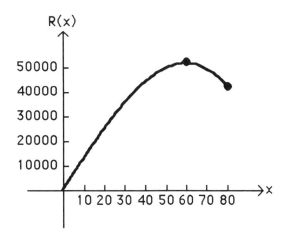

61. $P(x) = \dfrac{2x}{1 - x}$, $0 \le x < 1$

(A) $P'(x) = \dfrac{(1 - x)(2) - 2x(-1)}{(1 - x)^2} = \dfrac{2}{(1 - x)^2}$

$P'(x) > 0$ for $0 \le x < 1$.   Thus, P is increasing on $(0, 1)$.

(B) From (A), $P'(x) = 2(1 - x)^{-2}$.   Thus,

$P''(x) = -4(1 - x)^{-3}(-1) = \dfrac{4}{(1 - x)^3}$.

$P''(x) > 0$ for $0 \le x < 1$, and the graph of P is concave upward on $(0, 1)$.

(C) Since the domain of P is $[0, 1)$, there are no horizontal asymptotes. The denominator is 0 at $x = 1$ and the numerator is nonzero there. Thus, $x = 1$ is a vertical asymptote.

(D) $P(0) = \dfrac{2 \cdot 0}{1 - 0} = 0$.

Thus, the origin is both an x and a y intercept of the graph.

(E) The graph of P is:

x	$P(x)$
0	0
$\frac{1}{2}$	2
$\frac{3}{4}$	6

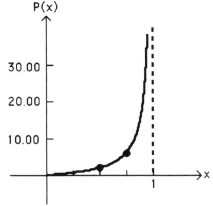

63. $C(n) = 3200 + 250n + 50n^2$, $0 < n < \infty$

(A) Average cost per year:

$$\overline{C}(n) = \frac{C(n)}{n} = \frac{3200}{n} + 250 + 50n, \ 0 < n < \infty$$

(B) Graph $\overline{C}(n)$:

Step 1. Use $\overline{C}(n)$:

Domain: $0 < n < \infty$

Intercepts: C intercept: None $(n > 0)$

n intercepts: $\dfrac{3200}{n} + 250 + 50n > 0$ on $(0, \infty)$;

there are no n intercepts.

Asymptotes: For large n, $C(n) = \dfrac{3200}{n} + 250 + 50n \approx 250 + 50n$.

Thus, $y = 250 + 50n$ is an oblique asymptote. As $n \to 0$, $\overline{C}(n) \to \infty$; thus, $n = 0$ is a vertical asymptote.

Step 2. Use $\overline{C}'(n)$:

$$\overline{C}'(n) = -\frac{3200}{n^2} + 50 = \frac{50n^2 - 3200}{n^2} = \frac{50(n^2 - 64)}{n^2}$$

$$= \frac{50(n - 8)(n + 8)}{n^2}, \ 0 < n < \infty$$

Critical value: $n = 8$

Sign chart for $\overline{C}'$:

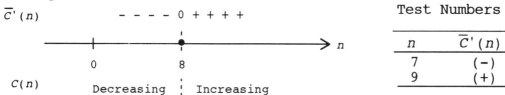

Test Numbers	
n	$\overline{C}'(n)$
7	$(-)$
9	$(+)$

Thus, $\overline{C}$ is decreasing on $(0, 8)$ and increasing on $(8, \infty)$; $n = 8$ is a local minimum.

Step 3: Use $\overline{C}''(n)$:

$$\overline{C}''(n) = \frac{6400}{n^3}, \ 0 < n < \infty$$

$\overline{C}''(n) > 0$ on $(0, \infty)$. Thus, the graph of $\overline{C}$ is concave upward on $(0, \infty)$.

Step 4. Sketch the graph of $\overline{C}$:

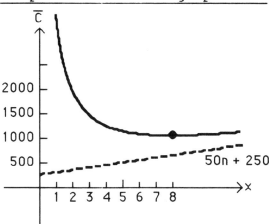

(C) The average cost per year is a minimum when $n = 8$ years.

65. $C(x) = 1000 + 5x + \frac{1}{10}x^2$, $0 < x < \infty$.

(A) The average cost function is: $\overline{C}(x) = \frac{1000}{x} + 5 + \frac{1}{10}x$.

Now, $\overline{C}'(x) = -\frac{1000}{x^2} + \frac{1}{10} = \frac{x^2 - 10,000}{10x^2} = \frac{(x + 100)(x - 100)}{10x^2}$

Sign chart for $\overline{C}'$:

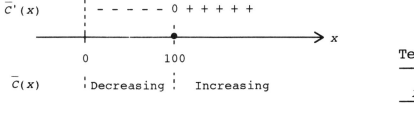

Test Numbers	
x	$\overline{C}'(x)$
1	≈ -1000 $(-)$
101	$\approx \frac{1}{500}$ $(+)$

Thus, $\overline{C}$ is decreasing on $(0, 100)$ and increasing on $(100, \infty)$; $\overline{C}$ has a minimum at $x = 100$.

Since $\overline{C}''(x) = \frac{2000}{x^3} > 0$ for $0 < x < \infty$, the graph of $\overline{C}$ is concave upward on $(0, \infty)$. The line $x = 0$ is a vertical asymptote and the line $y = 5 + \frac{1}{10}x$ is an oblique asymptote for the graph of $\overline{C}$.

The marginal cost function is $C'(x) = 5 + \frac{1}{5}x$.

The graphs of $\overline{C}$ and C' are:

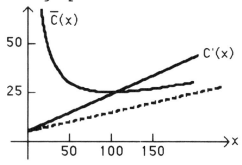

(B) The minimum average cost is:

$$\overline{C}(100) = \frac{1000}{100} + 5 + \frac{1}{10}(100) = 25$$

67. $C(t) = \dfrac{0.14t}{t^2 + 1}$

<u>Step 1. Use $C(t)$</u>:

Domain: $t \geq 0$, i.e., $[0, \infty)$

Intercepts: y intercept: $C(0) = 0$

$\qquad\qquad\quad$ t intercepts: $\dfrac{0.14t}{t^2 + 1} = 0$

$\qquad\qquad\qquad\qquad\qquad\qquad t = 0$

Asymptotes:

<u>Horizontal asymptote</u>: $\displaystyle\lim_{t\to\infty} \dfrac{0.14t}{t^2 + 1} = \lim_{t\to\infty} \dfrac{0.14t}{t^2\left(1 + \dfrac{1}{t^2}\right)} = \lim_{t\to\infty} \dfrac{0.14}{t\left(1 + \dfrac{1}{t^2}\right)} = 0$

Thus, $y = 0$ (the t axis) is a horizontal asymptote.

<u>Vertical asymptotes</u>: Since $t^2 + 1 > 0$ for all t, there are no vertical asymptotes.

<u>Step 2. Use $C'(t)$</u>:

$C'(t) = \dfrac{(t^2 + 1)(0.14) - 0.14t(2t)}{(t^2 + 1)^2} = \dfrac{0.14(1 - t^2)}{(t^2 + 1)^2} = \dfrac{0.14(1 - t)(1 + t)}{(t^2 + 1)^2}$

Critical values on $[0, \infty)$: $t = 1$

Sign chart for C':

Test Numbers	
t	$C'(t)$
0	$(+)$
2	$(-)$

$C'(t)$ $\quad\quad + + + + + 0 - - - - -$

$\qquad\qquad\qquad\qquad 0 \qquad\quad 1$

$C(t) \qquad$ Increasing $\vdots$ Decreasing

Thus, C is increasing on $(0, 1)$ and decreasing on $(1, \infty)$; C has a maximum value at $t = 1$.

Step 3. Use $C''(t)$:

$$C''(t) = \frac{(t^2 + 1)^2(-0.28t) - 0.14(1 - t^2)(2)(t^2 + 1)(2t)}{(t^2 + 1)^4}$$

$$= \frac{(t^2 + 1)(-0.28t) - 0.56t(1 - t^2)}{(t^2 + 1)^3} = \frac{0.28t^3 - 0.84t}{(t^2 + 1)^3}$$

$$= \frac{0.28t(t^2 - 3)}{(t^2 + 1)^3} = \frac{0.28t(t - \sqrt{3})(t + \sqrt{3})}{(t^2 + 1)^3}, \quad 0 \le t < \infty$$

Partition numbers for C'' on $[0, \infty)$: $t = \sqrt{3}$

Sign chart for C'':

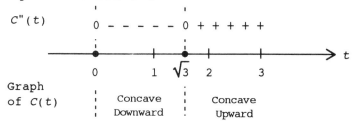

Test Numbers		
t	$C''(t)$	
1	-0.07	$(-)$
2	≈ 0.005	$(+)$

Thus, the graph of C is concave downward on $(0, \sqrt{3})$ and concave upward on $(\sqrt{3}, \infty)$; the graph has an inflection point at $t = \sqrt{3}$.

Step 4. Sketch the graph of $C(t)$:

t	$C(t)$
0	0
1	0.07
$\sqrt{3}$	≈ 0.06

69. $N(t) = \dfrac{5t + 20}{t} = 5 + 20t^{-1}, \quad 1 \le t \le 30$

Step 1. Use $N(t)$:

Domain: $1 \le t \le 30$, or $[1, 30]$.

Intercepts: There are no t or N intercepts.

Asymptotes: Since N is defined only for $1 \le t \le 30$, there are no horizontal asymptotes. Also, since $t \ne 0$ on $[1, 30]$, there are no vertical asymptotes.

Step 2. Use $N'(t)$:

$$N'(t) = -20t^{-2} = \frac{-20}{t^2}, \quad 1 \le t \le 30$$

Since $N'(t) < 0$ for $1 \le t \le 30$, N is decreasing on $(1, 30)$; N has no local extrema.

Step 3. Use $N''(t)$:

$N''(t) = \dfrac{40}{t^3}$, $1 \leq t \leq 30$

Since $N''(t) > 0$ for $1 \leq t \leq 30$, the graph of N is concave upward on $(1, 30)$.

Step 4. Sketch the graph of N:

t	$N(t)$
1	25
5	9
10	7
30	5.67

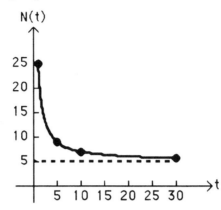

EXERCISE 10-4

Things to remember:

1. A function f continuous on a closed interval $[a, b]$ assumes both an absolute maximum and an absolute minimum on that interval. Absolute extrema (if they exist) must always occur at critical values or at endpoints.

2. STEPS FOR FINDING ABSOLUTE EXTREMA:

   To find the absolute maximum and absolute minimum of a function f on the closed interval $[a, b]$:

   (a) Verify that f is continuous on $[a, b]$.

   (b) Determine the critical values of f on the open interval (a, b).

   (c) Evaluate f at the endpoints a and b and at the critical values found in (b).

   (d) The absolute maximum $f(x)$ on $[a, b]$ is the largest of the values found in step (c).

   (e) The absolute minimum $f(x)$ on $[a, b]$ is the smallest of the values found in step (c).

3. SECOND-DERIVATIVE TEST FOR ABSOLUTE EXTREMA:

   Suppose f is continuous and has only one critical value c on the interval I, and suppose $f'(c) = 0$. Then $f(c)$ is the absolute minimum of f if $f''(c) > 0$, or $f(c)$ is the absolute maximum of f if $f''(c) < 0$; the test fails if $f''(c) = 0$.

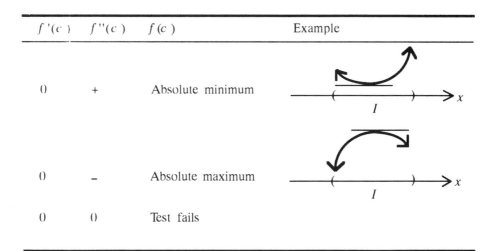

$f'(c)$	$f''(c)$	$f(c)$	Example
0	+	Absolute minimum	
0	−	Absolute maximum	
0	0	Test fails	

4. STRATEGY FOR SOLVING APPLIED OPTIMIZATION PROBLEMS

Step 1: Introduce variables and a function f, including the domain I of f, and then construct a mathematical model of the form:

Maximize (or minimize) f on the interval I

Step 2: Find the absolute maximum (or minimum) value of f on the interval I and the value(s) of x where this occurs.

Step 3: Use the solution to the mathematical model to answer the questions asked in the application.

1. $f(x) = x^2 - 4x + 5$, $I = (-\infty, \infty)$
$f'(x) = 2x - 4 = 2(x - 2)$
$x = 2$ is the *only critical value* on I, and $f(2) = 2^2 - 4 \cdot 2 + 5 = 1$.
$f''(x) = 2$
$f''(2) = 2 > 0$. Therefore, $f(2) = 1$ is the absolute minimum.
The function does not have an absolute maximum.

3. $f(x) = 10 + 8x - x^2$, $I = (-\infty, \infty)$
$f'(x) = 8 - 2x = 2(4 - x)$
$x = 4$ is the *only critical value* on I, and $f(4) = 10 + 32 - 16 = 26$.
$f''(x) = -2$
$f''(4) = -2 < 0$. Therefore, $f(4) = 26$ is the absolute maximum.
The function does not have an absolute minimum.

5. $f(x) = 1 - x^3$, $I = (-\infty, \infty)$
$f'(x) = -3x^2$
$x = 0$ is the *only critical value* on I, and $f(0) = 1 - 0^3 = 1$.
$f''(x) = -6x$
$f''(0) = 0$. Therefore, the test fails.
Since $f'(x) = -3x^2 < 0$ on $(-\infty, 0)$ and on $(0, \infty)$, f is decreasing on I and f does not have any absolute extrema.

7. $f(x) = 24 - 2x - \dfrac{8}{x}$, $x > 0$

$f'(x) = -2 + \dfrac{8}{x^2} = \dfrac{-2x^2 + 8}{x^2} = \dfrac{-2(x^2 - 4)}{x^2} = \dfrac{-2(x + 2)(x - 2)}{x^2}$

Critical values: $x = 2$ [Note: $x = -2$ is not a critical value, since the domain of f is $x > 0$.]

$f''(x) = -\dfrac{16}{x^3}$ and $f''(2) = -\dfrac{16}{8} = -2 < 0$.

Thus, the absolute maximum value of f is $f(2) = 24 - 2 \cdot 2 - \dfrac{8}{2} = 16$.

9. $f(x) = 5 + 3x + \dfrac{12}{x^2}$, $x > 0$

$f'(x) = 3 - \dfrac{24}{x^3} = \dfrac{3x^3 - 24}{x^3} = \dfrac{3(x^3 - 8)}{x^3} = \dfrac{3(x - 2)(x^2 + 2x + 4)}{x^3}$

Critical value: $x = 2$

$f''(x) = \dfrac{72}{x^4}$ and $f''(2) = \dfrac{72}{2^4} = \dfrac{72}{16} > 0$.

Thus, the absolute minimum of f is: $f(2) = 5 + 3(2) + \dfrac{12}{2^2} = 14$.

11. $f(x) = x^3 - 6x^2 + 9x - 6$
$f'(x) = 3x^2 - 12x + 9 = 3(x^2 - 4x + 3) = 3(x - 3)(x - 1)$
Critical values: $x = 1, 3$

(A) On the interval $[-1, 5]$: $f(-1) = -1 - 6 - 9 - 6 = -22$
$\qquad\qquad\qquad\qquad\qquad\quad f(1) = 1 - 6 + 9 - 6 = -2$
$\qquad\qquad\qquad\qquad\qquad\quad f(3) = 27 - 54 + 27 - 6 = -6$
$\qquad\qquad\qquad\qquad\qquad\quad f(5) = 125 - 150 + 45 - 6 = 14$
Thus, the absolute maximum of f is $f(5) = 14$, and the absolute minimum of f is $f(-1) = -22$.

(B) On the interval $[-1, 3]$: $f(-1) = -22$
$\qquad\qquad\qquad\qquad\qquad\quad f(1) = -2$
$\qquad\qquad\qquad\qquad\qquad\quad f(3) = -6$
Absolute maximum of f: $f(1) = -2$
Absolute minimum of f: $f(-1) = -22$

(C) On the interval $[2, 5]$: $f(2) = 8 - 24 + 18 - 6 = -4$
$\qquad\qquad\qquad\qquad\qquad\quad f(3) = -6$
$\qquad\qquad\qquad\qquad\qquad\quad f(5) = 14$
Absolute maximum of f: $f(5) = 14$
Absolute minimum of f: $f(3) = -6$

13. $f(x) = (x - 1)(x - 5)^3 + 1$
$f'(x) = (x - 1)3(x - 5)^2 + (x - 5)^3$
$\qquad\;\, = (x - 5)^2(3x - 3 + x - 5)$
$\qquad\;\, = (x - 5)^2(4x - 8)$
Critical values: $x = 2, 5$

(A) Interval [0, 3]: $f(0) = (-1)(-5)^3 + 1 = 126$
$f(2) = (2 - 1)(2 - 5)^3 + 1 = -26$
$f(3) = (3 - 1)(3 - 5)^3 + 1 = -15$

Absolute maximum of f: $f(0) = 126$
Absolute minimum of f: $f(2) = -26$

(B) Interval [1, 7]: $f(1) = 1$
$f(2) = -26$
$f(5) = 1$
$f(7) = (7 - 1)(7 - 5)^3 + 1 = 6 \cdot 8 + 1 = 49$

Absolute maximum of f: $f(7) = 49$
Absolute minimum of f: $f(2) = -26$

(C) Interval [3, 6]: $f(3) = (3 - 1)(3 - 5)^3 + 1 = -15$
$f(5) = 1$
$f(6) = (6 - 1)(6 - 5)^3 + 1 = 6$

Absolute maximum of f: $f(6) = 6$
Absolute minimum of f: $f(3) = -15$

15. Let one length $= x$ and the other $= 10 - x$.
Since neither length can be negative, we have $x \geq 0$ and $10 - x \geq 0$, or $x \leq 10$. We want the maximum value of the product $x(10 - x)$, where $0 \leq x \leq 10$.

Let $f(x) = x(10 - x) = 10x - x^2$; domain $I = [0, 10]$
$f'(x) = 10 - 2x$; $x = 5$ is the only critical value
$f''(x) = -2$
$f''(5) = -2 < 0$
Thus, $f(5) = 25$ is the absolute maximum; divide the line in half.

17. Let one number $= x$ and other other $= x + 30$.
$f(x) = x(x + 30) = x^2 + 30x$; domain $I = (-\infty, \infty)$
$f'(x) = 2x + 30$; $x = -15$ is the only critical value
$f''(x) = 2$
$f''(-15) = 2 > 0$
Thus, the absolute minimum of f occurs at $x = -15$. The numbers, then, are -15 and $-15 + 30 = 15$.

19. Let $x =$ the length of the rectangle
and $y =$ the width of the rectangle.
Then, $2x + 2y = 100$
$x + y = 50$
$y = 50 - x$

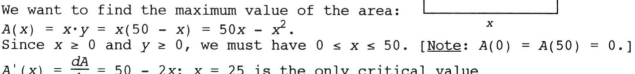

We want to find the maximum value of the area:
$A(x) = x \cdot y = x(50 - x) = 50x - x^2$.
Since $x \geq 0$ and $y \geq 0$, we must have $0 \leq x \leq 50$. [<u>Note</u>: $A(0) = A(50) = 0$.]
$A'(x) = \dfrac{dA}{dx} = 50 - 2x$; $x = 25$ is the only critical value
Now, $A'' = -2$ and $A''(25) = -2 < 0$. Thus, $A(25)$ is the absolute maximum.
The maximum area is $A(25) = 25(50 - 25) = 625$ cm^2, which means that the rectangle is actually a square with sides measuring 25 cm each.

21. (A) Revenue $R(x) = x \cdot p(x) = x\left(200 - \frac{x}{30}\right) = 200x - \frac{x^2}{30}$, $0 \leq x \leq 6,000$

$R'(x) = 200 - \frac{2x}{30} = 200 - \frac{x}{15}$

Now $R'(x) = 200 - \frac{x}{15} = 0$

implies $x = 3000$.

$R''(x) = -\frac{1}{15} < 0$.

Thus, $R''(3000) = -\frac{1}{15} < 0$ and we conclude that R has an absolute maximum at $x = 3000$. The maximum revenue is

$R(3000) = 200(3000) - \frac{(3000)^2}{30} = \$300,000$

(B) Profit $P(x) = R(x) - C(x) = 200x - \frac{x^2}{30} - (72,000 + 60x)$

$= 140x - \frac{x^2}{30} - 72,000$

$P'(x) = 140 - \frac{x}{15}$

Now $140 - \frac{x}{15} = 0$ implies $x = 2,100$.

$P''(x) = -\frac{1}{15}$ and $P''(2,100) = -\frac{1}{15} < 0$. Thus, the maximum profit occurs when 2,100 television sets are produced. The maximum profit is

$P(2,100) = 140(2,100) - \frac{(2,100)^2}{30} - 72,000 = \$75,000$

the price that the company should charge is

$p(2,100) = 200 - \frac{2,100}{30} = \$130$ for each set.

(C) If the government taxes the company \$5 for each set, then the profit $P(x)$ is given by

$P(x) = 200x - \frac{x^2}{30} - (72,000 + 60x) - 5x$

$= 135x - \frac{x^2}{30} - 72,000$.

$P'(x) = 135 - \frac{x}{15}$.

Now $135 - \frac{x}{15} = 0$ implies $x = 2,025$.

$P''(x) = -\frac{1}{15}$ and $P''(2,025) = -\frac{1}{15} < 0$. Thus, the maximum profit in this case occurs when 2,025 television sets are produced. The maximum profit is

$P(2,025) = 135(2,025) - \frac{(2,025)^2}{30} - 72,000$

$= \$64,687.50$

and the company should charge $p(2,025) = 200 - \frac{2,025}{30}$

$= \$132.50$ per set.

23. Let x = number of dollar increases in the rate per day. Then
$200 - 5x$ = total number of cars rented and $30 + x$ = rate per day.
Total income = (total number of cars rented)(rate)
$$y(x) = (200 - 5x)(30 + x), \ 0 \le x \le 40$$
$$y'(x) = (200 - 5x)(1) + (30 + x)(-5)$$
$$= 200 - 5x - 150 - 5x$$
$$= 50 - 10x$$
$$= 10(5 - x)$$
Thus, $x = 5$ is the only critical value and $y(5) = (200 - 25)(30 + 5) = $
6125.
$y''(x) = -10$
$y''(5) = -10 < 0$
Therefore, the absolute maximum income is $y(5) = \$6125$ when the rate is
$35 per day.

25. Let x = number of additional trees planted per acre. Then
$30 + x$ = total number of trees per acre and $50 - x$ = yield per tree.
Yield per acre = (total number of trees per acre)(yield per tree)
$$y(x) = (30 + x)(50 - x), \ 0 \le x \le 20$$
$$y'(x) = (30 + x)(-1) + (50 - x)$$
$$= 20 - 2x$$
$$= 2(10 - x)$$
The only critical value is $x = 10$, $y(10) = 40(40) = 1600$ pounds per acre.
$y''(x) = -2$
$y''(10) = -2 < 0$
Therefore, the absolute maximum yield is $y(10) = 1600$ pounds per acre
when the number of trees per acre is 40.

27. Volume = $V(x) = (12 - 2x)(8 - 2x)x, \ 0 \le x \le 4$
$$= 96x - 40x^2 + 4x^3$$
$$V'(x) = 96 - 80x + 12x^2$$
$$= 4(24 - 20x + 3x^2)$$
We solve $24 - 20x + 3x^2 = 0$ by using the
quadratic formula:

$$x = \frac{20 \pm \sqrt{400 - 4 \cdot 24 \cdot 3}}{6} = \frac{10 \pm 2\sqrt{7}}{3}$$
Thus, $x = \dfrac{10 - 2\sqrt{7}}{3} \approx 1.57$ is the only critical value on the interval
$[0, 4]$.
$V''(x) = -80 + 24x$
$V''(1.57) = -80 + 24(1.57) < 0$
Therefore, a square with a side of length $x = 1.57$ inches should be cut
from each corner to obtain the maximum volume.

29. Area = 800 square feet = xy (1)
Cost = $18x + 6(2y + x)$
From (1), we have $y = \dfrac{800}{x}$.

Hence, cost $C(x) = 18x + 6\left(\dfrac{1600}{x} + x\right)$, or

$$C(x) = 24x + \frac{9600}{x}, \quad x > 0,$$

$$C'(x) = 24 - \frac{9600}{x^2} = \frac{24(x^2 - 400)}{x^2} = \frac{24(x - 20)(x + 20)}{x^2}.$$

Therefore, $x = 20$ is the only critical value.

$$C''(x) = \frac{19,200}{x^3}$$

$$C''(20) = \frac{19,200}{8000} > 0. \quad \text{Therefore, } x = 20 \text{ for the}$$

minimum cost.
The dimensions of the fence are shown in the
diagram at the right.

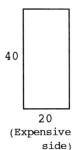

40

20
(Expensive
side)

31. Let x = number of books produced each printing. Then, the number of
    printings = $\frac{50,000}{x}$.

    Cost = $C(x)$ = cost of storage + cost of printing
    $$= \frac{x}{2} + \frac{50,000}{x}(1000), \quad x > 0$$

    [Note: $\frac{x}{2}$ is the average number in storage each day.]

    $$C'(x) = \frac{1}{2} - \frac{50,000,000}{x^2} = \frac{x^2 - 100,000,000}{2x^2} = \frac{(x + 10,000)(x - 10,000)}{2x^2}$$

    Critical value: $x = 10,000$

    $$C''(x) = \frac{100,000,000}{x^3}$$

    $$C''(10,000) = \frac{100,000,000}{(10,000)^3} > 0$$

    Thus, the minimum cost occurs when $x = 10,000$ and the number of
    printings is $\frac{50,000}{10,000} = 5$.

33. (A) Let the cost to lay the pipe on the land be 1 unit; then the cost to
    lay the pipe in the lake is 1.4 units.
    $$C(x) = \text{total cost} = (1.4)\sqrt{x^2 + 25} + (1)(10 - x), \quad 0 \le x \le 10$$
    $$= (1.4)(x^2 + 25)^{1/2} + 10 - x$$

    $$C'(x) = (1.4)\frac{1}{2}(x^2 + 25)^{-1/2}(2x) - 1$$
    $$= (1.4)x(x^2 + 25)^{-1/2} - 1$$
    $$= \frac{1.4x - \sqrt{x^2 + 25}}{\sqrt{x^2 + 25}}$$

    $C'(x) = 0$ when $1.4x - \sqrt{x^2 + 25} = 0$ or $1.96x^2 = x^2 + 25$
    $$.96x^2 = 25$$
    $$x^2 = \frac{25}{.96} = 26.04$$
    $$x = \pm 5.1$$

    Thus, the critical value is $x = 5.1$.

$$C''(x) = (1.4)(x^2 + 25)^{-1/2} + (1.4)x\left(-\frac{1}{2}\right)(x^2 + 25)^{-3/2}2x$$

$$= \frac{1.4}{(x^2 + 25)^{1/2}} - \frac{(1.4)x^2}{(x^2 + 25)^{3/2}} = \frac{35}{(x^2 + 25)^{3/2}}$$

$$C''(5.1) = \frac{35}{[(5.1)^2 + 25]^{3/2}} > 0$$

Thus, the cost will be a minimum when $x = 5.1$.
Note that: $C(0) = (1.4)\sqrt{25} + 10 = 17$

$\quad\quad\quad\quad C(5.1) = (1.4)\sqrt{51.01} + (10 - 5.1) = 14.9$

$\quad\quad\quad\quad C(10) = (1.4)\sqrt{125} = 15.65$

Thus, the absolute minimum occurs when $x = 5.1$ miles.

(B) $C(x) = (1.1)\sqrt{x^2 + 25} + (1)(10 - x)$, $0 \le x \le 10$

$$C'(x) = \frac{(1.1)x - \sqrt{x^2 + 25}}{\sqrt{x^2 + 25}}$$

$C'(x) = 0$ when $1.1x - \sqrt{x^2 + 25} = 0$ or $(1.21)x^2 = x^2 + 25$

$$.21x^2 = 25$$

$$x^2 = \frac{25}{.21} = 119.05$$

$$x = \pm 10.91$$

Critical value: $x = 10.91 > 10$, i.e., there are no critical values
on the interval $[0, 10]$. Now,

$\quad C(0) = (1.1)\sqrt{25} + 10 = 15.5$,

$C(10) = (1.1)\sqrt{125} \approx 12.30$.

Therefore, the absolute minimum occurs when $x = 10$ miles.

35. $C(t) = 30t^2 - 240t + 500$, $0 \le t \le 8$

$C'(t) = 60t - 240$; $t = 4$ is the only critical value.

$C''(t) = 60$

$C''(4) = 60 > 0$

Now, $C(0) = 500$

$\quad\quad C(4) = 30(4)^2 - 240(4) + 500 = 20$,

$\quad\quad C(8) = 30(8)^2 - 240(8) + 500 = 500$.

Thus, 4 days after a treatment, the concentration will be minimum; the minimum concentration is 20 bacteria per cm^3.

37. Let x = the number of mice ordered in each order. Then the number of orders = $\frac{500}{x}$.

$$C(x) = \text{Cost} = \frac{x}{2} \cdot 4 + \frac{500}{x}(10) \quad \begin{array}{l}[\underline{\text{Note}}: \text{Cost} = \text{cost of feeding} + \text{cost of order,} \\ \frac{x}{2} \text{ is the average number of mice at any one time.}]\end{array}$$

$$C(x) = 2x + \frac{5000}{x}, \quad 0 < x \le 500$$

$$C'(x) = 2 - \frac{5000}{x^2} = \frac{2x^2 - 5000}{x^2} = \frac{2(x^2 - 2500)}{x^2} = \frac{2(x + 50)(x - 50)}{x^2}$$

Critical value: $x = 50$ (-50 is not a critical value, since the domain of C is $x > 0$.

$$C''(x) = \frac{10,000}{x^3} \quad \text{and} \quad C''(50) = \frac{10,000}{50^3} > 0$$

Therefore, the minimum cost occurs when 50 mice are ordered each time. The total number of orders is $\frac{500}{50} = 10$.

39. $H(t) = 4t^{1/2} - 2t, \ 0 \le t \le 2$
$H'(t) = 2t^{-1/2} - 2$
Thus, $t = 1$ is the only critical value.
Now, $H(0) = 4 \cdot 0^{1/2} - 2(0) = 0,$
$\qquad H(1) = 4 \cdot 1^{1/2} - 2(1) = 2,$
$\qquad H(2) = 4 \cdot 2^{1/2} - 4 \approx 1.66.$
Therefore, $H(1)$ is the absolute maximum, and after one month the maximum height will be 2 feet.

41. $N(t) = 30 + 12t^2 - t^3, \ 0 \le t \le 8$
The rate of increase $= R(t) = N'(t) = 24t - 3t^2$, and
$R'(t) = N''(t) = 24 - 6t.$
Thus, $t = 4$ is the only critical value of $R(t)$.
Now, $R(0) = 0,$
$\qquad R(4) = 24 \cdot 4 - 3 \cdot 4^2 = 48,$
$\qquad R(8) = 24 \cdot 8 - 3 \cdot 8^2 = 0.$
Therefore, the absolute maximum value of R occurs when $t = 4$; the maximum rate of increase will occur four years from now.

EXERCISE 10-5

Things to remember:

1. THE NUMBER e

    The irrational number e is defined by
    $$e = \lim_{n \to \infty} \left(1 + \frac{1}{n}\right)^n$$
    or alternatively,
    $$e = \lim_{s \to 0} (1 + s)^{1/s}$$
    $$e = 2.7182818\ldots$$

2. CONTINUOUS COMPOUND INTEREST

    $A = Pe^{rt}$
    where P = Principal
    $\qquad r$ = Annual nominal interest rate compounded continuously
    $\qquad t$ = Time in years
    $\qquad A$ = Amount at time t

1. $A = \$1000e^{0.1t}$

When $t = 2$, $A = \$1000e^{(0.1)2} = \$1000e^{0.2} = \$1221.40$.

When $t = 5$, $A = \$1000e^{(0.1)5} = \$1000e^{0.5} = \$1628.72$.

When $t = 8$, $A = \$1000e^{(0.1)8} = \$1000e^{0.8} = \$2225.54$

3. $2 = e^{0.06t}$

Take the natural log of both sides of this equation

$\ln(e^{0.06t}) = \ln 2$

$0.06t \ln e = \ln 2$

$\qquad 0.06t = \ln 2 \qquad (\ln e = 1)$

$\qquad\quad t = \dfrac{\ln 2}{0.06} \approx 11.55$

5. $3 = e^{0.1t}$

$\ln(e^{0.1t}) = \ln 3$

$\quad 0.1t = \ln 3$

$\qquad t = \dfrac{\ln 3}{0.1} \approx 10.99$

7. $2 = e^{5r}$

$\ln(e^{5r}) = \ln 2$

$\qquad 5r = \ln 2$

$\qquad r = \dfrac{\ln 2}{5} \approx 0.14$

9.

n	$\left(1 + \dfrac{1}{n}\right)^n$
10	2.59374
100	2.70481
1000	2.71692
10,000	2.71815
100,000	2.71827
1,000,000	2.71828
10,000,000	2.71828
↓	↓
∞	$e = 2.7182818\ldots$

11. The graphs of $y_1 = \left(1 + \dfrac{1}{n}\right)^n$, $y_2 = 2.718281828 \approx e$, and $y_3 = \left(1 + \dfrac{1}{n}\right)^{n+1}$ for $0 \le n \le 20$ are given below:

13. $A = Pe^{rt} = \$20,000e^{0.12(8.5)}$

$\qquad\qquad = \$20,000e^{1.02}$

$\qquad\qquad \approx \$55,463.90$

15. $A = Pe^{rt}$

$\$20,000 = Pe^{0.07(10)} = Pe^{0.7}$

Therefore,

$P = \dfrac{\$20,000}{e^{0.7}} = \$20,000e^{-0.7}$

$\qquad\qquad \approx \$9931.71$

17. $30,000 = 20,000e^{4r}$

$\qquad e^{4r} = 1.5$

$\qquad 4r = \ln 1.5$

$\qquad r = \dfrac{\ln 1.5}{4} \approx 0.1014 \quad \text{or} \quad 10.14\%$

19. $P = 10,000e^{-0.08t}$, $0 \le t \le 50$

(A)

t	0	10	20	30	40	50
P	10,000	4493.30	2019	907.18	407.62	183.16

The graph of P is shown below.

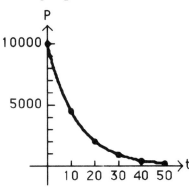

(B) $\lim\limits_{t \to \infty} 10,000e^{-0.08t} = 0$

21.
$$2P = Pe^{0.25t}$$
$$e^{0.25t} = 2$$
$$\ln(e^{0.25t}) = \ln 2$$
$$0.25t = \ln 2$$
$$t = \frac{\ln 2}{0.25} \approx 2.77 \text{ years}$$

23.
$$2P = Pe^{r(5)}$$
$$\ln(e^{5r}) = \ln 2$$
$$5r = \ln 2$$
$$r = \frac{\ln 2}{5} \approx .1386 \text{ or } = 13.86\%$$

25. $A = Pe^{rt}$. Let $A = 2P$. Then we have:
$2P = Pe^{rt}$ or $e^{rt} = 2$
Thus, $\ln e^{rt} = \ln 2$ and $rt = \ln 2$.
Therefore, $t = \dfrac{\ln 2}{r}$.

27. $2P_0 = P_0e^{0.02t}$ or $e^{0.02t} = 2$
Thus, $\ln(e^{0.02t}) = \ln 2$
and $0.02t = \ln 2$.
Therefore, $t = \dfrac{\ln 2}{0.02} \approx 34.66$ years.

29.
$$2P_0 = P_0e^{r(20)}$$
$$e^{20r} = 2$$
$$\ln(e^{20r}) = \ln 2$$
$$20r = \ln 2$$
$$r = \frac{\ln 2}{20} \approx 0.0347 \text{ or } 3.47\%$$

31.
$$Q = Q_0e^{-0.0004332t}$$
$$\frac{1}{2}Q_0 = Q_0e^{-0.0004332t}$$
$$e^{-0.0004332t} = \frac{1}{2}$$
$$\ln(e^{-0.0004332t}) = \ln\left(\frac{1}{2}\right) = \ln 1 - \ln 2$$
$$-0.0004332t = -\ln 2 \quad (\ln 1 = 0)$$
$$t = \frac{\ln 2}{0.0004332}$$
$$\approx \frac{0.6931}{0.0004332} \approx 1599.95$$

Thus, the half-life of radium is approximately 1600 years.

33.

$$Q = Q_0 e^{rt} \quad (r < 0)$$

$$\frac{1}{2} Q_0 = Q_0 e^{r(30)}$$

$$e^{30r} = \frac{1}{2}$$

$$\ln(e^{30r}) = \ln\left(\frac{1}{2}\right) = \ln 1 - \ln 2$$

$$30r = -\ln 2 \quad (\ln 1 = 0)$$

$$r = \frac{-\ln 2}{30} \approx \frac{-0.6931}{30}$$

$$\approx -0.0231$$

Thus, the continuous compound rate of decay of the cesium isotope is approximately -0.0231.

35.

$$A = Pe^{rt}$$

$$1.68 \times 10^{14} = 5 \times 10^9 e^{0.02t}$$

$$e^{0.02t} = \frac{1.68 \times 10^{14}}{5 \times 10^9} = 33{,}600$$

$$\ln(e^{0.02t}) = \ln(33{,}600)$$

$$0.02t = \ln 33{,}600$$

$$t = \frac{\ln 33{,}600}{0.02} \approx 521.11$$

Thus, there will be one square yard of land per person in approximately 521 years.

EXERCISE 10-6

Things to remember:

1. DERIVATIVES OF THE NATURAL LOGARITHMIC AND EXPONENTIAL FUNCTIONS

$$D_x \ln x = \frac{1}{x} \qquad D_x e^x = e^x$$

1. $f(x) = 6e^x - 7 \ln x$

$$f'(x) = 6D_x e^x - 7D_x \ln x = 6e^x - 7\left(\frac{1}{x}\right) = 6e^x - \frac{7}{x}$$

3. $f(x) = 2x^e + 3e^x$

$$f'(x) = 2D_x x^e + 3D_x e^x = 2ex^{e-1} + 3e^x$$

[<u>Note</u>: $e \approx 2.71828$ is a constant and so we use the power rule on the first term.]

5. $f(x) = \ln x^5 = 5 \ln x$ (Property of logarithms)

$$f'(x) = 5D_x \ln x = 5\left(\frac{1}{x}\right) = \frac{5}{x}$$

7. $f(x) = (\ln x)^2$

$$f'(x) = 2(\ln x)D_x \ln x \quad \text{(Power rule for functions)}$$

$$= 2(\ln x)\frac{1}{x} = \frac{2 \ln x}{x}$$

9. $f(x) = x^4 \ln x$

$$f'(x) = x^4 D_x \ln x + \ln x \, D_x x^4 \quad \text{(Product rule)}$$

$$= x^4\left(\frac{1}{x}\right) + (\ln x)4x^3 = x^3 + 4x^3 \ln x = x^3(1 + 4 \ln x)$$

11. $f(x) = x^3 e^x$

   $f'(x) = x^3 D_x e^x + e^x D_x x^3$ (Product rule)

        $= x^3 e^x + e^x 3x^2 = x^2 e^x (x + 3)$

13. $f(x) = \dfrac{e^x}{x^2 + 9}$

   $f'(x) = \dfrac{(x^2 + 9) D_x e^x - e^x D_x (x^2 + 9)}{(x^2 + 9)^2}$ (Quotient rule)

        $= \dfrac{(x^2 + 9) e^x - e^x (2x)}{(x^2 + 9)^2} = \dfrac{e^x (x^2 - 2x + 9)}{(x^2 + 9)^2}$

15. $f(x) = \dfrac{\ln x}{x^4}$

   $f'(x) = \dfrac{x^4 D_x \ln x - \ln x\, D_x x^4}{(x^4)^2}$ (Quotient rule)

        $= \dfrac{x^4 \left(\dfrac{1}{x}\right) - (\ln x) 4x^3}{x^8} = \dfrac{1 - 4 \ln x}{x^5}$

17. $f(x) = (x + 2)^3 \ln x$

   $f'(x) = (x + 2)^3 D_x \ln x + (\ln x) D_x (x + 2)^3$

        $= (x + 2)^3 \left(\dfrac{1}{x}\right) + (\ln x)[3(x + 2)^2 (1)]$

        $= 3(x + 2)^2 \ln x + \dfrac{(x + 2)^3}{x} = (x + 2)^2 \left[3 \ln x + \dfrac{x + 2}{x}\right]$

19. $f(x) = (x + 1)^3 e^x$

   $f'(x) = (x + 1)^3 D_x e^x + e^x D_x (x + 1)^3$

        $= (x + 1)^3 e^x + e^x (3)(x + 1)^2 (1)$

        $= (x + 1)^2 e^x [x + 1 + 3] = (x + 1)^2 (x + 4) e^x$

21. $f(x) = \dfrac{x^2 + 1}{e^x}$

   $f'(x) = \dfrac{e^x D_x (x^2 + 1) - (x^2 + 1) D_x e^x}{(e^x)^2} = \dfrac{e^x (2x) - (x^2 + 1) e^x}{e^{2x}} = \dfrac{2x - x^2 - 1}{e^x}$

23. $f(x) = x(\ln x)^3$

   $f'(x) = x D_x (\ln x)^3 + (\ln x)^3 D_x x$

        $= x(3)(\ln x)^2 \left(\dfrac{1}{x}\right) + (\ln x)^3 (1) = (\ln x)^2 [3 + \ln x]$

25. $f(x) = (4 - 5e^x)^3$

   $f'(x) = 3(4 - 5e^x)^2 (-5e^x) = -15e^x (4 - 5e^x)^2$

27. $f(x) = \sqrt{1 + \ln\ x} = (1 + \ln\ x)^{1/2}$

$$f'(x) = \frac{1}{2}(1 + \ln\ x)^{-1/2}\left(\frac{1}{x}\right)$$

$$= \frac{1}{2x(1 + \ln\ x)^{1/2}} = \frac{1}{2x\sqrt{1 + \ln\ x}}$$

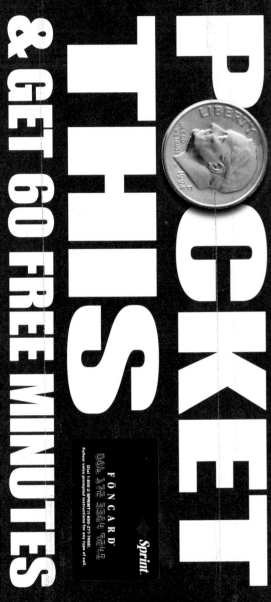

$- D_x e^x = xe^x + e^x - e^x = xe^x$

$\ln\ x\ D_x 2x^2 - D_x x^2 = 2x^2\left(\frac{1}{x}\right) + 4x\ \ln\ x - 2x = 4x\ \ln\ x$

$x = 1$ has an equation of the form

$1) = e$, and $m = f'(1) = e$. Thus, we have:

$y = ex$

$x = e$ has an equation of the form

$e) = \ln\ e = 1$, and $m = f'(e) = \frac{1}{e}$. Thus, we have:

$y = \frac{1}{e}x$

$x > 0$

$- (\ln\ x)D_x x$

$\ln\ x)(1) = 3 - \ln\ x, \ x > 0$

$x) = 3 - \ln\ x = 0$

$\ln\ x = 3$

$x = e^3$

nly critical value of f on $(0,\ \infty)$.

$\ln\ x) = -\frac{1}{x}$

ximum value at $x = e^3$, and $f(e^3) = 4e^3 - e^3\ \ln\ e^3$

.086 is the absolute maximum of f.

39. $f(x) = \dfrac{e^x}{x}$, $x > 0$

$$f'(x) = \frac{xD_x e^x - e^x D_x x}{x^2} = \frac{xe^x - e^x(1)}{x^2} = \frac{e^x(x-1)}{x^2}, \quad x > 0$$

Critical values: $f'(x) = \dfrac{e^x(x-1)}{x^2} = 0$

$$e^x(x-1) = 0$$

$$x = 1 \ [\underline{\text{Note}}: e^x \neq 0 \text{ for all } x.]$$

Thus, $x = 1$ is the only critical value of f on $(0, \infty)$.
Sign chart for f' [<u>Note</u>: This approach is a litle easier than calculating $f''(x)$]:

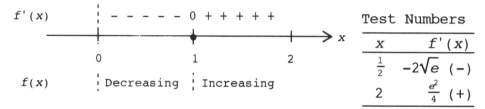

x	$f'(x)$
$\frac{1}{2}$	$-2\sqrt{e}\ (-)$
2	$\frac{e^2}{4}\ (+)$

By the first derivative test, f has a minimum value at $x = 1$; $f(1) = \dfrac{e}{1}$
$= e \approx 2.718$ is the absolute minimum of f.

41. $f(x) = \dfrac{1 + 2\ln x}{x}$, $x > 0$

$$f'(x) = \frac{xD_x(1 + 2\ln x) - (1 + 2\ln x)D_x x}{x^2} = \frac{x\left(\frac{2}{x}\right) - (1 + 2\ln x)(1)}{x^2}$$

$$= \frac{1 - 2\ln x}{x^2}, \quad x > 0$$

Critical values: $f'(x) = \dfrac{1 - 2\ln x}{x^2} = 0$

$$1 - 2\ln x = 0$$

$$\ln x = \frac{1}{2}$$

$$x = e^{1/2} = \sqrt{e} \approx 1.65$$

Sign chart for f':

f'(x) | + + + + + 0 - - - - - → x

0   1 √e   2   3

f(x) | Increasing | Decreasing

Test Numbers

x	$f'(x)$
1	$1\ (+)$
e	$-\frac{1}{e^2}\ (-)$

By the first derivative test, f has a maximum value at $x = \sqrt{e}$;

$$f(\sqrt{e}) = \frac{1 + 2\ln\sqrt{e}}{\sqrt{e}} = \frac{1 + 2\left(\frac{1}{2}\right)}{\sqrt{e}} = \frac{2}{\sqrt{e}} \approx 1.213 \text{ is the absolute maximum of } f.$$

43. $f(x) = 1 - e^x$

Step 1. Use $f(x)$:
(A) Domain: All real numbers, $(-\infty, \infty)$.

(B) Intercepts: y intercept: $f(0) = 1 - e^0 = 0$
x intercept: $1 - e^x = 0$
$$e^x = 1$$
$$x = 0$$

(C) Asymptotes:
Horizontal asymptote: $\lim\limits_{x \to -\infty} (1 - e^x) = 1 - \lim\limits_{x \to -\infty} e^x = 1 - 0 = 1$

$$[\lim\limits_{x \to \infty}(1 - e^x) = -\infty]$$

Thus, $y = 1$ is a horizontal asymptote.
Vertical asymptotes: There are no vertical asymptotes.

Step 2. Use $f'(x)$:
$f'(x) = -e^x$
Critical values: $f'(x) = -e^x$ is continuous and nonzero (negative) for all x; there are no critical values.
Partition numbers: There are no partition numbers. Since $f'(x) < 0$ for all x, f is decreasing on $(-\infty, \infty)$; f has no local extrema.

Step 3. Use $f''(x)$:
$f''(x) = -e^x < 0$ for all x
Thus, the graph of f is concave downward on $(-\infty, \infty)$.

Step 4. Sketch the graph of f:

x	$f(x)$
0	0
1	$1 - e \approx -1.718$
-1	$1 - \dfrac{1}{e}$

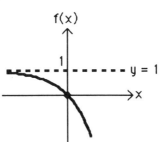

45. $f(x) = x - \ln x$

Step 1. Use $f(x)$:
(A) Domain: All positive real numbers, $(0, \infty)$.
[Note: $\ln x$ is defined only for positive numbers.]

(B) Intercepts: y intercept: There is no y intercept; $f(0) = 0 - \ln(0)$ is not defined.
x intercept: $x - \ln x = 0$
$$\ln x = x$$
Since the graph of $y = \ln x$ is below the graph of $y = x$, there are no solutions to this equation; there are no x intercepts.

(C) Asymptotes:

Horizontal asymptote: None

Vertical asymptotes: Since $\lim\limits_{x \to 0^+} \ln x = -\infty$, $\lim\limits_{x \to 0^+} (x - \ln x) = \infty$.

Thus, $x = 0$ is a vertical asymptote for $f(x) = x - \ln x$.

Step 2. Use $f'(x)$:

$f'(x) = 1 - \dfrac{1}{x} = \dfrac{x - 1}{x}$, $x > 0$

Critical values: $\dfrac{x - 1}{x} = 0$

$x = 1$

Partition numbers: $x = 1$

Sign chart for $f'(x) = \dfrac{x - 1}{x}$:

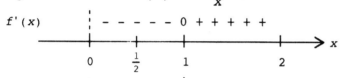

Test Numbers	
x	$f'(x)$
$\frac{1}{2}$	-1 $(-)$
2	$\frac{1}{2}$ $(+)$

$f(x)$ Decreasing   Increasing

Thus, f is decreasing on $(0, 1)$ and increasing on $(1, \infty)$; f has a local minimum at $x = 1$.

Step 3. Use $f''(x)$:

$f''(x) = \dfrac{1}{x^2}$, $x > 0$

Thus, $f''(x) > 0$ and the graph of f is concave upward on $(0, \infty)$.

Step 4. Sketch the graph of f:

x	$f(x)$
0.1	≈ 2.4
1	1
10	≈ 7.7

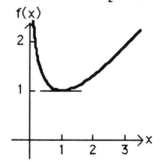

47. $f(x) = (3 - x)e^x$

Step 1. Use $f(x)$:

(A) Domain: All real numbers, $(-\infty, \infty)$.

(B) Intercepts: y intercept: $f(0) = (3 - 0)e^0 = 3$

x intercept: $(3 - x)e^x = 0$

$3 - x = 0$

$x = 3$

(C) Asymptotes:

Horizontal asymptote: Consider the behavior of f as $x \to \infty$ and as $x \to -\infty$.

Using the following tables,

x	-1	-10	-20
$f(x)$	1.47	0.00059	0.000000047

x	5	10
$f(x)$	-296.83	$-154,185.26$

we conclude that $\lim\limits_{x \to -\infty} f(x) = 0$ and $\lim\limits_{x \to \infty} f(x)$ does not exist. Because of the first limit, $y = 0$ is a horizontal asymptote.

Vertical asymptotes: There are no vertical asymptotes.

Step 2. Use $f'(x)$:

$f'(x) = (3 - x)e^x + e^x(-1) = (2 - x)e^x$

Critical values: $(2 - x)e^x = 0$

$$x = 2 \quad [\underline{\text{Note}}: e^x > 0]$$

Partition numbers: $x = 2$

Sign chart for f':

```
 f'(x)      + + + + + 0 - - - - -
       ┼────┼────●────┼──────────→ x
       0    1    2    3
 
 f(x)        Increasing ┊ Decreasing
```

Test Numbers	
x	$f'(x)$
0	2 (+)
3	$-e^3$ (−)

Thus, f is increasing on $(-\infty, 2)$ and decreasing on $(2, \infty)$; f has a local maximum at $x = 2$.

Step 3. Use $f''(x)$:

$f''(x) = (2 - x)e^x + e^x(-1) = (1 - x)e^x$

Partition number for f'': $x = 1$

Sign chart for f'':

```
 f''(x)       + + + + 0 - - - -
       ────────┼───●───┼───────→ x
               0   1   2
 Graph         Concave ┊ Concave
 of f          Upward  ┊ Downward
```

Test Numbers	
x	$f''(x)$
0	1 (+)
2	$-e^2$ (−)

Thus, the graph of f is concave upward on $(-\infty, 1)$ and concave downward on $(1, \infty)$; the graph has an inflection point at $x = 1$.

Step 4. Sketch the graph of f:

x	$f(x)$
0	3
2	$e^2 \approx 7.4$
3	0

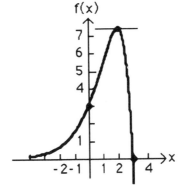

49. $f(x) = x^2 \ln x$.

Step 1. Use $f(x)$:

(A) Domain: All positive numbers, $(0, \infty)$.

(B) Intercepts: y intercept: There is no y intercept.

x intercept: $x^2 \ln x = 0$

$\ln x = 0$

$x = 1$

(C) Asymptotes: Consider the behavior of f as $x \to \infty$ and as $x \to 0$. It is
clear that $\lim\limits_{x \to \infty} f(x)$ does not exist; f is unbounded as x approaches ∘

The following table indicates that f approaches 0 as x approaches 0.

x	1	0.1	0.01	0.001
$f(x)$	0	−0.023	−0.00046	−0.000007

Thus, there are no vertical or horizontal asymptotes.

Step 2. Use $f'(x)$:

$f'(x) = x^2\left(\dfrac{1}{x}\right) + (\ln x)(2x) = x(1 + 2 \ln x)$

Critical values: $x(1 + 2 \ln x) = 0$

$1 + 2 \ln x = 0$ [Note: $x > 0$]

$\ln x = -\dfrac{1}{2}$

$x = e^{-1/2} = \dfrac{1}{\sqrt{e}} \approx 0.6065$

Partition number: $x = \dfrac{1}{\sqrt{e}} \approx 0.6065$

Sign chart for f':

Test Numbers	
x	$f'(x)$
$\frac{1}{2}$	≈ -0.19 (−)
1	1 (+)

$f'(x)$ $- - 0 + + + + +$ plotted at 0, $\frac{1}{\sqrt{e}}$, 1, 2 along x

$f(x)$ Decreasing ¦ Increasing

Thus, f is decreasing on $(0, e^{-1/2})$ and increasing on $(e^{-1/2}, \infty)$; f has
a local minimum at $x = e^{-1/2}$.

Step 3. Use $f''(x)$:

$f''(x) = x\left(\dfrac{2}{x}\right) + (1 + 2 \ln x) = 3 + 2 \ln x$

Partition number for f'': $3 + 2 \ln x = 0$

$\ln x = -\dfrac{3}{2}$

$x = e^{-3/2} \approx 0.2231$

Sign chart for f'':

Test Numbers	
x	$f''(x)$
$\frac{1}{10}$	≈ -1.61 (−)
1	3 (+)

$f''(x)$ $- - - 0 + + + +$ plotted at 0, $e^{-3/2}$, 1 along x

Graph
of f Concave ¦ Concave
              Downward ¦ Upward

Thus, the graph of f is concave downward on $(0, e^{-3/2})$ and concave upward on $(e^{-3/2}, \infty)$; the graph has an inflection point at $x = e^{-3/2}$.

Step 4. Sketch the graph of f:

x	$f(x)$
$e^{-3/2}$	≈ -0.075
$e^{-1/2}$	≈ -0.18
1	0

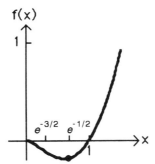

51. $f(x) = e^x - 2x^2 \qquad -\infty < x < \infty$

$f'(x) = e^x - 4x$

The graphs of $f(x)$ and $f'(x)$ are shown at the right.

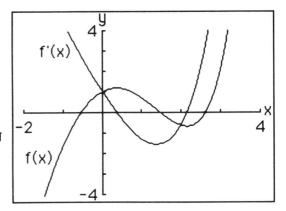

Critical values:

   Solve $f'(x) = e^x - 4x = 0$

To two decimal places, $x = 0.36$ and $x = 2.15$

Increasing/Decreasing: $f(x)$ is increasing on $(-\infty, 0.36)$ and on $(2.15, \infty)$; $f(x)$ is decreasing on $(0.36, 2.15)$

Local extrema: $f(x)$ has a local maximum at $x = 0.36$ and a local minimum at $x = 2.15$

53. $f(x) = 20 \ln x - e^x \qquad 0 < x < \infty$

$f'(x) = \dfrac{20}{x} - e^x$

The graphs of $f(x)$ and $f'(x)$ are shown at the right.

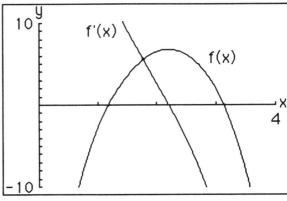

Critical values:

   Solve $f'(x) = \dfrac{20}{x} - e^x = 0$

To two decimal places, $x = 2.21$

Increasing/Decreasing: $f(x)$ is increasing on $(0, 2.21)$ and decreasing on $(2.21, \infty)$

Local extrema: $f(x)$ has a local maximum at $x = 2.21$

55. Demand: $p = 5 - \ln x, \; 5 \le x \le 50$

Revenue: $R = xp = x(5 - \ln x) = 5x - x \ln x$

Cost: $C = x(1) = x$

Profit = Revenue - Cost: $P = 5x - x \ln x - x$

$$\text{or} \quad P(x) = 4x - x \ln x$$

$$P'(x) = 4 - x\left(\frac{1}{x}\right) - \ln x$$

$$= 3 - \ln x$$

Critical value(s): $P'(x) = 3 - \ln x = 0$

$$\ln x = 3$$
$$x = e^3$$

$$P''(x) = -\frac{1}{x} \quad \text{and} \quad P''(e^3) = -\frac{1}{e^3} < 0.$$

Since $x = e^3$ is the only critical value and $P''(e^3) < 0$, the maximum weekly profit occurs when $x = e^3 \approx 20.09$ and the price $p = 5 - \ln(e^3) = 2$. Thus, the hot dogs should be sold at \$2.

57. Cost: $C(x) = 600 + 100x - 100 \ln x, \quad x \geq 1$

Average cost: $\overline{C}(x) = \frac{600}{x} + 100 - \frac{100}{x} \ln x$

$$\overline{C}'(x) = \frac{-600}{x^2} - \frac{100}{x^2} + \frac{100 \ln x}{x^2} = \frac{-700 + 100 \ln x}{x^2}, \quad x \geq 1$$

Critical value(s): $\overline{C}'(x) = \frac{-700 + 100 \ln x}{x^2} = 0$

$$-700 + 100 \ln x = 0$$
$$\ln x = 7$$
$$x = e^7$$

$$\overline{C}''(x) = \frac{x^2 \dfrac{100}{x} - (-700 + 100 \ln x)(2x)}{x^4}$$
$$= \frac{100x + 1400x - 200x \ln x}{x^4} = \frac{1500 - 200 \ln x}{x^3}$$

$$\overline{C}''(e^7) = \frac{1500 - 200 \ln(e^7)}{e^{21}} = \frac{100}{e^{21}} > 0$$

Since $x = e^7$ is the only critical value and $\overline{C}''(e^7) > 0$, the minimum average cost is

$$\overline{C}(e^7) = \frac{600}{e^7} + 100 - \frac{100}{e^7} \ln(e^7) = \frac{600}{e^7} + 100 - \frac{700}{e^7} = 100 - \frac{100}{e^7} \approx 99.91$$

Thus, the minimal average cost is approximately \$99.91.

59. Demand: $p = 10e^{-x} = \frac{10}{e^x}, \quad 0 \leq x \leq 2$

Revenue: $R(x) = xp = 10xe^{-x} = \frac{10x}{e^x}$

(A) $R'(x) = \frac{e^x(10) - 10xe^x}{e^{2x}} = \frac{10 - 10x}{e^x}, \quad 0 \leq x \leq 2$

Critical value(s): $\frac{10 - 10x}{e^x} = 0$

$$10 - 10x = 0$$
$$x = 1$$

$$R''(x) = \frac{e^x(-10) - (10 - 10x)e^x}{e^{2x}} = \frac{-20 + 10x}{e^x}$$

$$R''(1) = -\frac{10}{e} < 0$$

Now, $R(0) = 0$

$$R(1) = \frac{10}{e} \approx 3.68 \quad \text{Absolute maximum}$$

$$R(2) = \frac{20}{e^2} \approx 2.71$$

Thus, the maximum weekly revenue occurs at price $p = \frac{10}{e} \approx \$3.68$.

The maximum weekly revenue is $R(1) = 3.68$ thousand dollars, or $\$3680$.

(B) The sign chart for R' is:

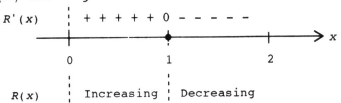

Test Numbers

x	$f'(x)$
0	10 (+)
2	$-\frac{10}{e^2}$ (−)

Thus, R is increasing on $(0, 1)$ and decreasing on $(1, 2)$; the maximum value of R occurs at $x = 1$, as noted in (A).

$R''(x) = 10e^{-x}(x - 2) < 0$ on $(0, 2)$
Thus, the graph of R is concave downward on $(0, 2)$. The graph is shown at the right.

x	$R(x)$
0	0
1	3.68
2	2.71

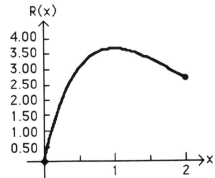

61. $P(x) = 17.5(1 + \ln x)$, $10 \le x \le 100$

$$P'(x) = \frac{17.5}{x}$$

$$P'(40) = \frac{17.5}{40} \approx 0.44$$

$$P'(90) = \frac{17.5}{90} \approx 0.19$$

Thus, at the 40 pound weight level, blood pressure would increase at the rate of 0.44 mm of mercury per pound of weight gain; at the 90 pound weight level, blood pressure would increase at the rate of 0.19 mm of mercury per pound of weight gain.

63. $C(t) = 4.35e^{-t} = \frac{4.35}{e^t}$, $0 \le t \le 5$

(A) $C'(t) = \frac{-4.35e^t}{e^{2t}} = \frac{-4.35}{e^t} = -4.35e^{-t}$

$C'(1) = -4.35e^{-1} \approx -1.60$
$C'(4) = -4.35e^{-4} \approx -0.08$
Thus, after one hour, the concentration is decreasing at the rate of 1.60 mg/ml per hour; after four hours, the concentration is decreasing at the rate of 0.08 mg/ml per hour.

(B) $C'(t) = -4.35e^{-t} < 0$ on $(0, 5)$

Thus, C is decreasing on $(0, 5)$; there are no local extrema.

$$C''(t) = \frac{4.35e^t}{e^{2t}} = \frac{4.35}{e^t} = 4.35e^{-t} > 0 \text{ on } (0, 5)$$

Thus, the graph of C is concave upward on $(0, 5)$. The graph of C is shown at the right.

t	$C(t)$
0	4.35
1	1.60
4	0.08
5	0.03

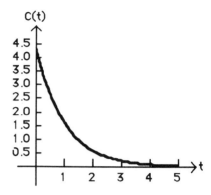

65. $R = k \ln(S/S_0)$

$\quad = k[\ln S - \ln S_0]$

$\dfrac{dR}{dS} = \dfrac{k}{S}$

EXERCISE 10-7

Things to remember:

1. COMPOSITE FUNCTIONS

    A function h is the composite of functions f and g (in this order) if

    $\quad h(x) = f[g(x)]$.

    The domain of h is the set of all numbers x in the domain of g such that $g(x)$ is in the domain of f.

2. CHAIN RULE

    If $y = f(u)$ and $u = g(x)$, define the composite function
    $\quad y = h(x) = f[g(x)]$.
    Then
    $$\frac{dy}{dx} = \frac{dy}{du} \cdot \frac{du}{dx}$$
    provided $\dfrac{dy}{du}$ and $\dfrac{du}{dx}$ exist.

    Or, equivalently, $h'(x) = f'[g(x)]g'(x)$ provided $f'[g(x)]$ and $g'(x)$ exist.

$\underline{3}$. GENERAL DERIVATIVE RULES

    (a) $D_x[f(x)]^n = n[f(x)]^{n-1}f'(x)$

    (b) $D_x \ln[f(x)] = \dfrac{1}{f(x)} \cdot f'(x)$

    (c) $D_x e^{f(x)} = e^{f(x)} \cdot f'(x)$

$\underline{4}$. OTHER LOGARITHMIC AND EXPONENTIAL FUNCTIONS

$$D_x[\log_b x] = \frac{1}{\ln b} \cdot \frac{1}{x}, \quad b \neq e$$
$$D_x[b^x] = b^x \ln b, \quad b \neq e$$

---

1. Let $u = g(x) = 2x + 5$ and $f(u) = u^3$. Then $y = f(u) = u^3$.

3. Let $u = g(x) = 2x^2 + 7$ and $f(u) = \ln u$. Then $y = f(u) = \ln u$.

5. Let $u = g(x) = x^2 - 2$ and $f(u) = e^u$. Then $y = f(u) = e^u$.

7. $y = u^2$, $u = 2 + e^x$. Thus, $y = (2 + e^x)^2$ and $\dfrac{dy}{dx} = \dfrac{dy}{du} \cdot \dfrac{du}{dx} = 2u(e^x)$
$$= 2e^x(2 + e^x).$$

9. $y = e^u$, $u = 2 - x^4$. Thus, $y = e^{2-x^4}$ and $\dfrac{dy}{dx} = \dfrac{dy}{du} \cdot \dfrac{du}{dx} = e^u(-4x^3)$
$$= -4x^3 e^{2-x^4}.$$

11. $y = \ln u$, $u = 4x^5 - 7$, so $y = \ln(4x^5 - 7)$ and
$$\frac{dy}{dx} = \frac{dy}{du} \cdot \frac{du}{dx} = \frac{1}{u}(20x^4) = \frac{1}{4x^5 - 7}(20x^4) = \frac{20x^4}{4x^5 - 7}$$

13. $D_x \ln(x - 3) = \dfrac{1}{x - 3}(1)$ (using $\underline{3}$b)
$$= \frac{1}{x - 3}$$

15. $D_t \ln(3 - 2t) = \dfrac{1}{3 - 2t}(-2)$ (using $\underline{3}$b)
$$= \frac{-2}{3 - 2t}$$

17. $D_x 3e^{2x} = 3D_x e^{2x} = 3e^{2x}(2)$ (using $\underline{3}$c)
$$= 6e^{2x}$$

19. $D_t 2e^{-4t} = 3D_t e^{-4t} = 2e^{-4t}(-4) = -8e^{-4t}$

21. $D_x 100e^{-0.03x} = 100D_x e^{-0.03x} = 100e^{-0.03x}(-0.03) = -3e^{-0.03x}$

23. $D_x \ln(x + 1)^4 = D_x 4 \ln(x + 1) = 4D_x \ln(x + 1) = 4\dfrac{1}{x + 1}(1) = \dfrac{4}{x + 1}$

25. $D_x(2e^{2x} - 3e^x + 5) = 2D_x e^{2x} - 3D_x e^x + D_x 5 = 2e^{2x}(2) - 3e^x = 4e^{2x} - 3e^x$

27. $D_x e^{3x^2-2x} = e^{3x^2-2x}(6x - 2) = (6x - 2)e^{3x^2-2x}$

29. $D_t \ln(t^2 + 3t) = \dfrac{1}{t^2 + 3t}(2t + 3) = \dfrac{2t + 3}{t^2 + 3t}$

31. $D_x \ln(x^2 + 1)^{1/2} = D_x \dfrac{1}{2}\ln(x^2 + 1) = \dfrac{1}{2}D_x \ln(x^2 + 1)$

$$= \dfrac{1}{2}\left(\dfrac{1}{x^2 + 1}\right)(2x) = \dfrac{x}{x^2 + 1}$$

33. $D_t[\ln(t^2 + 1)]^4 = 4[\ln(t^2 + 1)]^3 \dfrac{1}{t^2 + 1}(2t) = \dfrac{8t}{t^2 + 1}[\ln(t^2 + 1)]^3$

35. $D_x(e^{2x} - 1)^4 = 4(e^{2x} - 1)^3[e^{2x}(2)] = 8e^{2x}(e^{2x} - 1)^3$

37. $D_x \dfrac{e^{2x}}{x^2 + 1} = \dfrac{(x^2 + 1)D_x e^{2x} - e^{2x}D_x(x^2 + 1)}{(x^2 + 1)^2} = \dfrac{(x^2 + 1)e^{2x}(2) - e^{2x}(2x)}{(x^2 + 1)^2}$

$$= \dfrac{2e^{2x}(x^2 - x + 1)}{(x^2 + 1)^2}$$

39. $D_x(x^2 + 1)e^{-x} = (x^2 + 1)D_x e^{-x} + e^{-x}D_x(x^2 + 1)$

$$= (x^2 + 1)e^{-x}(-1) + e^{-x}(2x) = e^{-x}(2x - x^2 - 1)$$

41. $D_x e^{-x} \ln x = e^{-x}D_x \ln x + \ln x \, D_x e^{-x} = e^{-x}\left(\dfrac{1}{x}\right) + (\ln x)(e^{-x})(-1)$

$$= \dfrac{e^{-x}}{x} - e^{-x} \ln x$$

43. $D_x \dfrac{1}{\ln(1 + x^2)} = D_x[\ln(1 + x^2)]^{-1} = -1[\ln(1 + x^2)]^{-2}D_x \ln(1 + x^2)$

$$= -[\ln(1 + x^2)]^{-2}\dfrac{1}{1 + x^2}(2x) = \dfrac{-2x}{(1 + x^2)[\ln(1 + x^2)]^2}$$

45. $D_x \sqrt[3]{\ln(1 - x^2)} = D_x[\ln(1 - x^2)]^{1/3} = \dfrac{1}{3}[\ln(1 - x^2)]^{-2/3}D_x \ln(1 - x^2)$

$$= \dfrac{1}{3}[\ln(1 - x^2)]^{-2/3}\dfrac{1}{1 - x^2}(-2x) = \dfrac{-2x}{3(1 - x^2)[\ln(1 - x^2)]^{2/3}}$$

47. $f(x) = 1 - e^{-x}$

Step 1. Use $f(x)$:

(A) Domain: All real numbers, $(-\infty, \infty)$.

(B) Intercepts: y intercept: $f(0) = 1 - e^{-0} = 0$

$\qquad\qquad\qquad x$ intercept: $1 - e^{-x} = 0$

$$e^{-x} = 1$$

$$x = 0$$

(C) Asymptotes:

Horizontal asymptote: $\displaystyle\lim_{x \to \infty}(1 - e^{-x}) = \lim_{x \to \infty}\left(1 - \dfrac{1}{e^x}\right) = 1$

$\qquad\qquad\qquad\quad \displaystyle\lim_{x \to -\infty}(1 - e^{-x})$ does not exist.

$\qquad\qquad\qquad\quad y = 1$ is a horizontal asymptote.

Vertical asymptotes: There are no vertical asymptotes.

Step 2. Use $f'(x)$:

$f'(x) = -e^{-x}(-1) = e^{-x}$

Since $e^{-x} > 0$ for all x, f is increasing on $(-\infty, \infty)$; there are no local extrema.

Step 3. Use $f''(x)$:

$f''(x) = e^{-x}(-1) = -e^{-x}$

Since $-e^{-x} < 0$ for all x, the graph of f is concave downward on $(-\infty, \infty)$.

Step 4. Sketch the graph of f:

x	$f(x)$
0	0
-1	≈ -1.72
1	≈ 0.63

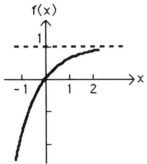

49. $f(x) = \ln(1 - x)$

Step 1. Use $f(x)$:

(A) Domain: All real numbers x such that $1 - x > 0$, i.e., $x < 1$ or $(-\infty, 1)$.

(B) Intercepts: y intercept: $f(0) = \ln(1 - 0) = \ln 1 = 0$

$\quad\quad\quad\quad\quad\quad x$ intercepts: $\ln(1 - x) = 0$
$\quad\quad\quad\quad\quad\quad\quad\quad\quad\quad\quad 1 - x = 1$
$\quad\quad\quad\quad\quad\quad\quad\quad\quad\quad\quad\quad\quad x = 0$

(C) Asymptotes:
$\quad$ Horizontal asymptote: $\lim\limits_{x \to -\infty} f(x) = \lim\limits_{x \to -\infty} \ln(1 - x)$ does not exist.

$\quad\quad\quad\quad\quad\quad\quad\quad\quad\quad$ Thus, there are no horizontal asymptotes.

$\quad$ Vertical asymptote: From the table,

x	0.9	0.99	0.99999	0.9999999	$\to$ 1
$f(x)$	-2.30	-4.61	-11.51	-16.12	$\to -\infty$

We conclude that $x = 1$ is a vertical asymptote.

Step 2. Use $f'(x)$:

$f'(x) = \dfrac{1}{1 - x}(-1),\ x < 1$

$\quad\quad = \dfrac{1}{x - 1}$

Now, $f'(x) = \dfrac{1}{x - 1} < 0$ on $(-\infty, 1)$.

Thus, f is decreasing on $(-\infty, 1)$; there are no critical values and no local extrema.

Step 3. Use $f''(x)$:

$f'(x) = (x - 1)^{-1}$

$f''(x) = -1(x - 1)^{-2} = \dfrac{-1}{(x - 1)^2}$

Since $f''(x) = \dfrac{-1}{(1-x)^2} < 0$ on $(-\infty, 1)$, the graph of f is concave downward on $(-\infty, 1)$; there are no inflection points.

Step 4. Sketch the graph of f:

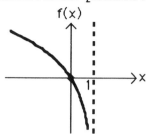

x	$f(x)$
0	0
-2	$\approx$ 1.10
.9	$\approx$ -2.30

51. $f(x) = e^{-(1/2)x^2}$

Step 1. Use $f(x)$:

(A) Domain: All real numbers, $(-\infty, \infty)$.

(B) Intercepts: y intercept: $f(0) = e^{-(1/2)0} = e^0 = 1$

   x intercepts: Since $e^{-(1/2)x^2} \neq 0$ for all x, there are no x intercepts.

(C) Asymptotes: $\displaystyle\lim_{x \to \infty} f(x) = \lim_{x \to \infty} e^{-(1/2)x^2} = \lim_{x \to \infty} \dfrac{1}{e^{(1/2)x^2}} = 0$

   $\displaystyle\lim_{x \to -\infty} f(x) = \lim_{x \to -\infty} e^{-(1/2)x^2} = \lim_{x \to -\infty} \dfrac{1}{e^{(1/2)x^2}} = 0$

Thus, $y = 0$ is a horizontal asymptote.

Since
$f(x) = e^{-(1/2)x^2} = \dfrac{1}{e^{(1/2)x^2}}$ and $g(x) = e^{(1/2)x^2} \neq 0$ for all x, there are no vertical asymptotes.

Step 2. Use $f'(x)$:

$f'(x) = e^{-(1/2)x^2}(-x) = -xe^{-(1/2)x^2}$
Critical values: $-xe^{-(1/2)x^2} = 0$
   $x = 0$
Partition numbers: $x = 0$
Sign chart for f':

	Test Numbers

x	$f'(x)$
-1	$\frac{1}{e^{1/2}}$ (+)
1	$-\frac{1}{e^{1/2}}$ (-)

$f'(x)$ $\quad$ + + + + + 0 - - - - -
(number line with -1, 0, 1)
$f(x)$ $\quad$ Increasing | Decreasing

Thus, f is increasing on $(-\infty, 0)$ and decreasing on $(0, \infty)$; f has a local maximum at $x = 0$.

Step 3. Use $f''(x)$:

$$f''(x) = -xe^{-(1/2)x^2}(-x) - e^{-(1/2)x^2}$$
$$= e^{-(1/2)x^2}(x^2 - 1) = e^{-(1/2)x^2}(x - 1)(x + 1)$$

Partition numbers for f'': $e^{-(1/2)x^2}(x - 1)(x + 1) = 0$
$$(x - 1)(x + 1) = 0$$
$$x = -1, \ 1$$

Sign chart for f'':

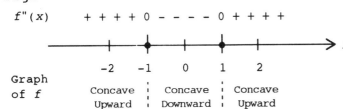

Test Numbers

x	$f''(x)$
-2	$\frac{3}{e^2}$ (+)
0	-1 (−)
2	$\frac{3}{e^2}$ (+)

Thus, the graph of f is concave upward on $(-\infty, -1)$ and on $(1, \infty)$; the graph of f is concave downward on $(-1, 1)$; the graph has inflection points at $x = -1$ and at $x = 1$.

Step 4. Sketch the graph of f:

x	$f(x)$
0	1
-1	≈ 0.61
1	≈ 0.61

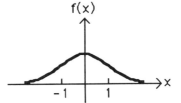

53. $y = 1 + w^2$, $w = \ln u$, $u = 2 + e^x$
Thus, $y = 1 + [\ln(2 + e^x)]^2$ and
$$\frac{dy}{dx} = 2[\ln(2 + e^x)]\left(\frac{1}{2 + e^x}\right)(e^x) = \frac{2e^x \ \ln(2 + e^x)}{2 + e^x}$$

55. $D_x \log_2(3x^2 - 1) = \frac{1}{\ln 2} \cdot \frac{1}{3x^2 - 1} \cdot 6x = \frac{1}{\ln 2} \cdot \frac{6x}{3x^2 - 1}$

57. $D_x 10^{x^2+x} = 10^{x^2+x}(\ln 10)(2x + 1) = (2x + 1)10^{x^2+x} \ln 10$

59. Let $u = f(x)$ and $y = \ln u$. Then, by the chain rule,
$$\frac{dy}{dx} = \frac{dy}{du} \cdot \frac{du}{dx} = \frac{1}{u} \cdot f'(x) = \frac{1}{f(x)} \cdot f'(x).$$

61. Price: $p = 100e^{-0.05x}$, $x \geq 0$
Revenue: $R(x) = xp = 100xe^{-0.05x}$
$$R'(x) = 100xe^{-0.05x}(-0.05) + 100e^{-0.05x}$$
$$= 100e^{-0.05x}(1 - 0.05x)$$
Critical value(s): $R'(x) = 100e^{-0.05x}(1 - 0.05x) = 0$
$$1 - 0.05x = 0$$
$$x = 20$$
$$R''(x) = 100e^{-0.05x}(-0.05) + (1 - 0.05x)100e^{-0.05x}(-0.05)$$
$$= 100e^{-0.05x}(0.0025x - 0.1)$$
$$R''(20) = -100e^{-1}(0.05) = \frac{-5}{e} < 0$$

Since $x = 20$ is the only critical value and $R''(20) < 0$, the production level that maximizes the revenue is 20 units. The maximum revenue is $R(20) = 20(36.79) = 735.80$ or \$735.80, and the price is $p(20) = 36.79$ or \$36.79 each.

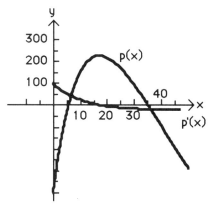

63. The cost function $C(x)$ is given by
$$C(x) = 400 + 6x$$
and the revenue function $R(x)$ is
$$R(x) = xp = 100xe^{-0.05x}$$
The profit function $P(x)$ is
$$P(x) = R(x) - C(x)$$
$$= 100xe^{-0.05x} - 400 - 6x$$
and $P'(x) = 100e^{-0.05x} - 5xe^{-0.05x} - 6$
We graph $y = P(x)$ and $y = P'(x)$ in the viewing rectangle $0 \le x \le 50$, $-400 \le y \le 300$

Critical value: Solve $P'(x) = (100 - 5x)e^{-0.05x} - 6 = 0$
To the nearest integer, $x = 17$.
$P(x)$ is increasing on $(0, 17)$ and decreasing on $(17, \infty)$; $P(x)$ has a maximum at $x = 17$. Thus, the maximum profit $P(17) = \$224.61$ is realized at a production level of 17 units at a price of \$42.74 per unit.

65. $S(t) = 300{,}000e^{-0.1t}$, $t \ge 0$
$S'(t) = 300{,}000e^{-0.1t}(-0.1) = -30{,}000e^{-0.1t}$
The rate of depreciation after one year is:
$S'(1) = -30{,}000e^{-0.1} \approx -\$27{,}145.12$ per year.
The rate of depreciation after five years is:
$S'(5) = -30{,}000e^{-0.5} \approx -\$18{,}195.92$ per year.
The rate of depreciation after ten years is:
$S'(10) = -30{,}000e^{-1} \approx -\$11{,}036.38$ per year.

67. Revenue: $R(t) = 200{,}000(1 - e^{-0.03t})$, $t \ge 0$
Cost: $C(t) = 4000 + 3000t$, $t \ge 0$
Profit: $P(t) = R(t) - C(t) = 200{,}000(1 - e^{-0.03t}) - (4000 + 3000t)$
$$= 200{,}000(1 - e^{-0.03t}) - 3000t - 4000$$
(A) $P'(t) = -200{,}000e^{-0.03t}(-0.03) - 3000 = 6000e^{-0.03t} - 3000$
Critical value(s): $P'(t) = 6000e^{-0.03t} - 3000 = 0$
$$e^{-0.03t} = \frac{1}{2}$$
$$-0.03t = \ln\left(\frac{1}{2}\right) = -\ln 2$$
$$t = \frac{\ln 2}{0.03} \approx 23$$
$$P''(t) = 6000e^{-0.03t}(-0.03) = -180e^{-0.03t}$$
$$P''(23) = -180e^{-0.69} < 0$$

Since $t = 23$ is the only critical value and $P''(23) < 0$, 23 days of TV promotion should be used to maximize profits. The maximum profit is:

$P(23) = 200,000(1 - e^{-0.03(23)}) - 3000(23) - 4000 \approx \$26,685$

The proportion of people buying the disk after t days is:

$p(t) = 1 - e^{-0.03t}$

Thus,

$p(23) = 1 - e^{-0.03(23)} \approx 0.50$ or approximately 50%.

(B) From A, the sign chart for P' is:

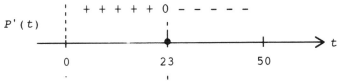

Test Numbers	
t	$P'(t)$
0	3000 (+)
50	-1661.22 (-)

$P(t)$:   Increasing : Decreasing

Thus, P is increasing on $(0, 23)$ and decreasing on $(23, \infty)$; P has a maximum at $t = 23$.
Since $P''(t) = -180e^{-0.03t} < 0$ on $(0, \infty)$, the graph of P is concave downward on $(0, \infty)$; $P(0) = -4000$ and $P(50) \approx 0$.
The graph of P is shown at the right.

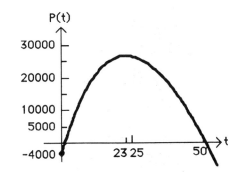

69. $P(x) = 40 + 25 \ln(x + 1)$ $0 \le x \le 65$

$P'(x) = 25\left(\dfrac{1}{x + 1}\right)(1) = \dfrac{25}{x + 1}$

$P'(10) = \dfrac{25}{11} \approx 2.27$

$P'(30) = \dfrac{25}{31} \approx 0.81$

$P'(60) = \dfrac{25}{61} \approx 0.41$

Thus, the rate of change of pressure at the end of 10 years is 2.27 millimeters of mercury per year; at the end of 30 years the rate of change is 0.81 millimeters of mercury per year; at the end of 60 years the rate of change is 0.41 millimeters of mercury per year.

71. $A(t) = 5000 \cdot 2^{2t}$

$A'(t) = 5000 \cdot 2^{2t}(2)(\ln 2) = 10,000 \cdot 2^{2t}(\ln 2)$

$A'(1) = 10,000 \cdot 2^2(\ln 2) = 40,000 \ln 2$

    $\approx 27,726$ rate of change of bacteria at the end of the first hour.

$A'(5) = 10,000 \cdot 2^{2 \cdot 5}(\ln 2) = 10,000 \cdot 2^{10}(\ln 2)$

    $\approx 7,097,827$ rate of change of bacteria at the end of the fifth hour.

73. $N(n) = 1,000,000e^{-0.09(n-1)}$, $1 \le n \le 20$

There are no asymptotes and no intercepts.

Using the first derivative:

$N'(n) = 1,000,000e^{-0.09(n-1)}(-0.09)$

$\quad\quad = -90,000e^{-0.09(n-1)} < 0$, $1 \le n \le 20$

Thus, N is decreasing on $(0, 20)$.

Using the second derivative:

$N''(n) = -90,000e^{-0.09(n-1)}(-0.09)$

$\quad\quad = 8100e^{-0.09(n-1)} > 0$, $1 \le n \le 20$

Thus, the graph of N is concave upward on $(0, 20)$.

The graph of N is:

n	N
1	1,000,000
20	$\approx$ 180,866

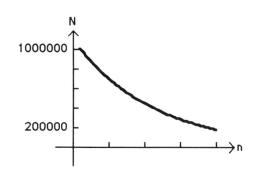

EXERCISE 10-8   CHAPTER REVIEW

1. The function f is increasing on (a, c_1), (c_3, c_6).

2. $f'(x) < 0$ on (c_1, c_3), (c_6, b).

3. The graph of f is concave downward on (a, c_2), (c_4, c_5), (c_7, b).

4. A local minimum occurs at $x = c_3$.

5. The absolute maximum occurs at $x = c_6$.

6. $f'(x)$ appears to be zero at $x = c_1$, c_3, c_5.

7. $f'(x)$ does not exist at $x = c_6$.

8. $x = c_2$, c_4, c_5, c_7 are points of inflection

9.

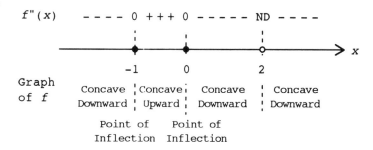

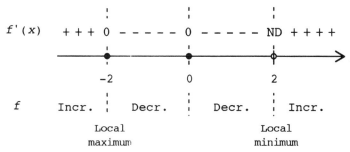

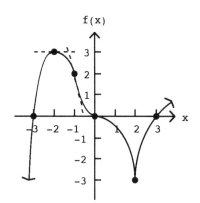

$f'(x)$ $+ + + 0 - - - - - 0 - - - - -$ ND $+ + + +$

f Incr.   Decr.   Decr.   Incr.

Local maximum   Local minimum

Using this information together with the points $(-3, 0)$, $(-2, 3)$, $(-1, 2)$, $(0, 0)$, $(2, -3)$, $(3, 0)$ on the graph, we have

10. $f(x) = x^4 + 5x^3$

$f'(x) = 4x^3 + 15x^2$

$f''(x) = 12x^2 + 30x$

11. $y = 3x + \dfrac{4}{x}$

$y' = 3 - \dfrac{4}{x^2}$

$y'' = \dfrac{8}{x^3}$

12. $A(t) = 2000e^{0.09t}$

$A(5) = 2000e^{0.09(5)} = 2000e^{0.45} \approx 3136.62$ or 3136.62

$A(10) = 2000e^{0.09(10)} = 2000e^{0.9} \approx 4919.21$ or 4919.21

$A(20) = 2000e^{0.09(20)} = 2000e^{1.8} \approx 12,099.29$ or $12,099.29$

13. $D_x(2 \ln x + 3e^x) = 2D_x \ln x + 3D_x e^x = \dfrac{2}{x} + 3e^x$

14. $D_x e^{2x-3} = e^{2x-3} D_x(2x - 3)$ (by the chain rule)

$\qquad = 2e^{2x-3}$

15. $y = \ln(2x + 7)$

$y' = \dfrac{1}{2x + 7}(2)$ (by the chain rule)

$\quad = \dfrac{2}{2x + 7}$

16. $y = \ln u$, where $u = 3 + e^x$.

(A) $y = \ln[3 + e^x]$

(B) $\dfrac{dy}{dx} = \dfrac{dy}{du} \cdot \dfrac{du}{dx} = \dfrac{1}{u}(e^x) = \dfrac{1}{3 + e^x}(e^x) = \dfrac{e^x}{3 + e^x}$

17. $x^2 - x < 12$ or $x^2 - x - 12 < 0$

Let $f(x) = x^2 - x - 12 = (x + 3)(x - 4)$.   Then f is continuous for all x and $f(-3) = f(4) = 0$.   Thus, $x = -3$ and $x = 4$ are partition numbers.

$f(x)$ $+ + + + +$ $- - - - - -$ $+ + + +$

Test Numbers

x	$f(x)$
-4	8 (+)
0	-12 (−)
5	8 (+)

Thus, $x^2 - x < 12$ for: $-3 < x < 4$ or $(-3, 4)$.

18. $\dfrac{x-5}{x^2+3x} > 0$ or $\dfrac{x-5}{x(x+3)} > 0$

Let $f(x) = \dfrac{x-5}{x(x+3)}$. Then f is discontinuous at $x = 0$ and $x = -3$, and $f(5) = 0$. Thus, $x = -3$, $x = 0$, and $x = 5$ are partition numbers.

Test Numbers	
x	$f(x)$
-4	$-\frac{9}{4}$ $(-)$
-1	3 $(+)$
1	-1 $(-)$
6	$\frac{1}{54}$ $(+)$

Thus, $\dfrac{x-5}{x^2+3x} > 0$ for $-3 < x < 0$ or $x > 5$

or $(-3, 0) \cup (5, \infty)$.

19. $f(x) = x^3 - 18x^2 + 81x$

(A) Domain: f is defined for all real numbers.

(B) Intercepts: y intercept: $f(0) = 0^3 - 18(0)^2 + 81(0) = 0$

$\qquad x$ intercepts: $x^3 - 18x^2 + 81x = 0$

$\qquad\qquad\qquad\quad x(x^2 - 18x + 81) = 0$

$\qquad\qquad\qquad\qquad\quad x(x - 9)^2 = 0$

$\qquad\qquad\qquad\qquad\qquad\quad x = 0, 9$

(C) Since f is a polynomial, there are no horizontal or vertical asymptotes.

20. $f(x) = x^3 - 18x^2 + 81x$
$f'(x) = 3x^2 - 36x + 81 = 3(x^2 - 12x + 27) = 3(x - 3)(x - 9)$

(A) Critical values: $x = 3$, $x = 9$

(B) Partition numbers: $x = 3$, $x = 9$

(C) Sign chart for f':

Test Numbers	
x	$f'(x)$
0	81 $(+)$
5	-24 $(-)$
10	21 $(+)$

Thus, f is increasing on $(-\infty, 3)$ and on $(9, \infty)$; f is decreasing on $(3, 9)$.

(D) There is a local maximum at $x = 3$ and a local minimum at $x = 9$.

21. $f'(x) = 3x^2 - 36x + 81$

$f''(x) = 6x - 36 = 6(x - 6)$

Thus, $x = 6$ is a partition number for f''.

(A) Sign chart for f'':

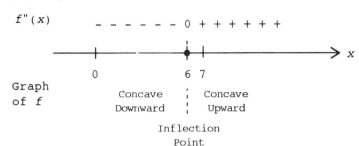

x	f''(x)
0	−36 (−)
7	6 (+)

Test Numbers

Thus, the graph of f is concave downward on $(-\infty, 6)$ and concave upward on $(6, \infty)$.

(B) The point $x = 6$ is an inflection point.

22. The graph of f:

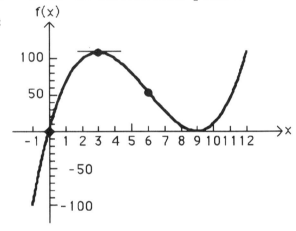

23. $f(x) = \dfrac{3x}{x + 2}$

(A) The domain of f is all real numbers except $x = -2$.

(B) Intercepts: y intercept: $f(0) = \dfrac{3(0)}{0 + 2} = 0$

x intercepts: $\dfrac{3x}{x + 2} = 0$

$3x = 0$

$x = 0$

(C) Asymptotes:

<u>Horizontal asymptotes</u>: $\lim\limits_{x \to \infty} \dfrac{3x}{x + 2} = \lim\limits_{x \to \infty} \dfrac{3x}{x\left(1 + \dfrac{2}{x}\right)} = \lim\limits_{x \to \infty} \dfrac{3}{1 + \dfrac{2}{x}} = 3$

Thus, $y = 3$ is a horizontal asymptote.

<u>Vertical asymptote(s)</u>: The denominator is 0 at $x = -2$ and the numerator is nonzero at $x = -2$. Thus, $x = -2$ is a vertical asymptote.

24. $f(x) = \dfrac{3x}{x + 2}$

$f'(x) = \dfrac{(x + 2)(3) - 3x(1)}{(x + 2)^2} = \dfrac{6}{(x + 2)^2}$

(A) Critical values: $f'(x) = \dfrac{6}{(x + 2)^2} \neq 0$ for all x ($x \neq -2$).

   Thus, f does not have any critical values.

(B) Partition numbers: $x = -2$ is a partition number for f'.

(C) Sign chart for f':

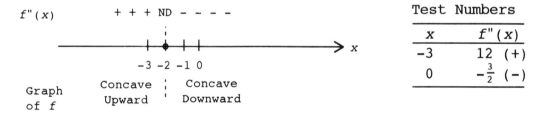

Test Numbers	
x	$f'(x)$
-3	6 $(+)$
0	$\frac{3}{2}$ $(+)$

Thus, f is increasing on $(-\infty, -2)$ and on $(-2, \infty)$.

(D) $f'(x) > 0$ for all x ($x \neq -2$). Thus, from (C), f does not have any local extrema.

25. $f'(x) = \dfrac{6}{(x + 2)^2} = 6(x + 2)^{-2}$

$f''(x) = -12(x + 2)^{-3} = \dfrac{-12}{(x + 2)^3}$

(A) Partition numbers for f'': $x = -2$
   Sign chart for f'':

Test Numbers	
x	$f''(x)$
-3	12 $(+)$
0	$-\frac{3}{2}$ $(-)$

The graph of f is concave upward on $(-\infty, -2)$ and concave downward on $(-2, \infty)$.

(B) The graph of f does not have any inflection points.

26. The graph of f is:

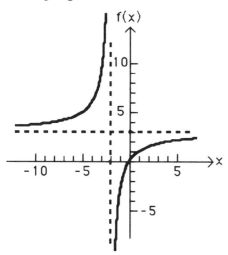

27. $f(x) = x^3 - 6x^2 - 15x + 12$
$f'(x) = 3x^2 - 12x - 15$
$3x^2 - 12x - 15 = 0$
$3(x^2 - 4x - 5) = 0$
$3(x - 5)(x + 1) = 0$
Thus, $x = -1$ and $x = 5$ are critical values of f.
$\quad f''(x) = 6x - 12$
Now, $f''(-1) = 6(-1) - 12 = -18 < 0$. Thus, f has a local maximum at $x = -1$.
Also, $f''(5) = 6(5) - 12 = 18 > 0$ and f has a local minimum at $x = 5$.

28. $y = f(x) = x^3 - 12x + 12$, $-3 \leq x \leq 5$
$f'(x) = 3x^2 - 12$
Critical values: f' is defined for all x:
$f'(x) = 3x^2 - 12 = 0$
$\quad\quad 3(x^2 - 4) = 0$
$\quad 3(x - 2)(x + 2) = 0$
Thus, the critical values of f are: $x = -2$, $x = 2$.
$f(-3) = (-3)^3 - 12(-3) + 12 = 21$
$f(-2) = (-2)^3 - 12(-2) + 12 = 28$
$\quad f(2) = 2^3 - 12(2) + 12 = -4$ Absolute minimum
$\quad f(5) = 5^3 - 12(5) + 12 = 77$ Absolute maximum

29. $y = f(x) = x^2 + \dfrac{16}{x^2}$, $x > 0$

$f'(x) = 2x - \dfrac{32}{x^3} = \dfrac{2x^4 - 32}{x^3} = \dfrac{2(x^4 - 16)}{x^3} = \dfrac{2(x - 2)(x + 2)(x^2 + 4)}{x^3}$

$f''(x) = 2 + \dfrac{96}{x^4}$

The only critical value of f in the interval $(0, \infty)$ is $x = 2$.
Since

$f''(2) = 2 + \dfrac{96}{2^4} = 8 > 0$,

$f(2) = 8$ is the absolute minimum of f on $(0, \infty)$.

30. $f(x) = \dfrac{x}{x^2 + 9}$

$$\lim_{x \to \infty} f(x) = \lim_{x \to \infty} \frac{x}{x^2 + 9} = \lim_{x \to \infty} \frac{\dfrac{1}{x}}{1 + \dfrac{9}{x^2}} = 0$$

Thus, $y = 0$, or the x axis, is a horizontal asymptote. Since $x^2 + 9 \neq 0$ for all x, there are no vertical asymptotes.

31. $f(x) = \dfrac{x^3}{x^2 - 9}$

$$\lim_{x \to \infty} f(x) = \lim_{x \to \infty} \frac{x^3}{x^2 - 9} = \lim_{x \to \infty} \frac{1}{\dfrac{1}{x} - \dfrac{9}{x^3}}$$

This limit does not exist. Thus, there are no horizontal asymptotes. Since the denominator $x^2 - 9 = (x + 3)(x - 3) = 0$ when $x = -3$ and when $x = 3$, and since the numerator $x^3 \neq 0$ at these values, $x = -3$ and $x = 3$ are vertical asymptotes.

32. $y = 100e^{-0.1x}$

Step 1. Use $f(x)$:
(A) Domain: All real numbers, $(-\infty, \infty)$.

(B) Intercepts: y intercept: $f(0) = 100e^{-0.1(0)} = 100$
x intercept: Since $100e^{-0.1x} \neq 0$ for all x, there are no x intercepts.

(C) Asymptotes:
$$\lim_{x \to \infty} 100e^{-0.1x} = \lim_{x \to \infty} \frac{100}{e^{0.1x}} = 0$$
$$\lim_{x \to -\infty} 100e^{-0.1x} \text{ does not exist.}$$
Thus, $y = 0$ is a horizontal asymptote. There are no vertical asymptotes.

Step 2. Use $f'(x)$:
$y' = 100e^{-0.1x}(-0.1)$
$\quad = -10e^{-0.1x} < 0$ on $(-\infty, \infty)$
Thus, y is decreasing on $(-\infty, \infty)$; there are no local extrema.

Step 3. Use $f''(x)$:
$y'' = -10e^{-0.1x}(-0.1)$
$\quad = e^{-0.1x} > 0$ on $(-\infty, \infty)$
Thus, the graph of f is concave upward on $(-\infty, \infty)$; there are no inflection points.

Step 4. Sketch the graph of f:

x	y
0	100
−1	≈ 110
10	≈ 37

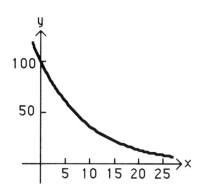

33. $D_z[(\ln z)^7 + \ln z^7] = D_z[\ln z]^7 + D_z 7 \ln z$

$$= 7[\ln z]^6 D_z \ln z + 7 D_z \ln z$$

$$= 7[\ln z]^6 \frac{1}{z} + \frac{7}{z}$$

$$= \frac{7(\ln z)^6 + 7}{z} = \frac{7[(\ln z)^6 + 1]}{z}$$

34. $D_x x^6 \ln x = x^6 D_x \ln x + (\ln x) D_x x^6$

$$= x^6 \left(\frac{1}{x}\right) + (\ln x) 6x^5 = x^5(1 + 6 \ln x)$$

35. $D_x\left(\dfrac{e^x}{x^6}\right) = \dfrac{x^6 D_x e^x - e^x D_x x^6}{(x^6)^2} = \dfrac{x^6 e^x - 6x^5 e^x}{x^{12}} = \dfrac{xe^x - 6e^x}{x^7} = \dfrac{e^x(x-6)}{x^7}$

36. $y = \ln(2x^3 - 3x)$

$$y' = \frac{1}{2x^3 - 3x}(6x^2 - 3) = \frac{6x^2 - 3}{2x^3 - 3x}$$

37. $f(x) = e^{x^3 - x^2}$

$$f'(x) = e^{x^3 - x^2}(3x^2 - 2x)$$

$$= (3x^2 - 2x)e^{x^3 - x^2}$$

38. $y = e^{-2x} \ln 5x$

$$\frac{dy}{dx} = e^{-2x}\left(\frac{1}{5x}\right)(5) + (\ln 5x)(e^{-2x})(-2)$$

$$= e^{-2x}\left(\frac{1}{x} - 2 \ln 5x\right) = \frac{1 - 2x \ln 5x}{xe^{2x}}$$

39. $f(x) = 6x^2 - x^3 + 8$

$f'(x) = 12x - 3x^2$

$f''(x) = 12 - 6x$

Now, $f''(x)$ is defined for all x and $f''(x) = 12 - 6x = 0$ implies $x = 2$. Thus, f' has a critical value at $x = 2$. Since this is the only critical value of f' and $(f'(x))'' = f'''(x) = -6$ so that $f'''(2) = -6 < 0$, $f'(2) = 12$ is the absolute maximum of f'. The graph is shown at the right.

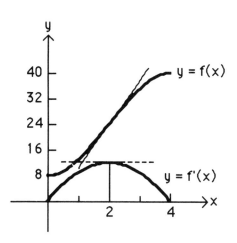

40. Let $x > 0$ be one of the numbers. Then $\dfrac{400}{x}$ is the other number. Now, we

have:

$$S(x) = x + \frac{400}{x}, \quad x > 0,$$

$$S'(x) = 1 - \frac{400}{x^2} = \frac{x^2 - 400}{x^2} = \frac{(x - 20)(x + 20)}{x^2}$$

Thus, $x = 20$ is the only critical value of S on $(0, \infty)$.

$$S''(x) = \frac{800}{x^3} \quad \text{and} \quad S''(20) = \frac{800}{8000} = \frac{1}{10} > 0$$

Therefore, $S(20) = 20 + \dfrac{400}{20} = 40$ is the absolute minimum sum, and this

occurs when each number is 20.

41. $f(x) = (x - 1)^3(x + 3)$

<u>Step 1. Use $f(x)$:</u>
(A) Domain: All real numbers.
(B) Intercepts: y intercept: $f(0) = (-1)^3(3) = -3$
$\qquad\qquad\qquad$ x intercepts: $(x - 1)^3(x + 3) = 0$
$\qquad\qquad\qquad\qquad\qquad\qquad\qquad x = 1, -3$
(C) Asymptotes: Since f is a polynomial (of degree 4), the graph of f
$\qquad$ has no asymptotes.

<u>Step 2. Use $f'(x)$:</u>
$$\begin{aligned} f'(x) &= (x - 1)^3(1) + (x + 3)(3)(x - 1)^2(1) \\ &= (x - 1)^2[(x - 1) + 3(x + 3)] \\ &= 4(x - 1)^2(x + 2) \end{aligned}$$
Critical values: $x = -2$, $x = 1$
Partition numbers: $x = -2$, $x = 1$
Sign chart for f':

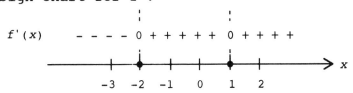

Test Numbers	
x	$f'(x)$
-3	-64 (-)
0	8 (+)
2	16 (+)

Thus, f is decreasing on $(-\infty, -2)$; f is increasing on $(-2, 1)$ and
$(1, \infty)$; f has a local minimum at $x = -2$.
<u>Step 3. Use $f''(x)$:</u>
$$\begin{aligned} f''(x) &= 4(x - 1)^2(1) + 4(x + 2)(2)(x - 1)(1) \\ &= 4(x - 1)[(x - 1) + 2(x + 2)] \\ &= 12(x - 1)(x + 1) \end{aligned}$$
Partition numbers for f'': $x = -1$, $x = 1$.

Sign chart for f'':

$f''(x)$ $\quad$ + + + + 0 - - - - 0 + + + +

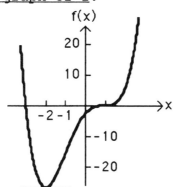

$\qquad$ -2 $\quad$ -1 $\quad$ 0 $\quad$ 1 $\quad$ 2

Graph
of f $\qquad$ Concave $\vdots$ Concave $\vdots$ Concave
$\qquad\qquad$ Upward $\vdots$ Downward $\vdots$ Upward

Test Numbers	
x	$f''(x)$
-2	36 (+)
0	-12 (-)
2	36 (+)

Thus, the graph of f is concave upward on $(-\infty, -1)$ and on $(1, \infty)$; the graph of f is concave downward on $(-1, 1)$; the graph has inflection points at $x = -1$ and at $x = 1$.

Step 4. Sketch the graph of f:

x	$f(x)$
-2	-27
0	-3
1	0

42. $f(x) = 11x - 2x \ln x, \ x > 0$

$f'(x) = 11 - 2x\left(\dfrac{1}{x}\right) - (\ln x)(2)$

$\qquad = 11 - 2 - 2 \ln x = 9 - 2 \ln x, \ x > 0$

Critical value(s): $f'(x) = 9 - 2 \ln x = 0$

$\qquad\qquad\qquad\qquad\qquad 2 \ln x = 9$

$\qquad\qquad\qquad\qquad\qquad\ \ \ln x = \dfrac{9}{2}$

$\qquad\qquad\qquad\qquad\qquad\qquad x = e^{9/2}$

$f''(x) = -\dfrac{2}{x}$ and $f''(e^{9/2}) = -\dfrac{2}{e^{9/2}} < 0$

Since $x = e^{9/2}$ is the only critical value, and $f''(e^{9/2}) < 0$, f has an absolute maximum at $x = e^{9/2}$. The absolute maximum is:

$f(e^{9/2}) = 11e^{9/2} - 2e^{9/2}\ln(e^{9/2})$

$\qquad\quad = 11e^{9/2} - 9e^{9/2}$

$\qquad\quad = 2e^{9/2} \approx 180.03$

43. $f(x) = 10xe^{-2x}, \ x > 0$

$f'(x) = 10xe^{-2x}(-2) + 10e^{-2x}(1) = 10e^{-2x}(1 - 2x), \ x > 0$

Critical value(s): $f'(x) = 10e^{-2x}(1 - 2x) = 0$

$\qquad\qquad\qquad\qquad\qquad\quad 1 - 2x = 0$

$\qquad\qquad\qquad\qquad\qquad\qquad\quad x = \dfrac{1}{2}$

$f''(x) = 10e^{-2x}(-2) + 10(1 - 2x)e^{-2x}(-2)$

$\qquad = -20e^{-2x}(1 + 1 - 2x)$

$$= -40e^{-2x}(1 - x)$$

$$f''\left(\frac{1}{2}\right) = -20e^{-1} < 0$$

Since $x = \frac{1}{2}$ is the only critical value, and $f''\left(\frac{1}{2}\right) = -20e^{-1} < 0$, f has an absolute maximum at $x = \frac{1}{2}$.   The absolute maximum of f is:

$$f\left(\frac{1}{2}\right) = 10\left(\frac{1}{2}\right)e^{-2(1/2)}$$

$$= 5e^{-1} \approx 1.84$$

44. $f(x) = 5 - 5e^{-x}$

<u>Step 1. Use $f(x)$</u>:

(A) Domain: All real numbers, $(-\infty, \infty)$.

(B) Intercepts: y intercept: $f(0) = 5 - 5e^{-0} = 0$

$\qquad\qquad\qquad\quad$ x intercepts: $5 - 5e^{-x} = 0$

$$e^{-x} = 1$$
$$x = 0$$

(C) Asymptotes:

$$\lim_{x \to \infty} (5 - 5e^{-x}) = \lim_{x \to \infty}\left(5 - \frac{5}{e^x}\right) = 5$$

$\displaystyle\lim_{x \to -\infty} (5 - 5e^{-x})$ does not exist.

Thus, $y = 5$ is a horizontal asymptote.

Since $f(x) = 5 - \dfrac{5}{e^x} = \dfrac{5e^x - 5}{e^x}$ and $e^x \ne 0$ for all x, there are no vertical asymptotes.

<u>Step 2. Use $f'(x)$</u>:

$f'(x) = -5e^{-x}(-1) = 5e^{-x} > 0$ on $(-\infty, \infty)$

Thus, f is increasing on $(-\infty, \infty)$; there are no local extrema.

<u>Step 3. Use $f''(x)$</u>:

$f''(x) = -5e^{-x} < 0$ on $(-\infty, \infty)$.

Thus, the graph of f is concave downward on $(-\infty, \infty)$; there are no inflection points.

<u>Step 4. Sketch the graph of f</u>:

x	$f(x)$
0	0
-1	-8.59
2	4.32

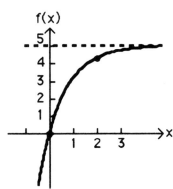

45. $f(x) = x^3 \ln x$

Step 1. Use $f(x)$:

(A) Domain: all positive real numbers, $(0, \infty)$.

(B) Intercepts: y intercept: Since $x = 0$ is not in the domain, there is no y intercept.

x intercepts: $x^3 \ln x = 0$

$\ln x = 0$

$x = 1$

(C) Asymptotes:

$\lim\limits_{x \to \infty} (x^3 \ln x)$ does not exist.

It can be shown that $\lim\limits_{x \to 0^+} (x^3 \ln x) = 0$. Thus, there are no horizontal or vertical asymptotes.

Step 2. Use $f'(x)$:

$f'(x) = x^3 \left(\dfrac{1}{x} \right) + (\ln x) 3x^2$

$= x^2 [1 + 3 \ln x], \quad x > 0$

Critical values: $x^2 [1 + 3 \ln x] = 0$

$1 + 3 \ln x = 0 \quad$ (since $x > 0$)

$\ln x = -\dfrac{1}{3}$

$x = e^{-1/3} \approx 0.72$

Partition numbers: $x = e^{-1/3}$

Sign chart for f':

x	$f'(x)$
0.5	-0.27 (-)
1	1 (+)

Thus, f is decreasing on $(0, e^{-1/3})$ and increasing on $(e^{-1/3}, \infty)$; f has a local minimum at $x = e^{-1/3}$.

Step 3. Use $f''(x)$:

$f''(x) = x^2 \left(\dfrac{3}{x} \right) + (1 + 3 \ln x) 2x$

$= x(5 + 6 \ln x), \quad x > 0$

Partition numbers: $x(5 + 6 \ln x) = 0$

$5 + 6 \ln x = 0$

$\ln x = -\dfrac{5}{6}$

$x = e^{-5/6} \approx 0.43$

Sign chart for f'':

$f''(x)$ $\quad$ – – – – – 0 + + + + +

$$\xrightarrow{\hspace{3cm}} x$$

$\quad\quad$ 0 $\quad\quad\quad e^{-5/6}$ $\quad$ .5 $\quad\quad$ 1

Graph $\quad$: $\quad$ Concave $\quad$: $\quad$ Concave
of f $\quad$: $\quad$ Downward $\quad$: $\quad$ Upward

Test Numbers	
x	$f''(x)$
.2	-0.93 (–)
1	5 (+)

Thus, the graph of f is concave downward on $(0, e^{-5/6})$ and concave upward on $(e^{-5/6}, \infty)$; the graph has an inflection point at $x = e^{-5/6}$.

<u>Step 4.</u> <u>Sketch the graph of f:</u>

x	$f(x)$
$e^{-5/6}$	-0.07
$e^{-1/3}$	-0.12
1	0

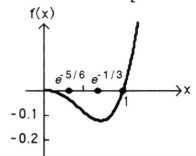

46. $y = w^3$, $w = \ln u$, $u = 4 - e^x$

(A) $y = [\ln(4 - e^x)]^3$

(B) $\dfrac{dy}{dx} = \dfrac{dy}{dw} \cdot \dfrac{dw}{du} \cdot \dfrac{du}{dx}$

$\quad = 3w^2 \cdot \dfrac{1}{u} \cdot (-e^x) = 3[\ln(4 - e^x)]^2 \left(\dfrac{1}{4 - e^x}\right)(-e^x)$

$\quad = \dfrac{-3e^x[\ln(4 - e^x)]^2}{4 - e^x}$

47. $y = 5^{x^2-1}$

$y' = 5^{x^2-1}(\ln 5)(2x) = 2x5^{x^2-1}(\ln 5)$

48. $D_x \log_5(x^2 - x) = \dfrac{1}{x^2 - x} \cdot \dfrac{1}{\ln 5} \cdot D_x(x^2 - x) = \dfrac{1}{\ln 5} \cdot \dfrac{2x - 1}{x^2 - x}$

49. $D_x\sqrt{\ln(x^2 + x)} = D_x[\ln(x^2 + x)]^{1/2} = \dfrac{1}{2}[\ln(x^2 + x)]^{-1/2} D_x \ln(x^2 + x)$

$\quad = \dfrac{1}{2}[\ln(x^2 + x)]^{-1/2} \dfrac{1}{x^2 + x} D_x(x^2 + x)$

$\quad = \dfrac{1}{2}[\ln(x^2 + x)]^{-1/2} \cdot \dfrac{2x + 1}{x^2 + x} = \dfrac{2x + 1}{2(x^2 + x)[\ln(x^2 + x)]^{1/2}}$

50. $P(x) = 150x - \dfrac{x^2}{40} - 50{,}000$, $0 \le x \le 5000$

First, compute the critical values:

$P'(x) = 150 - \dfrac{x}{20}$

Now, $150 - \dfrac{x}{20} = 0$ when $x = 3000$. Thus, $x = 3000$ is the critical value.

The absolute maximum or minimum can occur at $x = 0$, 3000, or 5000.

$P(0) = -50,000$

$P(3000) = 150(3000) - \dfrac{(3000)^2}{40} - 50,000 = 175,000$

$P(5000) = 150(5000) - \dfrac{(5000)^2}{40} - 50,000 = 75,000$

Thus, the maximum occurs at a production level of $x = 3000$, and the maximum profit is $P(3000) = 175,000$.

51.

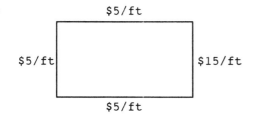

Let x be the length and y the width of the rectangle.

(A) $C(x, y) = 5x + 5x + 5y + 15y = 10x + 20y$

Also, Area $A = xy = 5000$, so $y = \dfrac{5000}{x}$

and $C(x) = 10x + \dfrac{100,000}{x}$, $x \geq 0$

Now, $C'(x) = 10 - \dfrac{100,000}{x^2}$ and

$10 - \dfrac{100,000}{x^2} = 0$ implies $10x^2 = 100,000$

$$x^2 = 10,000$$
$$x = \pm 100$$

Thus, $x = 100$ is the critical value.

Now, $C''(x) = \dfrac{200,000}{x^3}$ and $C''(100) = \dfrac{200,000}{1,000,000} = 0.2 > 0$

and the most economical (i.e. least cost) fence will have

dimensions: length $x = 100$ feet and width $y = \dfrac{5000}{100} = 50$ feet.

(B) We want to maximize $A = xy$ subject to
$C(x, y) = 10x + 20y = 3000$ or $x = 300 - 2y$
Thus, $A = y(300 - 2y) = 300y - 2y^2$, $0 \leq y \leq 150$.
Now, $A'(y) = 300 - 4y$ and
$300 - 4y = 0$ implies $y = 75$.
Therefore, $y = 75$ is the critical value.

Now, $A''(y) = -4$ and $A''(75) = -4 < 0$. Thus, A has an absolute maximum when $y = 75$. Therefore the dimensions of the rectangle that will enclose maximum area are:
length $x = 300 - 2(75) = 150$ feet and width $y = 75$ feet.

52. $C(x) = 4000 + 10x + \dfrac{1}{10}x^2$, $x > 0$

Average cost $= \overline{C}(x) = \dfrac{4000}{x} + 10 + \dfrac{1}{10}x$

Marginal cost $= C'(x) = 10 + \dfrac{2}{10}x = 10 + \dfrac{1}{5}x$

The graph of $C'(x)$ is a straight line with slope $\frac{1}{5}$ and y intercept 10.

$$\overline{C}'(x) = \frac{-4000}{x^2} + \frac{1}{10} = \frac{-40,000 + x^2}{10x^2} = \frac{(x + 200)(x - 200)}{10x^2}$$

Thus, $\overline{C}'(x) < 0$ on $(0, 200)$ and $\overline{C}'(x) > 0$ on $(200, \infty)$. Therefore, $\overline{C}(x)$ is decreasing on $(0, 200)$, increasing on $(200, \infty)$, and a minimum occurs at $x = 200$.

$$\text{Min } \overline{C}(x) = \overline{C}(200) = \frac{4000}{200} + 10 + \frac{1}{10}(200) = 50$$

$$\overline{C}''(x) = \frac{8000}{x^3} > 0 \text{ on } (0, \infty).$$

Therefore, the graph of $\overline{C}(x)$ is concave upward on $(0, \infty)$.

Using this information and point-by-point plotting (use a calculator), the graphs of $C'(x)$ and $\overline{C}(x)$ are as shown in the diagram at the right.

The line $y = \frac{1}{10}x + 10$ is an oblique asymptote for $y = \overline{C}(x)$.

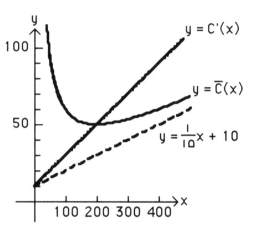

53. Let x = the number of dollars increase in the nightly rate, $x \geq 0$. Then $200 - 4x$ rooms will be rented at $(40 + x)$ dollars per room. [Note: Since $200 - 4x \geq 0$, $x \leq 50$.] The cost of service for $200 - 4x$ rooms at $8 per room is $8(200 - 4x)$. Thus:

Gross profit: $P(x) = (200 - 4x)(40 + x) - 8(200 - 4x)$
$= (200 - 4x)(32 + x)$
$= 6400 + 72x - 4x^2$, $0 \leq x \leq 50$

$P'(x) = 72 - 8x$
Critical value: $72 - 8x = 0$
$x = 9$

Now, $P(0) = 6400$
$P(9) = 6724$ Absolute maximum
$P(50) = 0$

Thus, the maximum gross profit is $6724 and this occurs at $x = 9$, i.e., the rooms should be rented at $49 per night.

54. Let x = number of times the company should order. Then, the number of discs per order = $\frac{7200}{x}$. The average number of unsold discs is given by:

$$\frac{7200}{2x} = \frac{3600}{x}$$

Total cost: $C(x) = 5x + 0.2\left(\frac{3600}{x}\right)$, $x > 0$

$$C(x) = 5x + \frac{720}{x}$$

$$C'(x) = 5 - \frac{720}{x^2} = \frac{5x^2 - 720}{x^2} = \frac{5(x^2 - 144)}{x^2}$$
$$= \frac{5(x + 12)(x - 12)}{x^2}$$

Critical value: $x = 12$ [Note: $x > 0$, so $x = -12$ is not a critical value.]

$C''(x) = \dfrac{1440}{x^3}$ and $C''(12) = \dfrac{1440}{12^3} > 0$

Therefore, $C(x)$ is a minimum when $x = 12$.

55. (A) The compound interest formula is: $A = P(1 + r)^t$.   Thus, the time for P to double when $r = 0.05$ and interest is compounded annually can be found by solving
$2P = P(1 + 0.05)^t$ or $2 = (1.05)^t$ for t.
$\ln(1.05)^t = \ln 2$
$t \ln(1.05) = \ln 2$
$t = \dfrac{\ln 2}{\ln(1.05)} \approx 14.2$ years

(B) The continuous compound interest formula is: $A = Pe^{rt}$.   Proceeding as above, we have
$2P = Pe^{0.05t}$ or $e^{0.05t} = 2$.
Therefore, $0.05t = \ln 2$ and
$t = \dfrac{\ln 2}{.05} \approx 13.9$ years

56. $A(t) = 100e^{0.1t}$
$A'(t) = 100(0.1)e^{0.1t} = 10e^{0.1t}$
$A'(1) = 11.05$ or   \$11.05 per year
$A'(10) = 27.18$ or   \$27.18 per year

57. $R(x) = xp(x) = 1000xe^{-0.02x}$
$R'(x) = 1000[xD_x e^{-0.02x} + e^{-0.02x}D_x x]$
$\quad = 1000[x(-0.02)e^{-0.02x} + e^{-0.02x}]$
$\quad = (1000 - 20x)e^{-0.02x}$

58. From Problem 57,
$R'(x) = (1000 - 20x)e^{-0.02x}$
Critical value(s): $R'(x) = (1000 - 20x)e^{-0.02x} = 0$
$\qquad\qquad\qquad\qquad\qquad 1000 - 20x = 0$
$\qquad\qquad\qquad\qquad\qquad\qquad\quad x = 50$
$R''(x) = (1000 - 20x)e^{-0.02x}(-0.02) + e^{-0.02x}(-20)$
$\quad = e^{-0.02x}[0.4x - 20 - 20]$
$\quad = e^{-0.02x}(0.4x - 40)$
$R''(50) = e^{-0.02(50)}[0.4(50) - 40] = -20e^{-1} < 0$

Since $x = 50$ is the only critical value and $R''(50) < 0$, R has an absolute maximum at a production level of 50 units.   The maximum revenue is

$R(50) = 1000(50)e^{-0.02(50)} = 50,000e^{-1} \approx 18,394$ or $\$18,394$.
The price per unit at the production level of 50 units is
$p(50) = 1000e^{-0.02(50)} = 1000e^{-1} \approx 367.88$ or $\$367.88$.

59. $R(x) = 1000xe^{-0.02x}$, $0 \le x \le 100$

<u>Step 1. Use $R(x)$</u>:

(A) Domain: $0 \le x \le 100$ or $[0, 100]$

(B) Intercepts: y intercept: $R(0) = 0$

x intercepts: $100xe^{-0.02x} = 0$

$x = 0$

(C) Asymptotes: There are no horizontal or vertical asymptotes.

<u>Step 2. Use $R'(x)$</u>:

From Problems 54 and 55, $R'(x) = (1000 - 20x)e^{-0.02x}$ and $x = 50$ is a critical value.
Sign chart for R':

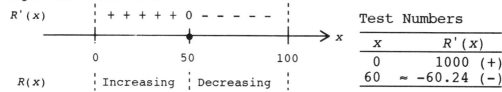

x	$R'(x)$
0	1000 (+)
60	$\approx$ -60.24 (-)

Test Numbers

Thus, R is increasing on $(0, 50)$ and decreasing on $(50, 100)$; R has a maximum at $x = 50$.

<u>Step 3. Use $R''(x)$</u>:

$R''(x) = (0.4x - 40)e^{-0.02x} < 0$ on $(0, 100)$
Thus, the graph of R is concave downward on $(0, 100)$.

<u>Step 4. Sketch the graph of R</u>:

x	$R(x)$
0	0
50	18,394
100	13,533

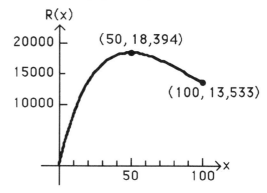

60. Cost: $C(x) = 200 + 50x - 50 \ln x$, $x \ge 1$

Average cost: $\overline{C} = \dfrac{C(x)}{x} = \dfrac{200}{x} + 50 - \dfrac{50}{x} \ln x$, $x \ge 1$

$\overline{C}'(x) = \dfrac{-200}{x^2} - \dfrac{50}{x}\left(\dfrac{1}{x}\right) + (\ln x)\dfrac{50}{x^2} = \dfrac{50(\ln x - 5)}{x^2}$, $x \ge 1$

Critical value(s): $\overline{C}'(x) = \dfrac{50(\ln x - 5)}{x^2} = 0$

$\ln x = 5$

$x = e^5$

Sign chart for $\overline{C}'$:

Test Numbers

x	$\overline{C}'(x)$
1	-250 $(-)$
e^6	$\frac{50}{e^{12}}$ $(+)$

By the first derivative test, $\overline{C}$ has a local minimum at $x = e^5$. Since this is the only critical value of $\overline{C}$, $\overline{C}$ has as absolute minimum at $x = e^5$. Thus, the minimal average cost is:

$$\overline{C}(e^5) = \frac{200}{e^5} + 50 - \frac{50}{e^5} \ln(e^5)$$

$$= 50 - \frac{50}{e^5} \approx 49.66 \quad \text{or} \quad \$49.66$$

61. $C(t) = 20t^2 - 120t + 800, \quad 0 \le t \le 9$
$C'(t) = 40t - 120 = 40(t - 3)$
Critical value: $t = 3$
$C''(t) = 40$ and $C''(3) = 40 > 0$
Therefore, a local minimum occurs at $t = 3$.
$C(3) = 20(3^2) - 120(3) + 800 = 620$ Absolute minimum
$C(0) = 800$
$C(9) = 20(81) - 120(9) + 800 = 1340$
Therefore, the bacteria count will be at a minimum three days after a treatment.

62. $C(t) = 5e^{-0.3t}$
$C'(t) = 5e^{-0.3t}(-0.3) = -1.5e^{-0.3t}$
After one hour, the rate of change of concentration is
$C'(1) = -1.5e^{-0.3(1)} = -1.5e^{-0.3} \approx -1.111$ mg/ml per hour.
After five hours, the rate of change of concentration is
$C'(5) = -1.5e^{-0.3(5)} = -1.5e^{-1.5} \approx -0.335$ mg/ml per hour.

63. $N = 10 + 6t^2 - t^3, \quad 0 \le t \le 5$
$\frac{dN}{dt} = 12t - 3t^2$
Now, find the critical values of the rate function $R(t)$:
$R(t) = \frac{dN}{dt} = 12t - 3t^2$
$R'(t) = \frac{dR}{dt} = \frac{d^2N}{dt^2} = 12 - 6t$
Critical value: $t = 2$
$R''(t) = -6$ and $R''(2) = -6 < 0$
$R(0) = 0$
$R(2) = 12$ Absolute maximum
$R(5) = -15$
Therefore, $R(t)$ has an absolute maximum at $t = 2$. The rate of increase will be a maximum after two years.

64. $N(t) = 10(1 - e^{-0.4t})$

(A) $N'(t) = -10e^{-0.4t}(-0.4) = 4e^{-0.4t}$

$N'(1) = 4e^{-0.4(1)} = 4e^{-0.4} \approx 2.68$.

Thus, learning is increasing at the rate of 2.68 units per day afte 1 day.

$N'(5) = 4e^{-0.4(5)} = 4e^{-2} \approx 0.54$

Thus, learning is increasing at the rate of 0.54 units per day afte 5 days.

(B) From (A), $N'(t) = 4e^{-0.4t} > 0$ on $(0, 10)$. Thus, N is increasing on $(0, 10)$.

$N''(t) = 4e^{-0.4t}(-0.4) = -1.6e^{-0.4t} < 0$ on $(0, 10)$.

Thus, the graph of N is concave downward on $(0, 10)$. The graph of N is:

t	$N(t)$
0	0
5	8.65
10	9.82

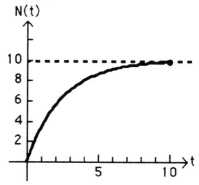

Things to remember:

$\underline{1}$. A function $F(x)$ is an ANTIDERIVATIVE of $f(x)$ if $F'(x) = f(x)$.

$\underline{2}$. The INDEFINITE INTEGRAL of $f(x)$, denoted

$$\int f(x)\,dx,$$

represents all antiderivatives of $f(x)$ and is given by

$$\int f(x)\,dx = F(x) + C$$

where $F(x)$ is any antiderivative of $f(x)$ and C is an arbitrary constant. The symbol $\int$ is called an INTEGRAL SIGN, the function $f(x)$ is called the INTEGRAND, and C is called the CONSTANT OF INTEGRATION.

$\underline{3}$. Indefinite integration and differentiation are reverse operations (except for the addition of the constant of integration). This is expressed symbolically by:

(a) $D_x\left(\int f(x)\,dx \right) = f(x)$

(b) $\int F'(x)\,dx = F(x) + C$

$\underline{4}$. INDEFINITE INTEGRAL FORMULAS:

(a) $\int k\ dx = kx + C$, k constant

(b) $\int x^n dx = \dfrac{x^{n+1}}{n+1} + C$, $n \neq -1$

(c) $\int e^x dx = e^x + C$

(d) $\int \dfrac{dx}{x} = \ln|x| + C$, $x \neq 0$

INDEFINITE INTEGRATION PROPERTIES:

   (a) $\displaystyle\int kf(x)\,dx = k\int f(x)\,dx,\ k$ constant

   (b) $\displaystyle\int [f(x) \pm g(x)]\,dx = \int f(x)\,dx \pm \int g(x)\,dx$

---

1. $\displaystyle\int 7\,dx = 7x + C$ [using $\underline{4}$(a)] <u>Check</u>: $(7x + C)' = 7$

3. $\displaystyle\int x^6\,dx = \frac{x^{6+1}}{6+1} + C$ [using $\underline{4}$(b)]

   $= \dfrac{x^7}{7} + C$ <u>Check</u>: $\left(\dfrac{x^7}{7} + C\right)' = x^6$

5. $\displaystyle\int 8t^3\,dt = 8\int t^3\,dt$ [using $\underline{5}$(a)]

   $= 8\,\dfrac{t^{3+1}}{3+1} + C = 2t^4 + C$ <u>Check</u>: $(2t^4 + C)' = 8t^3$

7. $\displaystyle\int (2u + 1)\,du = \int 2u\,du + \int 1\,du$ [using $\underline{5}$(b)]

   $= 2\,\dfrac{u^2}{2} + u + C = u^2 + u + C$ <u>Check</u>: $(u^2 + u + C)' = 2u + 1$

9. $\displaystyle\int (3x^2 + 2x - 5)\,dx = \int 3x^2\,dx + \int 2x\,dx - \int 5\,dx$

   $= 3\displaystyle\int x^2\,dx + 2\int x\,dx - \int 5\,dx$

   $= 3\,\dfrac{x^3}{3} + 2\,\dfrac{x^2}{2} - 5x + C = x^3 + x^2 - 5x + C$

   <u>Check</u>: $(x^3 + x^2 - 5x + C)' = 3x^2 + 2x - 5$

11. $\displaystyle\int (s^4 - 8s^5)\,ds = \int s^4\,ds - \int 8s^5\,ds = \int s^4\,ds - 8\int s^5\,ds$

   $= \dfrac{s^5}{5} - 8\,\dfrac{s^6}{6} + C = \dfrac{s^5}{5} - \dfrac{4s^6}{3} + C$

   <u>Check</u>: $\left(\dfrac{s^5}{5} - \dfrac{4s^6}{3} + C\right)' = s^4 - 8s^5$

13. $\displaystyle\int 3e^t\,dt = 3\int e^t\,dt$ [using $\underline{4}$(c)]

   $= 3e^t + C$ <u>Check</u>: $(3e^t + C)' = 3e^t$

15. $\int 2z^{-1}dz = 2 \int \frac{1}{z} dz = 2 \ln|z| + C$ [using 4(d)]

Check: $(2 \ln|z| + C)' = \frac{2}{z}$

17. $\frac{dy}{dx} = 200x^4$

$y = \int 200x^4 dx = 200 \int x^4 dx = 200 \frac{x^5}{5} + C = 40x^5 + C$

19. $\frac{dP}{dx} = 24 - 6x$

$P = \int (24 - 6x) dx = \int 24 \; dx - \int 6x \; dx = \int 24 \; dx - 6 \int x \; dx$

$= 24x - \frac{6x^2}{2} + C = 24x - 3x^2 + C$

21. $\frac{dy}{du} = 2u^5 - 3u^2 - 1$

$y = \int (2u^5 - 3u^2 - 1) du = \int 2u^5 du - \int 3u^2 du - \int 1 \; du$

$= 2 \int u^5 du - 3 \int u^2 du - \int du$

$= \frac{2u^6}{6} - \frac{3u^3}{3} - u + C = \frac{u^6}{3} - u^3 - u + C$

23. $\frac{dy}{dx} = e^x + 3$

$y = \int (e^x + 3) dx = \int e^x dx + \int 3 \; dx = e^x + 3x + C$

25. $\frac{dx}{dt} = 5t^{-1} + 1$

$x = \int (5t^{-1} + 1) dt = \int 5t^{-1} dt + \int 1 \; dt = 5 \int \frac{1}{t} dt + \int dt$

$= 5 \ln|t| + t + C$

27. $\int 6x^{1/2} dx = 6 \int x^{1/2} dx = 6 \frac{x^{(1/2)+1}}{\frac{1}{2} + 1} + C = \frac{6x^{3/2}}{\frac{3}{2}} + C = 4x^{3/2} + C$

Check: $(4x^{3/2} + C)' = 4\left(\frac{3}{2}\right)x^{1/2} = 6x^{1/2}$

29. $\int 8x^{-3} dx = 8 \int x^{-3} dx = 8 \frac{x^{-3+1}}{-3 + 1} + C = \frac{8x^{-2}}{-2} + C = -4x^{-2} + C$

Check: $(-4x^{-2} + C)' = -4(-2)x^{-3} = 8x^{-3}$

31. $\int \dfrac{du}{\sqrt{u}} = \int \dfrac{du}{u^{1/2}} = \int u^{-1/2} du = \dfrac{u^{(-1/2)+1}}{-\dfrac{1}{2}+1} + C = \dfrac{u^{1/2}}{\dfrac{1}{2}} + C$

$$= 2u^{1/2} + C \text{ or } 2\sqrt{u} + C$$

Check: $(2u^{1/2} + C)' = 2\left(\dfrac{1}{2}\right)u^{-1/2} = \dfrac{1}{u^{1/2}} = \dfrac{1}{\sqrt{u}}$

33. $\int \dfrac{dx}{4x^3} = \dfrac{1}{4}\int x^{-3} dx = \dfrac{1}{4}\cdot\dfrac{x^{-2}}{-2} + C = \dfrac{-x^{-2}}{8} + C$

Check: $\left(\dfrac{-x^{-2}}{8} + C\right)' = \dfrac{1}{8}(-2)(-x^{-3}) = \dfrac{1}{4}x^{-3} = \dfrac{1}{4x^3}$

35. $\int \dfrac{du}{2u^5} = \dfrac{1}{2}\int u^{-5} du = \dfrac{1}{2}\cdot\dfrac{u^{-4}}{-4} + C = \dfrac{-u^{-4}}{8} + C$

Check: $\left(\dfrac{-u^{-4}}{8} + C\right)' = \dfrac{-1}{8}(-4)u^{-5} = \dfrac{u^{-5}}{2} = \dfrac{1}{2u^5}$

37. $\int\left(3x^2 - \dfrac{2}{x^2}\right)dx = \int 3x^2 dx - \int \dfrac{2}{x^2} dx$

$$= 3\int x^2 dx - 2\int x^{-2} dx = 3\cdot\dfrac{x^3}{3} - \dfrac{2x^{-1}}{-1} + C = x^3 + 2x^{-1} + C$$

Check: $(x^3 + 2x^{-1} + C)' = 3x^2 - 2x^{-2} = 3x^2 - \dfrac{2}{x^2}$

39. $\int\left(10x^4 - \dfrac{8}{x^5} - 2\right)dx = \int 10x^4 dx - \int 8x^{-5} dx - \int 2\ dx$

$$= 10\int x^4 dx - 8\int x^{-5} dx - \int 2\ dx$$

$$= \dfrac{10x^5}{5} - \dfrac{8x^{-4}}{-4} - 2x + C = 2x^5 + 2x^{-4} - 2x + C$$

Check: $(2x^5 + 2x^{-4} - 2x + C)' = 10x^4 - 8x^{-5} - 2 = 10x^4 - \dfrac{8}{x^5} - 2$

41. $\int\left(3\sqrt{x} + \dfrac{2}{\sqrt{x}}\right)dx = 3\int x^{1/2} dx + 2\int x^{-1/2} dx$

$$= \dfrac{3x^{3/2}}{\dfrac{3}{2}} + \dfrac{2x^{1/2}}{\dfrac{1}{2}} + C = 2x^{3/2} + 4x^{1/2} + C$$

Check: $(2x^{3/2} + 4x^{1/2} + C)' = 2\left(\dfrac{3}{2}\right)x^{1/2} + 4\left(\dfrac{1}{2}\right)x^{-1/2}$

$$= 3x^{1/2} + 2x^{-1/2} = 3\sqrt{x} + \dfrac{2}{\sqrt{x}}$$

43. $\int \left(\sqrt[3]{x^2} - \dfrac{4}{x^3}\right) dx = \int x^{2/3} dx - 4 \int x^{-3} dx = \dfrac{x^{5/3}}{\frac{5}{3}} - \dfrac{4x^{-2}}{-2} + C$

$$= \dfrac{3x^{5/3}}{5} + 2x^{-2} + C$$

Check: $\left(\dfrac{3}{5} x^{5/3} + 2x^{-2} + C\right)' = \dfrac{3}{5}\left(\dfrac{5}{3}\right)x^{2/3} + 2(-2)x^{-3}$

$$= x^{2/3} - 4x^{-3} = \sqrt[3]{x^2} - \dfrac{4}{x^3}$$

45. $\int \dfrac{e^x - 3x}{4} dx = \int \left(\dfrac{e^x}{4} - \dfrac{3x}{4}\right)dx = \dfrac{1}{4}\int e^x dx - \dfrac{3}{4}\int x \; dx$

$$= \dfrac{1}{4}e^x - \dfrac{3}{4}\cdot\dfrac{x^2}{2} + C = \dfrac{1}{4}e^x - \dfrac{3x^2}{8} + C$$

Check: $\left(\dfrac{1}{4}e^x - \dfrac{3x^2}{8} + C\right)' = \dfrac{1}{4}e^x - \dfrac{6x}{8} = \dfrac{1}{4}e^x - \dfrac{3}{4}x$

47. $\int (2z^{-3} + z^{-2} + z^{-1})dz = 2\int z^{-3}dz + \int z^{-2}dz + \int \dfrac{1}{z}dz$

$$= \dfrac{2z^{-2}}{-2} + \dfrac{z^{-1}}{-1} + \ln|z| + C = -z^{-2} - z^{-1} + \ln|z| + C$$

Check: $(-z^{-2} - z^{-1} + \ln|z| + C)' = 2z^{-3} + z^{-2} + \dfrac{1}{z}$

49. $\dfrac{dy}{dx} = 2x - 3$

$y = \int (2x - 3) dx = 2\int x \; dx - \int 3 \; dx = \dfrac{2x^2}{2} - 3x + C = x^2 - 3x + C$

Given $y(0) = 5$: $5 = 0^2 - 3(0) + C$. Hence, $C = 5$ and $y = x^2 - 3x + 5$.

51. $C'(x) = 6x^2 - 4x$

$C(x) = \int (6x^2 - 4x) dx = 6\int x^2 dx - 4\int x \; dx = \dfrac{6x^3}{3} - \dfrac{4x^2}{2} + C = 2x^3 - 2x^2 + C$

Given $C(0) = 3000$: $3000 = 2(0^3) - 2(0^2) + C$. Hence, $C = 3000$ and
$C(x) = 2x^3 - 2x^2 + 3000$.

53. $\dfrac{dx}{dt} = \dfrac{20}{\sqrt{t}}$

$x = \int \dfrac{20}{\sqrt{t}} dt = 20 \int t^{-1/2} dt = 20\dfrac{t^{1/2}}{\frac{1}{2}} + C = 40\sqrt{t} + C$

Given $x(1) = 40$: $40 = 40\sqrt{1} + C$ or $40 = 40 + C$. Hence, $C = 0$ and
$x = 40\sqrt{t}$.

55. $\dfrac{dy}{dx} = 2x^{-2} + 3x^{-1} - 1$

$$y = \int (2x^{-2} + 3x^{-1} - 1)\,dx = 2\int x^{-2}dx + 3\int x^{-1}dx - \int dx$$

$$= \dfrac{2x^{-1}}{-1} + 3\,\ln|x| - x + C = \dfrac{-2}{x} + 3\,\ln|x| - x + C$$

Given $y(1) = 0$: $0 = -\dfrac{2}{1} + 3\,\ln|1| - 1 + C$. Hence, $C = 3$ and

$y = -\dfrac{2}{x} + 3\,\ln|x| - x + 3$.

57. $\dfrac{dx}{dt} = 4e^{t} - 2$

$$x = \int (4e^{t} - 2)\,dt = 4\int e^{t}dt - \int 2\,dt = 4e^{t} - 2t + C$$

Given $x(0) = 1$: $1 = 4e^{0} - 2(0) + C = 4 + C$. Hence, $C = -3$ and
$x = 4e^{t} - 2t - 3$.

59. $\dfrac{dy}{dx} = 4x - 3$

$$y = \int (4x - 3)\,dx = 4\int x\,dx - \int 3\,dx = \dfrac{4x^{2}}{2} - 3x + C = 2x^{2} - 3x + C$$

Given $y(2) = 3$: $3 = 2\cdot 2^{2} - 3\cdot 2 + C$. Hence, $C = 1$ and $y = 2x^{2} - 3x + 1$.

61. $\displaystyle\int \dfrac{2x^{4} - x}{x^{3}}\,dx = \int \left(\dfrac{2x^{4}}{x^{3}} - \dfrac{x}{x^{3}}\right)dx$

$$= 2\int x\,dx - \int x^{-2}dx = \dfrac{2x^{2}}{2} - \dfrac{x^{-1}}{-1} + C = x^{2} + x^{-1} + C$$

63. $\displaystyle\int \dfrac{x^{5} - 2x}{x^{4}}\,dx = \int \left(\dfrac{x^{5}}{x^{4}} - \dfrac{2x}{x^{4}}\right)dx$

$$= \int x\,dx - 2\int x^{-3}dx = \dfrac{x^{2}}{2} - \dfrac{2x^{-2}}{-2} + C = \dfrac{x^{2}}{2} + x^{-2} + C$$

65. $\displaystyle\int \dfrac{x^{2}e^{x} - 2x}{x^{2}}\,dx = \int \left(\dfrac{x^{2}e^{x}}{x^{2}} - \dfrac{2x}{x^{2}}\right)dx = \int e^{x}dx - 2\int x^{-1}dx = e^{x} - 2\,\ln|x| + C$

67. $\dfrac{dM}{dt} = \dfrac{t^{2} - 1}{t^{2}}$

$$M = \int \dfrac{t^{2} - 1}{t^{2}}\,dt = \int \left(\dfrac{t^{2}}{t^{2}} - \dfrac{1}{t^{2}}\right)dt = \int dt - \int t^{-2}dt = t - \dfrac{t^{-1}}{-1} + C = t + \dfrac{1}{t} + C$$

Given $M(4) = 5$: $5 = 4 + \dfrac{1}{4} + C$ or $C = 5 - \dfrac{17}{4} = \dfrac{3}{4}$. Hence, $M = t + \dfrac{1}{t} + \dfrac{3}{4}$.

69. $\dfrac{dy}{dx} = \dfrac{5x + 2}{\sqrt[3]{x}}$

$$y = \int \dfrac{5x + 2}{\sqrt[3]{x}}\,dx = \int \left(\dfrac{5x}{x^{1/3}} + \dfrac{2}{x^{1/3}}\right)dx = 5\int x^{2/3}\,dx + 2\int x^{-1/3}\,dx$$

$$= \dfrac{5x^{5/3}}{\dfrac{5}{3}} + \dfrac{2x^{2/3}}{\dfrac{2}{3}} + C = 3x^{5/3} + 3x^{2/3} + C$$

Given $y(1) = 0$: $0 = 3 \cdot 1^{5/3} + 3 \cdot 1^{2/3} + C$. Hence, $C = -6$ and
$y = 3x^{5/3} + 3x^{2/3} - 6$.

71. $p'(x) = -\dfrac{10}{x^2}$

$$p(x) = \int -\dfrac{10}{x^2}\,dx = -10\int x^{-2}\,dx = \dfrac{-10x^{-1}}{-1} + C = \dfrac{10}{x} + C$$

Given $p(1) = 20$: $20 = \dfrac{10}{1} + C = 10 + C$. Hence, $C = 10$ and

$$p(x) = \dfrac{10}{x} + 10.$$

73. $P'(x) = 50 - 0.04x$

$$P(x) = \int (50 - 0.04x)\,dx = \int 50\,dx - 0.04 \int x\,dx$$

$$= 50x - 0.04\,\dfrac{x^2}{2} + C = 50x - 0.02x^2 + C$$

Given $P(0) = 0$: $0 = 50 \cdot 0 - 0.02(0)^2 + C$. Hence, $C = 0$ and
$P(x) = 50x - 0.02x^2$. The profit on 100 units of production is:
$P(100) = 50(100) - 0.02(100)^2$
$\qquad\quad = 5000 - 200$
$\qquad\quad = \$4800$

75. $R'(x) = 100 - \dfrac{1}{5}x$

$$R(x) = \int \left(100 - \dfrac{1}{5}x\right)dx = \int 100\,dx - \dfrac{1}{5}\int x\,dx$$

$$= 100x - \dfrac{1}{5}\left(\dfrac{x^2}{2}\right) + C = 100x - \dfrac{1}{10}x^2 + C$$

Given $R(0) = 0$: $0 = 100(0) - \dfrac{1}{10} \cdot 0^2 + C = C$. Hence, $C = 0$ and

$R(x) = 100x - \dfrac{1}{10}x^2$. Since

$R(x) = xp(x) = x\left(100 - \dfrac{1}{10}x\right)$,

we have:

$p(x) = 100 - \dfrac{1}{10}x$

$p(700) = 100 - \dfrac{1}{10}(700) = 100 - 70 = 30$ or $p(700) = \$30$

77. $S'(t) = -25t^{2/3}$

$$S(t) = \int S'(t)\,dt = \int -25t^{2/3}\,dt = -25\int t^{2/3}\,dt = -25\,\frac{t^{5/3}}{\frac{5}{3}} + C = -15t^{5/3} + C$$

Given $S(0) = 2000$: $-15(0)^{5/3} + C = 2000$. Hence, $C = 2000$ and
$S(t) = -15t^{5/3} + 2000$. Now, we want to find t such that $S(t) = 800$,
that is:

$$-15t^{5/3} + 2000 = 800$$
$$-15t^{5/3} = -1200$$
$$t^{5/3} = 80$$

and $\qquad\qquad t = 80^{3/5} \approx 14$

Thus, the company should manufacture the computer for 14 months.

79. $L'(x) = g(x) = 2400x^{-1/2}$

$$L(x) = \int g(x)\,dx = \int 2400x^{-1/2}\,dx = 2400\int x^{-1/2}\,dx = 2400\,\frac{x^{1/2}}{\frac{1}{2}} + C$$

$$= 4800\,x^{1/2} + C$$

Given $L(16) = 19{,}200$: $19{,}200 = 4800(16)^{1/2} + C = 19{,}200 + C$. Hence,
$C = 0$ and $L(x) = 4800x^{1/2}$.
$L(25) = 4800(25)^{1/2} = 4800(5) = 24{,}000$ labor hours.

81. $\dfrac{dW}{dh} = 0.0015h^2$

$$W = \int 0.0015h^2\,dh = 0.0015\int h^2\,dh = 0.0015\,\frac{h^3}{3} + C = 0.0005h^3 + C$$

Given $W(60) = 108$: $108 = 0.0005(60)^3 + C$ or $108 = 108 + C$.
Hence, $C = 0$ and $W(h) = 0.0005h^3$. Now $5'10'' = 70''$ and
$W(70) = 0.0005(70)^3 = 171.5$ lb.

83. $\dfrac{dN}{dt} = 400 + 600\sqrt{t}$, $0 \le t \le 9$

$$N = \int (400 + 600\sqrt{t})\,dt = \int 400\,dt + 600\int t^{1/2}\,dt$$

$$= 400t + 600\,\frac{t^{3/2}}{\frac{3}{2}} + C = 400t + 400t^{3/2} + C$$

Given $N(0) = 5000$: $5000 = 400(0) + 400(0)^{3/2} + C$. Hence, $C = 5000$ and
$N(t) = 400t + 400t^{3/2} + 5000$.
$N(9) = 400(9) + 400(9)^{3/2} + 5000 = 3600 + 10{,}800 + 5000 = 19{,}400$

Things to remember:

1. GENERAL INDEFINITE INTEGRAL FORMULAS

(a) $\int [f(x)]^n f'(x)\,dx = \dfrac{[f(x)]^{n+1}}{n+1} + C, \ n \neq -1$

(b) $\int e^{f(x)} f'(x)\,dx = e^{f(x)} + C$

(c) $\int \dfrac{1}{f(x)} f'(x)\,dx = \ln|f(x)| + C$

(d) $\int u^n\,du = \dfrac{u^{n+1}}{n+1} + C, \ n \neq -1$

(e) $\int e^u\,du = e^u + C$

(f) $\int e^{au}\,du = \dfrac{1}{a} e^{au} + C, \ a \neq 0$ constant

(g) $\int \dfrac{1}{u}\,du = \ln|u| + C$

These formulas are valid if *u* is an independent variable or if *u* is a function of another variable and *du* is its differential with respect to that variable.

2. INTEGRATION BY SUBSTITUTION

(a) Select a substitution that appears to simplify the integrand. In particular, try to select *u* so that *du* is a factor in the integrand.

(b) Express the integrand entirely in terms of *u* and *du*, completely eliminating the original variable and its differential.

(c) Evaluate the new integral, if possible.

(d) Express the antiderivative found in Step (c) in terms of the original variable.

---

1. $\int (x^2 - 4)^5 2x\,dx$

Let $u = x^2 - 4$, then $du = 2x\,dx$ and

$\int (x^2 - 4)^5 2x\,dx = \int u^5\,du = \dfrac{u^6}{6} + C$ [using formula 1(d)]

$\qquad\qquad = \dfrac{(x^2 - 4)^6}{6} + C$

Check: $D_x\left[\dfrac{(x^2-4)^6}{6} + C\right] = \dfrac{1}{6}(6)(x^2-4)^5(2x) = (x^2-4)^5 2x$

3. $\displaystyle\int e^{4x}4\ dx$

Let $u = 4x$, then $du = 4\ dx$ and

$$\int e^{4x}4\ dx = \int e^u du = e^u + C \quad \text{[using formula 1(e)]}$$
$$= e^{4x} + C$$

Check: $D_x[e^{4x} + C] = e^{4x}(4)$

5. $\displaystyle\int \frac{1}{2t + 3}\ 2\ dt$

Let $u = 2t + 3$, then $du = 2\ dt$ and

$$\int \frac{1}{2t + 3}\ 2\ dt = \int \frac{1}{u}\ du = \ln|u| + C \quad \text{[using formula 1(g)]}$$
$$= \ln|2t + 3| + C$$

Check: $D_t[\ln|2t + 3| + C] = \dfrac{1}{2t + 3}(2)$

7. $\displaystyle\int (3x - 2)^7 dx$

Let $u = 3x - 2$, then $du = 3\ dx$ and

$$\int (3x - 2)^7 dx = \int (3x - 2)^7 \frac{3}{3}\ dx = \frac{1}{3}\int (3x - 2)^7 3\ dx = \frac{1}{3}\int u^7 du$$
$$= \frac{1}{3}\cdot\frac{u^8}{8} + C = \frac{(3x - 2)^8}{24} + C$$

Check: $D_x\left[\dfrac{(3x - 2)^8}{24} + C\right] = \dfrac{1}{24}(8)(3x - 2)^7(3) = (3x - 2)^7$

9. Let $u = x^2 + 3$, then $du = 2x\ dx$.

$$\int (x^2 + 3)^7 x\ dx = \int (x^2 + 3)^7 \frac{2}{2} x\ dx = \frac{1}{2}\int (x^2 + 3)^7 2x\ dx$$
$$= \frac{1}{2}\int u^7 du = \frac{1}{2}\cdot\frac{u^8}{8} + C = \frac{u^8}{16} + C = \frac{(x^2 + 3)^8}{16} + C$$

Check: $D_x\left[\dfrac{(x^2 + 3)^8}{16} + C\right] = \dfrac{1}{16}(8)(x^2 + 3)^7(2x) = (x^2 + 3)^7 x$

11. Let $u = -0.5t$, then $du = -0.5\ dt$.

$$\int 10e^{-0.5t}dt = 10\int e^{-0.5t}dt = 10\int e^{-0.5t}\frac{(-0.5)}{-0.5}\ dt$$
$$= \frac{10}{-0.5}\int e^u du = -20e^u + C = -20e^{-0.5t} + C$$

Check: $D_t[-20e^{-0.5t} + C] = -20e^{-0.5t}(-0.5) = 10e^{-0.5t}$

13. Let $u = 10x + 7$, then $du = 10\ dx$.

$$\int \frac{1}{10x + 7}\, dx = \int \frac{1}{10x + 7} \cdot \frac{10}{10}\, dx = \frac{1}{10} \int \frac{1}{10x + 7}\, 10\ dx$$

$$= \frac{1}{10} \int \frac{1}{u}\, du = \frac{1}{10} \ln|u| + C = \frac{1}{10} \ln|10x + 7| + C$$

Check: $D_x\left[\frac{1}{10} \ln|10x + 7| + C\right] = \left(\frac{1}{10}\right) \frac{1}{10x + 7} (10) = \frac{1}{10x + 7}$

15. Let $u = 2x^2$, then $du = 4x\ dx$.

$$\int xe^{2x^2}dx = \int e^{2x^2} x \cdot \frac{4}{4}\, dx = \frac{1}{4} \int e^{2x^2}(4x)\, dx$$

$$= \frac{1}{4} \int e^u du = \frac{1}{4} e^u + C = \frac{1}{4} e^{2x^2} + C$$

Check: $D_x\left[\frac{1}{4} e^{2x^2} + C\right] = \frac{1}{4} e^{2x^2}(4x) = xe^{2x^2}$

17. Let $u = x^3 + 4$, then $du = 3x^2 dx$.

$$\int \frac{x^2}{x^3 + 4}\, dx = \int \frac{1}{x^3 + 4} \cdot \frac{3}{3} x^2 dx = \frac{1}{3} \int \frac{1}{x^3 + 4} (3x^2)\, dx$$

$$= \frac{1}{3} \int \frac{1}{u}\, du = \frac{1}{3} \ln|u| + C = \frac{1}{3} \ln|x^3 + 4| + C$$

Check: $D_x\left[\frac{1}{3} \ln|x^3 + 4| + C\right] = \left(\frac{1}{3}\right) \frac{1}{x^3 + 4} (3x^2) = \frac{x^2}{x^3 + 4}$

19. Let $u = 3t^2 + 1$, then $du = 6t\ dt$.

$$\int \frac{t}{(3t^2 + 1)^4}\, dt = \int (3t^2 + 1)^{-4} t\ dt = \int (3t^2 + 1)^{-4}\, \frac{6}{6} t\ dt$$

$$= \frac{1}{6} \int (3t^2 + 1)^{-4} 6t\ dt = \frac{1}{6} \int u^{-4} du$$

$$= \frac{1}{6} \cdot \frac{u^{-3}}{-3} + C = \frac{-1}{18} (3t^2 + 1)^{-3} + C$$

Check: $D_t\left[\frac{-1}{18} (3t^2 + 1)^{-3} + C\right] = \left(\frac{-1}{18}\right)(-3)(3t^2 + 1)^{-4}(6t) = \frac{t}{(3t^2 + 1)^4}$

21. Let $u = 4 - x^3$, then $du = -3x^2 dx$.

$$\int \frac{x^2}{(4 - x^3)^2}\, dx = \int (4 - x^3)^{-2} x^2 dx = \int (4 - x^3)^{-2}\left(\frac{-3}{-3}\right) x^2 dx$$

$$= \frac{-1}{3} \int (4 - x^3)^{-2} (-3x^2) \, dx = \frac{-1}{3} \int u^{-2} \, du = \frac{-1}{3} \cdot \frac{u^{-1}}{-1} + C$$

$$= \frac{1}{3} (4 - x^3)^{-1} + C$$

Check: $D_x \left[\frac{1}{3} (4 - x^3)^{-1} + C \right] = \frac{1}{3} (-1)(4 - x^3)^{-2} (-3x^2) = \dfrac{x^2}{(4 - x^3)^2}$

23. $\displaystyle\int x\sqrt{x + 4} \, dx$

Let $u = x + 4$, then $du = dx$ and $x = u - 4$.

$$\int x\sqrt{x + 4} \, dx = \int (u - 4) u^{1/2} \, du = \int (u^{3/2} - 4u^{1/2}) \, du$$

$$= \frac{u^{5/2}}{\frac{5}{2}} - \frac{4u^{3/2}}{\frac{3}{2}} + C = \frac{2}{5} u^{5/2} - \frac{8}{3} u^{3/2} + C$$

$$= \frac{2}{5} (x + 4)^{5/2} - \frac{8}{3} (x + 4)^{3/2} + C \quad \text{(since } u = x + 4\text{)}$$

Check: $D_x \left[\frac{2}{5} (x + 4)^{5/2} - \frac{8}{3} (x + 4)^{3/2} + C \right]$

$$= \frac{2}{5} \left(\frac{5}{2} \right) (x + 4)^{3/2} (1) - \frac{8}{3} \left(\frac{3}{2} \right) (x + 4)^{1/2} (1)$$

$$= (x + 4)^{3/2} - 4(x + 4)^{1/2} = (x + 4)^{1/2} [(x + 4) - 4] = x\sqrt{x + 4}$$

25. $\displaystyle\int \frac{x}{\sqrt{x - 3}} \, dx$

Let $u = x - 3$, then $du = dx$ and $x = u + 3$.

$$\int \frac{x}{\sqrt{x - 3}} \, dx = \int \frac{u + 3}{u^{1/2}} \, du = \int (u^{1/2} + 3u^{-1/2}) \, du = \frac{u^{3/2}}{\frac{3}{2}} + \frac{3u^{1/2}}{\frac{1}{2}} + C$$

$$= \frac{2}{3} u^{3/2} + 6u^{1/2} + C = \frac{2}{3} (x - 3)^{3/2} + 6(x - 3)^{1/2} + C$$

$$\text{(since } u = x - 3\text{)}$$

Check: $D_x \left[\frac{2}{3} (x - 3)^{3/2} + 6(x - 3)^{1/2} + C \right]$

$$= \frac{2}{3} \left(\frac{3}{2} \right) (x - 3)^{1/2} (1) + 6 \left(\frac{1}{2} \right) (x - 3)^{-1/2} (1)$$

$$= (x - 3)^{1/2} + \frac{3}{(x - 3)^{1/2}} = \frac{x - 3 + 3}{(x - 3)^{1/2}} = \frac{x}{\sqrt{x - 3}}$$

27. $\displaystyle\int x(x - 4)^9 \, dx$

Let $u = x - 4$, then $du = dx$ and $x = u + 4$.

$$\int x(x - 4)^9 \, dx = \int (u + 4) u^9 \, du = \int (u^{10} + 4u^9) \, du$$

$$= \frac{u^{11}}{11} + \frac{4u^{10}}{10} + C = \frac{(x - 4)^{11}}{11} + \frac{2}{5} (x - 4)^{10} + C$$

Check: $D_x \left[\dfrac{(x - 4)^{11}}{11} + \dfrac{2}{5}(x - 4)^{10} + C \right]$

$\quad = \dfrac{1}{11}(11)(x - 4)^{10}(1) + \dfrac{2}{5}(10)(x - 4)^9(1)$

$\quad = (x - 4)^9 [(x - 4) + 4] = x(x - 4)^9$

29. Let $u = 1 + e^{2x}$, then $du = 2e^{2x}dx$.

$\displaystyle\int e^{2x}(1 + e^{2x})^3 dx = \int (1 + e^{2x})^3 \dfrac{2}{2} e^{2x}dx = \dfrac{1}{2}\int (1 + e^{2x})^3 2e^{2x}dx$

$\qquad = \dfrac{1}{2}\int u^3 du = \dfrac{1}{2}\cdot\dfrac{u^4}{4} + C = \dfrac{1}{8}(1 + e^{2x})^4 + C$

Check: $D_x \left[\dfrac{1}{8}(1 + e^{2x})^4 + C \right] = \left(\dfrac{1}{8}\right)(4)(1 + e^{2x})^3 e^{2x}(2) = e^{2x}(1 + e^{2x})^3$

31. Let $u = 4 + 2x + x^2$, then $du = (2 + 2x)dx = 2(1 + x)dx$.

$\displaystyle\int \dfrac{1 + x}{4 + 2x + x^2} dx = \int \dfrac{1}{4 + 2x + x^2} \cdot \dfrac{2(1 + x)}{2} dx = \dfrac{1}{2}\int \dfrac{1}{4 + 2x + x^2} 2(1 + x)dx$

$\qquad = \dfrac{1}{2}\int \dfrac{1}{u} du = \dfrac{1}{2}\ln|u| + C = \dfrac{1}{2}\ln|4 + 2x + x^2| + C$

Check: $D_x \left[\dfrac{1}{2}\ln|4 + 2x + x^2| + C \right] = \left(\dfrac{1}{2}\right)\dfrac{1}{4 + 2x + x^2}(2 + 2x) = \dfrac{1 + x}{4 + 2x + x^2}$

33. Let $u = x^2 + x + 1$, then $du = (2x + 1)dx$.

$\displaystyle\int (2x + 1)e^{x^2+x+1}dx = \int e^u du = e^u + C = e^{x^2+x+1} + C$

Check: $D_x[e^{x^2+x+1} + C] = e^{x^2+x+1}(2x + 1)$

35. Let $u = e^x - 2x$, then $du = (e^x - 2)dx$.

$\displaystyle\int (e^x - 2x)^3(e^x - 2)dx = \int u^3 du = \dfrac{u^4}{4} + C = \dfrac{(e^x - 2x)^4}{4} + C$

Check: $D_x \left[\dfrac{(e^x - 2x)^4}{4} + C \right] = \dfrac{1}{4}(4)(e^x - 2x)^3(e^x - 2) = (e^x - 2x)^3(e^x - 2)$

37. Let $u = x^4 + 2x^2 + 1$, then $du = (4x^3 + 4x)dx = 4(x^3 + x)dx$.

$\displaystyle\int \dfrac{x^3 + x}{(x^4 + 2x^2 + 1)^4} dx = \int (x^4 + 2x^2 + 1)^{-4}\dfrac{4}{4}(x^3 + x)dx$

$$= \frac{1}{4} \int (x^4 + 2x^2 + 1)^{-4} 4(x^3 + x) \, dx$$

$$= \frac{1}{4} \int u^{-4} \, du = \frac{1}{4} \cdot \frac{u^{-3}}{-3} + C = \frac{-u^{-3}}{12} + C$$

$$= \frac{-(x^4 + 2x^2 + 1)^{-3}}{12} + C$$

Check: $D_x \left[-\frac{1}{12}(x^4 + 2x^2 + 1)^{-3} + C \right] = \left(-\frac{1}{12} \right)(-3)(x^4 + 2x^2 + 1)^{-4}(4x^3 + 4x)$

$$= (x^4 + 2x^2 + 1)^{-4}(x^3 + x)$$

39. Let $u = 3x^2 + 7$, then $du = 6x \, dx$.

$$\int x\sqrt{3x^2 + 7} \, dx = \int (3x^2 + 7)^{1/2} x \, dx = \int (3x^2 + 7)^{1/2} \frac{6}{6} x \, dx$$

$$= \frac{1}{6} \int u^{1/2} \, du = \frac{1}{6} \cdot \frac{u^{3/2}}{\frac{3}{2}} + C = \frac{1}{9}(3x^2 + 7)^{3/2} + C$$

Check: $D_x \left[\frac{1}{9}(3x^2 + 7)^{3/2} + C \right] = \frac{1}{9} \left(\frac{3}{2} \right)(3x^2 + 7)^{1/2}(6x) = x(3x^2 + 7)^{1/2}$

41. $\int x(x^3 + 2)^2 \, dx = \int x(x^6 + 4x^3 + 4) \, dx = \int (x^7 + 4x^4 + 4x) \, dx$

$$= \frac{x^8}{8} + \frac{4}{5}x^5 + 2x^2 + C$$

Check: $D_x \left[\frac{x^8}{8} + \frac{4}{5}x^5 + 2x^2 + C \right] = x^7 + 4x^4 + 4x$

$$= x(x^6 + 4x^3 + 4) = x(x^3 + 2)^2$$

43. $\int x^2(x^3 + 2)^2 \, dx$

Let $u = x^3 + 2$, then $du = 3x^2 \, dx$.

$$\int x^2(x^3 + 2)^2 \, dx = \int (x^3 + 2)^2 \frac{3x^2}{3} \, dx = \frac{1}{3} \int (x^3 + 2)^2 3x^2 \, dx$$

$$= \frac{1}{3} \int u^2 \, du = \frac{1}{3} \cdot \frac{u^3}{3} + C = \frac{1}{9} u^3 + C = \frac{1}{9}(x^3 + 2)^3 + C$$

Check: $D_x \left[\frac{1}{9}(x^3 + 2)^3 + C \right] = \frac{1}{9}(3)(x^3 + 2)^2(3x^2) = x^2(x^3 + 2)^2$

45. Let $u = 2x^4 + 3$, then $du = 8x^3 \, dx$.

$$\int \frac{x^3}{\sqrt{2x^4 + 3}} \, dx = \int (2x^4 + 3)^{-1/2} x^3 \, dx = \int (2x^4 + 3)^{-1/2} \frac{8}{8} x^3 \, dx$$

$$= \frac{1}{8} \int u^{-1/2} \, du = \frac{1}{8} \cdot \frac{u^{1/2}}{\frac{1}{2}} + C = \frac{1}{4}(2x^4 + 3)^{1/2} + C$$

Check: $D_x\left[\frac{1}{4}(2x^4 + 3)^{1/2} + C\right] = \frac{1}{4}\left(\frac{1}{2}\right)(2x^4 + 3)^{-1/2}(8x^3) = \frac{x^3}{(2x^4 + 3)^{1/2}}$

47. Let $u = \ln x$, then $du = \frac{1}{x}dx$.

$$\int \frac{(\ln x)^3}{x}dx = \int u^3 du = \frac{u^4}{4} + C = \frac{(\ln x)^4}{4} + C$$

Check: $D_x\left[\frac{(\ln x)^4}{4} + C\right] = \frac{1}{4}(4)(\ln x)^3 \cdot \frac{1}{x} = \frac{(\ln x)^3}{x}$

49. Let $u = \frac{-1}{x} = -x^{-1}$, then $du = \frac{1}{x^2}dx$.

$$\int \frac{1}{x^2}e^{-1/x}dx = \int e^u du = e^u + C = e^{-1/x} + C$$

Check: $D_x[e^{-1/x} + C] = e^{-1/x}\left(\frac{1}{x^2}\right) = \frac{1}{x^2}e^{-1/x}$

51. $\frac{dx}{dt} = 7t^2(t^3 + 5)^6$

Let $u = t^3 + 5$, then $du = 3t^2 dt$.

$$x = \int 7t^2(t^3 + 5)^6 dt = 7\int t^2(t^3 + 5)^6 dt = 7\int (t^3 + 5)^6 \frac{3}{3}t^2 dt$$

$$= \frac{7}{3}\int u^6 du = \frac{7}{3}\cdot\frac{u^7}{7} + C = \frac{1}{3}(t^3 + 5)^7 + C$$

53. $\frac{dy}{dt} = \frac{3t}{\sqrt{t^2 - 4}}$

Let $u = t^2 - 4$, then $du = 2t\ dt$.

$$y = \int \frac{3t}{(t^2 - 4)^{1/2}}dt = 3\int (t^2 - 4)^{-1/2}t\ dt = 3\int (t^2 - 4)^{-1/2}\frac{2}{2}t\ dt$$

$$= \frac{3}{2}\int u^{-1/2}du = \frac{3}{2}\cdot\frac{u^{1/2}}{\frac{1}{2}} + C = 3(t^2 - 4)^{1/2} + C$$

55. $\frac{dp}{dx} = \frac{e^x + e^{-x}}{(e^x - e^{-x})^2}$

Let $u = e^x - e^{-x}$, then $du = (e^x + e^{-x})dx$.

$$p = \int \frac{e^x + e^{-x}}{(e^x - e^{-x})^2}dx = \int (e^x - e^{-x})^{-2}(e^x + e^{-x})dx = \int u^{-2}du$$

$$= \frac{u^{-1}}{-1} + C = -(e^x - e^{-x})^{-1} + C$$

57. Let $v = au$, then $dv = a\ du$.

$$\int e^{au}du = \int e^{au}\frac{a}{a}\ du = \frac{1}{a}\int e^{au}a\ du = \frac{1}{a}\int e^v dv = \frac{1}{a}e^v + C = \frac{1}{a}e^{au} + C$$

<u>Check</u>: $D_u\left[\frac{1}{a}e^{au} + C\right] + \frac{1}{a}e^{au}(a) = e^{au}$

59. $p'(x) = \dfrac{-6000}{(3x + 50)^2}$

Let $u = 3x + 50$, then $du = 3\ dx$.

$$p(x) = \int\frac{-6000}{(3x + 50)^2}dx = -6000\int(3x + 50)^{-2}dx = -6000\int(3x + 50)^{-2}\frac{3}{3}dx$$

$$= -2000\int u^{-2}du = -2000\cdot\frac{u^{-1}}{-1} + C = \frac{2000}{3x + 50} + C$$

Given $p(150) = 4$:

$$4 = \frac{2000}{(3\cdot 150 + 50)} + C$$

$$4 = \frac{2000}{500} + C$$

$$C = 0$$

Thus, $p(x) = \dfrac{2000}{3x + 50}$.

Now, $2.50 = \dfrac{2000}{3x + 50}$

$$2.50(3x + 50) = 2000$$
$$7.5x + 125 = 2000$$
$$7.5x = 1875$$
$$x = 250$$

Thus, the demand is 250 bottles when the price is $2.50.

61. $C'(x) = 12 + \dfrac{500}{x + 1}$, $x > 0$

$$C(x) = \int\left(12 + \frac{500}{x + 1}\right)dx = \int 12\ dx + 500\int\frac{1}{x + 1}dx \quad (u = x + 1,\ du = dx)$$

$$= 12x + 500\ \ln(x + 1) + C$$

Now, $C(0) = 2000$. Thus, $C(x) = 12x + 500\ \ln(x + 1) + 2000$. The average cost is:

$$\overline{C}(x) = 12 + \frac{500}{x}\ln(x + 1) + \frac{2000}{x}$$

and

$$\overline{C}(1000) = 12 + \frac{500}{1000}\ln(1001) + \frac{2000}{1000} = 12 + \frac{1}{2}\ln(1001) + 2$$

$$\approx 17.45 \text{ or } \$17.45 \text{ per pair of shoes}$$

63. $S'(t) = 10 - 10e^{-0.1t}$, $0 \le t \le 24$

$$S(t) = \int (10 - 10e^{-0.1t})\,dt = \int 10\,dt - 10\int e^{-0.1t}\,dt$$

$$= 10t - \frac{10}{-0.1}\int e^{-0.1t}(-0.1)\,dt = 10t + 100e^{-0.1t} + C$$

Given $S(0) = 0$:
$0 = 0 + 100e^0 + C$
$C = -100$
Thus, $S(t) = 10t + 100e^{-0.1t} - 100$, $0 \le t \le 24$.
The total sales for the first 12 months is given by:
$S(12) = 10(12) + 100e^{-0.1(12)} - 100$
$\qquad = 20 + 100e^{-1.2}$
$\qquad \approx 50$ or $50 million

65. $Q(t) = \int R(t)\,dt = \int \left(\frac{100}{t+1} + 5\right)dt = 100\int \frac{1}{t+1}\,dt + \int 5\,dt$

$$= 100\ \ln(t+1) + 5t + C$$

Given $Q(0) = 0$:
$0 = 100\ \ln(1) + 0 + C$
Thus, $C = 0$ and $Q(t) = 100\ \ln(t+1) + 5t$, $0 \le t \le 20$.
$Q(9) = 100\ \ln(9+1) + 5(9) = 100\ \ln 10 + 45 \approx 275$ thousand barrels.

67. $W(t) = \int W'(t)\,dt = \int 0.2e^{0.1t}\,dt = \frac{0.2}{0.1}\int e^{0.1t}(0.1)\,dt = 2e^{0.1t} + C$

Given $W(0) = 2$:
$2 = 2e^0 + C$.
Thus, $C = 0$ and $W(t) = 2e^{0.1t}$.
The weight of the culture after 8 hours is given by:
$W(8) = 2e^{0.1(8)} = 2e^{0.8} \approx 4.45$ grams.

69. $\dfrac{dN}{dt} = \dfrac{-2000t}{1 + t^2}$, $0 \le t \le 10$

Let $u = 1 + t^2$, then $du = 2t\ dt$.

$$N = \int \frac{-2000t}{1 + t^2}\,dt = \frac{-2000}{2}\int \frac{2t}{1 + t^2}\,dt = -1000\int \frac{1}{u}\,du$$

$$= -1000\ \ln|u| + C = -1000\ \ln(1 + t^2) + C$$

Given $N(0) = 5000$:
$5000 = -1000\ \ln(1) + C$
Hence, $C = 5000$ and $N(t) = 5000 - 1000\ \ln(1 + t^2)$, $0 \le t \le 10$.
Now, $N(10) = 5000 - 1000\ \ln(1 + 10^2)$
$\qquad\qquad = 5000 - 1000\ \ln(101)$
$\qquad\qquad \approx 385$ bacteria per milliliter.

71. $N'(t) = 6e^{-0.1t}$, $0 \le t \le 15$

$$N(t) = \int N'(t)\,dt = \int 6e^{-0.1t}\,dt = 6\int e^{-0.1t}\,dt$$

$$= \frac{6}{-0.1}\int e^{-0.1t}(-0.1)\,dt = -60e^{-0.1t} + C$$

Given $N(0) = 40$:

$40 = -60e^0 + C$

Hence, $C = 100$ and $N(t) = 100 - 60e^{-0.1t}$, $0 \le t \le 15$.
The number of words per minute after completing the course is:
$N(15) = 100 - 60e^{-0.1(15)} = 100 - 60e^{-1.5} \approx 87$ words per minute.

73. $\frac{dE}{dt} = 5000(t + 1)^{-3/2}$, $t \ge 0$

Let $u = t + 1$, then $du = dt$

$$E = \int 5000(t + 1)^{-3/2}\,dt = 5000\int (t + 1)^{-3/2}\,dt = 5000\int u^{-3/2}\,du$$

$$= 5000\frac{u^{-1/2}}{-\frac{1}{2}} + C = -10{,}000(t + 1)^{-1/2} + C$$

$$= \frac{-10{,}000}{\sqrt{t + 1}} + C$$

Given $E(0) = 2000$:

$$2000 = \frac{-10{,}000}{\sqrt{1}} + C$$

Hence, $C = 12{,}000$ and $E(t) = 12{,}000 - \frac{10{,}000}{\sqrt{t + 1}}$.

The projected enrollment 15 years from now is:

$$E(15) = 12{,}000 - \frac{10{,}000}{\sqrt{15 + 1}} = 12{,}000 - \frac{10{,}000}{\sqrt{16}} = 12{,}000 - \frac{10{,}000}{4}$$

$$= 9500 \text{ students}$$

EXERCISE 11-3

Things to remember:

<u>1</u>. DEFINITE INTEGRAL

The DEFINITE INTEGRAL of a continuous function f over an
interval from $x = a$ to $x = b$ is the net change of an
antiderivative of f over the interval. That is, if $F(x)$ is an
antiderivative of $f(x)$, then

$$\int_a^b f(x)\,dx = F(x)\Big|_a^b = F(b) - F(a) \text{ where } F'(x) = f(x)$$

Integrand: $f(x)$ Upper limit: b Lower limit: a

2. DEFINITE INTEGRAL PROPERTIES

(a) $\displaystyle\int_a^a f(x)\,dx = 0$

(b) $\displaystyle\int_a^b f(x)\,dx = -\int_b^a f(x)\,dx$

(c) $\displaystyle\int_a^b Kf(x)\,dx = K\int_a^b f(x)\,dx \qquad K$ is a constant

(d) $\displaystyle\int_a^b [f(x) \pm g(x)]\,dx = \int_a^b f(x)\,dx \pm \int_a^b g(x)\,dx$

(e) $\displaystyle\int_a^b f(x)\,dx = \int_a^c f(x)\,dx + \int_c^b f(x)\,dx$

---

1. $\displaystyle\int_2^3 2x\,dx = 2\cdot\frac{x^2}{2}\Big|_2^3 = 3^2 - 2^2 = 5$

3. $\displaystyle\int_3^4 5\,dx = 5x\Big|_3^4 = 5\cdot4 - 5\cdot3 = 5$

5. $\displaystyle\int_1^3 (2x - 3)\,dx = (x^2 - 3x)\Big|_1^3 = (3^2 - 3\cdot3) - (1^2 - 3\cdot1) = 2$

7. $\displaystyle\int_0^4 (3x^2 - 4)\,dx = (x^3 - 4x)\Big|_0^4 = (4^3 - 4\cdot4) - (0^3 - 4\cdot0) = 48$

9. $\displaystyle\int_{-3}^4 (4 - x^2)\,dx = \left(4x - \frac{x^3}{3}\right)\Big|_{-3}^4 = \left(4\cdot4 - \frac{4^3}{3}\right) - \left(4(-3) - \frac{(-3)^3}{3}\right)$

$$= 16 - \frac{64}{3} + 3 = -\frac{7}{3}$$

11. $\displaystyle\int_0^1 24x^{11}\,dx = 24\frac{x^{12}}{12}\Big|_0^1 = 2x^{12}\Big|_0^1 = 2\cdot1^{12} - 2\cdot0^{12} = 2$

13. $\displaystyle\int_0^1 e^{2x}\,dx = \frac{1}{2}e^{2x}\Big|_0^1 = \frac{1}{2}e^{2\cdot1} - \frac{1}{2}e^{2\cdot0} = \frac{1}{2}(e^2 - 1)$

15. $\displaystyle\int_1^{3.5} 2x^{-1}\,dx = 2\ln x\Big|_1^{3.5} = 2\ln 3.5 - 2\ln 1$

$$= 2\ln 3.5 \qquad (\text{Recall: } \ln 1 = 0)$$

17. $\displaystyle\int_1^2 (2x^{-2} - 3)\,dx = (-2x^{-1} - 3x)\Big|_1^2 = \left(-\frac{2}{x} - 3x\right)\Big|_1^2$

$$= -\frac{2}{2} - 3\cdot2 - \left(-\frac{2}{1} - 3\cdot1\right) = -7 - (-5) = -2$$

19. $\displaystyle\int_1^4 3\sqrt{x}\,dx = 3\int_1^4 x^{1/2}\,dx = 3\cdot\frac{2}{3}x^{3/2}\Big|_1^4 = 2x^{3/2}\Big|_1^4$

$$= 2\cdot4^{3/2} - 2\cdot1^{3/2} = 16 - 2 = 14$$

21. $\int_2^3 12(x^2 - 4)^5 x\, dx$. Consider the indefinite integral $\int 12(x^2 - 4)^5 x\, dx$.

Let $u = x^2 - 4$, then $du = 2x\, dx$.

$$\int 12(x^2 - 4)^5 x\, dx = 6\int (x^2 - 4)^5 2x\, dx = 6\int u^5 du$$

$$= 6\frac{u^6}{6} + C = u^6 + C = (x^2 - 4)^6 + C$$

Thus,

$$\int_2^3 12(x^2 - 4)^5 x\, dx = (x^2 - 4)^6 \Big|_2^3 = (3^2 - 4)^6 - (2^2 - 4)^6 = 5^6 = 15{,}625.$$

23. $\int_3^9 \frac{1}{x - 1}\, dx$

Let $u = x - 1$. Then $du = dx$ and $u = 8$ when $x = 9$, $u = 2$ when $x = 3$.

Thus,

$$\int_3^9 \frac{1}{x - 1}\, dx = \int_2^8 \frac{1}{u}\, du = \ln u \Big|_2^8 = \ln 8 - \ln 2 = \ln 4.$$

25. $\int_{-5}^{10} e^{-0.05x}\, dx$

Let $u = -0.05x$. Then $du = -0.05\, dx$ and $u = -0.5$ when $x = 10$, $u = 0.25$ when $x = -5$. Thus,

$$\int_{-5}^{10} e^{-0.05x}\, dx = -\frac{1}{0.05}\int_{-5}^{10} e^{-0.05x}(-0.05)\, dx = -\frac{1}{0.05}\int_{0.25}^{-0.5} e^u\, du$$

$$= -\frac{1}{0.05} e^u \Big|_{0.25}^{-0.5} = -\frac{1}{0.05}[e^{-0.5} - e^{0.25}]$$

$$= 20(e^{0.25} - e^{-0.5}) \approx 13.55$$

27. $\int_{-6}^0 \sqrt{4 - 2x}\, dx$

Consider the indefinite integral $\int \sqrt{4 - 2x}\, dx = \int (4 - 2x)^{1/2}\, dx$.

Let $u = 4 - 2x$, then $du = -2\, dx$.

$$\int (4 - 2x)^{1/2}\, dx = -\frac{1}{2}\int (4 - 2x)^{1/2}(-2)\, dx = -\frac{1}{2}\int u^{1/2}\, du$$

$$= -\frac{1}{2} \cdot \frac{u^{3/2}}{\frac{3}{2}} + C = -\frac{1}{3} u^{3/2} + C = -\frac{1}{3}(4 - 2x)^{3/2} + C$$

Thus,

$$\int_{-6}^0 (4 - 2x)^{1/2}\, dx = -\frac{1}{3}(4 - 2x)^{3/2} \Big|_{-6}^0 = -\frac{1}{3}[4^{3/2} - 16^{3/2}]$$

$$= -\frac{1}{3}(8 - 64) = \frac{56}{3} \approx 18.667$$

29. $\int_{-1}^{7} \dfrac{x}{\sqrt{x + 2}}\, dx$

Consider the indefinite integral $\int \dfrac{x}{\sqrt{x + 2}}\, dx = \int x(x + 2)^{-1/2} dx.$

Let $u = x + 2$, then $du = dx$ and $x = u - 2.$

$\int x(x + 2)^{-1/2} dx = \int (u - 2)u^{-1/2} du = \int (u^{1/2} - 2u^{-1/2})\, du$

$\qquad = \dfrac{u^{3/2}}{\frac{3}{2}} - \dfrac{2u^{1/2}}{\frac{1}{2}} + C = \dfrac{2}{3}(x + 2)^{3/2} - 4(x + 2)^{1/2} + C$

Thus,

$\int_{-1}^{7} \dfrac{x}{\sqrt{x + 2}}\, dx = \left[\dfrac{2}{3}(x + 2)^{3/2} - 4(x + 2)^{1/2} \right]\Big|_{-1}^{7}$

$\qquad = \dfrac{2}{3}(9)^{3/2} - 4(9)^{1/2} - \left(\dfrac{2}{3}(1)^{3/2} - 4(1)^{1/2} \right)$

$\qquad = \dfrac{2}{3}(27) - 12 - \left(\dfrac{2}{3} - 4 \right) = 6 + \dfrac{10}{3} = \dfrac{28}{3} \approx 9.333.$

31. $\int_{0}^{1} (e^{2x} - 2x)^2 (e^{2x} - 1)\, dx = \dfrac{1}{2}\int_{0}^{1} (e^{2x} - 2x)^2 (2e^{2x} - 2)\, dx$

$\qquad = \dfrac{1}{2} \cdot \dfrac{(e^{2x} - 2x)^3}{3}\Big|_{0}^{1}$ [Note : The integrand

$\qquad = \dfrac{1}{6}(e^{2x} - 2x)^3\Big|_{0}^{1}$ has the form $u^2 du$; an

$\qquad \qquad \qquad \qquad \qquad$ antiderivative is

$\qquad = \dfrac{1}{6}[(e^2 - 2)^3 - 1]$ $\dfrac{u^3}{3} = \dfrac{(e^{2x} - 2x)^3}{3}.$]

33. $\int_{-2}^{-1} (x^{-1} + 2x)\, dx = (\ln|x| + x^2)\Big|_{-2}^{-1}$

$\qquad = \ln|-1| + (-1)^2 - [\ln|-2| + (-2)^2]$

$\qquad = 1 - \ln 2 - 4$

$\qquad = -3 - \ln 2$

35. $\int_{2}^{3} x\sqrt{2x^2 - 3}\, dx = \int_{2}^{3} x(2x^2 - 3)^{1/2} dx$

$\qquad = \dfrac{1}{4}\int_{2}^{3} (2x^2 - 3)^{1/2} 4x\, dx$ [Note : The integrand has the

$\qquad = \dfrac{1}{4}\left(\dfrac{2}{3}\right)(2x^2 - 3)^{3/2}\Big|_{2}^{3}$ form $u^{1/2} du$; the antiderivative is $\dfrac{2}{3}u^{3/2} = \dfrac{2}{3}(2x^2 - 3)^{3/2}.$]

$\qquad = \dfrac{1}{6}[2(3)^2 - 3]^{3/2} - \dfrac{1}{6}[2(2)^2 - 3]^{3/2}$

$\qquad = \dfrac{1}{6}(15)^{3/2} - \dfrac{1}{6}(5)^{3/2} = \dfrac{1}{6}[15^{3/2} - 5^{3/2}]$

37. $\displaystyle\int_0^1 \frac{x - 1}{x^2 - 2x + 3}\, dx$

Consider the indefinite integral and let $u = x^2 - 2x + 3$.
Then $du = (2x - 2)dx = 2(x - 1)dx$.

$$\int \frac{x - 1}{x^2 - 2x + 3}\, dx = \frac{1}{2}\int \frac{2(x - 1)}{x^2 - 2x + 3}\, dx = \frac{1}{2}\int \frac{1}{u}\, du = \frac{1}{2}\ln|u| + C$$

Thus,

$$\int_0^1 \frac{x - 1}{x^2 - 2x + 3}\, dx = \frac{1}{2}\ln|x^2 - 2x + 3|\,\Big|_0^1$$

$$= \frac{1}{2}\ln 2 - \frac{1}{2}\ln 3 = \frac{1}{2}(\ln 2 - \ln 3) = \frac{1}{2}\ln\left(\frac{2}{3}\right)$$

39. $\displaystyle\int_{-1}^1 \frac{e^{-x} - e^x}{(e^{-x} + e^x)^2}\, dx$

Consider the indefinite integral and let $u = e^{-x} + e^x$.
Then $du = (-e^{-x} + e^x)dx = -(e^{-x} - e^x)dx$.

$$\int \frac{e^{-x} - e^x}{(e^{-x} + e^x)^2}\, dx = -\int \frac{-(e^{-x} - e^x)}{(e^{-x} + e^x)^2}\, dx = -\int u^{-2}\, du = \frac{-u^{-1}}{-1} + C = \frac{1}{u} + C$$

Thus,

$$\int_{-1}^1 \frac{e^{-x} - e^x}{(e^{-x} + e^x)^2}\, dx = \frac{1}{e^{-x} + e^x}\,\Big|_{-1}^1 = \frac{1}{e^{-1} + e^1} - \frac{1}{e^{-(-1)} + e^{-1}}$$

$$= \frac{1}{e^{-1} + e} - \frac{1}{e^{-1} + e} = 0$$

41. Total loss in value in the first 5 years:

$$V(5) - V(0) = \int_0^5 V'(t)\, dt = \int_0^5 500(t - 12)\, dt = 500\left(\frac{t^2}{2} - 12t\right)\Big|_0^5$$

$$= 500\left(\frac{25}{2} - 60\right) = -\$23,750$$

Total loss in value in the second 5 years:

$$V(10) - V(5) = \int_5^{10} V'(t)\, dt = \int_5^{10} 500(t - 12)\, dt = 500\left(\frac{t^2}{2} - 12t\right)\Big|_5^{10}$$

$$= 500\left[(50 - 120) - \left(\frac{25}{2} - 60\right)\right] = -\$11,250$$

43. (A) To find the useful life, set $C'(t) = R'(t)$ and solve for t.

$$\frac{1}{11}t = 5te^{-t^2}$$

$$e^{t^2} = 55$$

$$t^2 = \ln 55$$

$$t = \sqrt{\ln 55} \approx 2 \text{ years}$$

(B) The total profit accumulated during the useful life is:

$$P(2) - P(0) = \int_0^2 [R'(t) - C'(t)]dt = \int_0^2 \left(5te^{-t^2} - \frac{1}{11}t\right)dt$$

$$= \int_0^2 5te^{-t^2}dt - \int_0^2 \frac{1}{11}t\ dt$$

$$= -\frac{5}{2}\int_0^2 e^{-t^2}(-2t)dt - \frac{1}{11}\int_0^2 t\ dt$$

[Note: In the first integral, the integrand has the form $e^u du$, where $u = -t^2$; an antiderivative is $e^u = e^{-t^2}$.]

$$= -\frac{5}{2}e^{-t^2}\Big|_0^2 - \frac{1}{22}t^2\Big|_0^2$$

$$= -\frac{5}{2}e^{-4} + \frac{5}{2} - \frac{4}{22} = \frac{51}{22} - \frac{5}{2}e^{-4} \approx 2.272$$

Thus, the total profit is approximately $2,272.

45. $g(x) = 2400x^{-1/2}$ and $L'(x) = g(x)$.

The number of labor hours to assemble the 17th through the 25th control units is:

$$L(25) - L(16) = \int_{16}^{25} g(x)\,dx = \int_{16}^{25} 2400x^{-1/2}dx = 2400(2)x^{1/2}\Big|_{16}^{25}$$

$$= 4800x^{1/2}\Big|_{16}^{25} = 4800[25^{1/2} - 16^{1/2}] = 4800 \text{ labor hours}$$

47. Rate of production: $R(t) = \dfrac{100}{t+1} + 5,\ 0 \le t \le 20$

Total production from year N to year M is given by:

$$P = \int_N^M R(t)\,dt = \int_N^M \left(\frac{100}{t+1} + 5\right)dt = 100\int_N^M \frac{1}{t+1}\,dt + \int_N^M 5\ dt$$

$$= 100\ \ln|t+1|\ \Big|_N^M + 5t\Big|_N^M$$

$$= 100\ \ln(M+1) - 100\ \ln(N+1) + 5(M-N)$$

Thus, for total production during the first 10 years, let $M = 10$ and $N = 0$.

$P = 100\ \ln 11 - 100\ \ln 1 + 5(10 - 0)$
$\ = 100\ \ln 11 + 50 \approx 290$ thousand barrels

For the total production from the end of the 10th year to the end of the 20th year, let $M = 20$ and $N = 10$.

$P = 100\ \ln 21 - 100\ \ln 11 + 5(20 - 10)$
$\ = 100\ \ln 21 - 100\ \ln 11 + 50 \approx 115$ thousand barrels

49. $S'(t) = 10 - 10e^{-0.1t},\ 0 \le t \le 24$

The total sales during the first 12 months of the campaign is given by:

$$S(12) - S(0) = \int_0^{12} (10 - 10e^{-0.1t})\,dt = \int_0^{12} 10\ dt - 10\int_0^{12} e^{-0.1t}dt$$

$$= 10t\Big|_0^{12} - \frac{10}{-0.1}\int_0^{12} e^{-0.1t}(-0.1)\,dt \quad \text{(Let } u = -0.1t, \text{ then } du = -0.1dt.\text{)}$$

$$= 120 + 100e^{-0.1t}\Big|_0^{12}$$

$$= 120 + 100e^{-0.1(12)} - 100e^0$$
$$= 20 + 100e^{-1.2} \approx 50 \text{ or } \$50 \text{ million}$$

The total sales during the second twelve months is:

$$S(24) - S(12) = \int_{12}^{24} (10 - 10e^{-0.1t})\,dt = \int_{12}^{24} 10\,dt - 10\int_{12}^{24} e^{-0.1t}\,dt$$

$$= 10t\Big|_{12}^{24} + 100e^{-0.1t}\Big|_{12}^{24} = 120 + 100e^{-2.4} - 100e^{-1.2}$$

$$\approx 99 \text{ or } \$99 \text{ million}$$

51. Billions of cubic feet of wood that will be consumed from 1980 to 1990 is given by:

$$Q(20) - Q(10) = \int_{10}^{20} Q'(t)\,dt = \int_{10}^{20} (12 + 0.006t^2)\,dt = \left(12t + 0.006\frac{t^3}{3}\right)\Big|_{10}^{20}$$

$$= (12 \cdot 20 + 0.002 \cdot 20^3) - (12 \cdot 10 + 0.002 \cdot 10^3)$$
$$= 134 \text{ billion cubic feet}$$

53. $W'(t) = 0.2e^{0.1t}$

The weight increase during the first eight hours is given by:

$$W(8) - W(0) = \int_0^8 W'(t)\,dt = \int_0^8 0.2e^{0.1t}\,dt = 0.2\int_0^8 e^{0.1t}\,dt$$

$$= \frac{0.2}{0.1}\int_0^8 e^{0.1t}(0.1)\,dt \qquad \text{(Let } u = 0.1t, \text{ then } du = 0.1dt.)$$

$$= 2e^{0.1t}\Big|_0^8 = 2e^{0.8} - 2 \approx 2.45 \text{ grams}$$

The weight increase during the second eight hours, i.e., from the 8th hour through the 16th hour, is given by:

$$W(16) - W(8) = \int_8^{16} W'(t)\,dt = \int_8^{16} 0.2e^{0.1t}\,dt = 2e^{0.1t}\Big|_8^{16}$$

$$= 2e^{1.6} - 2e^{0.8} \approx 5.45 \text{ grams}$$

55. $N'(t) = 6e^{-0.1t}$, $0 \le t \le 15$

The improvement from the end of week K to the end of week M is given by:

$$N = \int_K^M N'(t)\,dt = \int_K^M 6e^{-0.1t}\,dt$$

$$= \frac{6}{-0.1}\int_K^M e^{-0.1t}(-0.1)\,dt \quad \text{[Let } u = -0.1t, \text{ then } du = -0.1dt.]$$

$$= -60e^{-0.1t}\Big|_K^M = -60e^{-0.M} + 60e^{-0.K}$$

Thus, for the improvement during the first five weeks, let $K = 0$ and $M = 5$. Then,
$N = -60e^{-0.5} + 60 \approx 24$ words per minute.
For the improvement during the second five weeks, let $K = 5$ and $M = 10$. Then,
$N = -60e^{-1} + 60e^{-0.5} \approx 14$ words per minute.
For the improvement during the last five weeks, let $K = 10$ and $M = 15$. Then,
$N = -60e^{-1.5} + 60e^{-1} \approx 9$ words per minute.

Things to remember:

1. AREA UNDER A CURVE

   If f is continuous and $f(x) \geq 0$ over the interval $[a, b]$, then the area bounded by $y = f(x)$, the x axis $(y = 0)$, and the vertical lines $x = a$ and $x = b$, is given exactly by:

   $$A = \int_a^b f(x)\,dx$$

   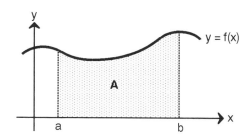

   If $f(x) \leq 0$ over the interval $[a, b]$, then the area of the region bounded by $y = f(x)$, the x axis, and the vertical lines $x = a$ and $x = b$ is given by

   $$\int_a^b [-f(x)]\,dx.$$

   Finally, if $f(x)$ is positive for some values of x and negative for others, the area between the graph of f and the x axis can be obtained by dividing $[a, b]$ into subintervals on which f is always positive or always negative, finding the area over each subinterval, and then summing these areas.

2. AREA BETWEEN TWO CURVES

   If f and g are continuous and $f(x) \geq g(x)$ over the interval $[a, b]$, then the area bounded by $y = f(x)$ and $y = g(x)$, for $a \leq x \leq b$, is given exactly by:

   $$A = \int_a^b [f(x) - g(x)]\,dx.$$

   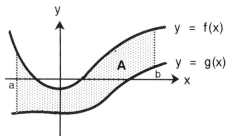

3. COEFFICIENT OF INEQUALITY FOR A LORENZ CURVE

   If $y = f(x)$ is the equation of a Lorenz curve, then the

   $$\text{coefficient of inequality} = 2\int_0^1 [x - f(x)]\,dx.$$

1. $A = \displaystyle\int_{1}^{3} (2x + 4)\,dx = \left(\dfrac{2x^2}{2} + 4x\right)\Big|_{1}^{3}$

$\quad = \left(x^2 + 4x\right)\Big|_{1}^{3} = (9 + 12) - (1 + 4)$

$\quad = 21 - 5 = 16$

3. $A = \displaystyle\int_{1}^{2} 3x^2\,dx = \dfrac{3x^3}{3}\Big|_{1}^{2} = x^3\Big|_{1}^{2}$

$\quad = 2^3 - 1 = 7$

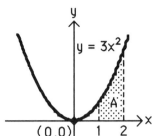

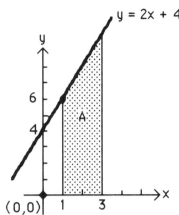

5. $A = \displaystyle\int_{-1}^{0} (x^2 + 2)\,dx = \left(\dfrac{x^3}{3} + 2x\right)\Big|_{-1}^{0}$

$\quad = 0 - \left(-\dfrac{1}{3} - 2\right) = \dfrac{7}{3}$

7. $A = \displaystyle\int_{-1}^{2} (4 - x^2)\,dx = \left(4x - \dfrac{x^3}{3}\right)\Big|_{-1}^{2}$

$\quad = \left(4 \cdot 2 - \dfrac{2^3}{3}\right) - \left(-4 + \dfrac{1}{3}\right)$

$\quad = \dfrac{16}{3} + \dfrac{11}{3} = 9$

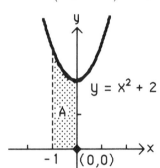

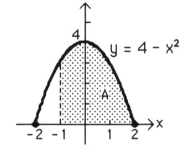

9. $A = \displaystyle\int_{-1}^{2} e^x\,dx = e^x\Big|_{-1}^{2}$

$\quad = e^2 - e^{-1}$

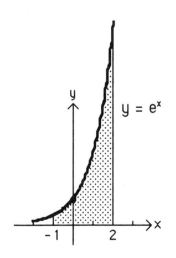

11. $A = \displaystyle\int_{0.5}^{1} \frac{1}{t}\,dt = \ln t \Big|_{0.5}^{1}$

$\quad = \ln 1 - \ln 0.5 = -\ln 0.5 \approx 0.6931$

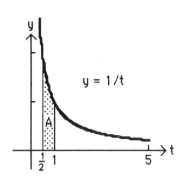

13. $A = \displaystyle\int_{-1}^{2} [12 - (-2x + 8)]\,dx = \int_{-1}^{2} (2x + 4)\,dx$

$\quad = \left(\dfrac{2x^2}{2} + 4x\right)\Big|_{-1}^{2} = (x^2 + 4x)\Big|_{-1}^{2} = (4 + 8) - (1 - 4)$

$\quad = 12 + 3 = 15$

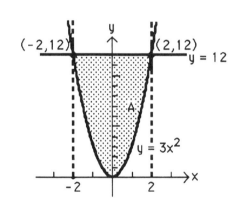

15. $A = \displaystyle\int_{-2}^{2} (12 - 3x^2)\,dx = \left(12x - \dfrac{3x^3}{3}\right)\Big|_{-2}^{2}$

$\quad = (12x - x^3)\Big|_{-2}^{2}$

$\quad = (12\cdot 2 - 2^3) - [12\cdot(-2) - (-2)^3]$

$\quad = 16 - (-16) = 32$

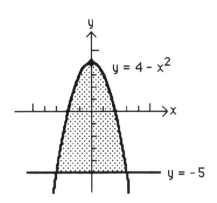

17. $(3, -5)$ and $(-3, -5)$ are the points of intersection.

$A = \displaystyle\int_{-3}^{3} [4 - x^2 - (-5)]\,dx$

$\quad = \displaystyle\int_{-3}^{3} (9 - x^2)\,dx = \left(9x - \dfrac{x^3}{3}\right)\Big|_{-3}^{3}$

$\quad = \left(9\cdot 3 - \dfrac{3^3}{3}\right) - \left(9(-3) - \dfrac{(-3)^3}{3}\right)$

$\quad = 18 + 18 = 36$

19. $A = \displaystyle\int_{-1}^{2} [(x^2 + 1) - (2x - 2)]\,dx$

$\qquad = \displaystyle\int_{-1}^{2} (x^2 - 2x + 3)\,dx = \left(\dfrac{x^3}{3} - x^2 + 3x\right)\Big|_{-1}^{2}$

$\qquad = \left(\dfrac{8}{3} - 4 + 6\right) - \left(-\dfrac{1}{3} - 1 - 3\right)$

$\qquad = 3 - 4 + 6 + 1 + 3 = 9$

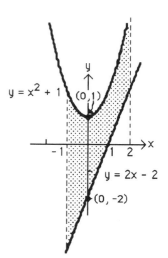

21. $A = A_1 + A_2 = \displaystyle\int_{-2}^{0} -x\,dx + \int_{0}^{1} -(-x)\,dx$

$\qquad = -\displaystyle\int_{-2}^{0} x\,dx + \int_{0}^{1} x\,dx$

$\qquad = -\dfrac{x^2}{2}\Big|_{-2}^{0} + \dfrac{x^2}{2}\Big|_{0}^{1}$

$\qquad = -\left(0 - \dfrac{(-2)^2}{2}\right) + \left(\dfrac{1^2}{2} - 0\right)$

$\qquad = 2 + \dfrac{1}{2} = \dfrac{5}{2}$

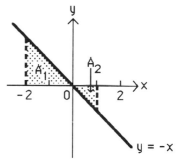

23. $A = \displaystyle\int_{1}^{2} \left[e^{0.5x} - \left(-\dfrac{1}{x}\right)\right]dx$

$\qquad = \displaystyle\int_{1}^{2} \left(e^{0.5x} + \dfrac{1}{x}\right)dx$

$\qquad = \left(\dfrac{e^{0.5x}}{0.5} + \ln|x|\right)\Big|_{1}^{2}$

$\qquad = 2e + \ln 2 - 2e^{0.5}$

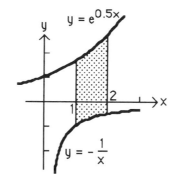

25. $A = A_1 + A_2 = \displaystyle\int_{0}^{2} -(x^2 - 4)\,dx + \int_{2}^{3} (x^2 - 4)\,dx$

$\qquad = \displaystyle\int_{0}^{2} (4 - x^2)\,dx + \int_{2}^{3} (x^2 - 4)\,dx$

$\qquad = \left(4x - \dfrac{x^3}{3}\right)\Big|_{0}^{2} + \left(\dfrac{x^3}{3} - 4x\right)\Big|_{2}^{3}$

$\qquad = \left(8 - \dfrac{8}{3}\right) + \left(\dfrac{27}{3} - 12\right) - \left(\dfrac{8}{3} - 8\right)$

$\qquad = 13 - \dfrac{16}{3} = \dfrac{39}{3} - \dfrac{16}{3} = \dfrac{23}{3}$

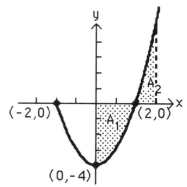

27. The graphs of $y = 10 - 2x$ and $y = 4 + 2x$, $0 \le x \le 4$, are shown at the right.

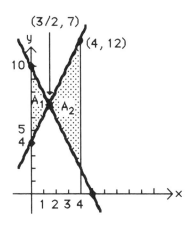

To find the point of intersection of the two lines, solve:

$$10 - 2x = 4 + 2x$$
$$-4x = -6$$
$$x = \frac{3}{2}$$

Substituting $x = \frac{3}{2}$ into either equation, we find $y = 7$. Now we have:

$$A = A_1 + A_2$$

$$= \int_0^{3/2} [(10 - 2x) - (4 + 2x)]dx + \int_{3/2}^4 [(4 + 2x) - (10 - 2x)]dx$$

$$= \int_0^{3/2} (6 - 4x)dx + \int_{3/2}^4 (4x - 6)dx$$

$$= (6x - 2x^2)\Big|_0^{3/2} + (2x^2 - 6x)\Big|_{3/2}^4$$

$$= 9 - \frac{9}{2} + (32 - 24) - \left(\frac{9}{2} - 9\right) = 17$$

29. The graphs of $y = 5x - x^2$ and $y = x + 3$, $0 \le x \le 5$ are shown at the right.

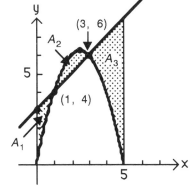

To find the points of intersection, solve:

$$5x - x^2 = x + 3$$
$$x^2 - 4x + 3 = 0$$
$$(x - 1)(x - 3) = 0$$
$$x = 1, \quad x = 3$$

Substituting into the equation $y = x + 3$, we get $y = 4$ when $x = 1$, and $y = 6$ when $x = 3$. Thus, we have:

$$A = A_1 + A_2 + A_3$$

$$= \int_0^1 [x + 3 - (5x - x^2)]dx + \int_1^3 [5x - x^2 - (x + 3)]dx$$

$$\qquad + \int_3^5 [x + 3 - (5x - x^2)dx$$

$$= \int_0^1 (x^2 - 4x + 3)dx + \int_1^3 (-x^2 + 4x - 3)dx + \int_3^5 (x^2 - 4x + 3)dx$$

$$= \left(\frac{x^3}{3} - 2x^2 + 3x\right)\Big|_0^1 + \left(-\frac{x^3}{3} + 2x^2 - 3x\right)\Big|_1^3 + \left(\frac{x^3}{3} - 2x^2 + 3x\right)\Big|_3^5$$

$$= \left(\frac{1}{3} - 2 + 3\right) + (-9 + 18 - 9) - \left(-\frac{1}{3} + 2 - 3\right) + \left(\frac{125}{3} - 50 + 15\right) - (9 - 18 + 9)$$

$$= \frac{4}{3} + \frac{4}{3} + \frac{20}{3} = \frac{28}{3}$$

31. The graphs of $y = x^2 + 2x + 3$ and $y = 2x + 4$ are shown at the right. To find the points of intersection, solve:

$$2x + 4 = x^2 + 2x + 3$$
$$x^2 = 1$$
$$x = \pm 1$$

Thus, $(1, 6)$ and $(-1, 2)$ are the points of intersection.

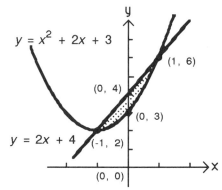

$$A = \int_{-1}^{1} [(2x + 4) - (x^2 + 2x + 3)]\,dx$$

$$= \int_{-1}^{1} (1 - x^2)\,dx = \left(x - \frac{x^3}{3}\right)\Big|_{-1}^{1}$$

$$= \left(1 - \frac{1}{3}\right) - \left(-1 + \frac{1}{3}\right) = \frac{2}{3} + \frac{2}{3} = \frac{4}{3}$$

33. The graphs are given at the right. To find the points of intersection, solve:

$$x^2 - 4x - 10 = 14 - 2x - x^2$$
$$2x^2 - 2x - 24 = 0$$
$$2(x^2 - x - 12) = 0$$
$$2(x - 4)(x + 3) = 0$$

Thus, the points of intersection are $(-3, 11)$ and $(4, -10)$.

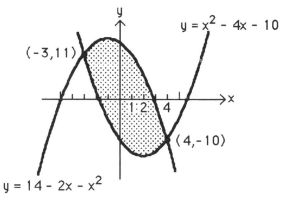

$$A = \int_{-3}^{4} [(14 - 2x - x^2) - (x^2 - 4x - 10)]\,dx$$

$$= \int_{-3}^{4} (-2x^2 + 2x + 24)\,dx = \left(-\frac{2}{3}x^3 + x^2 + 24x\right)\Big|_{-3}^{4}$$

$$= \frac{-2}{3}(4)^3 + (4)^2 + 24(4) - \left[-\frac{2}{3}(-3)^3 + (-3)^2 + 24(-3)\right]$$

$$= \frac{-2}{3}(64) + 16 + 96 - [18 + 9 - 72]$$

$$= \frac{-128}{3} + 157 = \frac{343}{3}$$

35. The graphs are given at the right. To find the points of intersection, solve:

$$x^3 = 4x$$
$$x^3 - 4x = 0$$
$$x(x^2 - 4) = 0$$
$$x(x + 2)(x - 2) = 0$$

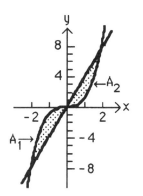

Thus, the points of intersection are $(-2, -8)$, $(0, 0)$, and $(2, 8)$.

$$A = A_1 + A_2 = \int_{-2}^{0} (x^3 - 4x)\,dx + \int_{0}^{2} (4x - x^3)\,dx$$

$$= \left(\frac{x^4}{4} - 2x^2\right)\Big|_{-2}^{0} + \left(2x^2 - \frac{x^4}{4}\right)\Big|_{0}^{2}$$

$$= 0 - \left[\frac{(-2)^4}{4} - 2(-2)^2\right] + \left[2(2^2) - \frac{2^4}{4}\right] - 0$$

$$= -4 + 8 + 8 - 4 = 8$$

37. The graphs are given at the right.
To find the points of intersection, solve:

$$x^3 - 3x^2 - 9x + 12 = x + 12$$
$$x^3 - 3x^2 - 10x = 0$$
$$x(x^2 - 3x - 10) = 0$$
$$x(x - 5)(x + 2) = 0$$
$$x = -2, \ x = 0, \ x = 5$$

Thus, $(-2, 10)$, $(0, 12)$, and $(5, 17)$ are the points of intersection.

$$A = A_1 + A_2$$

$$= \int_{-2}^{0} [x^3 - 3x^2 - 9x + 12 - (x + 12)]dx$$

$$+ \int_{0}^{5} [x + 12 - (x^3 - 3x^2 - 9x + 12)]dx$$

$$= \int_{-2}^{0} (x^3 - 3x^2 - 10x)dx + \int_{0}^{5} (-x^3 + 3x^2 + 10x)dx$$

$$= \left(\frac{x^4}{4} - x^3 - 5x^2\right)\Big|_{-2}^{0} + \left(-\frac{x^4}{4} + x^3 + 5x^2\right)\Big|_{0}^{5}$$

$$= -\left[\frac{(-2)^4}{4} - (-2)^3 - 5(-2)^2\right] + \left(\frac{-5^4}{4} + 5^3 + 5\cdot5^2\right) = 8 + \frac{375}{4} = \frac{407}{4}$$

39. $\displaystyle\int_{5}^{10} R(t)dt = \int_{5}^{10} \left(\frac{100}{t + 10} + 10\right)dt = 100\int_{5}^{10} \frac{1}{t + 10}dt + \int_{5}^{10} 10 \ dt$

$$= 100 \ \ln(t + 10)\Big|_{5}^{10} + 10t\Big|_{5}^{10}$$

$$= 100 \ \ln 20 - 100 \ \ln 15 + 10(10 - 5)$$

$$= 100 \ \ln 20 - 100 \ \ln 15 + 50 \approx 79$$

The total production from the end of the fifth year to the end of the tenth year is approximately 79 thousand barrels.

41. To find the useful life, set $R'(t) = C'(t)$ and solve for t:

$$9e^{-0.3t} = 2$$
$$e^{-0.3t} = \frac{2}{9}$$
$$-0.3t = \ln\frac{2}{9}$$
$$-0.3t \approx -1.5$$
$$t \approx 5 \text{ years}$$

$$\int_0^5 [R'(t) - C'(t)]dt] = \int_0^5 [9e^{-0.3t} - 2]dt$$

$$= 9\int_0^5 e^{-0.3t}dt - \int_0^5 2\,dt = \frac{9}{-0.3}e^{-0.3t}\Big|_0^5 - 2t\Big|_0^5$$

$$= -30e^{-1.5} + 30 - 10$$

$$= 20 - 30e^{-1.5} \approx 13.306$$

The total profit over the useful life of the game is approximately $13,306.

43. For 1935: $f(x) = x^{2.4}$

Coefficient of inequality $= 2\int_0^1 [x - f(x)] = 2\int_0^1 (x - x^{2.4})\,dx$

$$= 2\left(\frac{x^2}{2} - \frac{x^{3.4}}{3.4}\right)\Big|_0^1$$

$$= 2\left(\frac{1}{2} - \frac{1}{3.4}\right) \approx 0.412$$

For 1947: $g(x) = x^{1.6}$

Coefficient of inequality $= 2\int_0^1 [x - g(x)]\,dx = 2\int_0^1 (x - x^{1.6})\,dx$

$$= 2\left(\frac{x^2}{2} - \frac{x^{2.6}}{2.6}\right)\Big|_0^1$$

$$= 2\left(\frac{1}{2} - \frac{1}{2.6}\right) \approx 0.231$$

Interpretation: Income was more equally distributed in 1947.

45. For 1963: $f(x) = x^{10}$

Coefficient of inequality $= 2\int_0^1 [x - f(x)]\,dx = 2\int_0^1 (x - x^{10})\,dx$

$$= 2\left(\frac{x^2}{2} - \frac{x^{11}}{11}\right)\Big|_0^1$$

$$= 2\left(\frac{1}{2} - \frac{1}{11}\right) \approx 0.818$$

For 1983: $g(x) = x^{12}$

Coefficient of inequality $= 2\int_0^1 [x - g(x)]\,dx = 2\int_0^1 (x - x^{12})\,dx$

$$= 2\left(\frac{x^2}{2} - \frac{x^{13}}{13}\right)\Big|_0^1$$

$$= 2\left(\frac{1}{2} - \frac{1}{13}\right) \approx 0.846$$

Total assets were more equally distributed in 1963.

47. $W(t) = \displaystyle\int_0^{10} W'(t)\,dt = \int_0^{10} 0.3e^{0.1t}\,dt = 0.3\int_0^{10} e^{0.1t}\,dt$

$$= \frac{0.3}{0.1}e^{0.1t}\Big|_0^{10} = 3e^{0.1t}\Big|_0^{10} = 3e - 3 \approx 5.15$$

Total weight gain during the first 10 hours is approximately 5.15 grams.

49. $V = \displaystyle\int_2^4 \frac{15}{t}\,dt = 15\int_2^4 \frac{1}{t}\,dt = 15\,\ln t\,\Big|_2^4$

$$= 15\,\ln 4 - 15\,\ln 2 = 15\,\ln\!\left(\frac{4}{2}\right) = 15\,\ln 2 \approx 10$$

Average number of words learned during the second 2 hours is 10.

51. $y = \sqrt{x}, \; y = x^2 - 4x + 3$

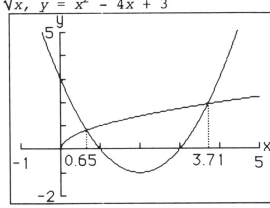

The graphs intersect at the points where $x = 0.65$ and $x = 3.71$

$A = \displaystyle\int_{0.65}^{3.71}[\sqrt{x} - (x^2 - 4x + 3)]\,dx = \int_{0.65}^{3.71}(x^{1/2} - x^2 + 4x - 3)\,dx$

$$= \left(\frac{2}{3}x^{3/2} - \frac{x^3}{3} + 2x^2 - 3x\right)\Big|_{0.65}^{3.71} = 4.99$$

53. $y = e^{2x}, \; y = 2x + 3$

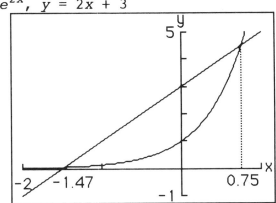

The graphs intersect at the points where $x = -1.47$ and $x = 0.75$

$A = \displaystyle\int_{-1.47}^{0.75}(2x + 3 - e^{2x})\,dx = \left(x^2 + 3x - \frac{1}{2}e^{2x}\right)\Big|_{-1.47}^{0.75} = 2.85$

55. $y = 2 + 6x - x^3$, $y = 0$ (the x-axis)

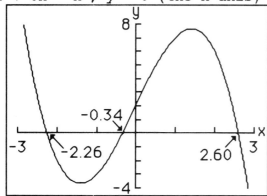

The x-intercepts are: -2.26, -0.34, and 2.60

$$A = A_1 + A_2 = \int_{-2.26}^{-0.34} -(2 + 6x - x^3)\,dx + \int_{-0.34}^{2.60} (2 + 6x - x^3)\,dx$$

$$= -\left(2x + 3x^2 - \frac{x^4}{4}\right)\Bigg|_{-2.26}^{-0.34} + \left(2x + 3x^2 - \frac{x^4}{4}\right)\Bigg|_{-0.34}^{2.60}$$

$$= 4.618 + 14.393 = 19.01$$

57. $y = \dfrac{6x}{x^2 + 1}$, $y = x - 1$

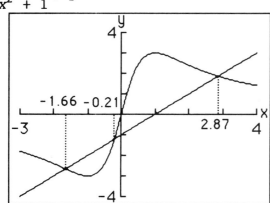

The graphs intersect at the points where $x = -1.66$, $x = -0.21$, and $x = 2.87$.

$$A = A_1 + A_2 = \int_{-1.66}^{-0.21} \left(x - 1 - \frac{6x}{x^2 + 1}\right)dx + \int_{-0.21}^{2.87} \left(\frac{6x}{x^2 + 1} - [x - 1]\right)dx$$

$$= \left[\frac{x^2}{2} - x - 3\ln(x^2 + 1)\right]\Bigg|_{-1.66}^{-0.21} + \left[3\ln(x^2 + 1) - \frac{x^2}{2} + x\right]\Bigg|_{-0.21}^{2.87}$$

$$= 1.035 + 5.524 = 6.56$$

Things to remember:

1. RECTANGLE RULE

   Divide the interval from $x = a$ to $x = b$ into n equal subintervals of length $\Delta x = [b - a]/n$. Let c_k be any point in the kth subinterval. Then

   $$\int_a^b f(x)\,dx \approx f(c_1)\Delta x + f(c_2)\Delta x + \cdots + f(c_n)\Delta x$$
   $$= \Delta x[f(c_1) + f(c_2) + \cdots + f(c_n)].$$

2. DEFINITION OF A DEFINITE INTEGRAL

   Let f be a continuous function defined on the closed interval $[a, b]$, and let:

   (a) $a = x_0 \le x_1 \le \ldots \le x_{n-1} \le x_n = b$

   (b) $\Delta x_k = x_k - x_{k-1}$ for $k = 1, 2, \ldots, n$

   (c) $\Delta x_k \to 0$ as $n \to \infty$

   (d) $x_{k-1} \le c_k \le x_k$ for $k = 1, 2, \ldots, n$

   Then

   $$\int_a^b f(x)\,dx = \lim_{n \to \infty}[f(c_1)\Delta x_1 + f(c_2)\Delta x_2 + \cdots + f(c_n)\Delta x_n]$$

   is called a DEFINITE INTEGRAL.

3. FUNDAMENTAL THEOREM OF CALCULUS

   Under the conditions stated in the definition of a definite integral:

   (Definition)
   $$\int_a^b f(x)\,dx = \lim_{n \to \infty}[f(c_1)\Delta x_1 + f(c_2)\Delta x_2 + \cdots + f(c_n)\Delta x_n]$$

   (Theorem)
   $$= F(b) - F(a) \quad \text{where } F'(x) = f(x)$$

   In particular, from the rectangle rule:
   $$\int_a^b f(x)\,dx = \lim_{n \to \infty}[f(c_1)\Delta x + f(c_2)\Delta x + \cdots + f(c_n)\Delta x]$$
   $$= \lim_{n \to \infty} \Delta x[f(c_1) + f(c_2) + \cdots + f(c_n)]$$
   $$= F(b) - F(a) \quad \text{where } F'(x) = f(x)$$

4. The AVERAGE VALUE of a continuous function f over $[a, b]$ is given by:

$$\frac{1}{b - a} \int_a^b f(x)\, dx$$

1. (A) $\Delta x = \dfrac{5 - 1}{2} = 2$

   Subintervals: $[1, 3]$, $[3, 5]$
   Midpoints: $c_1 = 2$, $c_2 = 4$
   Using the rectangle rule with $n = 2$:

   $$\int_1^5 3x^2\, dx \approx \Delta x[f(c_1) + f(c_2)] = 2[f(2) + f(4)]$$
   $$= 2[3 \cdot 2^2 + 3 \cdot 4^2] = 2(60) = 120$$

   (B) $\displaystyle \int_1^5 3x^2\, dx = x^3 \Big|_1^5 = 5^3 - 1^3 = 124$

3. (A) $\Delta x = \dfrac{5 - 1}{4} = 1$

   Subintervals: $[1, 2]$, $[2, 3]$,
   $\qquad\qquad\quad [3, 4]$, $[4, 5]$
   Midpoints: $c_1 = 1.5$, $c_2 = 2.5$,
   $\qquad\qquad c_3 = 3.5$, $c_4 = 4.5$
   Using the rectangle rule with $n = 4$:

   $$\int_1^5 3x^2\, dx \approx \Delta x[f(c_1) + f(c_2) + f(c_3) + f(c_4)]$$
   $$= 1[f(1.5) + f(2.5) + f(3.5) + f(4.5)]$$
   $$= 3(1.5)^2 + 3(2.5)^2 + 3(3.5)^2 + 3(4.5)^2 = 123$$

   (B) $\displaystyle \int_1^5 3x^2\, dx = x^3 \Big|_1^5 = 5^3 - 1^3 = 124$

5. (A) $\Delta x = \dfrac{4 - 0}{2} = 2$

   Subintervals: $[0, 2]$, $[2, 4]$
   Midpoints: $c_1 = 1$, $c_2 = 3$
   Using the rectangle rule with $n = 2$:

   $$\int_0^4 (4 - x^2)\, dx \approx \Delta x[f(c_1) + f(c_2)]$$
   $$= 2[f(1) + f(3)]$$
   $$= 2[(4 - 1^2) + (4 - 3^2)]$$
   $$= 2[3 - 5] = -4$$

   (B) $\displaystyle \int_0^4 (4 - x^2)\, dx = \left(4x - \frac{x^3}{3}\right) \Big|_0^4$

   $$= 4 \cdot 4 - \frac{4^3}{3}$$
   $$= 16 - \frac{64}{3}$$
   $$= -\frac{16}{3} \approx -5.33$$

7. (A) $\Delta x = \dfrac{4 - 0}{4} = 1$

   Subintervals: $[0, 1]$, $[1, 2]$
   $\qquad\qquad\quad [2, 3]$, $[3, 4]$
   Midpoints: $c_1 = 0.5$, $c_2 = 1.5$,
   $\qquad\qquad c_3 = 2.5$, $c_4 = 3.5$

   (B) $\displaystyle \int_0^4 (4 - x^2)\, dx = \left(4x - \frac{x^3}{3}\right) \Big|_0^4$

   $$= 16 - \frac{64}{3}$$
   $$= -5.33$$

Using the rectangle rule with $n = 4$:

$$\int_0^4 (4 - x^2)\,dx \approx \Delta x[f(c_1) + f(c_2) + f(c_3) + f(c_4)]$$

$$= 1[f(0.5) + f(1.5) + f(2.5) + f(3.5)]$$
$$= ([4 - (0.5)^2] + [4 - (1.5)^2] + [4 - (2.5)^2]$$
$$+ [4 - (3.5)^2])$$

$$= -5.00$$

9.

x	1	3	5	7
$f(x)$	4.5	3.2	2.4	1.6

Using the rectangle rule with $n = 4$, $\Delta x = 2$, and the values of f in the table, we have:

$$\int_0^8 f(x)\,dx \approx \Delta x[f(1) + f(3) + f(5) + f(7)]$$

$$= 2[4.5 + 3.2 + 2.4 + 1.6] = 23.4$$

11.

x	1.5	2.5	3.5	4.5
$f(x)$	12.5	16.7	15.4	10.7

Using the rectangle rule with $n = 4$, $\Delta x = 1$, and the values of f in the table, we have:

$$\int_1^5 f(x)\,dx \approx \Delta x[f(1.5) + f(2.5) + f(3.5) + f(4.5)]$$

$$= 1[12.5 + 16.7 + 15.4 + 10.7] = 55.3$$

13. Average value $= \dfrac{1}{b - a}\displaystyle\int_a^b f(x)\,dx = \dfrac{1}{10 - 0}\int_0^{10} (500 - 50x)\,dx$

$$= \frac{1}{10}\left(500x - \frac{50x^2}{2}\right)\Bigg|_0^{10} = \frac{1}{10}[500 \cdot 10 - 25(10)^2) - 0] = 250$$

15. $\dfrac{1}{2 - (-1)}\displaystyle\int_{-1}^2 (3t^2 - 2t)\,dt = \dfrac{1}{3}\left(3 \cdot \dfrac{t^3}{3} - 2 \cdot \dfrac{t^2}{2}\right)\Bigg|_{-1}^2$ (using **4**)

$$= \frac{1}{3}(t^3 - t^2)\Bigg|_{-1}^2$$

$$= \frac{1}{3}[(2^3 - 2^2) - ((-1)^3 - (-1)^2)] = \frac{1}{3}[6] = 2$$

17. $\dfrac{1}{8 - 1}\displaystyle\int_1^8 \sqrt[3]{x}\,dx = \dfrac{1}{7}\int_1^8 x^{1/3}\,dx = \dfrac{1}{7}\left(\dfrac{x^{4/3}}{4/3}\right)\Bigg|_1^8 = \dfrac{1}{7} \cdot \dfrac{3}{4}[(8)^{4/3} - (1)^{4/3}]$

$$= \frac{3}{28}(16 - 1) = \frac{45}{28} \approx 1.61$$

19. $\dfrac{1}{10 - 0}\displaystyle\int_0^{10} 4e^{-0.2x}\,dx = \dfrac{1}{10} \cdot 4 \cdot \dfrac{e^{-0.2x}}{-0.2}\Bigg|_0^{10}$

$$= -2e^{-0.2x}\Bigg|_0^{10} = -2e^{-2} + 2 = 2(1 - e^{-2}) \approx 1.73$$

21. $\dfrac{1}{1 - (-1)} \displaystyle\int_{-1}^{1} \dfrac{x + 1}{x^2 + 1} \, dx = \dfrac{1}{2} \displaystyle\int_{-1}^{1} \dfrac{x + 1}{x^2 + 1} \, dx$

$\Delta x = \dfrac{1 - (-1)}{4} = \dfrac{1}{2}$

Subintervals: $[-1, -0.5]$, $[-0.5, 0]$, $[0, 0.5]$, $[0.5, 1]$
Midpoints: $c_1 = -0.75$, $c_2 = -0.25$, $c_3 = 0.25$, $c_4 = 0.75$

Using the rectangle rule with $n = 4$:

$\dfrac{1}{2} \displaystyle\int_{-1}^{1} \dfrac{x + 1}{x^2 + 1} \, dx \approx \dfrac{1}{2} \cdot \dfrac{1}{2} [f(c_1) + f(c_2) + f(c_3) + f(c_4)]$

$\qquad\qquad = \dfrac{1}{4} [f(-0.75) + f(-0.25) + f(0.25) + f(0.75)]$

$\qquad\qquad = \dfrac{1}{4} (0.16 + 0.706 + 1.176 + 1.12) \approx 0.791$

23. $A = \displaystyle\int_{0}^{2} \ln(1 + x^3) \, dx$

$\Delta x = \dfrac{2 - 0}{4} = 0.5$

Subintervals: $[0, 0.5]$, $[0.5, 1]$, $[1, 1.5]$, $[1.5, 2]$
Midpoints: $c_1 = 0.25$, $c_2 = 0.75$, $c_3 = 1.25$, $c_4 = 1.75$

Using the rectangle rule with $n = 4$:

$\displaystyle\int_{0}^{2} \ln(1 + x^3) \, dx \approx 0.5[f(c_1) + f(c_2) + f(c_3) + f(c_4)]$

$\qquad\qquad = 0.5[f(0.25) + f(0.75) + f(1.25) + f(1.75)]$

$\qquad\qquad = 0.5[0.016 + 0.352 + 1.083 + 1.850] = 1.650$

25. $\displaystyle\int_{0}^{1} e^{-x^2} \, dx$

$\Delta x = \dfrac{1 - 0}{5} = 0.2$

Subintervals: $[0, 0.2]$, $[0.2, 0.4]$, $[0.4, 0.6]$, $[0.6, 0.8]$, $[0.8, 1.0]$
Midpoints: $c_1 = 0.1$, $c_2 = 0.3$, $c_3 = 0.5$, $c_4 = 0.7$, $c_5 = 0.9$

Using the rectangle rule with $n = 5$:

$\displaystyle\int_{0}^{1} e^{-x^2} \, dx \approx 0.2[f(c_1) + f(c_2) + f(c_3) + f(c_4) + f(c_5)]$

$\qquad\qquad = 0.2[f(0.1) + f(0.3) + f(0.5) + f(0.7) + f(0.9)]$

$\qquad\qquad = 0.2[0.990 + 0.914 + 0.779 + 0.613 + 0.445] = 0.748$

27. $\displaystyle\int_0^1 e^{-x^2}dx$

$$\Delta x = \frac{1 - 0}{10} = 0.1$$

Subintervals: $[0, 0.1]$, $[0.1, 0.2]$, $[0.2, 0.3]$, $[0.3, 0.4]$, $[0.4, 0.5]$
$[0.5, 0.6]$, $[0.6, 0.7]$, $[0.7, 0.8]$, $[0.8, 0.9]$, $[0.9, 1]$

Midpoints: $c_1 = 0.05$, $c_2 = 0.15$, $c_3 = 0.25$, $c_4 = 0.35$, $c_5 = 0.45$
$c_6 = 0.55$, $c_7 = 0.65$, $c_8 = 0.75$, $c_9 = 0.85$, $c_{10} = 0.95$

Using the rectangle rule with $n = 10$:

$$\int_0^1 e^{-x^2}dx \approx 0.1[f(c_1) + f(c_2) + f(c_3) + f(c_4) + f(c_5) + f(c_6) + f(c_7)$$
$$+ f(c_8) + f(c_9) + f(c_{10})]$$
$$= 0.1[f(0.05) + f(0.15) + f(0.25) + f(0.35) + f(0.45) + f(0.55)$$
$$+ f(0.65) + f(0.75) + f(0.85) + f(0.95)]$$
$$= 0.1[0.998 + 0.978 + 0.939 + 0.885 + 0.817 + 0.739 + 0.655$$
$$+ 0.570 + 0.486 + 0.406]$$
$$= 0.747$$

29. Average value of f' over $[a, b]$:

$$\frac{1}{b - a}\int_a^b f'(x)\,dx = \frac{1}{b - a}f(x)\Big|_a^b = \frac{f(b) - f(a)}{b - a}$$

31. (A) The inventory function is obtained by finding the equation of the line joining $(0, 600)$ and $(3, 0)$.

Slope: $m = \dfrac{0 - 600}{3 - 0} = -200$, y intercept: $b = 600$

Thus, the equation of the line is: $I = -200t + 600$

(B) The average of I over $[0, 3]$ is given by:

$$\frac{1}{3 - 0}\int_0^3 I(t)\,dt = \frac{1}{3}\int_0^3 (-200t + 600)\,dt = \frac{1}{3}(-100t^2 + 600t)\Big|_0^3$$
$$= \frac{1}{3}[-100(3^2) + 600(3) - 0] = \frac{900}{3} = 300 \text{ units}$$

33. Average cash reserve for the first quarter:

$$\frac{1}{3 - 0}\int_0^3 (1 + 12x - x^2)\,dx = \frac{1}{3}\left(x + \frac{12x^2}{2} - \frac{x^3}{3}\right)\Big|_0^3 = \frac{1}{3}\left(x + 6x^2 - \frac{x^3}{3}\right)\Big|_0^3$$
$$= \frac{1}{3}\left[3 + 6\cdot3^2 - \frac{3^3}{3}\right] = \frac{1}{3}[48] = 16 \text{ or } \$16,000$$

35. $C(x) = 60,000 + 300x$

(A) Average cost per unit: $\overline{C}(x) = \dfrac{C(x)}{x} = \dfrac{60,000}{x} + 300$

$\overline{C}(500) = \dfrac{60,000}{500} + 300 = 420 \text{ or } \$420$

(B) Average value: $\dfrac{1}{500 - 0} \displaystyle\int_0^{500} (60{,}000 + 300x)\,dx$

$$= \dfrac{1}{500}[60{,}000x + 150x^2] \Big|_0^{500}$$

$$= \dfrac{1}{500}[30{,}000{,}000 + 37{,}500{,}000] = 135{,}000 \text{ or } \$135{,}000$$

37. Continuous compound interest: $A = Pe^{rt}$
    Now, $P = \$100$, $r = 0.08$, $t = 5$.
    (A) $A = 100e^{0.08(5)}$
    $\quad\ = 100e^{0.40}$
    $\quad\ \approx 149.18 \text{ or } \$149.18$

    (B) Average amount: $\dfrac{1}{5 - 0} \displaystyle\int_0^5 100e^{0.08t}\,dt = 20 \int_0^5 e^{0.08t}\,dt$

    $$= \dfrac{20}{0.08} \int_0^5 e^{0.08t}(0.08)\,dt \quad u = 0.08t$$

    $$du = 0.08\,dt$$

    $$= 250e^{0.08t} \Big|_0^5$$

    $$= 250e^{0.40} - 250 = 122.96$$
    $$\text{or} \quad \$122.96$$

39. Average price:
    $$\dfrac{1}{30 - 20} \int_{20}^{30} 10(e^{0.02x} - 1)\,dx = \int_{20}^{30}(e^{0.02x} - 1)\,dx = \int_{20}^{30} e^{0.02x}\,dx - \int_{20}^{30} dx$$

    $$= \dfrac{1}{0.02} \int_{20}^{30} e^{0.02x}(0.02)\,dx - x \Big|_{20}^{30}$$

    $$= 50e^{0.02x} \Big|_{20}^{30} - (30 - 20)$$

    $$= 50e^{0.6} - 50e^{0.4} - 10 \approx 6.51 \text{ or } \$6.51$$

41. $S(t) = 20 - 10e^{-0.1t}$
    (A) Average number of hamburgers sold each day during the first week:

    $$\dfrac{1}{7 - 0} \int_0^7 (20 - 10e^{-0.1t})\,dt = \dfrac{1}{7} \int_0^7 20\,dt - \dfrac{10}{7} \int_0^7 e^{-0.1t}\,dt$$

    $$= \dfrac{20}{7}\,t \Big|_0^7 - \dfrac{10}{7(-0.1)} \int_0^7 e^{-0.1t}(-0.1)\,dt$$

    $$= 20 + \left(\dfrac{100}{7} e^{-0.1t}\right) \Big|_0^7 = 20 + \dfrac{100}{7} e^{-0.7} - \dfrac{100}{7}$$

    $$= \dfrac{1}{7}(40 + 100e^{-0.7})$$

    $$\approx 12.8 \text{ or } 12{,}800 \text{ hamburgers}$$

(B) Average number of hamburgers sold during the second week:

$$\frac{1}{14-7} \int_7^{14} (20 - 10e^{-0.1t})\,dt = \frac{1}{7} \int_7^{14} 20\,dt - \frac{10}{7} \int_7^{14} e^{-0.1t}\,dt$$

$$= \frac{20}{7} t \Big|_7^{14} + \frac{100}{7} \int_7^{14} e^{-0.1t}(-0.1)\,dt$$

$$= 20 + \left(\frac{100}{7} e^{-0.1t}\right) \Big|_7^{14}$$

$$= 20 + \frac{100}{7}(e^{-1.4} - e^{-0.7})$$

$$\approx 16.4 \text{ or } 16,400 \text{ hamburgers}$$

43. Let $P(t) = R(t) - C(t)$. Then the total accumulated profits over the five-year period are given by:

$$P(5) - P(0) = \int_0^5 P'(t)\,dt = \int_0^5 [R'(t) - C'(t)]\,dt$$

$$= \int_0^5 R'(t)\,dt - \int_0^5 C'(t)\,dt$$

Now, $C'(t) = 1500$ (constant). Therefore, we have:

$$\int_0^5 C'(t)\,dt = \int_0^5 1500\,dt = 1500t \Big|_0^5 = 7500$$

To approximate $\int_0^5 R'(t)\,dt$, let $n = 5$, $\Delta x = \frac{5-0}{5} = 1$, and $c_1 = \frac{1}{2}$,
$c_2 = \frac{3}{2}$, $c_3 = \frac{5}{2}$, $c_4 = \frac{7}{2}$, and $c_5 = \frac{9}{2}$.

Then, from the given graph, using the rectangle rule, we have:

$$\int_0^5 R'(t)\,dt \approx 1\left[R'\left(\frac{1}{2}\right) + R'\left(\frac{3}{2}\right) + R'\left(\frac{5}{2}\right) + R'\left(\frac{7}{2}\right) + R'\left(\frac{9}{2}\right)\right]$$

$$= 5000 + 4500 + 3500 + 2500 + 2000 = 17,500$$

Therefore, the total accumulated profits are (approximately):

$$P(5) - P(0) = \int_0^5 R'(t)\,dt - \int_0^5 C'(t)\,dt \approx 17,500 - 7500 = \$10,000$$

45.

x	300	900	1500	2100
$f(x)$	900	1700	1700	900

Using the rectangle rule with $n = 4$, $\Delta x = 600$, and the values of f in the table, we have:

$$\int_0^{2400} f(x)\,dx \approx \Delta x[f(c_1) + f(c_2) + f(c_3) + f(c_4)]$$

$$= 600[900 + 1700 + 1700 + 900]$$

$$= 600(5200) = 3,120,000 \text{ square feet}$$

47. Average temperature over time period $[0, 2]$ is given by:

$$\frac{1}{2-0}\int_0^2 C(t)\,dt = \frac{1}{2}\int_0^2 (t^3 - 2t + 10)\,dt = \frac{1}{2}\left(\frac{t^4}{4} - \frac{2t^2}{2} + 10t\right)\bigg|_0^2$$

$$= \frac{1}{2}(4 - 4 + 20) = 10° \text{ Celsius}$$

49. $n = 3$, $\Delta t = \dfrac{3-0}{3} = 1$, $c_1 = \dfrac{1}{2}$, $c_2 = \dfrac{3}{2}$, $c_3 = \dfrac{5}{2}$

Using the given graph and the rectangle rule, we have:

$$\int_0^3 R(t)\,dt \approx 1\left[R\!\left(\frac{1}{2}\right) + R\!\left(\frac{3}{2}\right) + R\!\left(\frac{5}{2}\right)\right] = 0.3 + 0.5 + 0.3 = 1.1$$

Thus, the total volume of air inhaled is approximately 1.1 liters.

51. $P(t) = \dfrac{8.4t}{t^2 + 49} + 0.1$, $0 \le t \le 24$

(A) Average fraction of people during the first seven months:

$$\frac{1}{7-0}\int_0^7 \left[\frac{8.4t}{t^2+49} + 0.1\right]dt = \frac{4.2}{7}\int_0^7 \frac{2t}{t^2+49}\,dt + \frac{1}{7}\int_0^7 0.1\,dt$$

$$= 0.6\,\ln(t^2 + 49)\bigg|_0^7 + \frac{0.1}{7}\,t\bigg|_0^7$$

$$= 0.6[\ln 98 - \ln 49] + 0.1$$

$$= 0.6\,\ln 2 + 0.1 \approx 0.516$$

(B) Average fraction of people during the first two years:

$$\frac{1}{24-0}\int_0^{24} \left[\frac{8.4t}{t^2+49} + 0.1\right]dt = \frac{4.2}{24}\int_0^{24} \frac{2t}{t^2+49}\,dt + \frac{1}{24}\int_0^{24} 0.1\,dt$$

$$= 0.175\,\ln(t^2 + 49)\bigg|_0^{24} + \frac{0.1}{24}\,t\bigg|_0^{24}$$

$$= 0.175[\ln 625 - \ln 49] + 0.1 \approx 0.546$$

53. $y = 1 + 12x - x^2$
$\bar{y} = 16$
$0 \le x \le 3$

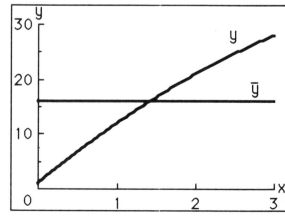

55. $y = 100e^{0.08x}$
$\bar{y} = 122.96$
$0 \le x \le 5$

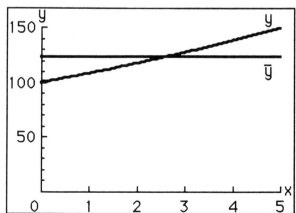

57. $y = 10(e^{0.02x} - 1)$
$\overline{y} = 6.51$
$20 \leq x \leq 30$

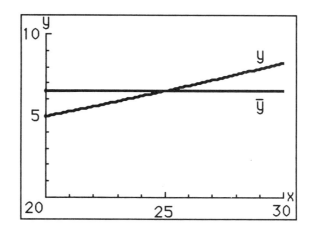

EXERCISE 11-6
_____

Things to remember:

1. TOTAL INCOME FOR A CONTINUOUS INCOME STREAM

   If $f(t)$ is the rate of flow of a continuous income stream, then the TOTAL INCOME produced during the time period from $t = a$ to $t = b$ is:

   $$\text{Total income} = \int_a^b f(t)\,dt$$

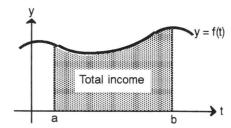

2. FUTURE VALUE OF A CONTINUOUS INCOME STREAM

   If $f(t)$ is the rate of flow of a continuous income stream, $0 \leq t \leq T$, and if the income is continuously invested at a rate r, compounded continuously, then the FUTURE VALUE, FV, at the end of T years is given by:

   $$FV = \int_0^T f(t)\,e^{r(T-t)}\,dt = e^{rT}\int_0^T f(t)\,e^{-rt}\,dt$$

   The future value of a continuous income stream is the total value of all money produced by the continuous income stream (income and interest) at the end of T years.

3. CONSUMERS' SURPLUS

   If $(\overline{x}, \overline{p})$ is a point on the graph of the price-demand equation $p = D(x)$ for a particular product, then the CONSUMERS' SURPLUS, CS, at a price level of $\overline{p}$ is

   $$CS = \int_0^{\overline{x}} [D(x) - \overline{p}]\,dx$$

which is the area between $p = \overline{p}$ and $p = D(x)$ from $x = 0$ to $x = \overline{x}$.

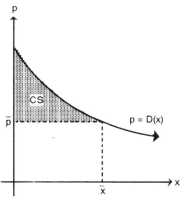

Consumer's surplus represents the total savings to consumers who are willing to pay a higher price for the product.

4. PRODUCERS' SURPLUS

If $(\overline{x}, \overline{p})$ is a point on the graph of the price-supply equation $p = S(x)$, then the PRODUCERS' SURPLUS, PS, at a price level of $\overline{p}$ is

$$PS = \int_0^{\overline{x}} [\overline{p} - S(x)]\,dx$$

which is the area between $p = \overline{p}$ and $p = S(x)$ from $x = 0$ to $x = \overline{x}$.

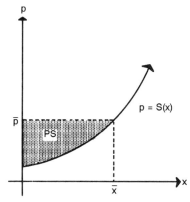

Producers' surplus represents the total gain to producers who are willing to supply units at a lower price.

5. EQUILIBRIUM PRICE AND EQUILIBRIUM QUANTITY

If $p = D(x)$ and $p = S(x)$ are the price-demand and the price-supply equations, respectively, for a product and if $(\overline{x}, \overline{p})$ is the point of intersection of these equations, then $\overline{p}$ is called the EQUILIBRIUM PRICE and $\overline{x}$ is called the EQUILIBRIUM QUANTITY.

1. $D(x) = 400 - \frac{1}{20}x$, $\bar{p} = 150$

First, find $\bar{x}$: $150 = 400 - \frac{1}{20}\bar{x}$

$$\bar{x} = 5000$$

$$CS = \int_0^{5000}\left[400 - \frac{1}{20}x - 150\right]dx = \int_0^{5000}\left(250 - \frac{1}{20}x\right)dx$$

$$= \left(250x - \frac{1}{40}x^2\right)\Big|_0^{5000} = \$625,000$$

3. $S(x) = 10 + \frac{1}{10}x + \frac{1}{3600}x^2$, $\bar{p} = 65$

First, find $\bar{x}$: $65 = 10 + \frac{1}{10}\bar{x} + \frac{1}{3600}\bar{x}^2$

$\bar{x}^2 + 360\bar{x} - 198,000 = 0$

$(\bar{x} - 300)(\bar{x} + 660) = 0$

Thus, $\bar{x} = 300$

$$PS = \int_0^{300}\left[65 - \left(10 + \frac{1}{10}x + \frac{1}{3600}x^2\right)\right]dx$$

$$= \int_0^{300}\left(55 - \frac{1}{10}x - \frac{1}{3600}x^2\right)dx = \left(55x - \frac{1}{20}x^2 - \frac{1}{10,800}x^3\right)\Big|_0^{300} = \$9500$$

5. $D(x) = 50 - \frac{1}{10}x$, $S(x) = 11 + \frac{1}{20}x$

Equilibrium price: $D(x) = S(x)$

$$50 - \frac{1}{10}x = 11 + \frac{1}{20}x$$

$$\frac{3}{20}x = 39$$

$$\bar{x} = 260$$

Thus, $\bar{p} = 50 - \frac{1}{10}(260) = 24$.

$$CS = \int_0^{260}\left[50 - \frac{1}{10}x - 24\right]dx = \int_0^{260}\left(26 - \frac{1}{10}x\right)dx = \left(26x - \frac{1}{20}x^2\right)\Big|_0^{260} = \$3380$$

$$PS = \int_0^{260}\left[24 - \left(11 + \frac{1}{20}x\right)\right]dx = \int_0^{260}\left(13 - \frac{1}{20}x\right)dx = \left(13x - \frac{1}{40}x^2\right)\Big|_0^{260} = \$1690$$

7. $D(x) = 80e^{-0.001x}$ and $S(x) = 30e^{0.001x}$

Equilibrium price: $D(x) = S(x)$

$$80e^{-0.001x} = 30e^{0.001x}$$

$$e^{0.002x} = \frac{8}{3}$$

$$0.002x = \ln\left(\frac{8}{3}\right)$$

$$\bar{x} = \frac{\ln\left(\frac{8}{3}\right)}{0.002} \approx 490$$

Thus, $\bar{p} = 30e^{0.001(490)} \approx 49$.

$$CS = \int_0^{490} [80e^{-0.001x} - 49]dx = \left(\frac{80e^{-0.001x}}{-0.001} - 49x\right)\Bigg|_0^{490}$$

$$= -80,000e^{-0.49} + 80,000 - 24,010 \approx \$6980$$

$$PS = \int_0^{490} [49 - 30e^{0.001x}]dx = \left(49x - \frac{30e^{0.001x}}{0.001}\right)\Bigg|_0^{490}$$

$$= 24,010 - 30,000(e^{0.49} - 1) \approx \$5041$$

9. The revenue R is given by $R(x) = xp$, where p is the price. Thus:

$$R(x) = x\left(100 - \frac{1}{10}x\right) = 100x - \frac{1}{10}x^2$$

Profit $P(x) = R(x) - C(x)$

$$= 100x - \frac{1}{10}x^2 - (5000 + 60x)$$

$$= -\frac{1}{10}x^2 + 40x - 5000, \quad x \geq 0$$

Now, $P'(x) = -\frac{1}{5}x + 40$

Critical value: $-\frac{1}{5}x + 40 = 0$

$$\frac{1}{5}x = 40$$

$$x = 200$$

Since $P''(x) = -\frac{1}{5} < 0$, $x = 200$ produces the maximum value of P. The price p when $x = 200$ is $p = 100 - \frac{1}{10}(200) = 80$.

$$CS = \int_0^{200} \left[100 - \frac{1}{10}x - 80\right]dx = \int_0^{200}\left(20 - \frac{1}{10}x\right)dx = \left(20x - \frac{1}{20}x^2\right)\Bigg|_0^{200}$$

$$= 20(200) - \frac{1}{20}(200)^2 = 2000$$

Thus, the consumers surplus at the production level 200 is \$2000.

11. $p = D(x) = 50 - \frac{1}{10}x$

$p = S(x) = 11 + \frac{1}{20}x$

$0 \leq x \leq 400, \ 0 \leq p \leq 50$

From Problem 5, $\bar{p} = 24$

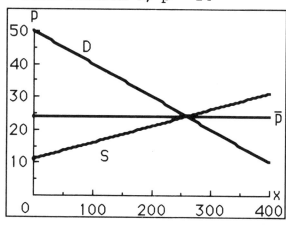

13. $p = D(x) = 80e^{-0.001x}$

$p = S(x) = 30e^{0.001x}$

$0 \leq x \leq 750, \ 0 \leq p \leq 100$

From Problem 7, $\bar{p} = 49$

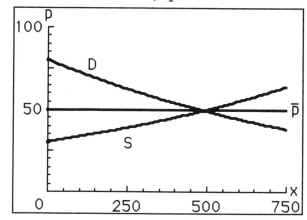

1. $\int (3t^2 - 2t)\,dt = 3\int t^2\,dt - 2\int t\,dt = 3\cdot\dfrac{t^3}{3} - 2\cdot\dfrac{t^2}{2} + C = t^3 - t^2 + C$

2. $\int_2^5 (2x - 3)\,dx = 2\int_2^5 x\,dx - 3\int_2^5 dx = x^2\Big|_2^5 - 3x\Big|_2^5$

$\qquad\qquad = (25 - 4) - (15 - 6) = 12$

3. $\int (3t^{-2} - 3)\,dt = 3\int t^{-2}\,dt - 3\int dt = 3\cdot\dfrac{t^{-1}}{-1} - 3t + C = -3t^{-1} - 3t + C$

4. $\int_1^4 x\,dx = \dfrac{x^2}{2}\Big|_1^4 = \dfrac{16}{2} - \dfrac{1}{2} = \dfrac{15}{2}$

5. $\int e^{-0.5x}\,dx = \dfrac{e^{-0.5x}}{-0.5} + C = -2e^{-0.5x} + C$

6. $\int_1^5 \dfrac{2}{u}\,du = 2\int_1^5 \dfrac{du}{u} = 2\ln u\Big|_1^5 = 2\ln 5 - 2\ln 1 = 2\ln 5$

7. $\dfrac{dy}{dx} = 3x^2 - 2$

$\qquad y = f(x) = \int (3x^2 - 2)\,dx$

$\qquad\quad f(x) = x^3 - 2x + C$

$\qquad\quad f(0) = C = 4$

$\qquad\quad f(x) = x^3 - 2x + 4$

8. The required area is given by:

$\int_{-1}^2 (3x^2 + 1)\,dx = (x^3 + x)\Big|_{-1}^2$

$\qquad\qquad = (8 + 2) - (-1 - 1)$

$\qquad\qquad = 12$

9. $\Delta x = 2$

Subintervals: $[1, 3]$, $[3, 5]$

Midpoints: $c_1 = 2$, $c_2 = 4$

Using the rectangle rule with $n = 2$:

$\int_1^5 (x^2 + 1)\,dx \approx \Delta x[f(c_1) + f(c_2)] = 2[f(2) + f(4)]$

$\qquad\qquad\qquad = 2[(2^2 + 1) + (4^2 + 1)] = 2(22) = 44$

10.

x	3	7	11	15
$f(x)$	1.2	3.4	2.6	0.5

Using the rectangle rule with $n = 4$, $\Delta x = 4$, and the values of f in the table, we have:

$\int_1^{17} f(x)\,dx \approx 4[f(3) + f(7) + f(11) + f(15)]$

$\qquad\qquad = 4[1.2 + 3.4 + 2.6 + 0.5] = 30.8$

11. Average value $= \dfrac{1}{2 - (-1)} \displaystyle\int_{-1}^{2} (6x^2 + 2x)\,dx = \dfrac{1}{3}[2x^3 + x^2]\Big|_{-1}^{2}$

$$= \dfrac{1}{3}\{2(2^3) + 2^2 - [2(-1)^3 + (-1)^2]\}$$

$$= \dfrac{1}{3}(16 + 4 + 2 - 1) = 7$$

12. $\displaystyle\int \sqrt[3]{6x - 5}\,dx = \int (6x - 5)^{1/3}dx = \dfrac{1}{6}\int (6x - 5)^{1/3}6\,dx$

$$= \dfrac{1}{6}\dfrac{(6x - 5)^{4/3}}{\frac{4}{3}} + C$$

$$= \dfrac{1}{8}(6x - 5)^{4/3} + C$$

13. $\displaystyle\int_{0}^{1} 10(2x - 1)^4 dx = 5\int_{0}^{1} (2x - 1)^4 2\,dx = \dfrac{5(2x - 1)^5}{5}\Big|_{0}^{1}$

$$= (2x - 1)^5\Big|_{0}^{1} = 1 - (-1)^5 = 2$$

14. $\displaystyle\int \left(\dfrac{2}{x^2} - 2xe^{x^2}\right)dx = 2\int x^{-2}dx - \int 2xe^{x^2}dx = \dfrac{2x^{-1}}{-1} - e^{x^2} + C$

$$= -2x^{-1} - e^{x^2} + C$$

15. $\displaystyle\int_{0}^{4} \sqrt{x^2 + 4}\,x\,dx = \int_{0}^{4} (x^2 + 4)^{1/2}x\,dx = \dfrac{1}{2}\int_{0}^{4} (x^2 + 4)^{1/2}2x\,dx$

$$= \dfrac{1}{2}\cdot\dfrac{(x^2 + 4)^{3/2}}{\frac{3}{2}}\Big|_{0}^{4} = \dfrac{(x^2 + 4)^{3/2}}{3}\Big|_{0}^{4} = \dfrac{(20)^{3/2} - 8}{3}$$

16. $\displaystyle\int (e^{-2x} + x^{-1})\,dx = \int e^{-2x}dx + \int \dfrac{1}{x}dx = -\dfrac{1}{2}\int e^{-2x}(-2)\,dx + \ln|x| + C$

$$= -\dfrac{1}{2}e^{-2x} + \ln|x| + C$$

17. $\displaystyle\int_{0}^{10} 10e^{-0.02x}dx = 10\int_{0}^{10} e^{-0.02x}dx = \dfrac{10}{-0.02}\int_{0}^{10} e^{-0.02x}(-0.02)\,dx$

$$= -500e^{-0.02x}\Big|_{0}^{10} = -500e^{-0.2} + 500 \approx 90.63$$

18. $\dfrac{dy}{dx} = 3x^{-1} - x^{-2}$

$$y = \int (3x^{-1} - x^{-2})\,dx = 3\int \dfrac{1}{x}dx - \int x^{-2}dx$$

$$= 3\ln|x| - \dfrac{x^{-1}}{-1} + C = 3\ln|x| + x^{-1} + C$$

Given $y(1) = 5$:
$5 = 3\ln 1 + 1 + C$ and $C = 4$
Thus, $y = 3\ln|x| + x^{-1} + 4$.

19. $\dfrac{dy}{dx} = 6x + 1$

$f(x) = y = \displaystyle\int (6x+1)\,dx = \dfrac{6x^2}{2} + x + C = 3x^2 + x + C$

We have $y = 10$ when $x = 2$: $\quad 3(2)^2 + 2 + C = 10$

$\qquad\qquad\qquad\qquad\qquad\qquad\quad C = 10 - 12 - 2 = -4$

Thus, the equation of the curve is $y = 3x^2 + x - 4$.

20. $\Delta x = \dfrac{1 - 0}{5} = 0.2$

Subintervals: $[0, 0.2]$, $[0.2, 0.4]$, $[0.4, 0.6]$, $[0.6, 0.8]$, $[0.8, 1]$
Midpoints: $c_1 = 0.1$, $c_2 = 0.3$, $c_3 = 0.5$, $c_4 = 0.7$, $c_5 = 0.9$
Using the rectangle rule with $n = 5$:

$\displaystyle\int_0^1 e^{2x^2}\,dx \approx 0.2[f(0.1) + f(0.3) + f(0.5) + f(0.7) + f(0.9)]$

$\qquad\qquad \approx 0.2[1.020 + 1.197 + 1.649 + 2.664 + 5.053] = 2.317$

21. Average value $= \dfrac{1}{9 - 1}\displaystyle\int_1^9 3x^{1/2}\,dx = \dfrac{3}{8}\left(\dfrac{2}{3}\right)x^{3/2}\Big|_1^9 = \dfrac{1}{4}x^{3/2}\Big|_1^9$

$\qquad\qquad\qquad = \dfrac{1}{4}(9)^{3/2} - \dfrac{1}{4}(1)^{3/2} = \dfrac{27}{4} - \dfrac{1}{4} = \dfrac{13}{2}$

22. $A = A_1 + A_2 = \displaystyle\int_{-2}^{2} -(x^2 - 4)\,dx + \int_{2}^{4} (x^2 - 4)\,dx$

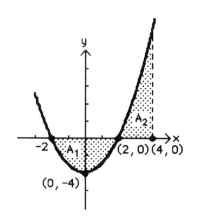

$\qquad\qquad\quad = \displaystyle\int_{-2}^{2} (4 - x^2)\,dx + \int_{2}^{4} (x^2 - 4)\,dx$

$\qquad\qquad\quad = \left(4x - \dfrac{x^3}{3}\right)\Big|_{-2}^{2} + \left(\dfrac{x^3}{3} - 4x\right)\Big|_{2}^{4}$

$\qquad\qquad\quad = \left[4(2) - \dfrac{2^3}{3}\right] - \left[4(-2) - \dfrac{(-2)^3}{3}\right]$

$\qquad\qquad\qquad + \dfrac{4^3}{3} - 4(4) - \left[\dfrac{2^3}{3} - 4(2)\right]$

$\qquad\qquad\quad = 8 - \dfrac{8}{3} - \left(-8 + \dfrac{8}{3}\right) + \dfrac{64}{3} - 16 - \left(\dfrac{8}{3} - 8\right)$

$\qquad\qquad\quad = 8 + \dfrac{40}{3} = \dfrac{64}{3}$

23. Let $u = 1 + x^2$, then $du = 2x\,dx$.

$\displaystyle\int_0^3 \dfrac{x}{1 + x^2}\,dx = \int_0^3 \dfrac{1}{1 + x^2}\dfrac{2}{2}x\,dx$

$\qquad\qquad\qquad = \dfrac{1}{2}\displaystyle\int_0^3 \dfrac{1}{1 + x^2}2x\,dx = \dfrac{1}{2}\ln(1 + x^2)\Big|_0^3$

$\qquad\qquad\qquad = \dfrac{1}{2}\ln 10 - \dfrac{1}{2}\ln 1 = \dfrac{1}{2}\ln 10$

24. Let $u = 1 + x^2$, then $du = 2x \, dx$.

$$\int_0^3 \frac{x}{(1 + x^2)^2} \, dx = \int_0^3 (1 + x^2)^{-2} \frac{2}{2} x \, dx = \frac{1}{2} \int_0^3 (1 + x^2)^{-2} 2x \, dx$$

$$= \frac{1}{2} \cdot \frac{(1 + x^2)^{-1}}{-1} \Big|_0^3 = \frac{-1}{2(1 + x^2)} \Big|_0^3 = -\frac{1}{20} + \frac{1}{2} = \frac{9}{20} = .45$$

25. Let $u = 2x^4 + 5$, then $du = 8x^3 dx$.

$$\int x^3 (2x^4 + 5)^5 dx = \int (2x^4 + 5)^5 \frac{8}{8} x^3 dx = \frac{1}{8} \int u^5 du$$

$$= \frac{1}{8} \cdot \frac{u^6}{6} + C = \frac{(2x^4 + 5)^6}{48} + C$$

26. Let $u = e^{-x} + 3$, then $du = -e^{-x} dx$.

$$\int \frac{e^{-x}}{e^{-x} + 3} \, dx = \int \frac{1}{e^{-x} + 3} \cdot \frac{(-1)}{(-1)} e^{-x} dx = -\int \frac{1}{u} \, du$$

$$= -\ln|u| + C = -\ln|e^{-x} + 3| + C = -\ln(e^{-x} + 3) + C$$

[<u>Note</u>: Absolute value not needed since $e^{-x} + 3 > 0$.]

27. Let $u = e^x + 2$, then $du = e^x dx$.

$$\int \frac{e^x}{(e^x + 2)^2} \, dx = \int (e^x + 2)^{-2} e^x dx = \int u^{-2} du$$

$$= \frac{u^{-1}}{-1} + C = -(e^x + 2)^{-1} + C = \frac{-1}{(e^x + 2)} + C$$

28. Let $u = \ln x$, then $du = \frac{1}{x} dx$.

$$\int \frac{(\ln x)^2}{x} \, dx = \int (\ln x)^2 \frac{1}{x} \, dx = \int u^2 du = \frac{u^3}{3} + C = \frac{(\ln x)^3}{3} + C$$

29. $\displaystyle\int x(x^3 - 1)^2 dx = \int x(x^6 - 2x^3 + 1) dx$ (square $x^3 - 1$)

$$= \int (x^7 - 2x^4 + x) dx = \frac{x^8}{8} - \frac{2x^5}{5} + \frac{x^2}{2} + C$$

30. Let $u = \sqrt{6 - x}$, then $u^2 = 6 - x$ or $x = 6 - u^2$ and $2u \, du = -dx$.

$$\int \frac{x}{\sqrt{6 - x}} \, dx = \int \frac{(6 - u^2)}{u} (-2u \, du) = -2 \int (6 - u^2) \, du$$

$$= -2\left(6u - \frac{u^3}{3}\right) + C = \frac{2}{3}(6 - x)^{3/2} - 12(6 - x)^{1/2} + C$$

31. $\displaystyle\int_0^7 x\sqrt{16 - x} \, dx.$ First consider the indefinite integral:

Let $u = \sqrt{16 - x}$, then $u^2 = 16 - x$ or $x = 16 - u^2$ and $2u \, du = -dx$.

$$\int x\sqrt{16 - x} \, dx = \int (16 - u^2) \cdot u(-2u \, du) = 2 \int (u^4 - 16u^2) \, du$$

$$= 2\left(\frac{u^5}{5} - \frac{16u^3}{3}\right) + C = \frac{2(16 - x)^{5/2}}{5} - \frac{32(16 - x)^{3/2}}{3} + C$$

$$\int_0^7 x\sqrt{16 - x}\, dx = \left[\frac{2(16 - x)^{5/2}}{5} - \frac{32(16 - x)^{3/2}}{3}\right]\Bigg|_0^7$$

$$= \frac{2\cdot 9^{5/2}}{5} - \frac{32\cdot 9^{3/2}}{3} - \left(\frac{2\cdot 16^{5/2}}{5} - \frac{32\cdot 16^{3/2}}{3}\right)$$

$$= \frac{2\cdot 3^5}{5} - \frac{32\cdot 3^3}{3} - \left(\frac{2\cdot 4^5}{5} - \frac{32\cdot 4^3}{3}\right)$$

$$= \frac{486}{5} - 288 - \left(\frac{2048}{5} - \frac{2048}{3}\right) = \frac{1234}{15} \approx 82.267$$

32.
$$\int_{-1}^1 x(x + 1)^4\, dx = \int_{-1}^1 x(x^4 + 4x^3 + 6x^2 + 4x + 1)\, dx$$

$$= \int_{-1}^1 (x^5 + 4x^4 + 6x^3 + 4x^2 + x)\, dx$$

$$= \left(\frac{x^6}{6} + \frac{4x^5}{5} + \frac{6x^4}{4} + \frac{4x^3}{3} + \frac{x^2}{2}\right)\Bigg|_{-1}^1$$

$$= \left(\frac{1}{6} + \frac{4}{5} + \frac{3}{2} + \frac{4}{3} + \frac{1}{2}\right) - \left(\frac{1}{6} - \frac{4}{5} + \frac{3}{2} - \frac{4}{3} + \frac{1}{2}\right)$$

$$= \frac{8}{5} + \frac{8}{3} = \frac{64}{15}$$

33. $\frac{dy}{dx} = 9x^2 e^{x^3}$, $f(0) = 2$

Let $u = x^3$, then $du = 3x^2 dx$.

$$y = \int 9x^2 e^{x^3}\, dx = 3\int e^{x^3}\cdot 3x^2\, dx = 3\int e^u du = 3e^u + C = 3e^{x^3} + C$$

Given $f(0) = 2$:

$2 = 3e^0 + C = 3 + C$

Hence, $C = -1$ and $y = f(x) = 3e^{x^3} - 1$.

34. The required area is given by:

$$A = A_1 + A_2 = \int_0^2 [(6 - x^2) - (x^2 - 2)]\, dx + \int_2^3 [(x^2 - 2) - (6 - x^2)]\, dx$$

$$= \int_0^2 (8 - 2x^2)\, dx + \int_2^3 (2x^2 - 8)\, dx$$

$$= \left(8x - \frac{2x^3}{3}\right)\Bigg|_0^2 + \left(\frac{2x^3}{3} - 8x\right)\Bigg|_2^3$$

$$= 16 - \frac{16}{3} + \frac{54}{3} - 24 - \left(\frac{16}{3} - 16\right)$$

$$= 8 + \frac{54}{3} - \frac{32}{3} = \frac{46}{3}$$

[Note: The required area has been shaded. The points of intersection of the two graphs are $(-2, 2)$ and $(2, 2)$.]

35. $P'(x) = 100 - 0.02x$

$$P(x) = \int (100 - 0.02x)\,dx = 100x - 0.02\frac{x^2}{2} + C = 100x - 0.01x^2 + C$$

$P(0) = 0 - 0 + C = 0$

$\qquad\qquad C = 0$

Thus, $P(x) = 100x - 0.01x^2$.

The profit on 10 units of production is given by:

$P(10) = 100(10) - 0.01(10)^2 = \$999$

36. The required definite integral is:

$$\int_0^{15} (60 - 4t)\,dt = (60t - 2t^2)\Big|_0^{15}$$

$$= 60(15) - 2(15)^2 = 450 \text{ or } 450{,}000 \text{ barrels}$$

The total production in 15 years is 450,000 barrels.

37. The total change in profit for a production change from 10 units per week to 40 units per week is given by:

$$\int_{10}^{40} \left(150 - \frac{x}{10}\right) dx = \left(150x - \frac{x^2}{20}\right)\Big|_{10}^{40}$$

$$= \left(150(40) - \frac{40^2}{20}\right) - \left(150(10) - \frac{10^2}{20}\right)$$

$$= 5920 - 1495 = \$4425$$

38. To find the useful life, set $R'(t) = C'(t)$:

$20e^{-0.1t} = 3$

$e^{-0.1t} = \dfrac{3}{20}$

$-0.1t = \ln\left(\dfrac{3}{20}\right) \approx -1.897$

$\qquad t = 18.97 \text{ or } 19 \text{ years}$

$$\text{Total profit} = \int_0^{19} [R'(t) - C'(t)]\,dt = \int_0^{19} (20e^{-0.1t} - 3)\,dt$$

$$= 20\int_0^{19} e^{-0.1t}\,dt - \int_0^{19} 3\,dt = \frac{20}{-0.1}\int_0^{19} e^{-0.1t}(-0.1)\,dt - \int_0^{19} 3\,dt$$

$$= -200e^{-0.1t}\Big|_0^{19} - 3t\Big|_0^{19}$$

$$= -200e^{-1.9} + 200 - 57 \approx 113.086 \text{ or } \$113{,}086$$

39. $S'(t) = 4e^{-0.08t}$, $0 \le t \le 24$. Therefore,

$$S(t) = \int 4e^{-0.08t}\,dt = \frac{4e^{-0.08t}}{-0.08} + C = -50e^{-0.08t} + C.$$

Now, $S(0) = 0$, so

$0 = -50e^{-0.08(0)} + C = -50 + C.$

Thus, $C = 50$, and $S(t) = 50(1 - e^{-0.08t})$ gives the total sales after t months.

Estimated sales after 12 months:

$S(12) = 50(1 - e^{-0.08(12)}) = 50(1 - e^{-0.96}) \approx 31 \text{ or } \$31 \text{ million}.$

To find the time to reach \$40 million in sales, solve

$40 = 50(1 - e^{-0.08t})$

for t.

$0.8 = 1 - e^{-0.08t}$

$e^{-0.08t} = 0.2$

$-0.08t = \ln(0.2)$

$t = \dfrac{\ln(0.2)}{-0.08} \approx 20$ months

40. Current Lorenz curve: $f(x) = \dfrac{1}{10}x + \dfrac{9}{10}x^2$

Coefficient of inequality $= 2\displaystyle\int_0^1 [x - f(x)]\,dx$

$$= 2\int_0^1 \left[x - \left(\frac{1}{10}x + \frac{9}{10}x^2\right)\right]dx = 2\int_0^1 \left(\frac{9}{10}x - \frac{9}{10}x^2\right)dx$$

$$= \frac{9}{5}\int_0^1 (x - x^2)\,dx = \frac{9}{5}\left(\frac{x^2}{2} - \frac{x^3}{3}\right)\Big|_0^1$$

$$= \frac{9}{5}\left(\frac{1}{2} - \frac{1}{3}\right) = \frac{9}{30} = 0.3$$

Projected Lorenz curve: $g(x) = x^{1.5}$

Coefficient of inequality $= 2\displaystyle\int_0^1 [x - g(x)]\,dx = 2\int_0^1 (x - x^{1.5})\,dx$

$$= 2\left(\frac{x^2}{2} - \frac{x^{2.5}}{2.5}\right)\Big|_0^1 = 2(0.5 - 0.4) = 0.2$$

Interpretation: Income will be more evenly distributed ten years from now.

41. Average inventory from $t = 3$ to $t = 6$:

$\dfrac{1}{6-3}\displaystyle\int_3^6 (10 + 36t - 3t^2)\,dt = \frac{1}{3}[10t + 18t^2 - t^3]\Big|_3^6$

$$= \frac{1}{3}[60 + 648 - 216 - (30 + 162 - 27)]$$

$$= 109 \text{ items}$$

42. $S(x) = 8(e^{0.05x} - 1)$

Average price over the interval $[40, 50]$:

$\dfrac{1}{50-40}\displaystyle\int_{40}^{50} 8(e^{0.05x} - 1)\,dx = \frac{8}{10}\int_{40}^{50}(e^{0.05x} - 1)\,dx$

$$= \frac{4}{5}\left[\frac{e^{0.05x}}{0.05} - x\right]\Big|_{40}^{50}$$

$$= \frac{4}{5}[20e^{2.5} - 50 - (20e^2 - 40)]$$

$$= 16e^{2.5} - 16e^2 - 8 \approx \$68.70$$

43. $D(x) = 70 - \frac{1}{5}x$, $S(x) = 13 + \frac{3}{2500}x^2$

(A) $\bar{p} = 50$. $D(\bar{x}) = 70 - \frac{1}{5}\bar{x} = 50$

$$-\frac{1}{5}\bar{x} = -20$$

$$\bar{x} = 100$$

Consumer's surplus $= \int_0^{100} [D(x) - \bar{p}]dx = \int_0^{100} \left(70 - \frac{1}{5}x - 50\right)dx$

$$= \int_0^{100} \left(20 - \frac{1}{5}x\right)dx = \left(20x - \frac{1}{10}x^2\right)\Big|_0^{100}$$

$$= 2000 - 1000 = \$1000$$

(B) $\bar{p} = 25$. $S(\bar{x}) = 13 + \frac{3}{2500}\bar{x}^2 = 25$

$$\frac{3}{2500}\bar{x}^2 = 12$$

$$\bar{x}^2 = 10,000$$

$$\bar{x} = 100$$

Producers' surplus $= \int_0^{100} [\bar{p} - S(x)]dx = \int_0^{100}\left[25 - \left(13 + \frac{3}{2500}x^2\right)\right]dx$

$$= \int_0^{100}\left(12 - \frac{3}{2500}x^2\right)dx = \left(12x - \frac{x^3}{2500}\right)\Big|_0^{100}$$

$$= 1200 - 400 = \$800$$

(C) Equilibrium price: $D(x) = S(x)$

$$70 - \frac{1}{5}x = 13 + \frac{3}{2500}x^2$$

$$\frac{3}{2500}x^2 + \frac{1}{5}x - 57 = 0$$

$$3x^2 + 500x - 142,500 = 0$$

$$(x - 150)(3x + 950) = 0$$

Thus, $\bar{x} = 150$ and the equilibrium price is $\bar{p} = 70 - \frac{1}{5}(150) = \$40$.

Consumers' surplus $= \int_0^{150} [D(x) - \bar{p}]dx = \int_0^{150}\left(70 - \frac{1}{5}x - 40\right)dx$

$$= \int_0^{150}\left(30 - \frac{1}{5}x\right)dx = \left(30x - \frac{1}{10}x^2\right)\Big|_0^{150} = \$2250$$

Producers' surplus $= \int_0^{150} [\bar{p} - S(x)]dx = \int_0^{150}\left[40 - \left(13 + \frac{3}{2500}x^2\right)\right]dx$

$$= \int_0^{150}\left(27 - \frac{3}{2500}x^2\right)dx = \left(27x - \frac{1}{2500}x^3\right)\Big|_0^{150}$$

$$= \$2700$$

44. $\dfrac{dA}{dt} = -5t^{-2}$, $1 \le t \le 5$

$A = \displaystyle\int -5t^{-2}dt = -5\int t^{-2}dt = -5 \cdot \dfrac{t^{-1}}{-1} + C = \dfrac{5}{t} + C$

Now $A(1) = \dfrac{5}{1} + C = 5$. Therefore, $C = 0$ and

$A(t) = \dfrac{5}{t}$

$A(5) = \dfrac{5}{5} = 1$

The area of the wound after 5 days is 1 cm^2.

45. Using the rectangle rule with $n = 6$, $\Delta x = \dfrac{12}{6} = 2$, $c_1 = 1$, $c_2 = 3$, $c_3 = 5$, $c_4 = 7$, $c_5 = 9$, $c_6 = 11$, and the values of C estimated from the graph, we have:

Average concentration

$= \dfrac{1}{12 - 0}\displaystyle\int_0^{12} C(t)\,dt = \dfrac{1}{12}\int_0^{12} C(t)\,dt$

$\approx \dfrac{1}{12} \cdot \Big(2[f(c_1) + f(c_2) + f(c_3) + f(c_4) + f(c_5) + f(c_6)]\Big)$

$= \dfrac{1}{6}[4 + 6 + 8 + 9 + 8 + 4] = \dfrac{39}{6} = 6.5$ parts per million

46. $N'(t) = 7e^{-0.1t}$ and $N(0) = 25$.

$N(t) = \displaystyle\int 7e^{-0.1t}dt = 7\int e^{-0.1t}dt = \dfrac{7}{-0.1}\int e^{-0.1t}(-0.1)\,dt$

$\qquad\qquad\qquad\qquad = -70e^{-0.1t} + C$, $0 \le t \le 15$

Given $N(0) = 25$: $25 = -70e^0 + C = -70 + C$

Hence, $C = 95$ and $N(t) = 95 - 70e^{-0.1t}$. The student would be expected to type $N(15) = 95 - 70e^{-0.1(15)} = 95 - 70e^{-1.5} \approx 79$ words per minute after completing the course.

EXERCISE A-1

Things to remember:

1. SEQUENCES

   A SEQUENCE is a function whose domain is a set of successive integers. If the domain of a given sequence is a finite set, then the sequence is called a FINITE SEQUENCE; otherwise, the sequence is an INFINITE SEQUENCE. In general, unless stated to the contrary or the context specifies otherwise, the domain of a sequence will be understood to be the set N of natural numbers.

2. NOTATION FOR SEQUENCES

   Rather than function notation $f(n)$, n in the domain of a given sequence f, subscript notation a_n is normally used to denote the value in the range corresponding to n, and the sequence itself is denoted $\{a_n\}$ rather than f or $f(n)$. The elements in the range, a_n, are called the TERMS of the sequence; a_1 is the first term, a_2 is the second term, and a_n is the nth term or general term.

3. SERIES

   Given a sequence $\{a_n\}$. The sum of the terms of the sequence, $a_1 + a_2 + a_3 + \cdots$ is called a SERIES. If the sequence is finite, the corresponding series is a FINITE SERIES; if the sequence is infinite, then the corresponding series is an INFINITE SERIES. Only finite series are considered in this section.

4. NOTATION FOR SERIES

   Series are represented using SUMMATION NOTATION. If $\{a_k\}$, $k = 1$, 2, ..., n is a finite sequence, then the series

   $$a_1 + a_2 + a_3 + \cdots + a_n$$

   is denoted

   $$\sum_{k=1}^{n} a_k.$$

   The symbol $\sum$ is called the SUMMATION SIGN and k is called the SUMMING INDEX.

5. ARITHMETIC MEAN

   If $\{a_k\}$, $k = 1$, 2, ..., n, is a finite sequence, then the ARITHMETIC MEAN $\bar{a}$ of the sequence is defined as

   $$\bar{a} = \frac{1}{n} \sum_{k=1}^{n} x_k.$$

1. $a_n = 2n + 3$; $a_1 = 2 \cdot 1 + 3 = 5$
$a_2 = 2 \cdot 2 + 3 = 7$
$a_3 = 2 \cdot 3 + 3 = 9$
$a_4 = 2 \cdot 4 + 3 = 11$

3. $a_n = \dfrac{n + 2}{n + 1}$; $a_1 = \dfrac{1 + 2}{1 + 1} = \dfrac{3}{2}$

$a_2 = \dfrac{2 + 2}{2 + 1} = \dfrac{4}{3}$

$a_3 = \dfrac{3 + 2}{3 + 1} = \dfrac{5}{4}$

$a_4 = \dfrac{4 + 2}{4 + 1} = \dfrac{6}{5}$

5. $a_n = (-3)^{n+1}$; $a_1 = (-3)^{1+1} = (-3)^2 = 9$
$a_2 = (-3)^{2+1} = (-3)^3 = -27$
$a_3 = (-3)^{3+1} = (-3)^4 = 81$
$a_4 = (-3)^{4+1} = (-3)^5 = -243$

7. $a_n = 2n + 3$; $a_{10} = 2 \cdot 10 + 3 = 23$

9. $a_n = \dfrac{n + 2}{n + 1}$; $a_{99} = \dfrac{99 + 2}{99 + 1} = \dfrac{101}{100}$

11. $\displaystyle\sum_{k=1}^{6} k = 1 + 2 + 3 + 4 + 5 + 6 = 21$

13. $\displaystyle\sum_{k=4}^{7} (2k - 3) = (2 \cdot 4 - 3) + (2 \cdot 5 - 3) + (2 \cdot 6 - 3) + (2 \cdot 7 - 3)$
$= 5 + 7 + 9 + 11 = 32$

15. $\displaystyle\sum_{k=0}^{3} \dfrac{1}{10^k} = \dfrac{1}{10^0} + \dfrac{1}{10^1} + \dfrac{1}{10^2} + \dfrac{1}{10^3} = 1 + \dfrac{1}{10} + \dfrac{1}{100} + \dfrac{1}{1000} = \dfrac{1111}{1000} = 1.111$

17. $a_1 = 5$, $a_2 = 4$, $a_3 = 2$, $a_4 = 1$, $a_5 = 6$. Here $n = 5$ and the arithmetic mean is given by:
$$\bar{a} = \dfrac{1}{5} \sum_{k=1}^{5} a_k = \dfrac{1}{5}(5 + 4 + 2 + 1 + 6) = \dfrac{18}{5} = 3.6$$

19. $a_1 = 96$, $a_2 = 65$, $a_3 = 82$, $a_4 = 74$, $a_5 = 91$, $a_6 = 88$, $a_7 = 87$, $a_8 = 91$, $a_9 = 77$, and $a_{10} = 74$. Here $n = 10$ and the arithmetic mean is given by:
$$\bar{a} = \dfrac{1}{10} \sum_{k=1}^{10} a_k = \dfrac{1}{10}(96 + 65 + 82 + 74 + 91 + 88 + 87 + 91 + 77 + 74)$$
$$= \dfrac{825}{10} = 82.5$$

21. $a_n = \dfrac{(-1)^{n+1}}{2^n}$; $a_1 = \dfrac{(-1)^2}{2^1} = \dfrac{1}{2}$

$a_2 = \dfrac{(-1)^3}{2^2} = -\dfrac{1}{4}$

$a_3 = \dfrac{(-1)^4}{2^3} = \dfrac{1}{8}$

$a_4 = \dfrac{(-1)^5}{2^4} = -\dfrac{1}{16}$

$a_5 = \dfrac{(-1)^6}{2^5} = \dfrac{1}{32}$

23. $a_n = n[1 + (-1)^n]$; $a_1 = 1[1 + (-1)^1] = 0$

$$a_2 = 2[1 + (-1)^2] = 4$$
$$a_3 = 3[1 + (-1)^3] = 0$$
$$a_4 = 4[1 + (-1)^4] = 8$$
$$a_5 = 5[1 + (-1)^5] = 0$$

25. $a_n = \left(-\dfrac{3}{2}\right)^{n-1}$; $a_1 = \left(-\dfrac{3}{2}\right)^0 = 1$

$$a_2 = \left(-\dfrac{3}{2}\right)^1 = -\dfrac{3}{2}$$
$$a_3 = \left(-\dfrac{3}{2}\right)^2 = \dfrac{9}{4}$$
$$a_4 = \left(-\dfrac{3}{2}\right)^3 = -\dfrac{27}{8}$$
$$a_5 = \left(-\dfrac{3}{2}\right)^4 = \dfrac{81}{16}$$

27. Given $-2, -1, 0, 1, \ldots$ The sequence is the set of successive integers beginning with -2. Thus, $a_n = n - 3$, $n = 1, 2, 3, \ldots$.

29. Given $4, 8, 12, 16, \ldots$ The sequence is the set of positive integer multiples of 4. Thus, $a_n = 4n$, $n = 1, 2, 3, \ldots$.

31. Given $\dfrac{1}{2}, \dfrac{3}{4}, \dfrac{5}{6}, \dfrac{7}{8}, \ldots$ The sequence is the set of fractions whose numerators are the odd positive integers and whose denominators are the even positive integers. Thus,

$$a_n = \dfrac{2n - 1}{2n}, \ n = 1, 2, 3, \ldots .$$

33. Given $1, -2, 3, -4, \ldots$ The sequence consists of the positive integers with alternating signs. Thus,

$$a_n = (-1)^{n+1}n, \ n = 1, 2, 3, \ldots .$$

35. Given $1, -3, 5, -7, \ldots$ The sequence consists of the odd positive integers with alternating signs. Thus,

$$a_n = (-1)^{n+1}(2n - 1), \ n = 1, 2, 3, \ldots .$$

37. Given $1, \dfrac{2}{5}, \dfrac{4}{25}, \dfrac{8}{125}, \ldots$ The sequence consists of the nonnegative integral powers of $\dfrac{2}{5}$. Thus,

$$a_n = \left(\dfrac{2}{5}\right)^{n-1}, \ n = 1, 2, 3, \ldots .$$

39. Given $x, x^2, x^3, x^4, \ldots$ The sequence is the set of positive integral powers of x. Thus, $a_n = x^n$, $n = 1, 2, 3, \ldots$.

41. Given x, $-x^3$, x^5, $-x^7$, ... The sequence is the set of positive odd integral powers of x with alternating signs. Thus,

$$a_n = (-1)^{n+1} x^{2n-1}, \quad n = 1, 2, 3, \ldots .$$

43. $\displaystyle\sum_{k=1}^{5} (-1)^{k+1}(2k-1)^2 = (-1)^2(2\cdot 1 - 1)^2 + (-1)^3(2\cdot 2 - 1)^2$
$$+ (-1)^4(2\cdot 3 - 1)^2 + (-1)^5(2\cdot 4 - 1)^2$$
$$+ (-1)^6(2\cdot 5 - 1)^2$$
$$= 1 - 9 + 25 - 49 + 81$$

45. $\displaystyle\sum_{k=2}^{5} \frac{2^k}{2k+3} = \frac{2^2}{2\cdot 2 + 3} + \frac{2^3}{2\cdot 3 + 3} + \frac{2^4}{2\cdot 4 + 3} + \frac{2^5}{2\cdot 5 + 3}$

$$= \frac{4}{7} + \frac{8}{9} + \frac{16}{11} + \frac{32}{13}$$

47. $\displaystyle\sum_{k=1}^{5} x^{k-1} = x^0 + x^1 + x^2 + x^3 + x^4 = 1 + x + x^2 + x^3 + x^4$

49. $\displaystyle\sum_{k=0}^{4} \frac{(-1)^k x^{2k+1}}{2k+1} = \frac{(-1)^0 x}{2\cdot 0 + 1} + \frac{(-1)x^3}{2\cdot 1 + 1} + \frac{(-1)^2 x^5}{2\cdot 2 + 1} + \frac{(-1)^3 x^7}{2\cdot 3 + 1} + \frac{(-1)^4 x^9}{2\cdot 4 + 1}$

$$= x - \frac{x^3}{3} + \frac{x^5}{5} - \frac{x^7}{7} + \frac{x^9}{9}$$

51. (A) $2 + 3 + 4 + 5 + 6 = \displaystyle\sum_{k=1}^{5} (k+1)$

    (B) $2 + 3 + 4 + 5 + 6 = \displaystyle\sum_{j=0}^{4} (j+2)$

53. (A) $1 - \dfrac{1}{2} + \dfrac{1}{3} - \dfrac{1}{4} = \displaystyle\sum_{k=1}^{4} \frac{(-1)^{k+1}}{k}$

    (B) $1 - \dfrac{1}{2} + \dfrac{1}{3} - \dfrac{1}{4} = \displaystyle\sum_{j=0}^{3} \frac{(-1)^j}{j+1}$

55. $2 + \dfrac{3}{2} + \dfrac{4}{3} + \ldots + \dfrac{n+1}{n} = \displaystyle\sum_{k=1}^{n} \frac{k+1}{k}$

57. $\dfrac{1}{2} - \dfrac{1}{4} + \dfrac{1}{8} - \ldots + \dfrac{(-1)^{n+1}}{2^n} = \displaystyle\sum_{k=1}^{n} \frac{(-1)^{k+1}}{2^k}$

59. $a_1 = 2$ and $a_n = 3a_{n-1} + 2$ for $n \geq 2$.
$a_1 = 2$
$a_2 = 3\cdot a_1 + 2 = 3\cdot 2 + 2 = 8$
$a_3 = 3\cdot a_2 + 2 = 3\cdot 8 + 2 = 26$
$a_4 = 3\cdot a_3 + 2 = 3\cdot 26 + 2 = 80$
$a_5 = 3\cdot a_4 + 2 = 3\cdot 80 + 2 = 242$

61. $a_1 = 1$ and $a_n = 2a_{n-1}$ for $n \geq 2$.

$a_1 = 1$

$a_2 = 2 \cdot a_1 = 2 \cdot 1 = 2$

$a_3 = 2 \cdot a_2 = 2 \cdot 2 = 4$

$a_4 = 2 \cdot a_3 = 2 \cdot 4 = 8$

$a_5 = 2 \cdot a_4 = 2 \cdot 8 = 16$

63. In $a_1 = \dfrac{A}{2}$, $a_n = \dfrac{1}{2}\left(a_{n-1} + \dfrac{A}{a_{n-1}}\right)$, $n \geq 2$, let $A = 2$. Then:

$a_1 = \dfrac{2}{2} = 1$

$a_2 = \dfrac{1}{2}\left(a_1 + \dfrac{A}{a_1}\right) = \dfrac{1}{2}(1 + 2) = \dfrac{3}{2}$

$a_3 = \dfrac{1}{2}\left(a_2 + \dfrac{A}{a_2}\right) = \dfrac{1}{2}\left(\dfrac{3}{2} + \dfrac{2}{3/2}\right) = \dfrac{1}{2}\left(\dfrac{3}{2} + \dfrac{4}{3}\right) = \dfrac{17}{12}$

$a_4 = \dfrac{1}{2}\left(a_3 + \dfrac{A}{a_3}\right) = \dfrac{1}{2}\left(\dfrac{17}{12} + \dfrac{2}{17/12}\right) = \dfrac{1}{2}\left(\dfrac{17}{12} + \dfrac{24}{17}\right) = \dfrac{577}{408} \approx 1.414216$

and $\sqrt{2} \approx 1.414214$

EXERCISE A-2

Things to remember:

1. A sequence of numbers a_1, a_2, a_3, $\ldots$, a_n, $\ldots$, is called an ARITHMETIC PROGRESSION if there is constant d, called the COMMON DIFFERENCE, such that

   $a_n - a_{n-1} = d$,

   that is,

   $a_n = a_{n-1} + d$

   for all $n > 1$.

2. If a_1, a_2, a_3, $\ldots$, a_n, $\ldots$, is an arithmetic progression with common difference d, then

   $a_n = a_1 + (n - 1)d$

   for all $n > 1$.

3. The sum S_n of the first n terms of an arithmetic progression a_1, a_2, a_3, $\ldots$, a_n, $\ldots$, with common difference d, is given by

   (a) $S_n = \dfrac{n}{2}[2a_1 + (n - 1)d]$

   or by

   (b) $S_n = \dfrac{n}{2}(a_1 + a_n)$.

---

1. (A) is an arithmetic progression; $a_2 - a_1 = a_3 - a_2 = 3$. Thus, $d = 3$, $a_4 = 14$, and $a_5 = 17$.

   (B) is not an arithmetic progression, since $a_2 - a_1 = 8 - 4 = 4 \neq a_3 - a_2 = 16 - 8 = 8$.

   (C) is not an arithmetic progression, since $-4 - (-2) = -2 \neq -8 - (-4) = -4$.

   (D) is an arithmetic progression; $a_2 - a_1 = a_3 - a_2 = -10$. Thus, $d = -10$, $a_4 = -22$, and $a_5 = -32$.

3. $a_2 = a_1 + d = 7 + 4 = 11$
   $a_3 = a_2 + d = 11 + 4 = 15$ (using 1)

5. $a_{21} = a_1 + (21 - 1)d = 2 + 20 \cdot 4 = 82$ (using 2)

   $S_{31} = \frac{31}{2}[2a_1 + (31 - 1)d] = \frac{31}{2}[2 \cdot 2 + 30 \cdot 4] = \frac{31}{2} \cdot 124 = 1922$ [using 3(a)]

7. Using 3(b), $S_{20} = \frac{20}{2}(a_1 + a_{20}) = 10(18 + 75) = 930$

9. $f(1) = -1$, $f(2) = 1$, $f(3) = 3$, ... This is an arithmetic progression with $a_1 = -1$, $d = 2$. Thus, using 3(a),

   $f(1) + f(2) + f(3) + \cdots + f(50) = \frac{50}{2}[2(-1) + 49 \cdot 2] = 25 \cdot 96 = 2400$

11. Let $a_1 = 13$, $d = 2$. Then, using 2, we can find n:
    $$67 = 13 + (n - 1)2 \quad \text{or} \quad 2(n - 1) = 54$$
    $$n - 1 = 27$$
    $$n = 28$$
    Therefore, using 3(b), $S_{28} = \frac{28}{2}[13 + 67] = 14 \cdot 80 = 1120$.

13. Consider the arithmetic progression with $a_1 = 1$, $d = 2$. This progression is the sequence of odd positive integers. Now, using 3(a), the sum of the first n odd positive integers is:

    $S_n = \frac{n}{2}[2 \cdot 1 + (n - 1)2] = \frac{n}{2}(2 + 2n - 2) = \frac{n}{2} \cdot 2n = n^2$

15. The yearly salaries from Firm A are: $24,000, $24,900, $25,800, ..., an arithmetic progression with $a_1 = 24,000$ and $d = 900$. Thus, in ten years Firm A will pay:

    $S_{10} = \frac{10}{2}[2(24,000) + 9(900)] = 5(56,100) = \$280,500$

    The salaries from Firm B are: $22,000, $23,300, $24,600, ..., an arithmetic progression with $a_1 = 22,000$ and $d = 1300$. Thus, in ten years Firm B will pay:

    $S_{10} = \frac{10}{2}[2(22,000) + 9(1300)] = 5(55,700) = \$278,500$

17. Consider the time line:

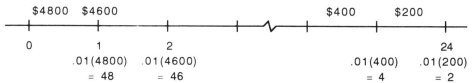

The total cost of the loan is $2 + 4 + 6 + \cdots + 46 + 48$. The terms form an arithmetic progression with $n = 24$, $a_1 = 2$, and $a_{24} = 48$. Thus, using $\underline{3}$(b):

$$S_{24} = \frac{24}{2}(2 + 48) = 24 \cdot 25 = \$600$$

EXERCISE A-3

Things to remember:

<u>1</u>. A sequence of numbers a_1, a_2, a_3, ..., a_n, ..., is called a GEOMETRIC PROGRESSION if there exists a nonzero constant r, called the COMMON RATIO, such that

$$\frac{a_n}{a_{n-1}} = r,$$

that is,

$$a_n = r a_{n-1}$$

for all $n > 1$.

<u>2</u>. If a_1, a_2, a_3, ..., a_n, ..., is a geometric progression with common ration r, then

$$a_n = a_1 r^{n-1}$$

for all $n > 1$.

<u>3</u>. The sum S_n of the first n terms of a geometric progression a_1, a_2, a_3, ..., a_n, ..., with common ration r, is given by:

$$S_n = \frac{a_1(r^n - 1)}{r - 1}, \quad r \neq 1,$$

or by

$$S_n = \frac{r a_n - a_1}{r - 1}, \quad r \neq 1.$$

<u>4</u>. If a_1, a_2, a_3, ..., a_n, ..., is a geometric progression with common ratio r having the property $|r| < 1$, then $S_\infty = \lim_{n \to \infty} S_n$ exists and is given by:

$$S_\infty = \frac{a_1}{1 - r}, \quad |r| < 1.$$

1. (A) is a geometric progression; $\frac{a_2}{a_1} = \frac{a_3}{a_2} = -2$. Thus, $r = -2$, $a_4 = -8$, $a_5 = 16$.

(B) is not a geometric progression, since $\frac{a_2}{a_1} = \frac{6}{7} \neq \frac{a_3}{a_2} = \frac{5}{6}$.

(C) is a geometric progression; $\frac{a_2}{a_1} = \frac{a_3}{a_2} = \frac{1}{2}$. Thus, $r = \frac{1}{2}$, $a_4 = \frac{1}{4}$, $a_5 = \frac{1}{8}$.

(D) is not a geometric progression, since $\frac{a_2}{a_1} = \frac{-4}{2} = -2 \neq \frac{a_3}{a_2} = \frac{-3}{2}$

3. $a_2 = a_1 r = 3(-2) = -6$
$a_3 = a_2 r = -6(-2) = 12$
$a_4 = a_3 r = 12(-2) = -24$ (using $\underline{1}$)

5. Using $\underline{3}$, $S_7 = \frac{-3 \cdot 729 - 1}{-3 - 1} = \frac{-2188}{-4} = 547$.

7. Using $\underline{2}$, $a_{10} = 100(1.08)^9 = 199.90$.

9. Using $\underline{2}$, $200 = 100r^8$. Thus, $r^8 = 2$ and $r = \sqrt[8]{2} \approx 1.09$.

11. Using $\underline{3}$, $S_{10} = \frac{500[(0.6)^{10} - 1]}{0.6 - 1} \approx 1242$,

$S_\infty = \frac{500}{1 - 0.6} = 1250$.

13. (A) $2 + 4 + 8 + \cdots$. Since $r = \frac{4}{2} = \frac{8}{4} = \cdots = 2$ and $|2| = 2 > 1$, the sum does not exist.

(B) $2, -\frac{1}{2}, \frac{1}{8}, \ldots$. In this case, $r = \frac{-\frac{1}{2}}{2} = \frac{\frac{1}{8}}{-\frac{1}{2}} = \cdots = -\frac{1}{4}$.

Since $|r| < 1$,
$S_\infty = \frac{2}{1 - \left(-\frac{1}{4}\right)} = \frac{2}{\frac{5}{4}} = \frac{8}{5} = 1.6$.

15. $f(1) = \frac{1}{2}$, $f(2) = \left(\frac{1}{2}\right)^2 = \frac{1}{4}$, $f(3) = \left(\frac{1}{2}\right)^3 = \frac{1}{8}$, $\ldots$. This is a geometric

progression with $a_1 = \frac{1}{2}$ and $r = \frac{1}{2}$. Thus, using $\underline{3}$:

$f(1) + f(2) + \cdots + f(10) = S_{10} = \frac{\frac{1}{2}\left[\left(\frac{1}{2}\right)^{10} - 1\right]}{\frac{1}{2} - 1} \approx 0.999$

17. This is a geometric progression with $a_1 = 3,500,000$ and $r = 0.7$. Thus, using $\underline{4}$:

$$S_\infty = \frac{3,500,000}{1 - 0.7} \approx \$11,670,000$$

19. This is a geometric progression with $a_1 = 20,000$ and $r = 1.05$. Using $\underline{2}$ and $\underline{3}$, respectively:

$$a_{10} = 20,000(1.05)^9 \approx \$31,027$$

$$S_{10} = \frac{20,000[(1.05)^{10} - 1]}{1.05 - 1} \approx 251,558$$

EXERCISE A-4

Things to remember:

$\underline{1}$. If n is a positive integer, then n FACTORIAL, denoted $n!$, is the product of the integers from 1 to n; that is,

$$n! = n \cdot (n - 1) \cdot \ldots \cdot 3 \cdot 2 \cdot 1.$$

Also, $0! = 1$.

$\underline{2}$. If n and r are nonnegative integers and $r \le n$, then:

$$C_{n,r} = \frac{n!}{r!(n - r)!}$$

$\underline{3}$. BINOMIAL FORMULA

For all positive integers n:

$$(a + b)^n = C_{n,0}a^n + C_{n,1}a^{n-1}b + C_{n,2}a^{n-2}b^2$$
$$+ \cdots + C_{n,n-1}ab^{n-1} + C_{n,n}b^n.$$

1. $6! = 6 \cdot 5 \cdot 4 \cdot 3 \cdot 2 \cdot 1 = 720$

3. $\dfrac{10!}{9!} = \dfrac{10 \cdot 9!}{9!} = 10$

5. $\dfrac{12!}{9!} = \dfrac{12 \cdot 11 \cdot 10 \cdot 9!}{9!} = 1320$

7. $\dfrac{5!}{2!3!} = \dfrac{5 \cdot 4 \cdot 3!}{2 \cdot 1 \cdot 3!} = 10$

9. $\dfrac{6!}{5!(6 - 5)!} = \dfrac{6 \cdot 5!}{5!1!} = 6$

11. $\dfrac{20!}{3!17!} = \dfrac{20 \cdot 19 \cdot 18 \cdot 17!}{3!17!}$
$$= \dfrac{20 \cdot 19 \cdot 18}{3 \cdot 2 \cdot 1} = 1140$$

13. $C_{5,3} = \dfrac{5!}{3!(5 - 3)!} = \dfrac{5!}{3!2!} = 10$ (see Problem 7)

15. $C_{6,5} = \dfrac{6!}{5!(6 - 5)!} = 6$ (see Problem 9)

17. $C_{5,0} = \dfrac{5!}{0!(5 - 0)!} = \dfrac{5!}{1 \cdot 5!} = 1$